-828-

WATER CONDITIONING FOR INDUSTRY

WATER CONDITIONING FOR INDUSTRY

SHEPPARD T. POWELL, *Consulting Engineer*
Member, American Institute of Consulting Engineers
Fellow and Honorary Member, American Society of Mechanical Engineers
Member, American Institute of Chemical Engineers
Honorary Member, American Water Works Association

FIRST EDITION

McGRAW-HILL BOOK COMPANY, INC.
New York Toronto London
1954

WATER CONDITIONING FOR INDUSTRY

Library of Congress Catalog Card Number: 53-11483

THE MAPLE PRESS COMPANY, YORK, PA.

To My Wife

RUBY IRENE

INSPIRER AND COWORKER

PREFACE

The enormous industrial expansion that has occurred in recent years has imposed a wide variety of specifications for water quality. To meet these requirements, many refined methods of water conditioning have been developed. It has been the author's aim in the preparation of this book to present in as nontechnical a form as possible the numerous processes by which water of the desired quality can be obtained. Many of the treatments which have broad application involve merely simple clarification, filtration, and softening. Others, however, demand water of the highest purity possible, thus requiring complex treating systems by which practically complete elimination of all soluble, insoluble, and gaseous compounds can be effected.

The subject matter in the several chapters has been arranged to indicate the progressive improvement in water quality obtainable, from simple removal of suspended solids to systems involving complex physicochemical principles. Some of these are still within the twilight zone of complete understanding. To cover this broad field of applied engineering practice adequately demands not only a review of the fundamentals upon which the processes are based but also a description of typical installations and operating techniques necessary to assure adequate and efficient performance.

Although the treatment processes covered have wide application in many industrial fields, emphasis has been placed on the preparation of water for high-pressure boilers and the solution of many related problems. Conditioning of water for steam generation is similar to that required for the manufacture of many products such as textiles, antibiotics, and photographic materials.

A number of chapters which have been included in the book may appear to be outside the scope of water conditioning. Such subject matter has been incorporated in the text because of its interrelation with water-conditioning requirements, with beneficial or detrimental effect on processing, and with maintaining equipment in good condition. It is for these reasons that there have been included chapters on cooling-tower design and operation, corrosion control, embrittlement of boiler steel, deaeration, and other subjects.

A comprehensive dissertation on the development of all mechanical and chemical arts should include historical information and contributions by others as a background for an evaluation of progress and for an understanding of the fundamental principles upon which the processes are based. Space requirements have limited the inclusion of such reviews. To offset this deficiency, there has been included at the end of each chapter a carefully selected bibliography to which the reader should refer for amplification of the subject under discussion. Where necessary, numerous additional references have been included as footnotes. There have also been included analytical procedures for testing, to guide necessary control measures.

The author has drawn heavily on his experience and that of his associates, which cover many years of practice in widely diversified water-conditioning procedures in a large group of industries. He has drawn, also, on the large mass of data which has been released in the technical literature or has been furnished by friends and associates.

It is impossible to acknowledge specifically all persons who furnished data, but an effort has been made to acknowledge and give credit to contributors by special reference to published data and other information which have been furnished to the author by correspondence.

The author wishes to acknowledge the valuable assistance rendered by a number of his associates who have contributed greatly to the preparation of the text, its arrangement, and its editing. Special mention in this respect is given to Dr. H. E. Bacon, Dr. E. L. Knoedler, L. G. von Lossberg, J. C. Booser, J. K. Rummel, A. Chesley Wilson, and Dr. J. B. Goodson. Acknowledgment is also given to the very valuable assistance of C. S. Ohrenschall, who supervised and personally prepared the large number of drawings and charts included in the text. The author wishes to acknowledge the valuable assistance rendered by Mrs. E. W. Harris, who typed most of the text and helped in the arrangement and preparation of the bibliographies. Lastly, acknowledgment is made of the helpful guidance given by the author's son, Robert F. Powell, and by Joseph C. McCabe, editor of *Combustion Magazine*, who reviewed the entire text and rendered valuable criticism and recommendations for changes, corrections, and additions to the manuscript.

SHEPPARD T. POWELL

CONTENTS

WATER CONDITIONING FOR INDUSTRY

1 NATURAL WATERS AND THEIR IMPURITIES

INTRODUCTION

Few engineering or industrial operations may be performed successfully without an adequate and properly conditioned water supply. Water of a given quality satisfactory for one set of conditions may be totally unfit for another purpose.

Experience has fairly well established definite specifications for water requirements in most industries and has indicated tolerances beyond which satisfactory performance is not obtainable.

Water purification satisfactory for municipal uses is frequently wholly unsuited for many industrial processes. It is obviously impracticable to condition the public supply for an entire community and to have the resulting product conform to the specific requirements of groups of industries. All that industries depending upon public water supplies may reasonably expect is that the raw product delivered shall fall within certain broad specifications. Refined processing for special usage is a function of private management.

Water is a tasteless, colorless, and odorless liquid. It is a chemical compound expressed by the formula H_2O, which indicates that it is the combination of two volumes of hydrogen and one volume of oxygen. Chemically pure water, however, is an extremely rare liquid and difficult to obtain, mostly because it is an almost universal solvent, and practically all substances are soluble in it to some degree. Because of this property, water is readily contaminated by substances with which it comes in contact.

On earth the primary source of water is the oceans, and all water eventually returns to these reservoirs. The cycle that water makes is similar to that of a huge distilling apparatus. Because of the heat of the sun, water vapor is passed upward from the water surfaces of the earth, eventually forming clouds. When brought in contact with cold air currents, these moisture-laden atmospheres are condensed and rain or snow results.

From the time water reaches the vapor stage until it is finally discharged again into the ocean, contamination is occurring (Fig. 1). The character of the contamination depends upon the territory over which the clouds form and where the rain or snow is precipitated. Rain water collected in rural districts, remote from the seaboard or at distances from industrial centers, contains only a small amount of salts, gases, acids, or

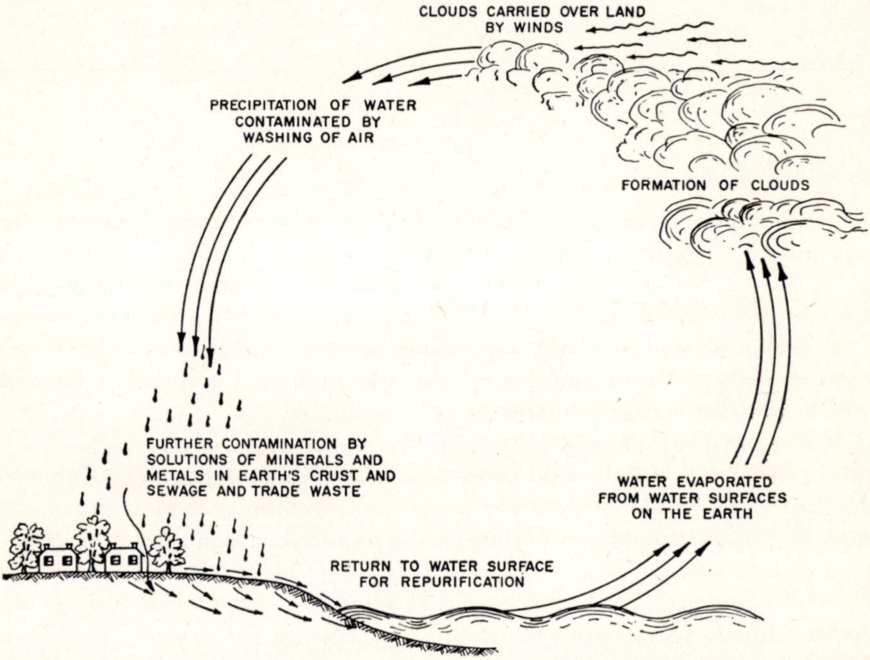

Fig. 1. Diagrammatic representation of the contamination and repurification of a natural water supply.

dust particles. The amount of these substances in rain water collected near cities is often relatively great (Table 1). The relative amounts of suspended and soluble solids deposited on land in a suburban and in an industrial area are shown in Fig. 2.

Of the water that falls on the earth's surface, approximately one-third flows back to the ocean by means of surface watercourses, while twice this volume is evaporated or sinks into the ground to reappear as springs and wells or eventually to reach the ocean by means of subterranean streams. Rain thus coming into contact with minerals and metals in the earth dissolves these materials. The solvent action of water upon mineral substances is increased greatly by the presence of gases dissolved in the

TABLE 1. ANALYSIS OF RELATIVELY PURE RAIN WATER*

Impurity	Ppm	Grains per U.S. gal†	Lb per in. of rain per acre†
Insoluble matter:			
Tar	0.41	0.023	0.092
Carbonaceous (other than tar)	8.0	0.466	1.807
Ash	16.0	0.932	3.614
Soluble matter:			
Loss on ignition	6.0	0.349	1.356
Ash	12.1	0.705	2.734
Total solids	41.0	2.39	9.271
Included in soluble matter:			
Sulfuric acid, as SO_4	8.0	0.466	1.807
Chlorine, as Cl	2.5	0.145	0.563
Ammonia, as NH_3	0.41	0.023	0.092

* After Sir Alexander Houston.
† Calculated by the author.

water. The most important gases contained in water are carbon dioxide, CO_2, and oxygen, O_2. The atmosphere contains approximately 0.04 per cent of carbon dioxide, and since this gas goes into solution readily in water, an appreciable amount of it is dissolved as the rain falls to the earth. The carbon dioxide content of water may be, and frequently is, increased by decomposing organic matter and by industrial wastes. The relatively weak solution of carbonic acid in rain water gradually dissolves the materials in the earth's crust, adding them to the dissolved oxygen and other constituents of rain water. As a result of these conditions, water obtained from rivers, lakes, springs, or other sources is not merely a simple chemical compound consisting of hydrogen and oxygen but is a complex solution containing many organic and inorganic substances.

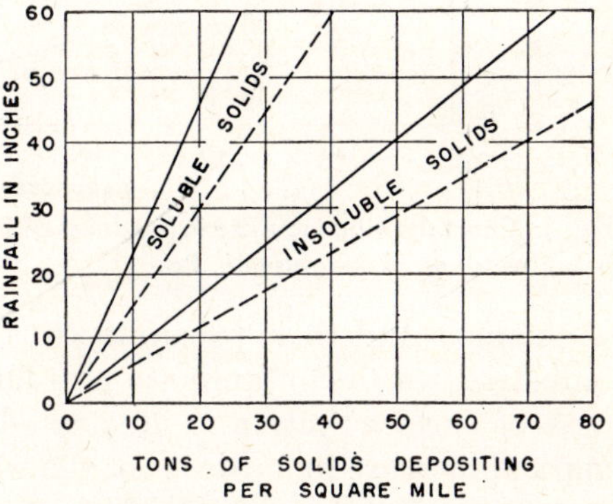

Fig. 2. Solids washed from the atmosphere in urban areas (dashed lines) and suburban areas (solid lines).

The chemical composition of any water reflects to a great extent the character of the geological territory from which it is obtained. Hard water is obtained from areas where limestone (calcium carbonate) or gypsum (calcium sulfate) is abundant; alkaline water, where alkaline deposits occur; and so on, through a large list of mineral or metallic substances. In addition to the constituents which go into solution in water, many surface streams contain much matter in suspension, owing to erosion of the watersheds or to the discharge of industrial wastes into the surface watercourses.

The quality of surface and underground waters fluctuates greatly from season to season, depending on variation in rainfall and other factors. During extreme droughts, such as occurred during 1930 and 1931, much difficulty in the operation of steam boilers occurred because of the marked increase in the mineral constituents of many water supplies. The magnitude of these changes depended upon the geographic location of the

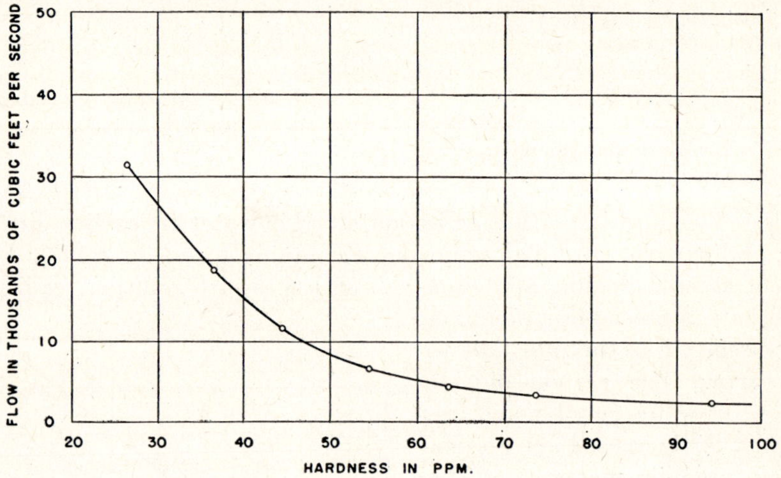

Fig. 3. Chart showing increase in hardness of Delaware River water as river flow decreases. Samples taken at Riegelsville, Pa.

watersheds and their proximity to tidal waters and upon domestic and industrial waste contamination. The influence of drought on the hardness of surface waters is illustrated in Fig. 3. This graph shows the marked increase in hardness of the water of the Delaware River resulting from reduced flow. The condition represented by this curve is typical of many other streams throughout the country. Further depreciation in the quality of water during extreme dry periods results from the influx of salt water into tidal streams during periods of flood tide. Such conditions detrimentally affect the performance of softening systems and result in marked increase in the operating costs of steam-generating units.

Contamination of raw water from dust in the atmosphere occurs to a much greater extent than is usually recognized. The nature and amount of such contamination varies with the locality where the precipitation occurs, being greatest in areas of industrial activity. The presence of dust in the atmosphere and the sources of such contamination which are constantly accumulating from the earth's surface are clearly described by Blacktin,[1*] who states as follows:

> The steady removal of dust over the earth's surface by the world-wind systems is so immense as to be unamenable to the comprehensive visual study of man. Rain plays an immense part in removing degraded solid matter, each drop more generally requiring a particle nucleus. Thus, A. Geikie[2] states: "Day by day the process is advancing. So far as we can tell it has never ceased since the first shower of rain fell upon the earth."

Atmospheric dust from industrial plants subsequently contaminating rain water is a factor in the final contamination of natural waters. Although it is of less importance than the contamination resulting from the solution of minerals on the earth surface and from industrial waste, it still plays a part in the physical and chemical composition of surface and underground waters.

SPECIFICATIONS FOR BOILER FEEDWATER

A perfect water for boiler-feed use is one which will not deposit any scale-forming substances, will not corrode metal of boilers or appurtenances, and will result in neither priming nor foaming. Such water is seldom obtainable except by artificial purification, whereby the impurities which may cause scale, corrosion, or other undesirable effects are removed or changed to less objectionable products. Although a great variety of mineral salts or acids may be present in natural waters, there are, however, relatively few of these constituents found in sufficient amounts to be troublesome. In Table 2 is given a list of the constituents occurring at times in water which may be objectionable when present in boiler feedwater. Before attempting to describe methods of treatment applicable to feedwater purification, it is essential to discuss the properties of some of the impurities encountered most frequently in feedwater supplies. These substances are present either as suspended matter or in solution. It is in the latter condition, however, that they generally cause the major difficulties in boiler operation.

SUSPENDED SOLIDS IN WATER

This term is applied to the heterogeneous matter which many waters contain. Suspended solids consist largely of mud, clay, sewage, and industrial waste products. Waters containing appreciable amounts of suspended solids are not well suited for boiler feed, since they result in deposits of mud within the boiler and may cause boiler water to foam and

* Superscript numbers refer to the items in the Bibliography at the end of the chapter.

TABLE 2. COMMON IMPURITIES IN WATER AND POSSIBLE EFFECT WHEN USED FOR BOILER-FEED PURPOSES

Constituent	Chemical formula	Principal source of contaminating substance	Possible effect when present in boiler waters	Treatment for removal from water
Suspended solids	....	Surface drainage, industrial wastes	Priming, foaming, sludge, or scale	Plain subsidence, coagulation, filtration, evaporation
Silica	SiO_2	Mineral deposits	Scale	Plain subsidence, coagulation, filtration, evaporation
Calcium carbonate	$CaCO_3$	Mineral deposits	Scale	Softening by chemicals, ion-exchange materials, evaporators
Calcium bicarbonate	$Ca(HCO_3)_2$	Mineral deposits	Scale	Softening by heaters, chemicals, ion-exchange materials, evaporators
Calcium sulfate	$CaSO_4$	Mineral deposits	Scale, corrosion	Softening by chemicals, ion-exchange materials, evaporators
Calcium chloride	$CaCl_2$	Mineral deposits	Scale	Softening by chemicals, ion-exchange materials, evaporators
Magnesium carbonate	$MgCO_3$	Mineral deposits	Scale	Softening by chemicals, ion-exchange materials, evaporators
Magnesium bicarbonate	$Mg(HCO_3)_2$	Mineral deposits	Scale	Softening by chemicals, ion-exchange materials, evaporators
Magnesium chloride	$MgCl_2$	Mineral deposits	Scale, corrosion	Softening by chemicals, ion-exchange materials, evaporators
Free acids	HCl, H_2SO_4	Mine drainage, industrial wastes	Corrosion	Neutralizing, followed at times by softening or evaporation
Sodium chloride	$NaCl$	Sewage, industrial wastes, mineral deposits	Inert, but may be corrosive under some conditions	Evaporation and demineralization by ion-exchange materials
Sodium carbonate	Na_2CO_3	Mineral deposits	Priming, foaming, embrittlement	Evaporation and demineralization by ion-exchange materials
Sodium bicarbonate	$NaHCO_3$	Mineral deposits	Priming, foaming, embrittlement	Evaporation and demineralization by ion-exchange materials
Carbonic acid	H_2CO_3	Absorption from the atmosphere, mineral deposits, decomposition of organic matter	Corrosion	De-aeration
Oxygen	O_2	Absorption from the atmosphere	Corrosion	De-aeration
Grease and oil	....	Industrial wastes	Corrosion, deposits, priming, foaming	Coagulation, filtration, evaporation
Organic matter and sewage	....	Domestic and industrial wastes	Corrosion, deposits, priming, foaming	Coagulation, filtration, evaporation

lay down scale. Where the condition is the result of sewage or industrial waste contamination, serious corrosion may also be experienced. Finely divided particles carried in suspension are frequently responsible for priming and foaming.

DISSOLVED SOLIDS AND GASES

The constituents contained in water that are most difficult to remove are those in solution. The soluble compounds are both organic and inorganic substances and are objectionable in varying degrees, as related to boiler-operation efficiency.

TYPES OF WATER

Waters are classified as "hard" or "soft," acid or alkaline, to describe their predominant characteristics reflected by the compounds they contain. The term hard water was coined to define its soap-consuming power, indicating quality with respect to its washing characteristics. Hardness of water is caused principally by calcium and magnesium salts in solution. However, other constituents also react to destroy soap and, therefore, are indicative of hardness. Among such compounds are iron, aluminum, mineral or organic acids, and some metals.

Total hardness,* as this term suggests, is a measure of calcium, magnesium, and all other soap-destroying compounds present in the supply.

Carbonate hardness, at one time identified as temporary hardness, represents carbonates and bicarbonates of calcium and magnesium.

Noncarbonate and permanent hardness are synonymous terms, indicating all hardness except that contributed by carbonates and bicarbonates.

Alkalinity of water is due to the presence of alkaline compounds. By testing water for alkalinity and hardness and comparing the results of alkalinity tests, it is possible to determine in what form the alkaline compounds are present in the water.†

SOLUBLE SOLIDS

Owing to the high solubility of sodium salts in water, they do not ordinarily form deposits or cause scale and are usually not objectionable except at extraordinarily high concentrations. The forms of soda present most frequently in water are sodium chloride, $NaCl$, sodium carbonate Na_2CO_3, sodium bicarbonate, $NaHCO_3$, sodium hydroxide, $NaOH$, sodium sulfate, Na_2SO_4, and sodium nitrate, $NaNO_3$.

* Analytical procedures for performing tests for this and other constituents and interpretation of the analytical data will be found in Chap. 24.

† Detailed instructions for making such tests and interpretation of the data are given in Chaps. 6 and 24.

Sodium chloride is common household salt. This form of sodium is found very widely distributed throughout the world and practically all waters contain some of it. It does not form scale at the concentrations usually present in boiler water. This salt is not particularly corrosive alone, but in the presence of magnesium salts, particularly magnesium sulfate, magnesium chloride is formed, thus giving a potentially corrosive substance.

Sodium bicarbonate is one of the most important salts found in boiler water. This substance, when heated, is reduced to sodium carbonate, commonly called soda ash. Bicarbonate of soda is highly soluble and is found in relatively high concentrations in many natural waters. Sodium carbonate is used extensively as a softening reagent to convert calcium sulfate into the carbonate form.

Sodium hydroxide (caustic soda) is always present in boiler water when the feedwater contains sodium carbonates or bicarbonates due to hydrolysis under high temperatures and pressures. This compound has high solubility and, therefore, does not form scale. Excessive concentration of caustic soda in boiler water causes many operating troubles. Caustic soda is known to attack steel, resulting in the characteristic cracking known as caustic embrittlement or crystallization of boiler steel. Under some conditions, it is responsible for active attack on steel. It has been claimed by Paul[3] that caustic soda may be responsible for active corrosion under some conditions.

Many natural waters contain sodium sulfate, and waters from softening plants of either the lime-soda or zeolite system frequently contain appreciable amounts of this salt. It does not form scale and in low concentrations is generally not objectionable. In high concentrations, it may be carried from the boiler with wet steam in sufficient amount to result in serious trouble. The author encountered a condition where sodium sulfate was discharged from a boiler to such an extent that it deposited in the main control valve of a turbine and caused the valve to stick closed.* Similar deposits were found on a number of the blades in the first few stages of the turbine.

GASES

There are several gases which at times may be dissolved in water. With the exception of carbon dioxide, oxygen, and ammonia, however, little difficulty is experienced from corrosion by gases. The gases referred to are responsible for the majority of the corrosion occurring in boilers, turbines, and auxiliary equipment. Carbon dioxide is actually

* It was not believed that the salt carried over selectively but that it was contained in the mixed boiler-water salines and was deposited selectively because of the solubility equilibria of sodium sulfate in the presence of other salts at a critical pressure.

an acid when combined with water (H_2CO_3) and as such, readily attacks metal. Dry oxygen has little, if any, direct action upon metal, but in solution in water it greatly accelerates corrosion, and much of the corrosion occurring in boilers may be corrected by the removal of this gas from the feedwater.

Many waters contain free mineral acids. It is evident that such supplies are highly injurious to boilers, even when the acidity is present in small amounts. Conditions of this kind are usually the result of mine drainage or contamination of the water source by industrial waste pollution.

ORGANIC MATERIAL

All waters contain some organic matter. The amount present in underground supplies is generally small, but surface waters usually contain relatively large amounts. Organic material present in boiler water is responsible for numerous operating difficulties. Most of the organic matter present in water is caused by the discharge of sanitary sewage and industrial wastes into the supplies.

Bibliography

1. Blacktin, S. Cyril: "Dust," pp. 7–8, The Sherwood Press, Cleveland, 1934.
2. Geikie, A.: "Textbook of Geology," 4th ed., The Macmillan Company, New York, 1903.
3. Paul, J. H.: "Boiler Chemistry and Feed Water Supplies," pp. 68–69, Longmans, Green & Co., Inc., New York, 1919.

2 REMOVAL OF SUSPENDED SOLIDS BY SETTLING WITHOUT CHEMICAL COAGULANTS

The oldest known method of water purification is the removal of suspended solids by subsidence. The clarification of muddy water by settling was practiced to some extent by ancient nations, particularly the Chinese, many centuries before the dawn of the Christian Era. It has been employed extensively by all peoples down to the present time because it is the least costly of all water-purification processes, whether the water is to be used for sanitary or industrial purposes.

Practically all surface waters contain some suspended matter which requires removal before the waters are suitable for industrial uses. Even where water is used merely for cooling, removal of gross amounts of suspended solids is necessary, and some form of treatment must be employed. The relative simplicity of plain sedimentation has always attracted attention, and the conditions which influence this phenomenon have been subject to critical studies. Scientifically, the problem of water clarification is somewhat involved, and the influence of the many contributing factors is not always apparent.

Many investigations have been made to establish the basic principles involved in the removal of solids from water by plain subsidence. The rate of settling of solid bodies in water depends upon the specific gravity, shape, and size of the particles, the viscosity of the water, and upon currents within the containing basin. Other things being equal, maximum removal of the coarser suspended solids in a specified time will result where the water is kept perfectly still. The removal of suspended matter under conditions of quiescent settling is shown in Fig. 1. However, in basins where the water remains entirely quiet, there is not always a perfect gradation of clarification from the top to the bottom, as might be expected. Very finely divided particles of matter may remain in suspension for weeks and are removed by subsidence only after they have coalesced. Spherical bodies settle more rapidly than those of any other shape and of the same specific gravity. The velocities at which sand particles are precipitated in still water have been determined by a number of investigators. Since the viscosity of water increases as the temperature decreases, the rate of settling will likewise decrease as the temperature of the water is reduced. A fall in temperature from 70 to 35°F, for example, results in approximately 45 per cent reduction in settling efficiency.

When water flows through reservoirs, basins, or tanks, it is not possible to effect the percentage of removal of suspended solids which may be obtained by holding water in a quiescent state, even though the same

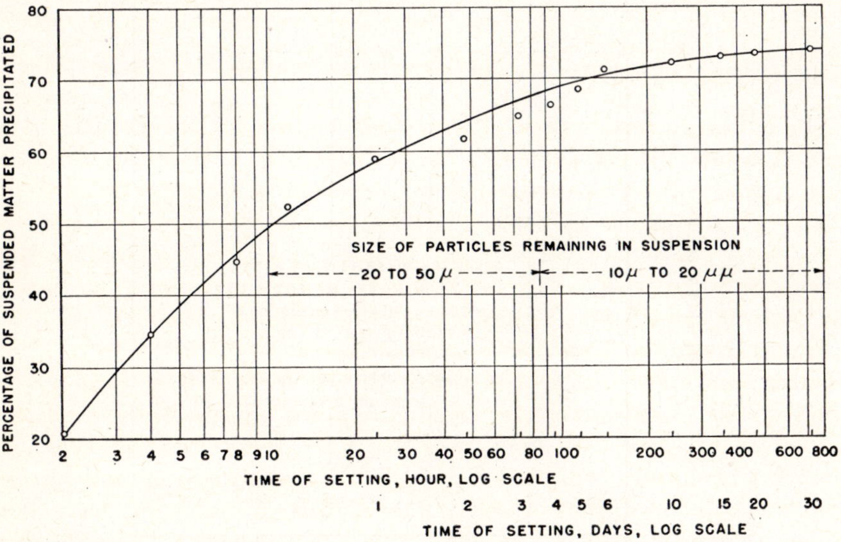

Fig. 1. Removal of suspended solids from water by subsidence in quiescent water at 62°F.

periods of detention are maintained in the basins. The horizontal velocity of the flowing water carries the particles along, thereby overcoming the tendency to deposit. Transporting capacity of water for suspended particles varies as the sixth power of its velocity.

Stokes' law states that the rate of fall of a small sphere in a viscous liquid ultimately reaches a constant value, depending upon the particle size and the viscosity of the medium. However, this law is not strictly applicable to the practical design of basins because of many factors which alter the theoretical concept. Even in a perfectly designed settling basin, wind action and slight changes in temperature as well as the kinetic energy of water entering the basin are all capable of producing motions sufficient to render theoretical assumptions erroneous. A mathematical treatment of Stokes' law, however, and an application of Allen Hazen's[1] classical studies bring out clearly certain pertinent facts which are often disregarded.

In a series of "Sedimentation Studies of American River Waters," Bull and Darby[2] made a number of interesting observations. It was pointed out that "turbidity is not simply a measure of the weight of suspended solids, but is also affected by their size. For example, a certain weight of suspended solids when very finely divided has a much greater turbidity or hiding power than the same weight of coarser solids. The suspended

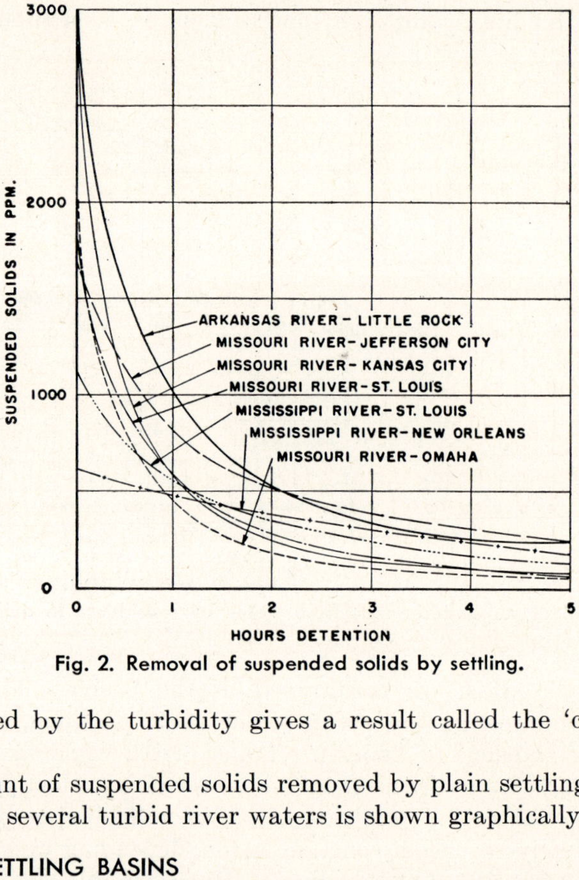

Fig. 2. Removal of suspended solids by settling.

solids divided by the turbidity gives a result called the 'coefficient of fineness.' "

The amount of suspended solids removed by plain settling vs. time of detention of several turbid river waters is shown graphically in Fig. 2.

TYPES OF SETTLING BASINS

Interpretations of the meanings of the terms "settling basin" and "sedimentation basin" vary, and they are frequently used synonymously.

Settling basins may be classified as being of either the continuous or the intermittent type. The former is generally used where the draft of water varies. Tanks operating intermittently are used to a great extent for the treatment of water at wayside railroad stations or at stationary steam-generating plants operated under widely varying load conditions. Frequently, fill-and-draw basins are arranged in pairs, one tank being in service supplying clarified water while the second one is being filled. A schematic sketch of such a design is shown in Fig. 3. The usual method of operation is to admit water at the bottom of the tank and, after a predetermined period of settling, to withdraw the clarified water by means of a "swing" pipe so arranged that the free end of the pipe will be slightly submerged below the surface of the water. Tanks of this kind can effect from 30 to 90 per cent removal of suspended solids. Their efficiency depends upon the amount of settleable solids in the raw water, the coarseness of the material, and the time of settling. As a rule, treatment of boiler feedwater in this way removes merely the gross suspended solids. With some water high in bicarbonates, these soluble salts may be reduced

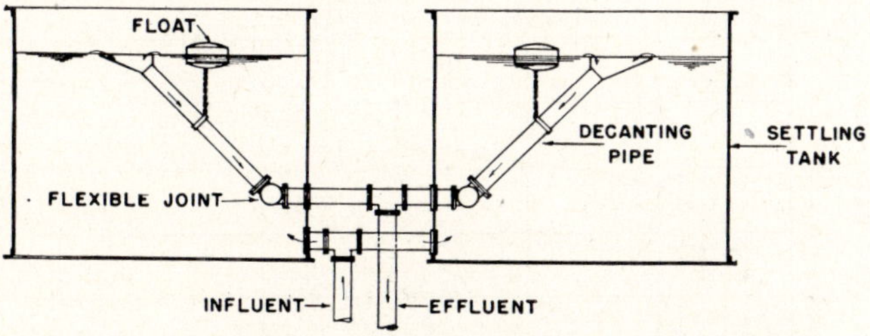

Fig. 3. Fill-and-draw tanks for intermittent sedimentation.

to normal carbonates because of the release of the carbon dioxide, in which case a slight degree of softening will occur with the removal of the insoluble calcium carbonate precipitate settling from the water. The iron content of some waters may be similarly reduced by conversion of soluble ferrous iron to the relatively insoluble ferric form. In the latter condition, the precipitate formed can be largely removed by plain subsidence.

BASIN DESIGN

Basins through which the water flows continuously may be either circular or rectangular. It is impracticable to design a basin in which all the suspended solids will be removed from the water during its passage through it. A basin may be designed, however, to eliminate eddy currents, scouring velocities, and other conditions which tend to prevent subsidence of the settleable material. Every basin should be designed with consideration to the physical characteristics of the turbidity of the water to be treated, and the velocity of flow through the basin should be maintained uniform. The inlet and take-off weirs should not be restricted, since this increases the velocity of the flowing liquid at adjacent sections of the tank. When this condition exists, currents and eddies are set up which prevent the suspended solids from settling.

The primary consideration in the design of a settling basin for the removal of the settleable solids should be the period of detention which will be available at the maximum rate of flow. The detention period which is incorporated in the design will depend on the coarseness of the suspended matter, the degree of purification desired, and the relative cost of the apparatus in comparison with other methods of clarification. From 4 to 24 hr of settling may be employed. The progressive clarification of the water as subsidence is continued is typified by the curve in

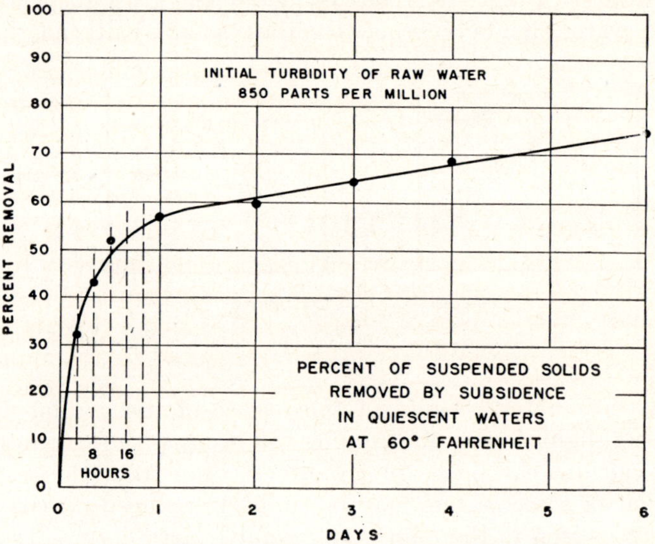

Fig. 4. Per cent of suspended solids removed by subsidence in quiescent waters at 60°F.

Fig. 4, which is based on experiments with water having an initial turbidity of 850 ppm. The per cent removal of suspended matter is plotted against the time of settling and shows a decreasing rate, indicating that the practical detention period is limited. It is doubtful that periods of settling in excess of one day are warranted for feedwater treatment, unless the basin is to be used as a storage reservoir.

It is customary to place baffles or stilling walls at the inlet side of the basin and at a short distance ahead of the effluent weir. These baffles should not be placed so near the outlet weir as to cause an increase in the velocity of the water passing under them. The depth of settling tanks or reservoirs is important where provision must be made to permit the building up of deep deposits between cleaning. Sludge or mud should never be permitted to accumulate in basins to such a depth that it will be picked up and carried along by the incoming water. No specified period for the removal of deposits may be designated, since each case will be governed by local conditions.

Although the theoretical concepts involved in the removal of suspended solids from water by plain subsidence are somewhat different from those relating to the removal of coagulated particles, the same fundamental principles of basin design apply in either case. All basins or reservoirs should be designed and constructed to assure uniform cross-sectional velocities of water passing through the units. An amplified discussion of these problems is given in Chaps. 3 and 4.

The number and arrangement of settling basins are governed largely by local conditions and the capacity of the plant. In small systems having a capacity of 1 mgd or less, usually only one basin will be provided when the basin is operated as a continuous-flow unit or two or more basins if the basins are designed for intermittent operation. In large plants, the basins are usually installed as multiple units. By such design, the flows through the units can be more adequately controlled than if one large basin were used.

Where basins are designed for manual cleaning, namely, to cut them out of service and remove the deposited solids intermittently, such operation demands multiple basins. However, if settled solids are discharged from the basin continuously, then there is less need for the multiple-basin design in small plants. Manual cleaning introduces a waste-disposal problem because large accumulations of sludge must be removed from the basin at one time. When a continuous removal of sludge is used, the amount wasted to the adjacent receiving stream is too small to create a nuisance.

VALUE OF PRELIMINARY SETTLING BASINS

The value of sedimentation, either alone or as an adjunct to other water-purification processes, is usually not recognized. In connection with boiler-feedwater conditioning, the removal of suspended solids in this manner is merely a preliminary step and, wherever employed, should be augmented by more complete treatment. It deserves consideration, however, in conjunction with filtration, softening, and other processes. It is the least costly method for eliminating settleable solids, and a large reduction in the cost of water-purification chemicals may often be effected by plain sedimentation prior to coagulation or softening.

The value of preliminary settling of water has been demonstrated many times in large and small plants.[3] Such treatment is particularly advantageous and economical when treating surface waters carrying excessive amounts of mud. Typical examples of such rivers are the Red, Mississippi, Missouri, Little Rock, and many others. Preliminary settling is also desirable as primary treatment of both large and small streams subjected to floods.

Too little consideration is given, generally, to this relatively simple method of feedwater treatment in designing a purification system to treat turbid water. Aside from the saving resulting from this form of pretreatment, much smoother control is possible than could be obtained otherwise. Basins or reservoirs for this kind of service should be designed carefully in respect to baffling, inlet and outlet weirs, and other essential details.

The determining factors in favor of preliminary settling basins are:

1. Reduction in the cost of coagulation and filtration, resulting from a decrease in the amount of chemicals required for these purposes.

2. Better clarification in coagulation and secondary basins, thereby reducing the load on filters.

3. Increased length of runs between cleaning of filters with reduction in wash water required.

4. Reduction of solids in evaporators where the equipment is supplied with waters that receive no other treatment.

5. Smoother plant operation where the characteristics of raw water fluctuate.

It is not possible to state in the abstract all situations in which preliminary settling is advisable. Where surface water supplies are employed which contain, at times, large amounts of suspended solids, this scheme has considerable merit. The ultimate decision as to the adoption of preliminary-subsidence basins will rest upon the saving to be effected and the flexibility of control of the system in comparison with the capital cost and operation of the equipment.

DISADVANTAGES OF PROLONGED STORAGE

There are certain disadvantages to prolonged storage of water exposed to sunlight. Algae and similar organisms develop prolifically in a relatively short time when water is exposed to sunlight and when the organic or chemical composition of the water will support such life. Certain forms of these organisms may entirely remove free carbon dioxide from the water and can consume a considerable amount of the half-bound carbon dioxide present. Under the latter conditions, bicarbonates are thereby reduced to normal carbonates, and the dissolved oxygen content of the supply may rise to a point of excessive supersaturation. When such organisms are present in great numbers, the chemical composition of the water may be so altered as to necessitate radical changes in the methods of operation or treatment procedures. Filter troubles are also frequently caused by these microscopic growths. It may be necessary to

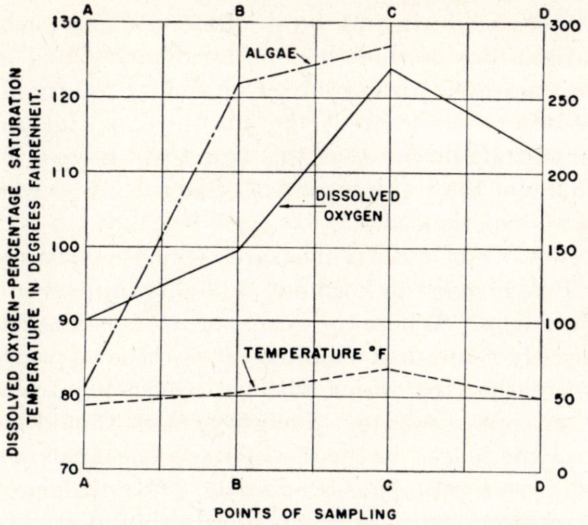

Fig. 5. Showing the effect of algae growths upon the dissolved oxygen in shallow reservoirs during a 24-hr period.

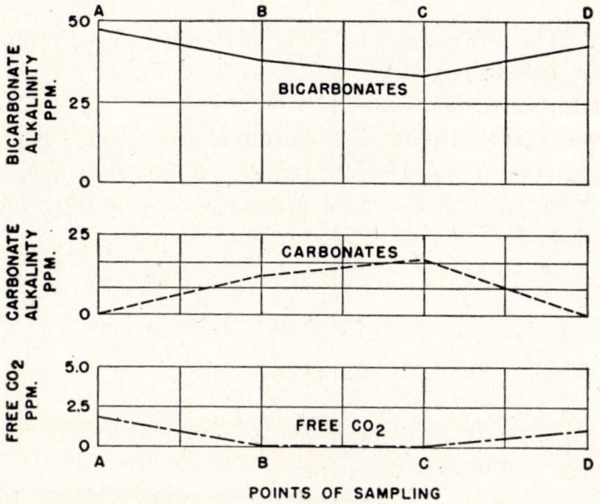

Fig. 6. Showing the effect of algae growths upon the free CO_2 and bicarbonates in a shallow reservoir during a 24-hr period.

employ copper sulfate treatments, chlorination, or other algaecidal treatments as control measures.

The author's[4] experience in this respect shows a typical example of the very active biological, chemical, and physical conditions of a surface water which can occur as a result of prolific development of certain microorganisms. This phenomenon is induced by photosynthesis and a favorable chemical environment.

The relative change during 24 hr in the concentration of dissolved oxygen, free carbon dioxide, bicarbonates, and carbonates as a result of algae growth in the water flowing through a shallow reservoir is shown in Figs. 5 and 6. It is obvious that changes in the chemical composition of water of the magnitude observed in this study will have a marked effect upon the adequacy of the routine control of any feedwater system where such problems are experienced.

There are a number of control measures which can be used to avoid such trouble. Basin covers to keep out sunlight will prevent the growth of many types of algae. Where the basins or reservoirs are so large as to prohibit these costly structures, elimination of these microorganisms can be effected by treating the water with chlorine, chloramines (chlorine and ammonia), or copper sulfate. Such treatment should be carried out only by persons experienced in the use of these chemicals and then only after a thorough investigation has been made. Overdosing of water used for boiler-feed purposes with these chemicals should be avoided, since such practices may result in serious operating difficulties.

Bibliography

1. Hazen, Allen: "The Filtration of Public Water Supplies," John Wiley & Sons, Inc., New York, 1910.
2. Bull, A. W., and G. M. Darby: *J. AWWA*, Vol. 19, No. 3, March, 1928.
3. Powell, Sheppard T.: "Boiler Feed-water Purification," McGraw-Hill Book Company, Inc., New York, 1927.
4. Powell, Sheppard T.: The Effect of Algae on Bicarbonates in Shallow Reservoirs, *J. AWWA*, Vol. 2, No. 4, December, 1915.

3 CHEMICAL COAGULATION: THEORY

PURPOSE OF COAGULATION

From ancient records it is known that the Egyptians used many processes to clarify the muddy water of the Nile River. Various types of coagulators were used, including almonds, beans, and alum.[1] The Chinese were credited, also, with the use of alum for water clarification. A great many accounts appear in ancient literature on this subject, all of which confirm the fact that the use of chemical coagulating materials was well known and fairly widely practiced for water treatment before the Christian Era.[2]

Surface waters normally encountered in nature contain varying proportions of relatively coarse suspended solids, finer solids that appear as turbidity and color, and certain other colloidal materials. Coagulation in water treatment will reduce the concentrations of these materials to limits that can be tolerated. When the water is used for steam generation, coagulation is employed usually to remove suspended solids and turbidity, the impurities most commonly present. However, situations exist where coagulation is practiced primarily to remove color and other materials from a water to render it suitable for boiler feed and miscellaneous uses.

Natural subsidence will remove relatively coarse suspended solids from water, but there is a definite practical limit to the length of the subsidence period. Much finely divided material, such as clays and similar substances, is generally present in surface waters.

THEORIES ADVANCED FOR FORMATION OF FLOC

There have been many concepts to account for the formation or inhibition of chemical flocculation as practiced in water-conditioning processes. Hay[3] has broken these down into two major classifications, namely, (1) the protogel and (2) the mechanistic concepts.

In the protogel concept it has been suggested that all of the water and impurities are at one time within a fibrillar network. As this is broken up and compacted by pressure differentials created in floc conditioning, the water must pass out through the interstices. As the floc particle first decreases in size and increases in apparent density, these pores become smaller and smaller until they reach such size that the viscosity of the water provides a resistance to flow which counterbalances further tendency for floc shrinkage. Then growth into larger flocs will provide the maximum toughness and best settling characteristics which can be attained under these conditions for formation. At low temperatures and high water viscosity the equilibrium will be reached with less dewatering of the floc and therefore less toughness and poorer settling.

In conjunction with mechanistic theories for floc formation, it must be recognized that the chemical and colloidal balance is of greatest significance. In a mathematical treatment, Camp and Stein[4] have assumed that two particles in stream flow will necessarily collide if calculations indicate that they can be brought within a distance equal to one-half the sum of their two diameters. Such calculations might hold for inert matter but not for electrically charged colloidal particles which frequently constitute a floc. Even with greatly reduced charges on the particles one must expect some orientation rather than random arrangements. The colloidal chemistry explanation for floc formation is too involved for detailed discussion at this time, but it can reasonably be said that it is very important in understanding the mechanism of floc formation as well as of practical significance in meeting problems of difficult coagulation. The benefits of colloidal balance for overcoming the low temperature problem are a case in point.

Regardless of which of these concepts is accepted, experience has demonstrated that satisfactory coagulation with maximum settling efficiency is attained only when all particles contained in the water, regardless of the physical or chemical condition, are agglomerated in relatively large masses having a specific gravity sufficiently high to cause them to deposit in a short period of time. To accomplish this favorable condition, the selection and doses of the coagulating chemical and the structure of the "floc" must be such as to effect satisfactory clarification.

COAGULATION OF COLLOIDS

Most waters can be clarified by the addition of alum or other coagulating materials followed by settling for a short period, usually from 2 to 4 hr. However, effective clarification is often retarded by a number of conditions which tend to inhibit the agglomeration of the suspended solids, thereby resulting in poor clarification. To condition such water properly requires a knowledge of physical and colloidal chemistry, since the removal of such substances is governed by fundamental chemical laws in these fields.[5,6]

A colloid is any substance that is sufficiently finely divided to form a relatively permanent suspension. The definition neglects basic classifications and qualifications which the interested reader will find in other works. In the present discussion, we are concerned primarily with suspensions of clay, silica, organic materials, and a few other constituents of such fineness that they completely resist removal by settling or settle out only after a long time.

In explaining the nature of colloids, it is customary to contrast them with true solutions and coarse, readily separable suspensions. A true solution, such as a sodium chloride brine, is homogeneous even down to atomic and molecular dimensions, that is, there is no evidence of a boundary between the minute units of the substances in the solution. However, the finest colloidal solution to be considered in this discussion consists of solid particles surrounded by liquid, each particle presenting a surface which is the boundary between solid and liquid. The existence of this boundary surface between the colloidal particles and the surrounding medium and the effects which are present as a consequence are the fundamental factors which account for the characteristic behavior of colloids. This is readily understood when it is realized that a substance in colloidal suspension has an enormous surface in relation to its total mass.

If a piece of an insoluble material like gravel should be crushed into small pieces and then pulverized until a minute state of subdivision is reached, the total surface area of the solid particles would tremendously exceed the surface area of the original material. If, for example, the piece of gravel before being broken up should have had a diameter of 10 mm and was then pulverized until the individual particles had a diameter of 0.00001 mm, assuming that the particles retain a spherical shape, the total surface area would be increased from less than 0.5 sq in. to 0.7 acre (Table 1).

When a liquid containing particles of this size is poured through filter paper, a large portion of the dispersed material goes through the paper as if it were truly dissolved. The filtrate may be perfectly clear and, to all outward appearances, it resembles a true solution. If a beam of light is passed through the beaker containing the filtrate, however, the light is reflected by the particles in suspension, and its path through the liquid can be seen clearly. The light diverges in such a manner as to produce a conical beam, and the liquid within the path of the beam will take on a milky or cloudy appearance because light is reflected by the colloidal particles. The reflection is known as "scattering" and occurs when the wave length of the light is materially shorter than the diameter of the reflecting particles. This path of the light is called a "Tyndall cone," and it is exhibited by all colloidal suspensions.

If a colloidal suspension is viewed under an ultramicroscope, which is simply an instrument for producing a Tyndall cone in such a manner that it may be seen under a microscope, the colloidal particles may be seen to possess Brownian movement. In other words, the particles in suspension will be seen to have a rapid vibratory motion or will appear to be dancing about in the field of vision. This motion is due to a bombardment of the particles by the molecules of the liquid medium, and the rapidity of the movement will increase as the size of the particles decreases.

TABLE 1. PROGRESSIVE EFFECT OF DECREASING SIZE OF SPHERES

	Diameter of particles, mm	Order of size	Number of spheres required for a volume equivalent to that of a sphere 10 mm in diameter	Total surface area	Time required for sphere of sp gr 2.65 to settle 1 ft
Suspensions	10	Gravel	1	0.487 sq in.	0.3 sec
Suspensions	1	Coarse sand	1,000	4.87 sq in.	3 sec
Suspensions	0.1	Fine sand	1,000,000	48.7 sq in.	38 sec
Suspensions	0.01	Silt	1,000,000,000	3.38 sq ft	33 min
Suspensions	0.001	Bacteria	1,000,000,000,000	33.8 sq ft	55 hr
Molecules or colloids	0.0001	Colloidal particles	1,000,000,000,000,000	3.8 sq yd	230 days
Molecules or colloids	0.00001	Colloidal particles	1,000,000,000,000,000,000	0.7 acre	6.3 years
Molecules or colloids	0.000001	Colloidal particles	1,000,000,000,000,000,000,000	7.0 acres	63 years minimum

As the size of the particles in suspension decreases, their settling rate diminishes. An approximation of the relative rates of settling of spheres of different sizes is shown in Table 1 and, generally speaking, a colloidal state can be said to exist where the effect of gravity is not sufficient to overcome the effects of other influencing factors. While there is no definitely fixed line of discrimination based on size alone, the following limits are given as a rough classification of matter contained within a liquid:

Suspensions, particles larger than 0.0001 mm
Colloidal solutions (hydrosols), 0.000001 to 0.0001 mm
Molecules of crystalloids, particles smaller than 0.000001 mm

Colloidal particles are electrically charged, and this is one of the principal factors that contribute to their remaining in a fine state of division. It has been given, therefore, as one reason why sedimentation of such particles is relatively difficult to accomplish. Because they have similar electrical charges, the particles repel one another, and they thus remain separated in a state of fine subdivision until the charges are neutralized.

It is apparent that colloidal conditions are largely controlled by surface phenomena. Consequently, the chemistry of colloids is largely a question of surface reactions which do not necessarily proceed in a stoichiometric* fashion.

It is more profitable to characterize colloids by certain typical behavior than by limits of particle size. Reference has been made to the slowness with which colloidal suspensions settle out, which is the most obvious characteristic. When a sample of boiler water is filtered for chemical testing and persists in remaining turbid, the suspended matter may, for practical purposes, be regarded as colloidal. Colloids resist ordinary methods of filtration.

To avoid false generalizations, it must be pointed out that colloids have no universal particle-size limits, that most colloidal suspensions in liquids would ultimately settle out, that they can be filtered out by special methods, and that some of them do not show the Tyndall effect. The Brownian movements do not occur to any appreciable extent in colloids whose particle size exceeds a certain value. For practical purposes, any finely divided suspended matter which resists separation by sedimentation and filtration may be regarded as colloidal, since the method of attack in water purification requires no sharper line of distinction.

The resistance of colloidal particles to agglomeration and subsequent settling is intensified by the presence of peptizing agents. Peptization involves the disintegration of colloidal aggregates by the adsorption of a peptizing agent on the surfaces of the particles. Liquids, ionic solutes or electrolytes, nonelectrolytes, and colloidal materials may serve as agents for peptizing various solids. A colloid that peptizes another colloidal aggregate is termed a "protective colloid." For this discussion, the significant peptizing agents are electrolytes and protective colloids, and the colloidal solutions encountered are termed "sols."

Coagulation of a colloid is simply the reverse of peptization and is effected by removing or neutralizing the stabilizing material. In the case of clay sols, the negatively charged clay particles can be coagulated if they can be caused to adsorb positively charged ions, such as, for instance, aluminum ions supplied by adding alum (aluminum sulfate). When adsorption of positive ions just balances the negative charges, the particles are neutral, and the sol is said to be in an isoelectric condition. The particles of clay will tend to coalesce in the absence of opposing forces and will then precipitate. In a sol of a particular composition, maximum precipitation will occur at a very definite concentration of hydrogen ions, which is indicated by the pH value. The pH value of maximum precipitation is dependent upon the composition of the sol. It is influenced by the concentrations and characters of the various ions in the sol.

* Pertaining to the art or process of calculating the combining weights, etc., of the elements.

A parallel phenomenon involving still more finely divided colloids is the removal of color from water. The color is due to a colloidal suspension of tannins and other materials of vegetable origin and is usually found in swamp water. Neutralization by trivalent aluminum or iron ions must be complete, and the coagulant demand is quite definite and may be high, as much as 4 to 6 grains per gal.

In some sols, it is possible to go beyond the isoelectric point and cause a reversal of charge on the colloidal particles by adding an excess of an adsorbed ion. For example, a stable sol in which the colloidal particles are negatively charged may be converted into a stable sol in which the particles are positively charged by the addition of an excess of strongly adsorbed positive ions.

Just as peptization can be reversed by the addition of electrolytic coagulants, so it is possible to coagulate, in some cases, with other colloids. A colloid which decreases the adsorption of the peptizing agent will thereby decrease the stability of the original suspension. Moreover, two oppositely charged colloids may tend to coagulate each other.

The problem of colloidal solutions and their reaction is often involved, and colloidal behavior in water purification or in other arts is much more complex than the simplified explanation given above. There is generally a great variety of ions and undissociated molecules present, including hydrous and anhydrous forms of the same substance and associations of two or more molecules, ions and molecules, etc. Moreover, considerable concentrations of suspended solids to be removed from the water are generally of greater than colloidal dimensions. The phenomena which account for the behavior of colloids, however, still exist in each of these cases to an extent depending on the fineness of the particles. Consequently, water treatment is properly guided by considerations of colloid chemistry.

CURRENT PRACTICE

When certain soluble metallic salts are added to artificially prepared or natural alkaline waters containing suspended and colloidal matter, chemical and physical changes take place and gelatinous substances are formed. This phenomenon is termed coagulation or flocculation. When the floc or coagulum first forms, it is in a minute state of subdivision, but owing to the structure and nature of the particles, it agglomerates and entrains and enmeshes other suspended solids in the water, thereby forming relatively larger bodies. Many writers have attempted to describe the structure of these coagula and have designated them as "spongelike," "open-mesh network," "jellylike," etc. An exact description is difficult, although in appearance and structure they resemble coarse sponges.

The primary function of the coagulant is to furnish highly charged ions that will effectively neutralize the electrical charges of most of the colloidal material present in the water and cause it to precipitate. In addition, the soluble coagulating chemical also undergoes rather intricate chemical reactions with certain definite compounds found in, or added to, the water. The latter reactions produce other insoluble compounds in the form of minute electrically charged particles. These compounds bring about the precipitation of still more of the colloidal matter. The material that is thus precipitated agglomerates to form the jellylike coagulum, or floc, described above.

The value of coagulants is due in part to the spongelike property of the floc, since this type of structure has large surface areas to which colloidal or semicolloidal particles of matter adhere. The addition of coagulants to water serves two purposes: it accelerates the subsidence of suspended matter, and it permits high rates of filtration which could not otherwise be attained successfully. When water which has been coagulated and settled passes through sand or other filter materials, the flocculated chemical remaining in the water is strained from the water and forms a gelatinous mat or cover on the surface of the filter medium. Owing to the spongelike structure of the floc, the water passes through it, but the suspended material is entrained and thus removed.

Although many materials are capable of coagulating solids contained in water, those most widely used are salts of aluminum and iron. Other coagulants employed are ammonia and potassium alum, chlorinated copperas, sodium aluminate, activated silica, bentonite clays, processed starches, and miscellaneous compounds.

ALUMINUM COMPOUNDS

Aluminum sulfate (filter alum) is more extensively used in water treatment than is any other coagulant now known. A brief résumé of the more important and probable reactions is necessary to indicate the value of such a process for water conditioning.

Filter Alum. When a solution of aluminum sulfate is added to a natural water, one or more chemical reactions may occur. Until recently, it was believed that the addition of an alum solution to water always produced a floc as the result of the formation of aluminum hydroxide. A number of investigators have studied the chemical phase of such a reaction and have demonstrated that the flocculation by means of alum solution may or may not be the result of forming aluminum hydroxide. The composition of the flocculent substance will depend upon the relative alkalinity or acidity of the water. If the water is close to neutrality (with methyl orange indicator), the floc will have a probable composition of $5Al_2O_3 \cdot 3SO_3$,[7] but when the water is sufficiently alkaline to react pink to phenolphthalein indicator, aluminum hydroxide, $Al(OH)_3$, is formed.

If the alkalinity of the water is due to calcium bicarbonate, the following reaction will occur:

$Al_2(SO_4)_3$ +	$3Ca(HCO_3)_2$ →	$3CaSO_4$ +	$2Al(OH)_3$ +	$6CO_2$
Aluminum sulfate (soluble)	Calcium bicarbonate (soluble)	Calcium sulfate (soluble)	Aluminum hydroxide (insoluble)	Carbon dioxide (soluble)

When the alkalinity is due to sodium salts in place of calcium in the above reactions, sodium sulfate will be formed instead of calcium sulfate. A similar reaction will take place with magnesium bicarbonate.

Filter alum is prepared by treating bauxite (aluminum oxide) with sulfuric acid. The product is usually purified to remove most of the inert material contained in the bauxite, although some grades of alum are used without removing such contaminants. The product is shipped in slab, lump, ground, or liquid form. Large shipments are delivered in bulk, but small and medium quantities are shipped in bags or barrels. When sold in liquid form, it is delivered in rubber-lined tank cars. The use of liquid alum is limited largely to water-treating plants in areas fairly close to the point of manufacture.

Alum in the form generally used has a chemical formula of $Al_2(SO_4)_3 \cdot 18H_2O$ and is available as an acid or basic material. The pure product contains 15.3 per cent aluminum oxide, Al_2O_3. Alum is applied to water by various equipment either in dry form or after being dissolved. When fed dry it must be in powder or granular form and put into solution before admixture with the water being treated.

When solution feeding is practiced, a stock solution is prepared and then fed through a measuring device in proportion to the volume of water requiring treatment.

Although alum is fairly soluble, it is preferable to heat the water used for dissolving the chemical and to agitate it to accelerate the rate of solution.

Ammonia and Potash Alums. These coagulants are not widely used for large plants but are employed primarily in "alum pots" where the coagulant is dissolved in a pot under pressure and the solution so formed is injected into the water. This type of equipment is used in the operation of pressure filters. Filter alum, owing to its lower solubility and other characteristics, is not well suited for use in pressure-pot feeders.

Sodium Aluminate. Sodium aluminate is now fairly widely used as a coagulant. This material is sold either in liquid form or as a powder. Sodium aluminate solutions contain various amounts of moisture and solids. The solids present in the product are sodium aluminate, sodium carbonate, caustic soda, and organic matter. Because of the presence of the sodium compounds, some softening action results when it is used with hard waters. It has a specific sphere of usefulness in the treatment of some types of waters (Figs. 1 and 2). It is of value for treating colored waters and soft waters which are usually corrosive after treatment with alum.

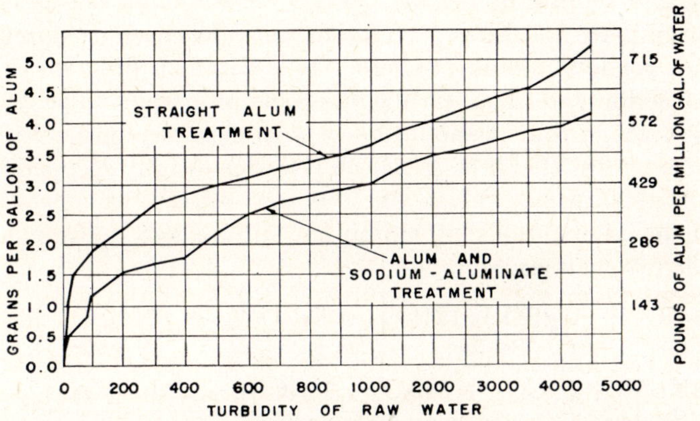

Fig. 1. Alum dose required for good coagulation with various turbidities in raw water in comparison with alum–sodium aluminate treatment. Data shown are actual operating results.

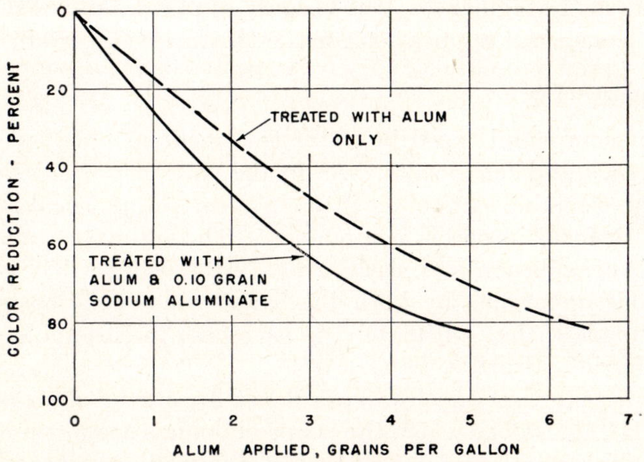

Fig. 2. Chart showing the per cent of color reduction effected by the application of alum in comparison with that by alum–sodium aluminate treatment.

Morse, Hechmer, and Powell[8] found that sodium aluminate was a helpful aid for coagulating certain suspended solids when applied in conjunction with aluminum sulfate. This type of coagulant has been used extensively as a supplementary reagent in water-softening processes to agglomerate finely divided suspended solids and also for softening, producing a lower residual hardness in some cases than can be obtained by lime and soda ash alone. It also has merit for reducing the silica content of some waters.

Sodium aluminate contains about 88 per cent $NaAlO_2$, the remaining constituents present in the product being sodium salts, principally sodium carbonate and sodium hydroxide. The excess alkalies are used to maintain the aluminate in solution. This chemical has been used in treatment compounds for low-pressure boilers. It is common practice to heat the solution of sodium aluminate while it is being prepared. The dry product is deliquescent, so that it should be kept in closed containers until used. It weighs between 50 and 60 lb per cu ft and is usually shipped in drums, bags, or barrels.

The chemical reactions of aluminum sulfate and sodium aluminate when added to water are:

$$\underset{\text{Soluble}}{6NaAlO_2} + \underset{\text{Soluble}}{Al_2(SO_4)_3 \cdot 18H_2O} \rightarrow \underset{\text{Insoluble}}{8Al(OH)_3} + \underset{\text{Soluble}}{3Na_2SO_4} + 6H_2O$$

$$\underset{\text{Soluble}}{2NaAlO_2} + \underset{\text{Soluble}}{CO_2 + 3H_2O} \rightarrow \underset{\text{Soluble}}{Na_2CO_3} + \underset{\text{Insoluble}}{2Al(OH)_3}$$

The floc is similar to that produced by aluminum sulfate when used alone, but the material has an alkaline reaction, whereas aluminum sulfate is an acid coagulant. It is because of this difference that sodium aluminate has special application for the treatment of certain waters.

IRON COAGULANTS

The iron compounds used in water treatment include several of the soluble ferrous and ferric salts. As in the case of the aluminum compounds, these soluble salts enter into chemical and physical reactions in the water to form insoluble compounds. The ultimate result of these reactions is coagulation and clarification of the water. The simplest of the reactions undergone by both the ferrous and ferric coagulants are presented to show that, ultimately, in each case the insoluble compound formed is ferric hydroxide, $Fe(OH)_3$.

One advantage that the iron coagulants have over alum is the wide range of pH values over which the iron compounds are precipitated. In studies on the formation of floc by means of ferric coagulants, by Bartow, Black, and Sansbury,[9] it was demonstrated that the point of optimum coagulation may be as low as pH 3.8 in the case of highly colored swamp waters, or it may be above 8.0 with other types of water. At times, the pH range within which good floc may be rapidly formed is relatively narrow, and the pH of the water must be carefully controlled. The zone of coagulation depends on the concentrations of various ions present in the water, the temperature, flocculating facilities, and other factors.

The curves in Fig. 3 show the completeness of flocculation of ferric iron and manganese in natural waters observed by the author[10] and by

Weston.[11] These data show the wide range of precipitation of iron and manganese in a number of natural waters.

Ferrous Sulfate. Ferrous sulfate was originally employed to clarify very turbid water and is still used extensively for this purpose. The material is known as copperas and has the chemical formula $FeSO_4 \cdot 7H_2O$.

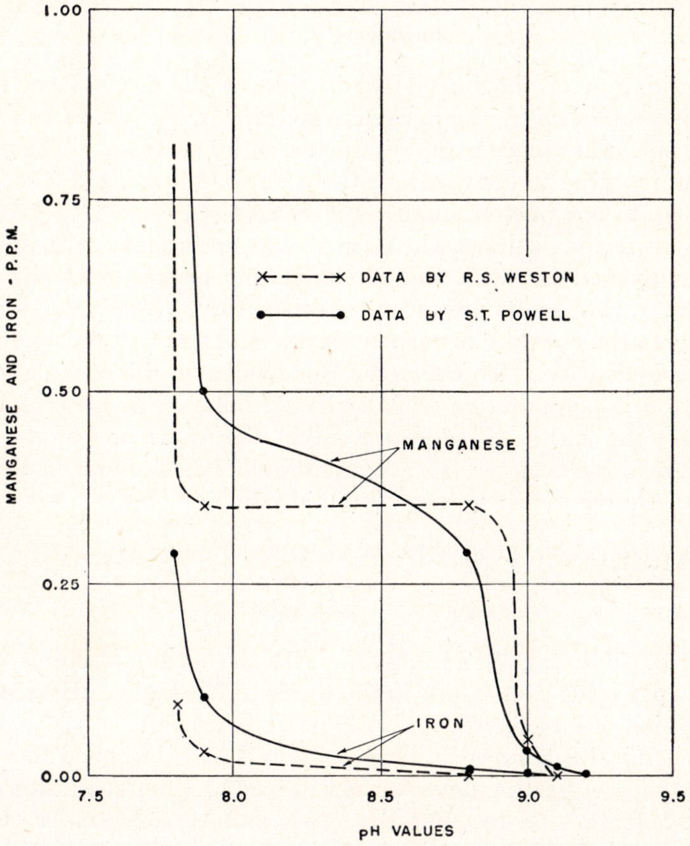

Fig. 3. Curves showing removal of manganese and iron at various pH values of the treated water.

Like aluminum sulfate, it reacts with the natural alkalinity of the water to form a floc. However, the initial floc that is formed is a relatively soluble compound, ferrous hydroxide, $Fe(OH)_2$. The reactions whereby this compound is formed are as follows:

$$\underset{\text{Soluble}}{FeSO_4} + \underset{\text{Soluble}}{Ca(HCO_3)_2} \rightarrow \underset{\text{Soluble}}{CaSO_4} + \underset{\text{Soluble}}{Fe(HCO_3)_2}$$

$$\underset{\text{Soluble}}{Fe(HCO_3)_2} \rightarrow \underset{\text{Partly soluble}}{Fe(OH)_2} + \underset{\text{Soluble}}{2CO_2}$$

Ferrous hydroxide is easily oxidized under certain conditions to ferric hydroxide, $Fe(OH)_3$, which is a very insoluble compound.* The following equation illustrates the reaction:

$$\underset{\substack{\text{Partly}\\ \text{soluble}}}{4Fe(OH)_2} + \underset{\substack{\text{Dissolved}\\ \text{from air}\\ \text{in water}}}{O_2} + \underset{\text{Water}}{2H_2O} \rightarrow \underset{\text{Insoluble}}{4Fe(OH)_3}$$

The oxidation of the ferrous hydroxide to the ferric form is fairly readily accomplished by the oxygen in solution in the water, and the rate at which this takes place is rapidly increased by aeration. The quantity of oxygen required to convert ferrous to ferric hydroxide is 0.03 parts of oxygen, O_2, to one part of copperas, $FeSO_4 \cdot 7H_2O$.

Although ferrous sulfate will form floc in accordance with the above equations by reacting with the alkalinity and oxygen normally present in the water, the results are often unsatisfactory. Such results are generally due to the presence of carbon dioxide and its effect on the solubility of ferrous hydroxide. To overcome this handicap, the coagulant is usually applied in combination with hydrated lime. The latter compound neutralizes the carbon dioxide present in the water and that resulting from the reaction between the coagulant and the alkalinity in the water. In effect, the reaction whereby ferrous hydroxide is produced becomes:

$$\underset{\text{Soluble}}{Fe(HCO_3)_2} + \underset{\substack{\text{Partly}\\ \text{soluble}}}{2Ca(OH)_2} \rightarrow \underset{\substack{\text{Partly}\\ \text{soluble}}}{Fe(OH)_2} + 2CaCO_3 + 2H_2O$$

An added advantage in adding lime with the coagulant stems from the increased pH value that results in the water. The higher pH value favors the oxidation of ferrous hydroxide to ferric hydroxide.

This chemical is shipped in bags or barrels and is applied to water by the same type of feeding devices used to feed aluminum sulfate.

Ferric Sulfate. In recent years ferric sulfate, sold under the trade names Ferrisul, Ferri-floc, and other identifications, has been widely used in the treatment of certain types of water. It has replaced ferrous sulfate in many cases. This coagulant reacts with the natural alkalinity in water to form insoluble ferric hydroxide and, therefore, does not require the addition of hydrated lime to complete the reaction.

The chemical reaction which takes place when ferric sulfate is added to a water containing calcium alkalinity is as follows:

* According to the data reported in Seidell's "Solubilities of Inorganic and Metal Organic Compounds," the solubilities of ferrous hydroxide and ferric hydroxide are 6.0 ppm and 0.000033 ppm, respectively. From these data, it is seen that ferric hydroxide is only about 1/180,000 as soluble as ferrous hydroxide.

$$Fe_2(SO_4)_3 + 3Ca(HCO_3)_2 \rightleftharpoons 2Fe(OH)_3 + 3CaSO_4 + 6CO_2$$

Soluble Soluble Insoluble Soluble Soluble

Ferric sulfate solutions are corrosive, requiring acidproof equipment for dissolving and applying to the water. Rubber- or lead-lined tanks, stainless-steel and Duriron construction are usually employed for such service.

It is important to control the ratio of the chemical to the water used for making the solution to avoid the deposition of hard insoluble deposits in the solution tanks. Ferric sulfate forms one hydrate, $Fe_2(SO_4)_3 \cdot H_2O$, which sets quickly, as plaster. It does not go into solution rapidly in cold water. The use of from two to four parts by weight of water to one part of ferric sulfate avoids the formation of the hard hydrate and conserves the heat of solution to aid in dissolving the material. Best practice is to furnish hot water proportioned to the flow of dry coagulant.

Ferric sulfate can be applied by means of dry feeding machines, and in dry form the product is not corrosive. Some ferric compounds are highly deliquescent, but some grades are treated with an inert material which coats the grains, thereby inhibiting absorption of the moisture.

The material is shipped in bags or paper-lined barrels.

Ferric Chloride and Chlorinated Copperas. These ferric compounds have been known and used for many years for coagulation of sewage sludges and, to a limited extent, for treatment of some water supplies used for sanitary and industrial uses. Ferric chloride, $FeCl_3 \cdot 6H_2O$, is widely employed for sewage treatment to improve the settling and filterability of sludge, but its use for water clarification is not extensive. However, ferric chloride is of value to produce coagulation at a low pH. Miller[12] found that satisfactory coagulation could be obtained at a pH of 5.0. Ferric chloride is shipped in drums or tank cars and is very corrosive, requiring special acid-resisting equipment for storage and feeding of the solution. It is because of its highly corrosive property that it is not so widely used as is ferric sulfate.

Chlorinated copperas, although not so efficient a coagulating material as some others, has considerable merit because its preparation can be carried out at the plant site. This is accomplished by chlorinating a solution of ferrous sulfate, both ferric sulfate and chloride being produced by the following reaction:

$$6FeSO_4 + 3Cl_2 \rightarrow 2Fe_2(SO_4)_3 + 2FeCl_3$$

In practice, the ratio of chlorine to the ferrous sulphate is 1:8. Hedgepeth and Olsen[13] reported its initial use for coagulating highly colored water at pH 4.9. Similarly, Enslow[14] specified the optimum zone of coagulation for swamp water containing a high color as that of pH values ranging from 3.8 to 4.7.

ACTIVATED SILICA

Although many investigators* had recognized the coagulating properties of certain forms of silica, the practical value of silica and of its preparation as an aid for coagulating turbid water was first placed on a firm scientific basis by the brilliant research studies of Baylis.[15] As a result of this work, a patent[16] was granted for a process for the preparation of an activated silica sol which, when added to water, improved the settling qualities of the coagulated solids. Other patents covering the preparation of the sol were granted to Graf and Schworm,[17] to Baker and Dedrick,[18] and to Hay.[19]

MISCELLANEOUS COAGULANTS

Many organic and inorganic materials, in addition to iron and aluminum compounds and activated silica, have been used to coagulate solids and improve settling rates. They have been used singly or in combination with other materials. Clay, bentonite, activated carbon, causticized starches,[20] and other compounds have been found to be helpful in the treatment of some supplies. In addition, such products as ethylcellulose and several proprietary compounds are said to aid other coagulants when used in relatively minute dosage.

CHLORINATION AS AN AID TO COAGULATION

Although chlorine is not a coagulant in the true sense of the word, it is employed frequently to assist in coagulation by conditioning the water to remove or destroy materials which inhibit floc formation. Such pretreatment is frequently used for waters contaminated with organic compounds, such as highly colored swamp water and those contaminated by sanitary or industrial wastes.

THE EFFECT OF pH ON COAGULATION

The amount of coagulant required to effect good clarification varies with the nature and amounts of suspended and soluble solids contained in the water. The fundamental work of Clark and Theriault, Miller, and others on the formation of floc by alum solution has proved that the acidity of the water, as measured by the hydrogen-ion concentration of the final mixture of alum and water, is of great importance in the production of floc. A portion of the summary of Clark and Theriault's study reported in the U.S. Public Health Report is as follows:

> When other possible factors are left out of consideration, the optimum conditions for floc formation will be found within a narrow zone of pH centered for

* For detailed information on these processes and instructions for preparation of the sols, the reader should refer to the patents and instructions given in the several papers listed in the Bibliography at the end of this chapter.

dilute solutions at pH 5.5. The more dilute the water in total salt content and the less alum added, the narrower becomes the pH zone with which optimum floc formation is to be found.

Although under controlled conditions maximum flocculation with alum is secured at pH 5.5, there are factors that may cause certain deviations from this point.[21,24] Further, maximum flocculation does not assure minimum solubility of residual ions remaining in the treated water.

In the comprehensive studies by Baylis[22] and by Bartow, Black, and Sansbury[9] on the flocculation of particular waters with alum and chlorinated copperas at various pH values, residual concentrations of aluminum and iron were determined in the treated water. These investigators reported that the minimum amount of residual aluminum occurred in the water investigated at approximately pH 6.3, whereas the minimum residual of iron in another water was found at a lower pH value of about 5.4. They also showed that there are two pH ranges in which iron is precipitated. In addition to the minimum residual at pH 5.4, another minimum is shown to exist in the alkaline region of pH values.[10,11] Hopkins[23] has demonstrated that this latter minimum occurs in the pH range 9.4 to 13.0.

Petersen and Bartow,[24] Black, Rice, and Bartow[21] and Bartow, Black, and Sansbury[9] have demonstrated that the pH range of optimum floc formation is broadened considerably by the presence of certain ions commonly found in natural waters. In general, ions of higher valence, such as sulfate, have been found to exert a greater influence in this respect than do the ions of lesser valence, such as chloride.

EFFECT OF TEMPERATURE ON COAGULATION

It is a well-known fact among filter operators that effective coagulation and the rate of floc formation are greatly influenced by the temperature of the water. As the temperature is lowered, the dosage of chemical for coagulation must be increased to insure properly formed floc. The detrimental effect of low temperature on floc formation can be minimized by adequate design of equipment with respect to thorough mixing preceding proper flocculation. Even when such desirable designs are available, more chemicals are required generally than when treating warmer water. This condition is demonstrated by the curves shown in Fig. 4, which indicate the effect of temperature on chemical requirement and also the variation in treatment required for water containing coarse, medium, and finely divided suspended solids.

EFFECT OF MIXING AND FLOCCULATION

One of the most important requisites for good coagulation and settling of suspended solids is the rapid and thorough mixing of the chemical and the water, followed by slow flocculation. Such treatment is necessary to ensure thorough admixture of the chemicals, thus avoiding the short-circuiting of water that has not received its aliquot proportion of the applied coagulant. After the mixing has been effected, the finely divided

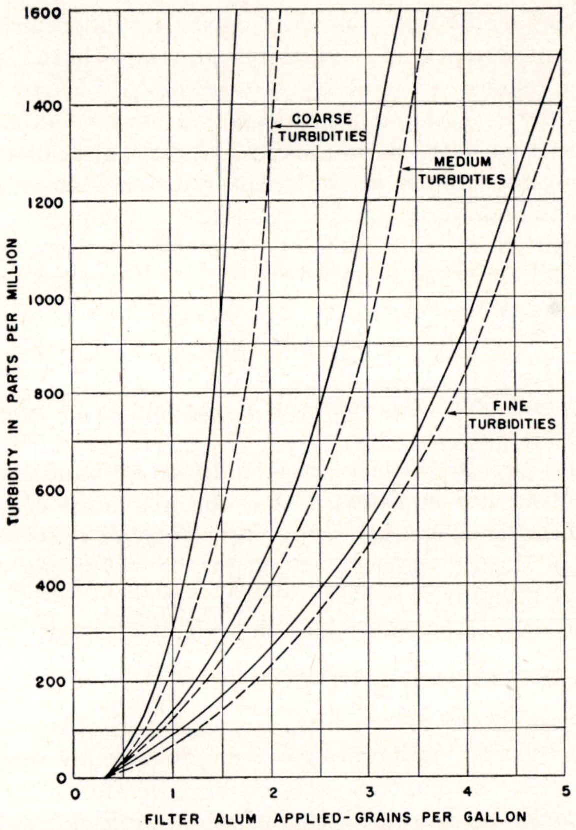

Fig. 4. Curves showing the approximate quantities of alum required for varying turbidities in raw water. Solid lines, above 60°F; dashed lines, below 50°F.

coagulum should be allowed to build up into larger masses to ensure maximum settling rates. Such treatment is best secured by imparting a rolling motion to the water as it passes through the flocculating chamber. Equipment for this purpose is described in the following chapter.

Bibliography

1. King, Thomson: Private communication.
2. Baker, M. N.: "The Quest for Pure Water," American Water Works Association, New York, 1948.

3. Hay, Harold R.: Improving Water Plant Operation with Activated Silica, Parts I, II, and III, *Water & Sewage Works*, September, October, and December, 1946.
4. Camp, T. R., and P. C. Stein: Velocity Gradients and Internal Work in Fluid Motion, *J. Boston Soc. Civil Engrs.*, **3** (4): 219–237 (October, 1943).
5. Thompson, M. deKay: "Theoretical and Applied Electrochemistry," The Macmillan Company, New York, 1930.
6. Zsigmondy, R.: "Chemistry of Colloids" (trans. by E. B. Spear), John Wiley & Sons, Inc., New York, 1917.
7. Williamson, F. S.: Basic Aluminum Sulfate, *J. Phys. Chem.*, **27**: 284 (1923).
8. Morse, R. B., Carl Hechmer, and S. T. Powell: Coagulation Studies at Washington Suburban Sanitary District, *J. Ind. Eng. Chem.*, **20**: 56 (January, 1928).
9. Bartow, Edward, A. P. Black, and W. E. Sansbury: Formation of Floc by Ferric Coagulants, *Ind. Eng. Chem.*, **25**: 898 (1933).
10. Powell, S. T.: Water As an Engineering and Industrial Material, Edgar Marburg Lecture, *ASTM Proc.*, **34** (II): 17 (1934).
11. Weston, R. S.: Manganese in Water, Its Occurrence and Removal, *J. AWWA*, **24**: 1272 (1932).
12. *U.S. Public Health Repts*, Feb. 2, 1923, pp. 191–197.
U.S. Public Health Service Reprint 862, p. 6.
13. Hedgepeth, L. L., and W. C. Olsen: Chlorinated Copperas—A New Coagulant, *J. AWWA*, **20**: 467 (1928).
14. Enslow, L. H.: *Proc. First Virginia Conf. Water Purification and Sewage Treatment*, 1929.
15. Baylis, J. R.: Silicates as Aids to Coagulation, *J. AWWA*, **29**: 1355 (1937); Effect of Silicate Treatment on Quality of Water, *Water Works & Sewerage*, **85**: 855 (1938).
16. Baylis, J. R.: U.S. Patent 2,217,466, Oct. 8, 1940.
17. Graf, A. V., and W. B. Schworm: U.S. Patent 2,234,285, Mar. 11, 1941.
18. Baker, C. L., and C. H. Dedrick: U.S. Patent 2,310,009, Feb. 2, 1943.
19. Hay, Harold R., U.S. Patent 2,444,774, July 6, 1948; U.S. Patent 2,567,285, Sept. 11, 1951.
20. Gardiner, G. R., and K. B. Ray: Flocculation and Clarification of Slimes with Organic Flocculents, *Am. Inst. Mining & Met. Engrs., Tech. Publ.* 1052, February, 1939; also *Mining Technol.*, May, 1939.
21. Black, A. P., Owen Rice, and Edward Bartow: Formation of Floc by Aluminum Sulfate, *Ind. Eng. Chem.*, **25**: 811 (1933).
22. Baylis, John R.: The Use of Acids with Alum in Water Purification and the Importance of Hydrogen-ion Concentration, *J. AWWA*, **10**: 365 (1923).
23. Hopkins, E. S.: Composition and Use of Ferric Hydroxide as a Coagulant, *Ind. Eng. Chem.*, **21**: 58 (1929).
24. Petersen, B. H., and Edward Bartow: Effects of Salts on the Rate of Coagulation and Optimum Precipitation of Alum Floc, *Ind. Eng. Chem.*, **20**: 51 (1928).

4 APPLICATION OF COAGULANTS, FLOCCULATION, AND SETTLING

Coagulation is almost always carried out as a stage of water purification preliminary to filtration. The complete sequence of basic operations for removal of turbidity, color, and other organic matter to prepare water for filtration is (1) addition of chemicals, (2) rapid mixing, (3) flocculation, and (4) settling.

The present chapter is limited to equipment for these four steps in the preparation of water for filtration. It should be noted, however, that the complete purification process usually includes, especially in the case of iron or manganese removal or softening, one or more of the following steps: (1) prechlorination, (2) aeration, (3) recarbonation, (4) acid treatment, (5) correction of pH of finished water, or (6) application of additives for corrosion or scale control.

Equipment for these operations is described in the appropriate chapters.

EQUIPMENT FOR FEEDING CHEMICALS USED FOR COAGULATION

Within recent years, most water-purification plants handling fairly large amounts of chemicals have employed equipment for measuring them in the dry state and dissolving them just prior to addition to the raw water. However, the original way of applying coagulating chemicals was to prepare a solution or slurry with the liquid in proportion to the water treated. This method is still the most widely used in small- to moderate-sized industrial water-treatment plants.

All coagulants, whether applied by dry or liquid feeders, are put into solution before being added to the water under treatment. When dry feed equipment is used, the measured chemical is discharged into a solution compartment and dissolved as rapidly as possible and then delivered to the raw water by pumps, ejectors, or gravity feed. Solution feeder tanks are equipped with motor-driven mixers or air or steam grids to aid dissolving. The chemical is weighed out according to a base charge (weight per inch of tank depth to be recharged) which will provide the desired dosage within the adjustment range of the feed control mechanism. Obviously, chemical-feeding equipment must be of a size to provide the desired dosage to the expected flow of water and to provide storage that will permit recharging not oftener than once a day or once a shift under maximum load conditions.

Many chemicals used for coagulating are corrosive, and equipment for feeding them must be constructed of acid-resisting materials. Lead- or rubber-lined or wood-stave tanks, stainless steel, monel, bronze, and other resisting materials are generally employed. Sodium aluminate, lime, soda ash, or caustic soda can be prepared and applied with iron or steel apparatus, pipes, and valves. Brass and copper should not be used as construction materials for equipment designed for alkaline service.

LIQUID FEEDERS

Dissolving Pots for Crystalline Chemicals. One of the earliest forms of chemical-feeding devices for alum solution was the alum pot, and the

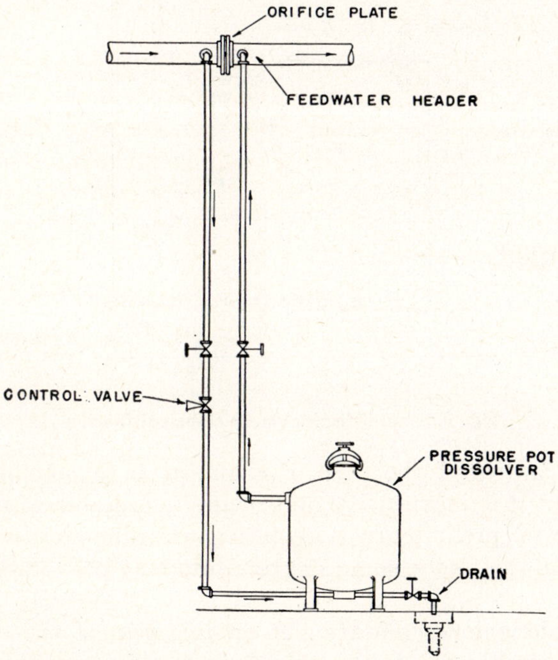

Fig. 1. Pressure pot for feeding crystal alum or sal soda.

same device is used for feeding sodium carbonate (Fig. 1). In this appliance, lumps of crystal potassium or ammonium alum or of other reagents are placed in a small pressure vessel, and solution is effected by passing a small fraction of the water continuously through it. The slip-stream for dissolving the chemicals is taken off the raw-water line just ahead of an orifice installed for this purpose and returned downstream from the orifice. A needle valve is used to adjust the rate of dissolving water flow and hence the rate of feed. This method of applying alum and soda is employed at the present time in many small pressure-filter installations where provision is not made for precoagulation. Since the exposed crystal surface slowly changes, the apparatus does not permit

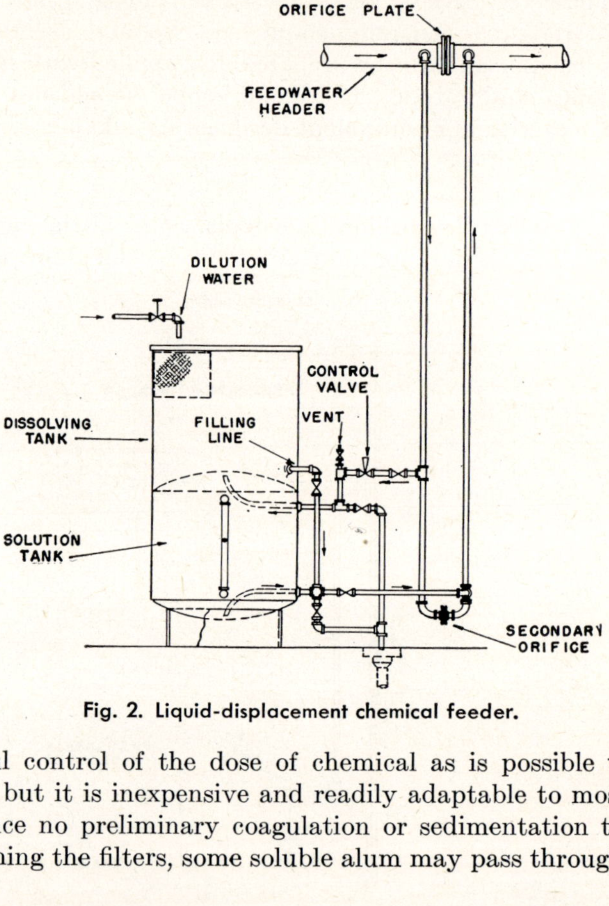

Fig. 2. Liquid-displacement chemical feeder.

such careful control of the dose of chemical as is possible with other appliances, but it is inexpensive and readily adaptable to most pressure filters. Since no preliminary coagulation or sedimentation takes place before reaching the filters, some soluble alum may pass through the filter bed.

Liquid Displacement Feeders. A modification of the dissolving-pot feeder is shown in Fig. 2. In this appliance, the alum or other coagulating chemical is first dissolved, and the solution is then discharged to the displacement tank. Because of its higher density, the chemical solution remains below the piston of raw water and is delivered into the main water supply by hydraulic displacement. It is necessary to make up the solution to a fairly high specific gravity to maintain separate layers. The boundary between the water and the solution is indicated by a density float in the gauge glass. The apparatus may be used for alum, soda ash, and other chemicals whose solutions are sufficiently heavy.

Water is bypassed around a primary orifice in the raw-water line, and its flow is therefore proportional to the main stream. To displace the solution from the feeder, water is bypassed around a secondary orifice in the slipstream, under control of a needle valve.

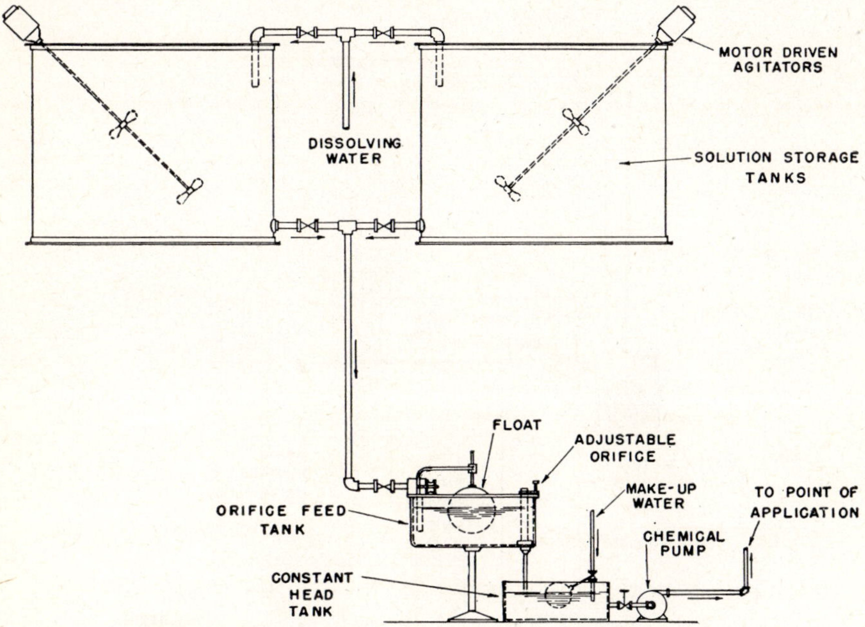

Fig. 3. Schematic diagram of orifice box chemical feeder.

Proportional feeding by these devices is inherently accurate; however, it is necessary to exercise great care to avoid mixing the two layers.

Orifice Box Feeder. This type of apparatus has been extensively used for many years and is offered by most vendors of water treating equipment. It is a simple device which permits applying a solution at a constant rate through a submerged orifice and under a fixed head maintained by a float-controlled valve. Two storage tanks are usually installed to permit feeding from one while the solution is being prepared in the other. The dosage can be regulated manually by positioning the orifice opening or automatically by electric control devices. A diagrammatic arrangement of the storage tanks and feeder is shown in Fig. 3.

The selection of the materials of construction will depend upon the chemical to be fed and must be of a type to resist corrosive attack. The small tanks are frequently of hard rubber or of porcelain-lined steel. Hard-rubber floats, valves, and orifices are available.

Feeders: Decanting or Swing-pipe. These feeders (Fig. 4) are used for proportioning chemical solutions. The swing pipe which skims off and delivers the solution is operated on a "stop-and-start" cycle actuated by an electric lowering mechanism interconnected to a proportioning meter. When a specific quantity of water passes through the meter, an electric contact starts the motor which rotates a drum over which a cable

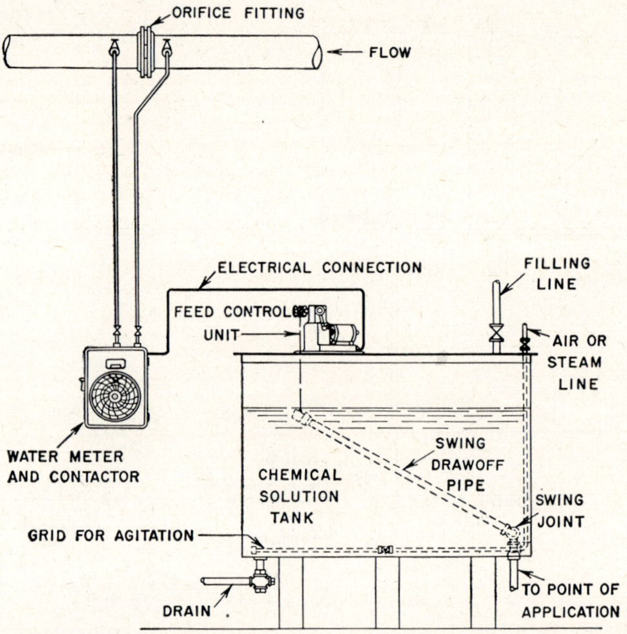

Fig. 4. Chemical-solution feeder: swing type. (*Courtesy of The Permutit Company.*)

is wound from which the swing pipe is suspended. The swing pipe is lowered a fixed distance, depending on the setting of the dosage control mechanism. Since the frequency of this cycle is governed by flow and the distance of lowering is manually adjustable, it is possible to feed any desired dosage accurately proportioned to water flow.

DRY FEEDERS

All appliances of this kind are classified as volumetric or gravimetric feeders, to identify the means by which the rate of feeding chemical is governed. The design and operation of dry feeders require that the chemical to be fed must be ground or powdered to pass through the control mechanism and to insure uniform flow.

Volumetric Dry Feeders. Volumetric dry feeders are designed to deliver chemicals in granular, lump, or powder form in a uniform flow which can be controlled with reasonable, but not precise, accuracy by setting the adjustment mechanism. The feeder shown in Fig. 5 has an oscillating throat which delivers onto a platform and moves a receiving tray back and forth with each motion, spilling off the dry chemical at the end of the stroke. The stroke length, which measures the rate of feed, is controlled by varying linkage to the connecting rod of the drive mechanism. Variation in speed or intermittent oper-

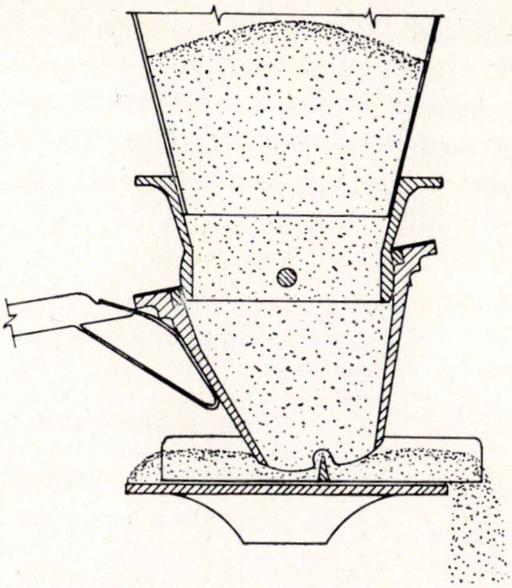

Fig. 5. Volumetric-type feeder operated by oscillating throat and feed tray. (*Courtesy of Omega Machine Co., Division of Builders Iron Foundry.*)

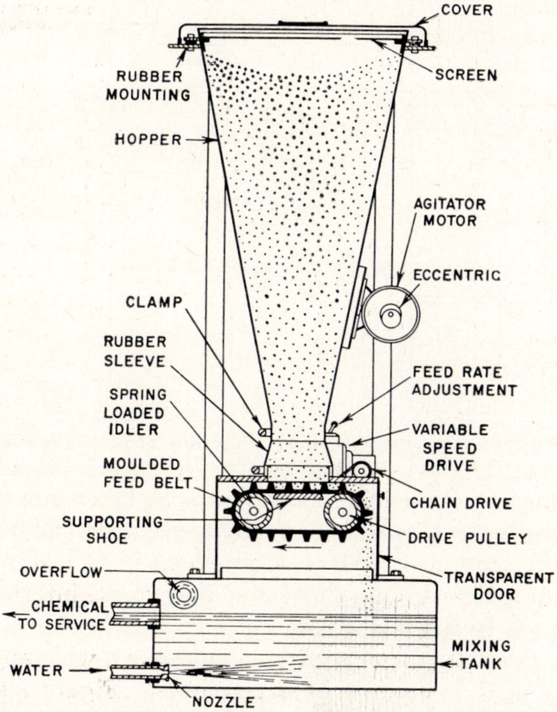

Fig. 6. Volumetric feeder delivering chemicals from a molded rubber belt. (*Courtesy of The Permutit Company.*)

ation or both make it possible to proportion the rate of feed to the flow of water being treated.

Development of volumetric feeders has been in the direction of more precise measurement of the bulk of solids being delivered. Numerous devices have been used, including those based on the screw-conveyer principle.

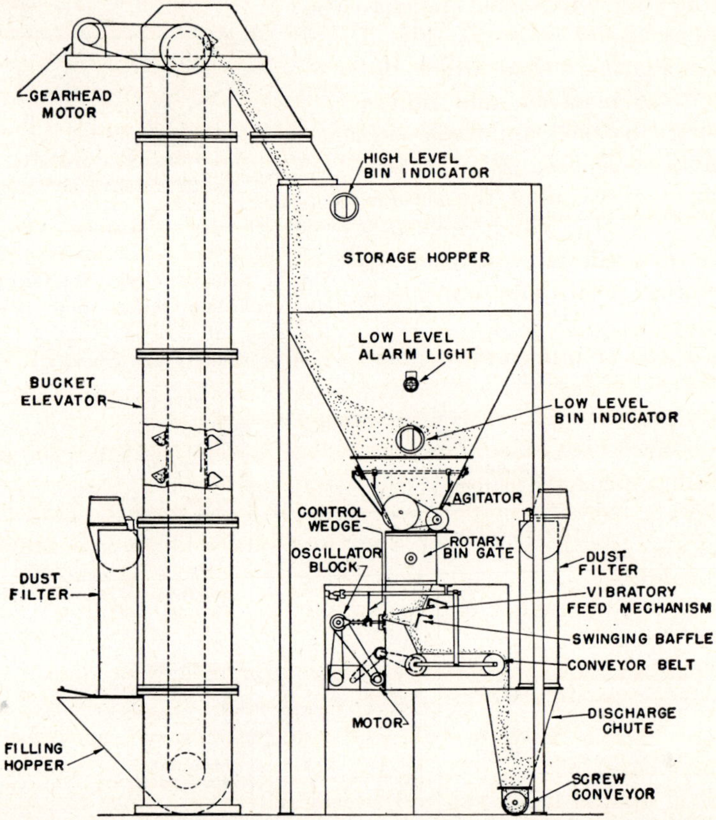

Fig. 7. Belt-type gravimetric feeder. *(Courtesy of Proportioneers, Inc.)*

Figure 6 shows a volumetric feeder in which the chemical is delivered by a molded rubber belt on which equally spaced vertical lugs form compartments of the same size. The feed rate is fixed by a stepless variable-speed drive. Impulses from a flowmeter may cause intermittent operation in proportion to the water flow.

Gravimetric Feeders. Units of this kind have a wide application and are extensively used, especially in large plants. Figure 7 shows a belt-type gravimetric feeder. The gravimetric control mechanism consists of a vibrating tray fed through a gate from a storage hopper built into the machine. The chemical is fed by this tray onto a short conveyor belt which is operated by a variable-speed drive. The belt is connected to, and suspended from, the beam of the weighing scale. A rubber wedge is attached to the beam of the scale and hangs freely, the positioning of the wedge depending upon the weight of material on the belt. As the rubber control wedge is raised or lowered by the scale into a V-shaped rubber bumper, the impact on the feeding tray is varied, thereby speeding up or reducing the rate of chemical delivery to the belt. This action maintains a constant weight of chemical on the belt, which is adjustable by setting the rider weights on the scale beam. The rate of chemical feed is accurately proportional to the weight of chemical on the belt and to the rate of belt travel, which may be proportioned to flow or made constant.

CHEMICAL-MIXING EQUIPMENT

After the coagulant is added to water, it must then be thoroughly and quickly mixed with the supply prior to flocculation. Mixing chambers are designed in a number of ways, depending upon the size, type, and purpose of the treating system. All of the following means have been used to effect mixing:

1. "Around-the-end" type, in which a rectangular basin is fitted with vertical baffles which cause the water to flow in a narrow but deep passage with 180-deg turns at the end.
2. "Up-and-down" type, in which the water is made to travel alternately to the surface and to the bottom by a succession of cross baffles in a flume or conduit.
3. Straight-flow conduit of considerable length, imposing high velocity and some turbulence.
4. Mechanical agitation by motor-driven paddles in circular or square tanks.
5. Combination of around-the-end and mechanical agitation in a rectangular basin.

A typical up-and-down type of baffling is shown in Fig. 8. Although such mixing compartments and those based on the around-the-end type have been used in the past and are still employed in some large plants, they are not so desirable as "flash mixers," in which the chemical and water are mixed rapidly by high-speed mechanically operated paddles or similar devices. The design of a high-speed mixer is shown in Fig. 9.

The time required for thorough mixing will depend upon the means provided. When the mixing is by baffled compartments, 5 to 20 min detention time is provided, whereas if flash mixing is used, the time may be from 3 to 15 sec. The time allotted for mixing should be governed by the secondary flocculation equipment. In general, only such mixing time should be allowed as to insure adequate admixture prior to floccula-

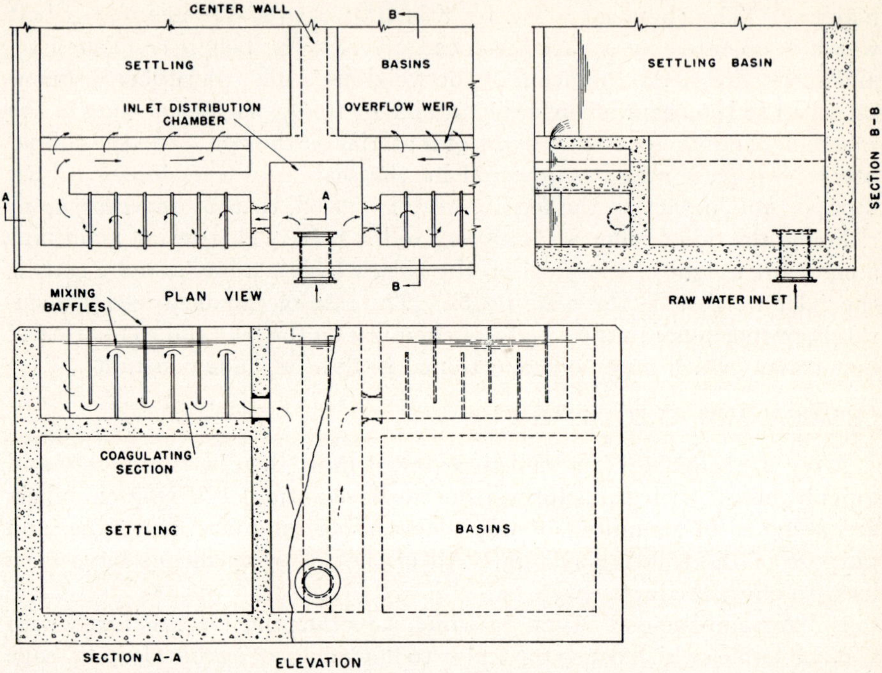

Fig. 8. Plan and sections of coagulating and settling basin showing the arrangement of mixing baffles and overflow weir.

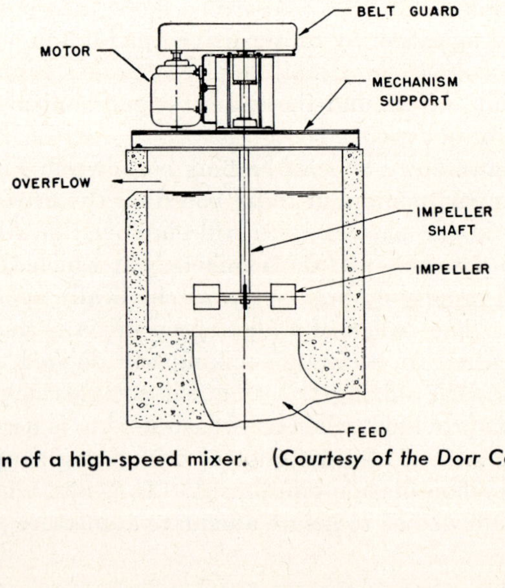

Fig. 9. Design of a high-speed mixer. *(Courtesy of the Dorr Company.)*

tion. If the mixing period is too long and the floc is well formed prior to flocculation, the coagulated particles may be broken up by the rotating mechanism in the flocculation chamber. When this occurs, flocculation is retarded.

FLOCCULATING EQUIPMENT

It was early recognized that development of large, dense, rapidly settling floc was promoted by following the rapid chemical mix with a period of slower mixing motion. This allows the forming floc particles and suspended matter to collide and grow larger. Before the widespread use of mechanical equipment, this was generally accomplished by a system of baffles similar in principle to those used for rapid chemical mixing but providing lower velocities suitable for flocculation. Practically all modern plants are constructed with motor-driven flocculating

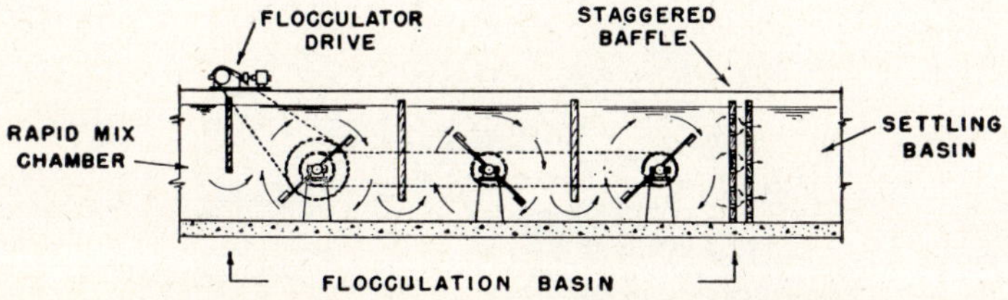

Fig. 10. Rotating flocculator mechanism: Dorr design.

equipment. Where conventional horizontal-flow settling basins are used, the flocculators are installed in a section ahead of the basin inlet. Special types of treatment tanks, to be described later, incorporate floc-forming and clarification facilities within the same basin.

A flocculator consists of motor-operated paddles rotated by a central shaft. The rotation is through a vertical or horizontal plane, depending on the specific design used. A number of flocculator designs are available, but all are based on the principle of rolling the water with its contained floc to increase the particle size, much as occurs when rolling a snowball.

The period required for flocculation depends on the design of the equipment used, the chemical characteristics of the water, and the amount and fineness of the suspended matter carried by the supply. It is generally conceded that flocculation time should be not less than 20 min nor more than 1 hr, although deviations from these limits are used by some designers.

The speed and size of the flocculator are governed by the paddle-tip velocity, which is the paddle-wheel diameter $\times \pi \times$ the revolutions per minute of rotation. The size and speed should be such as to limit the tip velocity to not more than 1.2 ft per sec. After leaving this compartment, the water should not be subjected to velocities higher than the flocculating velocity, otherwise the floc may be broken up and fail to re-form.

Figure 10 shows the principle of operation of a flocculator installation, and Fig. 11 is a typical installation of flocculators, which shows the arrangement of the baffles at the inlet end of a settling basin. The baffles placed between each set of flocculators are to prevent short-circuiting and to increase the travel time through the compartments.

Fig. 11. Photograph of a rotating paddle flocculator. *(Courtesy of the Dorr Company.)*

A novel type of flocculator, based on a walking-beam design, has recently been installed in some plants. The results obtained from the equipment have shown that its principle of operation is sound and that the degree of flocculation obtained is comparable to rotating units. A set of these flocculator units installed in a basin is shown in Fig. 12. The flocculation of the coagulum is obtained by an upward and downward travel similar to the action of a walking beam. To the moving mechanism are attached V-shaped troughs constructed of wooden slats. As the units move up and down, the finely divided floc particles are agglomerated into relatively large masses, as effected by rotating units. Flocculators of this kind have a specific sphere of usefulness, and they can be installed in existing basins requiring only minor reconstruction changes. Another feature of advantage in the construction of this equipment is that there are no bearings under water.

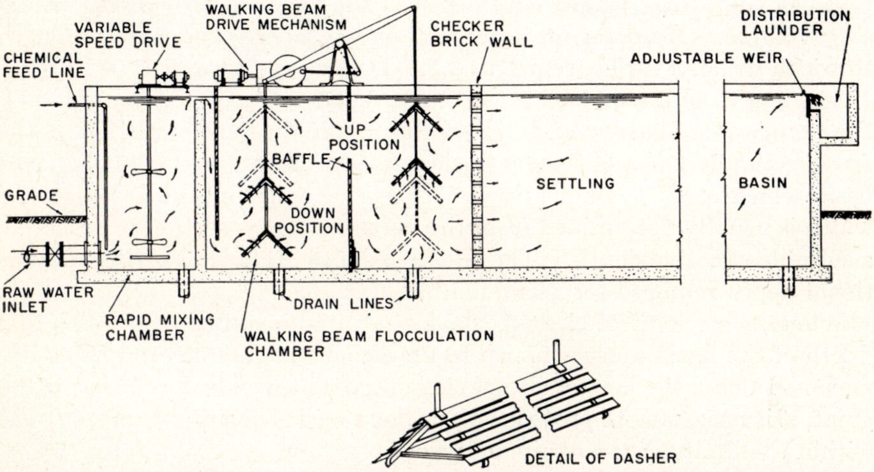

Fig. 12. Stuart walking-beam-type flocculator. (*Courtesy of the Stuart Corporation.*)

SETTLING BASINS

Settling facilities fall into two major classifications: (1) conventional settling basins providing a minimum of 2 hr of detention (usually more) and (2) rapid-flow treatment tanks providing 1 to 1½ hr of detention. The first group includes horizontal-flow basins; and the second type, for relatively small plants, includes round or square upflow units.

Design of Conventional Basins. Settling basins for large plants are generally rectangular, but small capacity units may be square or circular if space limitations require. It is customary to design the tanks for straight flow, although one or more round-the-end baffles may be included in the design to increase the length of travel of the water from the inlet to the outlets.

Retention Time. The time the water should be retained in the settling basins during passage from the inlet to the outlet will vary with the character of the raw water. Basin designs call for from 2 to 6 or 8 hr. On the average, most basins are designed for about 4 hr. If activated silica is used continuously, the detention periods may be shortened because of the rapid settling rates effected by the floc produced by this coagulant aid. Sludge produced by chemical softening requires less settling time than light and "fluffy" flocs. It is good practice to limit the maximum flow velocity through a settling basin to not more than 1 ft per min with most supplies, although some basins are designed to permit flow rates of 2.5 ft per min.

The volume of a standard settling basin is dependent upon two factors, the requirements for retention time in the basin, and the water flow in gallons per minute. There are no arbitrary standard relationships between the width, length, and depth of the basin to provide the necessary volume. However, in large rectangular basins, the ratio of length to width usually varies from 1 to 2.5. Depths vary from 10 to 20 ft.

The length of a standard straight-flow settling basin is determined by the permissible velocity of flow and the required retention period. The cross section is dependent upon the length and the required volume. In a basin which is not to be provided with a mechanical cleaning device, the width is usually determined from the necessary cross section by selecting a suitable effective depth. The effective depth excludes 3 or 4 ft of additional depth required for accumulation of sludge.

Where a mechanical cleaning device is to be installed, the width and depth of the basin must conform to the size of the manufacturer's equipment. In case the largest size of cleaning device yields depth that is too great, it is necessary to provide two or more basins of suitable proportions within the limits of the equipment sizes.

It is important to design the entrance and exit facilities of basins to secure maximum retention of water and to avoid short-circuiting. Numerous arrangements are used to effect this. A distribution baffle with staggered or checkerboard openings is often employed at the entrance to the basin. This arrangement provides uniform flow from the top to the bottom of the basin and thus avoids uneven flow rates throughout the basin.

The effluent or "take-off" weir should be carefully designed to ensure that each foot of weir will receive the same volume of water passing over it. To accomplish this, the weir must be absolutely level over its entire length. It is often desirable to install a weir in the form of a trough, thereby doubling the weir length and reducing the velocity over the weir edge. Two troughs of this type are installed at times to reduce still further the velocity over the weir.

Leveling long weirs, even when they are designed with adjustable weir plates, presents a problem. To overcome this difficulty, many tanks are now provided with troughs having serrated edges or a series of small rectangular weirs located as vertical slots. Others are provided with a series of holes placed in a uniform pattern in the bottom or sides of the take-off troughs.

Cleaning Deposits from Basins. It is imperative that deposits be removed from basins, either periodically or continuously, to ensure a satisfactory clarification of the water in its passage through such units. If large amounts of deposits are permitted to accumulate, the settling capacity is reduced. A more serious defect in operation occurs from decomposition of organic deposits, which results in gas formation. Such a condition creates eddy currents in the water as the gas bubbles rise through the main body of water in the tank. At times, this action is so violent as to cause large masses of deposited material to be carried to the surface, resulting in marked reduction in the clarification efficiency of the tank.

If settling basins are designed for periodic cleaning, it is preferable to provide enough additional capacity so that one basin may be taken out of service for cleaning without overloading the other unit or units in service. This type of cleaning operation can be avoided by equipping the basins with mechanically operated rakes which move the deposited sludge to a sump from which it is discharged to waste. In upflow basins designed to operate with a concentrated sludge blanket or slurry pool the control of the sludge mass is effected by blowing down continuously or semicontinuously, as conditions warrant. This operation is generally carried out by automatically controlled valves, supplemented by manual blowdown if needed.

The problem of sludge removal and its effect on receiving bodies of water is of growing importance and is related to stream pollution. The discharge of large volumes of sludge at one time into a small river or other surface water is likely to create sludge banks. Such a condition may in time restrict navigation, add to the frequency of dredging channels, and have a detrimental effect on fish and other aquatic life. This is more serious in large plants than in small ones, but if the available volume of dilution water is small in comparison with the amount of sludge dumped, even small plants may create a nuisance. Such problems are minimized when the basin deposits are discharged continuously, because the diluting water can more readily disperse the solids without creating a nuisance.

Round or Square Settling Tanks. Coagulation and sedimentation basins providing conventional retention periods for small to moderate flow capacity may be square or cylindrical with the flow predominantly upward rather than horizontal. It is more difficult to design properly and operate such tanks to obtain maximum subsidence efficiency owing to the trouble encountered in the placing of baffles and stilling walls and to the difficulty in controlling currents which may be set up. The influent may be a downcomer at the center of the tank and the effluent or take-off weir at the periphery.

One convenient arrangement of baffling is to have the influent pipe enter the tank at one side with a stilling wall placed as a chord and the center baffle placed at right angles to it. The opening between this baffle and the far side of the tank should be wide enough to avoid excessive velocities. One side of the baffle serves as the inlet and the opposite side as the outlet.

Many small installations of this type are built in steel tanks. The objection to metal tanks is the deterioration due to corrosion, although when the metal is properly protected, such tanks are quite satisfactory. Wood-stave tanks have frequently been used for reasons of economy and availability but have not always enjoyed these advantages under certain conditions of material shortages. Circular concrete tanks are employed at times, but where such construction is desired, it is preferable to build rectangular or square basins, since these may be designed to operate more efficiently.

RAPID-FLOW TREATMENT TANKS

In recent years a number of designs of clarification tanks have been used in which the removal of suspended solids takes place by upward filtration through a floating blanket of sludge. In others, the clarification

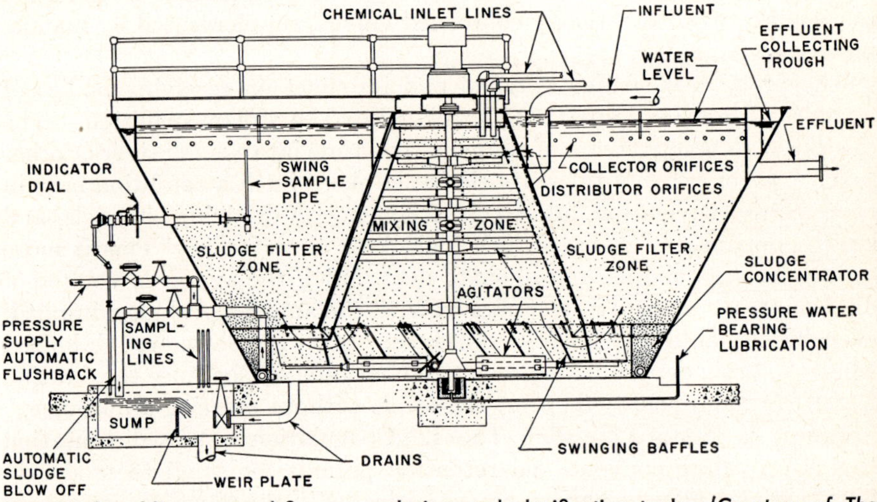

Fig. 13. Spaulding upward-flow coagulating and clarification tank. *(Courtesy of The Permutit Company.)*

is brought about by forced recirculation and intimate mixture of the prepared coagulated mass and upward flow, resulting in more effective clarification than may be secured in standard tanks depending only on removal of suspended matter by subsidence.

Spaulding Precipitator. The operation principle involved in this type of upward-flow coagulating and clarification tank is shown in Fig. 13. In this unit, the slurry in the central compartment is prepared by flocculating paddles and is discharged from the central compartment to the outer compartment, in which a predetermined level of the concentrated sludge is maintained. The water passes upward through the bed at controlled rates of flow, whereby clarification takes place. In general, the blanket of sludge serves as a floating filter bed. When properly designed, this type of unit can effect a high degree of clarification. The water containing that part of the floc which has not settled out passes upward through an annular opening into the outer part of the tank, which is formed by the central upright cone and the outer inverted cone. The cross section of this part of the tank increases as the elevation increases, and as the water flows upward, its velocity is reduced. At a certain elevation, the velocity is just sufficient to support the floc, which, therefore, forms a blanket at this point below a top layer of relatively clear water. The height of the sludge blanket is governed by the rate of flow through the plant and the withdrawal of excess sludge. The scrubbing of

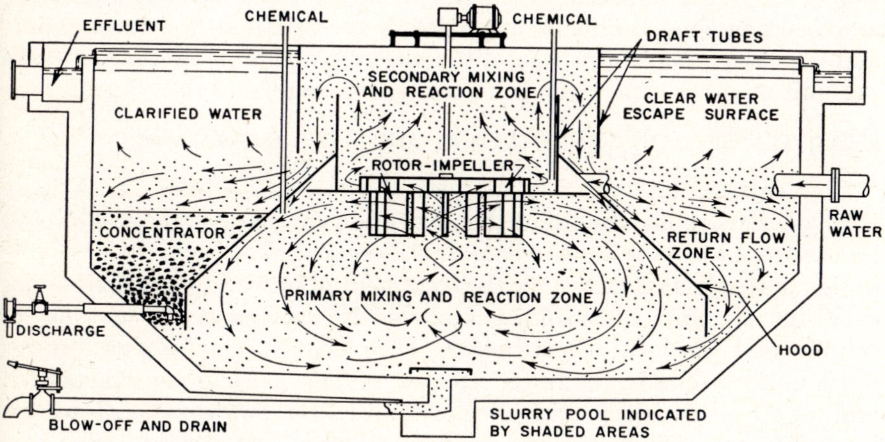

Fig. 14. Rapid-flow slurry-pool tank. (*Courtesy of Infilco Incorporated.*)

the water by the blanket of floc during the upward passage through the cone assures a completeness of reaction and retention of suspended matter that is not ordinarily obtainable in standard subsidence basins. The relative dimensions and angles shown here are those which were found to be best suited to the clarification only or for the softening of a particular water supply by means of lime and a coagulant. For other purification processes in which the characteristics of the sludge and density and viscosity of the water are different, it may be necessary to introduce wide variations in the rate of flow, angle of the cone, and mixing and sludge-removal facilities.

Accelator. Another type of tank, based on the rapid-flow principle but differing radically in design and operation, is shown in Fig. 14. In this tank, the chemicals are added directly to a relatively thick concentration of precipitated sludge near the bottom of the tank, and the raw water is treated with this mixture and is stirred at a higher velocity than is commonly employed in coagulating tanks. It is the purpose of this design to bring about the initial chemical reactions in the presence of sludge so that precipitation takes the form of accretion to existing particles rather than formation of new particles of much smaller size. From the primary mixing and reaction zone, the water flows upward into a final mixing and reaction zone, carrying with it a high concentration of sludge that is kept suspended by relatively rapid mixing. Flowing downward out of this area, the water enters an outer section of much larger diameter which is free from turbulence. The water then flows upward, which permits removal of the floc by settling in the compartment. Sludge settles out in a concentrator compartment covered with diagonal settling plates to furnish large area and is removed continuously. A larger portion of the water passes through a return zone back to the primary mixing and reaction zone. This recirculation is several times the throughput, and it is this feature which permits unusually high and fluctuating flows per unit of rising area without impairing the quality of the effluent.

RELATIVE MERITS OF CONVENTIONAL AND RAPID-FLOW TANKS

Conventional and rapid-flow types of clarification basins each have their definite application, and their selection for specific services should be carefully evaluated. In general, the conventional or horizontal-flow design of tank is best suited for the treatment of water supplies containing excessive amounts of turbidity and widely varying concentrations of soluble and insoluble constituents. The conventional design insures more uniform quality of water because of the longer detention period, and its operation is less critical than the rapid upflow units. However, standard conventional basins are generally more costly to construct and require elaborate sludge-removal facilities for best results. In some cases, mechanically operated desludging apparatus is not provided, and in this event, desludging must be done by intermittently removing the tanks from service, dewatering, and flushing the accumulated sludge to waste.

Wherever possible, mechanical desludging facilities should be provided because they avoid emptying basins intermittently with resulting loss of capacity and avoid dumping large amounts of mud to sewers or receiving bodies of water, whereas if the sludge is removed continuously in small quantities, these objectionable operating defects are eliminated. The conventional basin design imposes problems of ice and algae growth, which are more troublesome than those affecting rapid-flow tanks. Further, the conventional basins require more space per gallon of water treated.

In summary, conventional settling basins produce a more uniform quality of water, do not require so much detailed attention, and are better suited to treat water containing excessive quantities of turbidity.

Rapid-flow units have much merit and are well suited for clarification of many kinds of water. They are generally less costly to install than conventional basins and, when properly operated, can almost completely remove suspended solids. To accomplish these results it is necessary that the flow rates be low and that the slurry pool or floating sludge bed be satisfactorily maintained. When this is not done, water of poor quality will be produced.

This type of system has a distinct advantage over conventional units because of the ease of removing sludge, which is accomplished merely by opening blowdown valves. Some blowdown valves are operated manually, but for best control of the sludge level, the blowing down is automatically operated.

5 FILTRATION

The first successful municipal venture in this country for the treatment of water by filtration for municipal requirements was a small charcoal, sand, and gravel filter built in 1855 at Elizabeth, New Jersey. Up to the end of 1860, 136 water plants had been built in the United States and 10 in Canada.[1]

The rapid sand filtration process* was patented by John W. Hyatt,[2] of Newark, New Jersey, in the early 1880's. Ten years later, filtration of water for municipal and industrial use was on a sound engineering basis. Since that time, thousands of plants based on the Hyatt principle have been installed for conditioning water for sanitary and industrial uses.

THEORY OF FILTRATION

Removing suspended solids by passing water through a porous medium is primarily a mechanical action. Many particles will not pass through the interstices in the filter because of their size. This action alone, however, may not effect complete clarification. Finely divided solids are eliminated only by the formation of gelatinous films around the grains of filter media. In some filters this filming of sand grains is brought about by biological action. This is more pronounced in slow sand filters, where a slimy coating, technically called a *Schmutzdecke*, forms on the sand bed surface. Usually it is teeming with microscopic organisms that reinforce themselves to give a coating difficult to break. The mat they form aids clarification. A floc can be formed by a number of chemicals to create an artificial *Schmutzdecke*.

Just how far solids penetrate into a sand bed depends, to a great extent, upon the rate of filtration, the thoroughness of pretreatment, and the size and gradation of the filter medium. As time goes on, a properly

* This term was coined to distinguish it from the slow sand process, which uses no chemical coagulant. The rate of filtration through mechanical filters is much more rapid than through the slow sand type.

operated filter increases in efficiency. Flocculated matter fills the voids in the bed, and the artificial coating forms smaller interstices and aids in the removal of both soluble and suspended solids.

MECHANICAL FILTERS

Gravity Type. Rapid sand filters are either rectangular or circular and are made of wood, steel, or concrete. In the bottom is an underdraining system, which collects filtered water and distributes the backwash water. A bed of graded stone, coarse anthracite, or gravel is placed over the underdrain, and 24 to 36 in. of filter medium on top of the underdraining layer. Influent and effluent valves with controllers for maintaining uniform rate of flow of water through the apparatus, loss-of-head gauges, backwashing apparatus, and wash-water troughs or gutters complete the assembly.

Figure 1 shows a view of a rectangular filter unit. Water enters at the top and passes downward through the filter bed. The filtered water flows through the underdraining system and discharges from there into the clearwell, or filtered-water basin. To clean a unit, the filtering process is stopped and filtered water is discharged upward through the underdraining system and through the bed. Solid material collected on the surface of the bed is discharged into the wash-water troughs and then emptied into the sewer. In some units air is employed to agitate the filter medium. Wash water at high rates (15 to 30 gal of water per square foot of filter area) is generally used for agitation without air. Surface washing of the beds is often provided in addition to the upward flow of wash water. When an insufficient amount of water is available for washing the entire bed, it may be washed in sections. Sectional washing of filter beds is not employed except when the regular method of backwashing is not possible.

It is common practice to install coagulating and subsidence basins before the filter. This pretreatment serves two purposes: a large amount of suspended solids is removed by subsidence, reducing filter load, and the coagulated material coats the filter surface, preventing finely divided particles from penetrating the bed.

Pressure Type. Pressure filters are essentially the same as rapid-rate filters, except that the filter material is contained in a closed tank and the water is forced through the filter bed under fairly high pressure. This form of filtration has found favor in industries, especially for the treatment of feedwater, since the apparatus may be installed on pressure lines, thereby avoiding repumping.

Such units are built in two forms, vertical and horizontal. The former is adaptable primarily for filtering relatively small amounts of water, and the latter, where greater volumes are required. These two types of

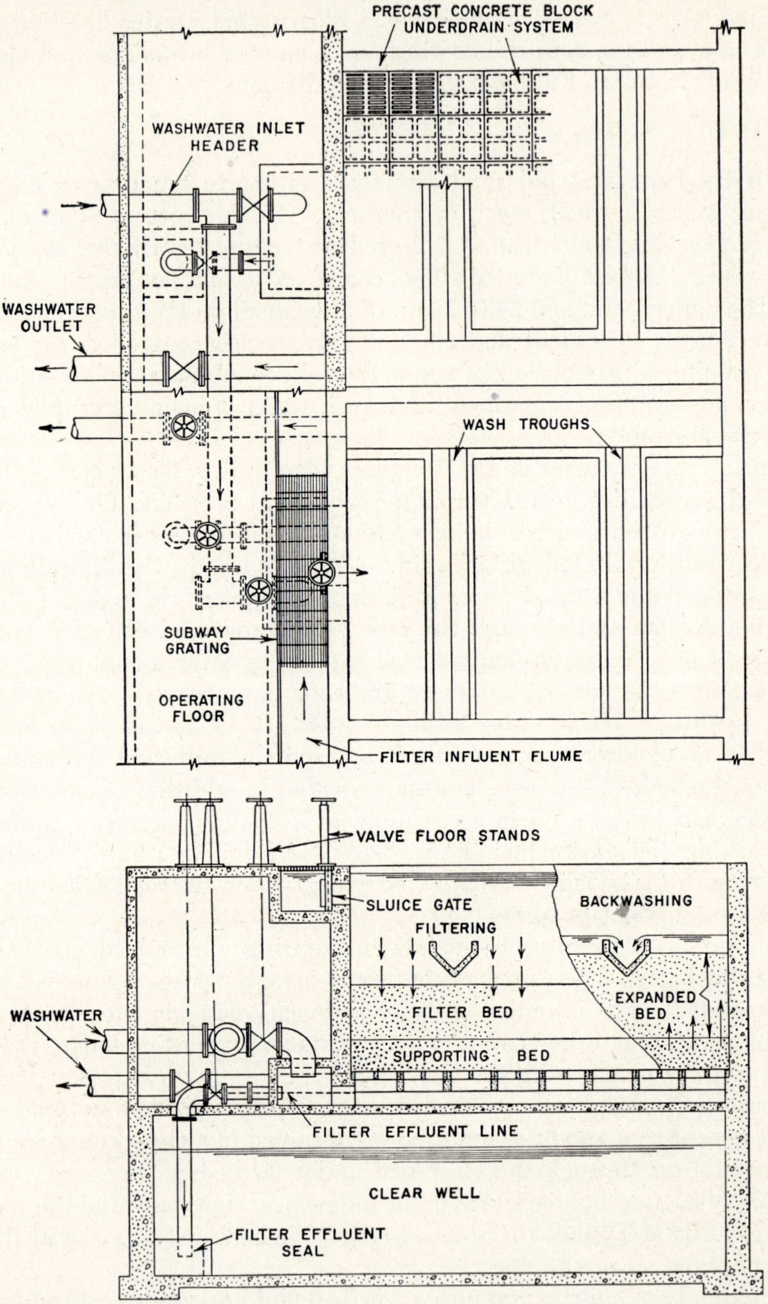

Fig. 1. Rapid sand filter (gravity type) showing arrangement of filter beds, valves, and clear-water well.

Fig. 2. Pressure filter equipped with hydraulic surface-washing mechanism. (*Courtesy of The Permutit Company.*)

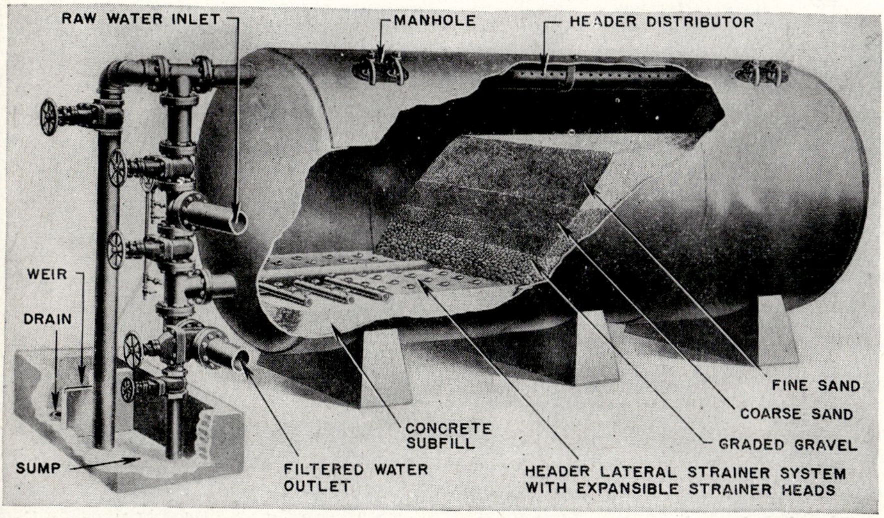

Fig. 3. Horizontal filter showing arrangement of underdraining system and filter medium. (*Courtesy of The Permutit Company.*)

filter are shown in Figs. 2 and 3. In some installations, one compartment of the filter shell is used as a settling chamber. This arrangement is of little value, since the time of detention of the water is so short that only a small amount of suspended solids is removed. Separate settling basins are necessary in the treatment of muddy water and should be used where the raw water contains a large amount of suspended matter.

In the operation of most pressure filters, the water enters the top of the unit and is distributed by means of a trough or perforated pipe and is filtered by passing downward through the filter medium. The filtered

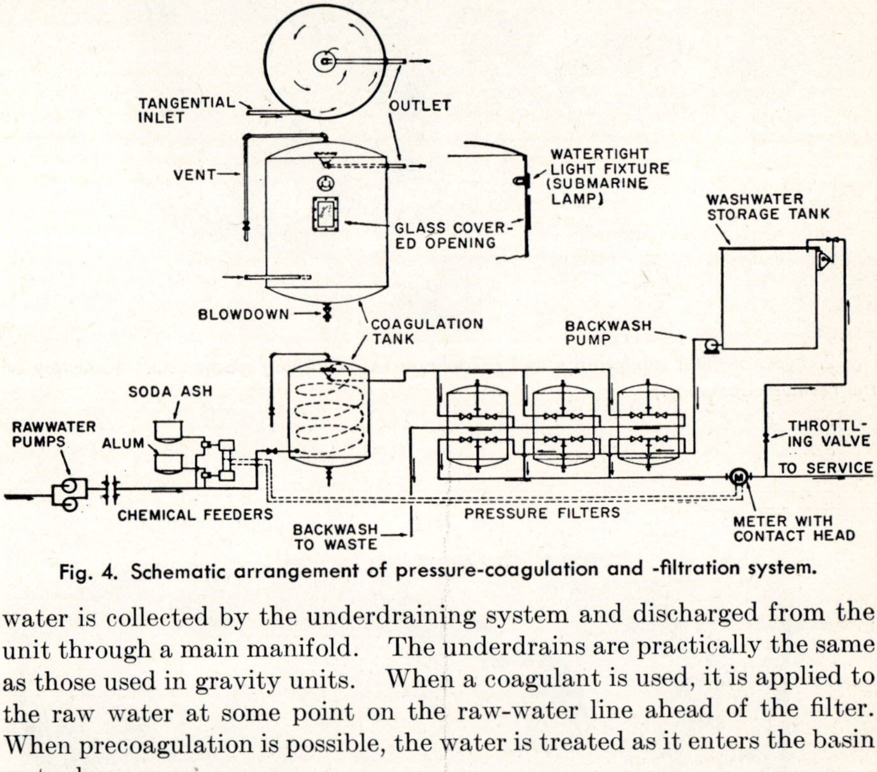

Fig. 4. Schematic arrangement of pressure-coagulation and -filtration system.

water is collected by the underdraining system and discharged from the unit through a main manifold. The underdrains are practically the same as those used in gravity units. When a coagulant is used, it is applied to the raw water at some point on the raw-water line ahead of the filter. When precoagulation is possible, the water is treated as it enters the basin or tank.

In some installations, a small coagulating tank is inserted in the line ahead of the filters. Such tanks will not serve to remove the coagulated material to any appreciable extent because the detention period is too short. However, a coagulating tank of this type is of value to give time for the chemical reaction to take place prior to reaching the filter beds. To this extent these tanks possess value. A typical arrangement of an installation of this type is shown in Fig. 4.

The depth and grading of the gravel and sand beds or other filter media are varied by different manufacturers. Depths of gravel from 8 to 20 in.

and of sand beds from 18 to 30 in. or more are employed. The majority of pressure filters are backwashed by passing the water upward through the filter at rates of from 7.5 to 15 gpm per sq ft of filter area. In the past, mechanical agitation of the filter medium was generally not used in pressure filters. At present, many units are being designed with supplementary devices for aid in removing surface accumulations. Compressed air and mechanically operated rakes are employed to some extent.

Vertical pressure filters range in size from 12 to 120 in. in diameter, and the horizontal filters are from 6 to 8 ft in diameter and from 8 to 25 ft in length. The shells are designed to withstand working pressures in

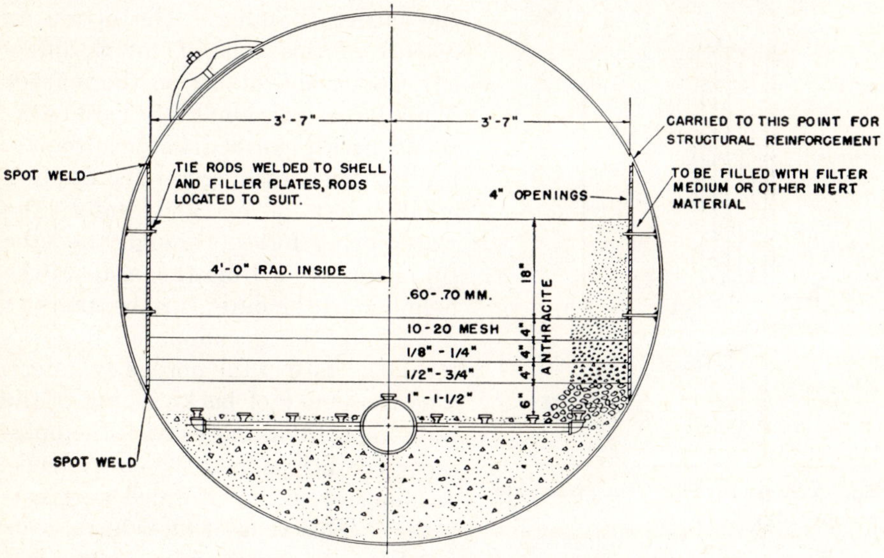

Fig. 5. Design for conversion of horizontal filters into vertical-type units by means of vertical steel filter plates.

accordance with the pressure of force mains supplying the raw water. It is common practice to have standard equipment to withstand working pressures of from 65 to 100 psi. Rules governing the choice of filter medium for pressure filters are the same as those which have been given for gravity installations.

Horizontal filters are not so satisfactory as vertical filters, owing to the tendency of the underdraining gravel to shift and form hills and valleys in the gravel or stone upon which the filter bed rests. When such failures occur, poor filtration results.

Defective operation of this kind has been overcome in some cases by installing vertical baffles in the tank, thereby converting the interior of the cylindrical shell into rectangular units. A drawing showing this type of reconstruction is shown in Fig. 5.

DIATOMACEOUS EARTH FILTERS

Recently, a number of manufacturers have designed filters, which are now fairly extensively used by industry, based on the principle of filtering the water through porous tubes or plates instead of beds of sand, anthracite coal, or other material. Figure 6 is a cut of a filter using porous tubes operated with or without a filter aid such as diatomaceous earth, diatomite, or other form of filter material.

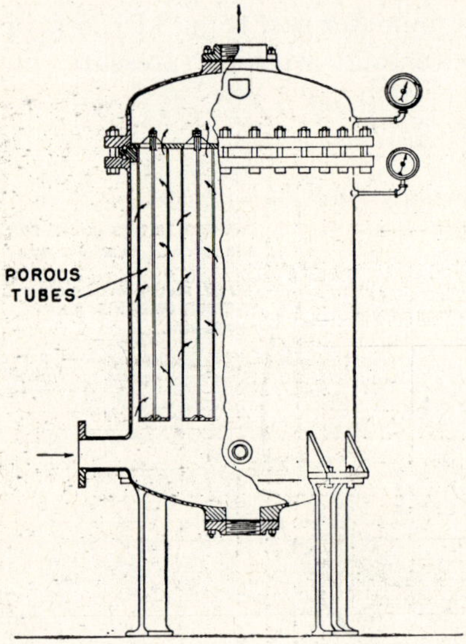

Fig. 6. Pressure filter equipped with porous tubes.

The water to be treated enters the shell and passes through the porous cylinders and is then conducted to the outlet chamber. The operating feature of this type of unit is different from some others on the market in that the individual tubes are backwashed with clarified water automatically. The washing is effected by a combination of air and water, the water being forced through the tube by means of air from the inside of the cylinder to the outer surface, thereby displacing the accumulated deposits on the surface of the porous cylinders. The frequency of backwashing of the individual tubes varies with the operating conditions, and the entire backwashing cycle is performed automatically. This type of filter has a wide application, and when properly adapted to meet the conditions of the raw water, it is capable of producing a water of a high degree of purity.

In a similar type of unit, water is filtered through porous cylinders coated with a slurry of diatomaceous earth and asbestos fibers which are applied to the water as it enters the filter unit. The raw supply with its dose of filter aid passes through the raw-water compartment and is deposited on the surface. This material thereby serves as the filter medium, effecting the desired clarification. The diatomaceous slurry has to be removed from time to time by backwashing. Removal of the deposited precoat, or filter cake, from the filter elements is effected by building up pressure on both sides of the elements. This is accomplished by closing the filter effluent valve and subsequently closing the inlet water valve. Following this operation, pressure is suddenly released on the outside of the element by opening the backwash water valve, causing a sudden flow of water at high rates through the element from the inside outward and thereby displacing the accumulated deposits of precoat material. The displaced cake drops to the bottom of the filter shell and then is forced out of the unit by a head of water in the shell over the backwash drain valve, which discharges to waste. This type of unit is efficient for oil removal and for other special services for preparation of boiler feedwater and for industrial processes.

PLATE-TYPE PRECOAT FILTERS

Another design of filter based on the precoat principle is shown in Fig. 7. These filters differ from the others described earlier in that the filtration takes place through perforated Monel plates over which filter

Fig. 7. Plate filter on which diatomaceous earth and asbestos are used. (Courtesy of Davis Filtration Equipment Co., Inc.)

material is deposited. Diatomaceous earth, activated carbon, asbestos, and special types of clay are employed to coat the plates. After the elements are precoated, the water flows through each side of the plate and is removed through its center section, then flowing to a header from which the water is discharged.

After the loss of head becomes too great to make possible the desired rated capacity of the filter, the inlet water is shut off, and the plates are removed as a unit. The photograph shows the plate arrangement and the method of removing the plate assembly from the filter shell.*

When the rack of plates has been removed, the spent filter coating is peeled off or otherwise removed from the plates. The units are again

* In recent installations, paper bags designed to fit over the leaves are used to assist in preventing break-through of the untreated water and to aid in complete removal of the deposited cake from the filter leaves during the cleaning operation.

inserted in the shell, and the filter is ready for operation. When the coating is being built up on the plates by the filter aid, the water is wasted until the effluent runs clear.

This type of unit differs from many of the other precoat filters in that no backwash water is used. For some conditions, this is an advantage, since it avoids the necessity of disposing of contaminated backwash water. These units have been used for the removal of oil from condensate and for other special water-processing requirements.

PREFLOC OIL-REMOVAL FILTERS

Filters for the removal of oil from condensate or other sources of water are now being fairly extensively used because of the necessity for eliminat-

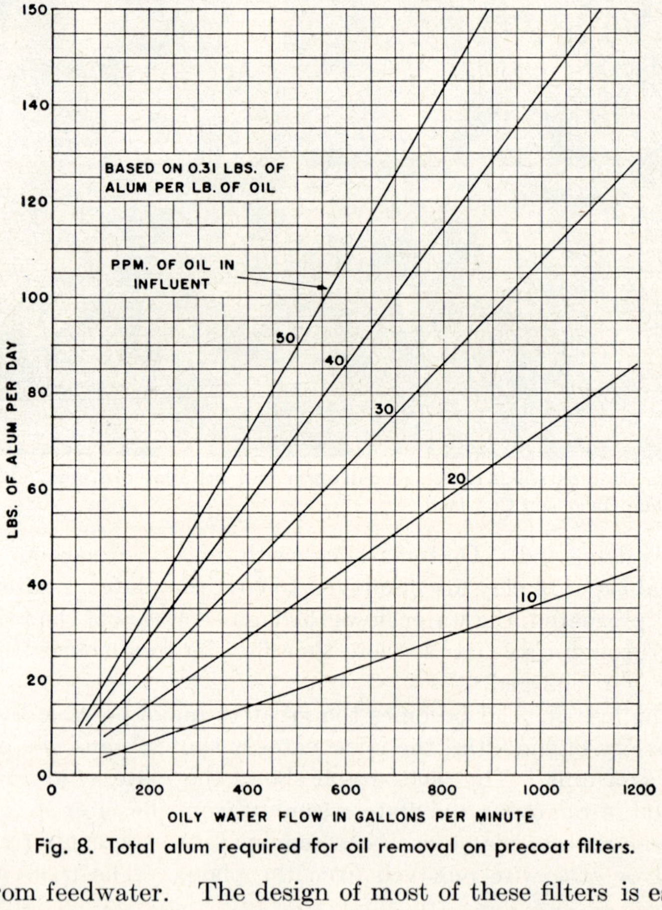

Fig. 8. Total alum required for oil removal on precoat filters.

ing oil from feedwater. The design of most of these filters is essentially the same as other types of pressure filters, but it is necessary to provide some form of agitation of the filter bed to insure proper cleaning of the

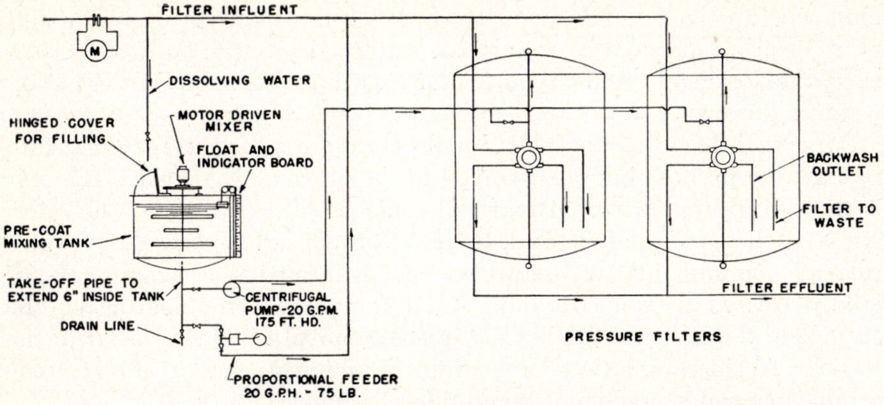

Fig. 9. Schematic arrangement of oil-removal filters and chemical-dosing apparatus.

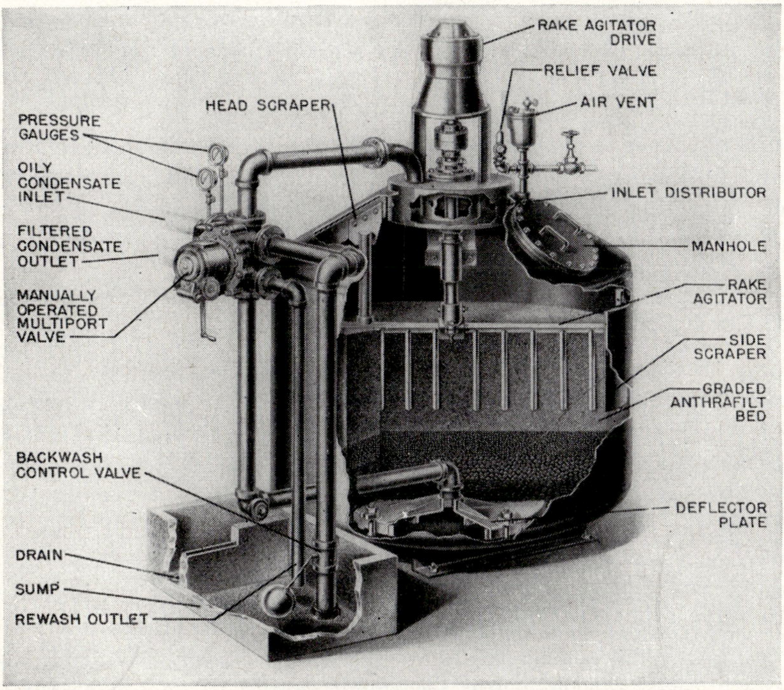

Fig. 10. Pressure-type oil-removal filter equipped with mechanically operated rotating rakes for cleaning sand. (*Courtesy of The Permutit Company.*)

unit. Further, the beds must be given a caustic wash from time to time in order to dislodge the oil which coats the grains of filter material.

It is common practice, also, in the operation of oil removal filters to precoat the surfaces of the filter bed by means of a preformed chemical floc. The prefloc is formed by the proper proportioning of a solution of alum with an alkali. The quantity of prefloc used depends on the amount of oil or other material carried in the water. From the author's experience, the dosage of prefloc required for treating various quantities of oily water is given in Fig. 8.

After each backwashing period and before the units are again cut into service, the prefloc slurry is pumped to the filters so that, when the unit is started up, the flocculent material coats the bed and enters the interstices at the surface of the bed, thereby preventing passage of oily water through the unit. After the prefloc coat is formed, the incoming water is also treated with a coagulant so as to maintain the coating on the surface of the bed and further aid in the removal of oil contained in the water. A diagrammatic sketch of an oil-removal system, showing the arrangement of a prefloc tank and filters, is shown in Fig. 9.

Figure 10 is a cutaway view of an oil-removal filter, showing the internal arrangement of sand and gravel and a cleaning device, motor-operated rakes. In some designs the rakes are replaced by rotating water jets.

COKE FILTERS

Units of this type are used to some extent either alone or as an adjunct to sand filters principally for the removal of iron and manganese from water containing objectionable amounts of these substances. The main purpose of these appliances is for aeration and as pretreatment of other forms of purification.

TYPES OF FILTER MATERIALS

Many substances have been employed and are being used for filtering industrial water supplies. These include sand, coke, charcoal, excelsior, marble chips, as well as magnetite, lava, anthracite coal, diatomaceous earth, asbestos, canvas, and many other substances. Larger installations, filtering cold boiler feedwater, generally employ sand or coal. Filter equipment for purifying industrial water is practically the same as that for sanitary purposes.

Sand. Sand is the most widely used medium for filtration of cold water. The choice of the right grade and sizing of sand and the depth of the bed are essential features in the design of a filter. Sand for use in filters employed for feedwater treatment should be chosen as carefully as that for use in the treatment of water for sanitary purposes. Pure quartz sand, free from loam, clay, or acid-soluble material, is the preferred material. The desirable size of sand varies with the kind and temperature of water to be treated, the rates at which the filter is to function, the depth of the bed, and other factors.

There has been a tendency within the past few years to use coarse sand, which formerly was believed to be unsatisfactory, in rapid sand filters. Some engineers have employed sand having an effective size* of 0.60 mm or larger. For average conditions of feedwater treatment, an effective size of from 0.40 to 0.45 mm is generally used.

The over-all efficiency of a filter medium, including the length of filter runs between washing periods, is obviously affected by the hydraulic gradation of the filtering material. The best results are obtained with material which is uniform in its dimensions.

Anthracite Coal. Anthracite coal has been used to a limited extent in municipal filtration for many years and has also been employed for the filtration of water from hot lime-soda softening systems. The material possesses a number of desirable characteristics and its use is rapidly increasing. Under some conditions, it is generally more efficient than sand, having a tendency to entrap certain kinds of suspended solids.[3–6]

The special merits of anthracite as a filter medium are due to its lower density, which is little more than half that of sand, and its irregular shape. These properties result in a porous bed which does not tend to crack and is easily backwashed. The area of the angular grains is greater per unit volume than the area of the more spherical sand grains. This facilitates entrapping floc and permits using a larger grain size without loss of filtering efficiency. Because of the coarse grain size, deeper parts of the bed are effective in filtration, higher filtration rates may be used, and longer runs are possible before excessive head loss is built up. The backwash rate required to accomplish satisfactory service is lower for anthracite than for the relatively heavier sand.

In certain applications, the chemical inertness of anthracite is of special value. Turner and Scott[4] state that iron and manganese deposits wash off the grains and that lime incrustations, common in filters receiving lime-softened water, do not form so readily on anthracite as on sand, leading to the inference that such deposits on sand are chemically combined with the silica. The most important chemical difference, however, is that strongly alkaline waters, especially when hot, leach silica from sand until it is completely dissolved. The low amount of silica in the proper grades of anthracite used for filtration makes this material preferable for such service. In boiler feedwater, any appreciable amount of silica is undesirable. Even a few parts per million of silica, such as might be leached from a siliceous filter medium, are undesirable.

* "Effective size" and "uniformity coefficient" are frequently used in examination of a sand to be used for filtration purposes. The term effective size was adopted by Hazen to apply to "the size of grain (in millimeters) such that 10 per cent, by weight, of the particles are smaller and 90 per cent larger than itself." Uniformity coefficient is the "term used to designate the ratio of the size of grain which has 60 per cent of the sample finer than itself to the size (effective size) which has 10 per cent finer than itself." (AWWA Manual, "Water Quality and Treatment," 2d ed., pp. 256, 257, American Water Works Association, New York, 1950.)

Anthracite is especially applicable to the following uses:

1. For replacement of other filter media in new or old filters, thereby increasing plant capacity, because of the higher filtration rates and lower backwash rates.

2. For filtering lime-softened or iron or manganese waters to minimize grain incrustations.

3. For filtering fine crystalline precipitates, such as calcium carbonate or phosphate from softeners. Anthracite coal may be used in a finer grain size than sand, calcite, or magnetite without imposing excessively low rates or high loss of head. Fine precipitates which would pass through other filtering media can sometimes be retained by anthracite.

4. For filtering alkaline boiler feedwater, either hot or cold, when the danger of contamination with silica must be carefully avoided.

5. For replacement of materials to improve backwashing and to overcome deficiencies due to inadequate pretreatment or other factors causing mud balls, upset gravel beds, etc.

When using anthracite for new installations or for replacement of other materials, the engineer should give attention to some of its inherent properties. The material cannot be substituted indiscriminately for materials of greater density. When designing a filtration system, it is good practice to provide a 30-in. bed of relatively coarse anthracite (0.60 to 0.70 mm) and sufficient freeboard height for from 50 to 100 per cent expansion of the bed.*

Backwash rates require careful consideration because of the lightness of the anthracite. This is especially important when the filters are backwashed with hot water.

One disadvantage of anthracite coal is that it is easily washed into the wash gutters and lost if too high rates of washing are used. When designing filters equipped with anthracite as the filter medium, the engineer must provide greater freeboard than for sand for specific wash-water rates. This necessitates greater depths of the gravity filter units and extra shell height for pressure installations.

While the depth and grading of supporting beds of gravel, coal, or other materials may vary in different plants, the operating success of a filter depends largely upon the gradation and condition of the supporting and distributing materials. A depth of 16 to 18 in. of graded material is used except where deeper beds are required to meet some specific condition. The individual layers should be from about 6 in. thick for the bottom layer of coarse material to 2 to 3 in. thick for the top and finer aggregate.

* The per cent of bed expansion during washing to ensure adequate cleaning of the bed depends on a number of design and operating factors. These include the adequacy of pretreatment of the water, the type and grading of filter medium, the temperature of the supply, and other conditions.

The different gradings used vary with the designing engineer's practice and experience. The general type of grading, however, is similar to that listed in the Appendix in Table 11.[7]

Since the gravel bed in a filter remains stationary, care should be taken, when placing the different layers, to insure satisfactory operation. The bottom layers should be laid by hand and then each succeeding layer carefully leveled off. Only clean, hard material should be used. Selected round gravel is preferred to other types of aggregate. In general, the deeper gravel beds yield better wash-water distribution, but little, if anything, is accomplished by having the depth exceed 18 or 20 in.

Magnetite. Frequently, existing filter shells are not deep enough to permit the desired depth of anthracite and still permit necessary expansion of the bed when backwashed. This problem may be overcome by substituting a shallow bed of magnetite, which is much heavier, for a supporting bed of gravel or coarse anthracite. A satisfactory depth of anthracite may then be laid over the magnetite.*

UNDERDRAINS

The essential requirement of filter underdrains is to withdraw filtered water from the filter area uniformly and to distribute the wash water in the same way without disturbing the gravel bed or other underdrain material used. It has been the practice for many years to use a center manifold with laterals tapped into it at right angles and spaced at equal distance over the floor of the filter. Strainers sometimes termed "strainer heads," "strainer cups," or "sand valves" are screwed into the laterals. Figures 11 and 12 show a typical strainer system and several types of strainers.

Many designers of filters use only perforated pipe in place of strainers. Such designs (Fig. 11) usually consist of a center manifold into which are tapped, on 6-in. centers, laterals laid in parallel lines running across the filter. These laterals are capped at the free end and tapped on the underside of the pipes with holes 7/32 in. in diameter. The holes are staggered at 45 deg and usually spaced about 3 in. apart. The size of the perforations and the spacing vary to meet special requirements. Where underdrains of this kind are employed, the size, depth, and grading of the gravel bed require special attention. Brass or other nonferrous metals may be used in place of ferrous metals, and the number, size, and spacing of holes in the collecting system may vary in accordance with requirements.

* Magnetite should be purchased under strict specifications to avoid the presence of sulfide. Inferior grades of magnetite can contaminate the water with hydrogen sulfide and other objectionable products.

Metal screen, perforated metal plates, wooden grids or lattice, false bottoms, and many other arrangements have been used. In tropical countries, false bottoms of interwoven strips of bamboo have been used with varying degrees of success. Underdrains based on the false-bottom

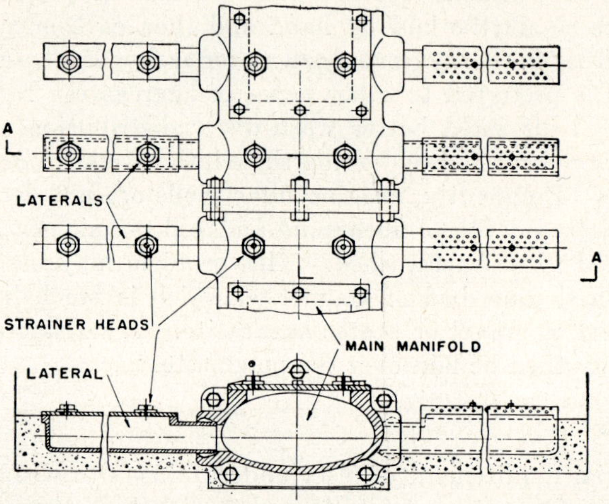

Fig. 11. Sectional view of a typical underdrain system for a rapid sand filter.

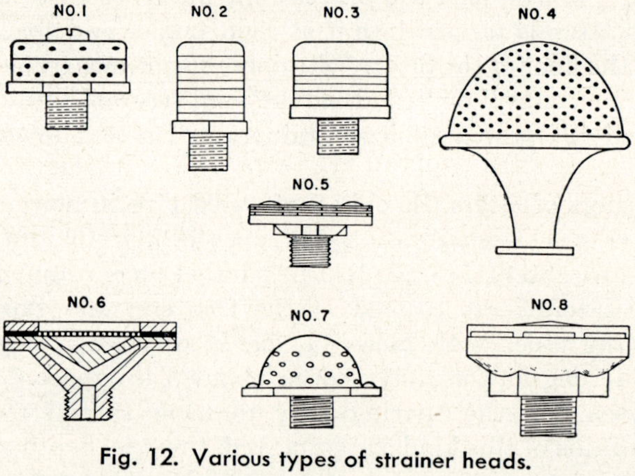

Fig. 12. Various types of strainer heads.

design have proved to be satisfactory and many systems of this type have been so constructed in recent years.

A satisfactory false bottom, consisting of slotted reinforced concrete slabs, has been used successfully in a number of plants. Individual 2- by 2-ft slabs are supported both longitudinally and crosswise, and the water used for washing has practically a free passage in the lengthwise direction. It is important in this type of underdraining system that velocity heads be kept sufficiently low to produce no interference with the distribution of the wash water. This requires designing the units to provide free and unobstructed passage of water under the false bottoms.

Another example of the simplified false-bottom type of underdrain consists of subway grating upon which the underdraining gravel rests. In circular units, the grating is placed in radial sections. The false flooring rests upon a steel framework supported from the bottom of the filter chamber.

Numerous arrangements of underdraining systems have been designed by engineers to meet specific requirements or to conform to local conditions.

POROUS-PLATE SYSTEMS

Porous plates, constructed of carborundum or other materials, may be used instead of standard underdrains. As long as the plates are clean, uniform filtering and backwashing result. When they become clogged, they require thorough cleaning, usually with acid. At times, the plates can be cleaned in place, while in other cases the unit has to be overhauled, the plates removed, cleaned, and replaced. The principal advantages of porous-plate designs are a saving of gravel or other supporting medium and shallower filter depths.

PREVENTION OF SHIFTING OF UNDERDRAINING GRAVEL OR COAL LAYERS

A serious operating problem which frequently occurs in filters is the shifting of the supporting medium in the lower part of filters. This difficulty is experienced in both gravity and pressure filters. Displacement of gravel or other supporting layers can be minimized by proper grading and care in backwashing. Too violent backwash or too high pressure on the backwash water mains is responsible in many cases for such defective operation. Entrained air can also upset the underdraining material.

WASH-WATER TROUGHS

Wash-water troughs, wash-water gutters, and similar facilities are used in both pressure and gravity filters for the dual purpose of collecting and removing wash water and for the distribution of water admitted to a filter. They should be placed sufficiently high above the surface of the filter material so that the latter will not be washed away during the cleaning period. The vertical distance between the surface of the filter bed and lip of the wash-water trough is called the "freeboard." The average height is about 2 ft, although greater and lesser distances are used. The greater the depth of the filter material, the greater will be the freeboard required in order to allow for adequate expansion during washing and to prevent loss of the material by washing it into the wash troughs or gutters. The size and specific gravity of the filtering material and the temperature of the wash water have very definite effects on the amount of expansion which will be obtained at given wash rates, and these factors must be taken into consideration in filter design. There should be sufficient trough capacity to carry all the wash water away from the unit without flooding. The troughs should be so placed that the maximum horizontal distance between their edges does not exceed 5 ft or at least so that the greatest horizontal travel of water does not exceed 3.5 ft.

Wash-water troughs are of various shapes and are made of cast iron, steel, or concrete. In pressure filters, sheet-steel troughs or perforated pipes are used.

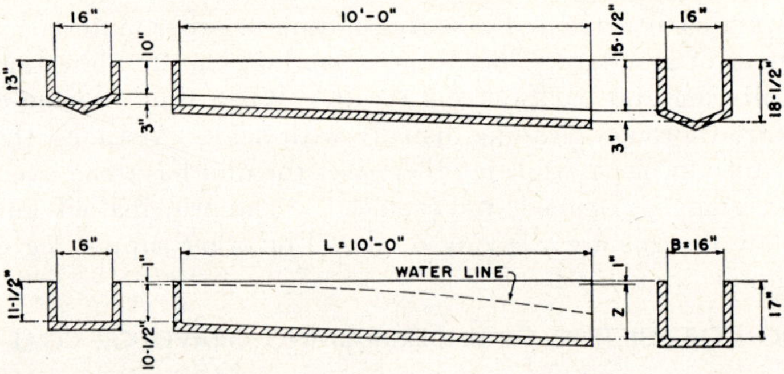

Fig. 13. Application of Miller-Streander formula for calculating the capacity of wash-water troughs.

The following formula for calculating, approximately, the discharge capacity of wash-water troughs, has been found convenient for designing such facilities:[8]

$$Q = 1.91\ BZ^{1.5}$$

where Q = total quantity of water received by trough, cfs
B = width of trough, ft
Z = vertical distance between inside bottom of lower end of trough and height of water at upper end, ft

This formula was originally developed for wash-water troughs of rectangular cross section, but it can be applied to troughs of any shape by assuming rectangular sections of equivalent area. An application of the formula to a concrete wash-water trough, 10 ft long and having a V-shaped bottom and capable of carrying approximately 4 cu ft per sec, is given in Fig. 13. In this example, a water line 1 in. below the upper end of the trough was assumed. Attention is called to the fact that, although the length of the trough is not a factor in the modified formula, adequate assumptions were made in C. N. Miller's* derivation of the original to make it applicable to the design of most wash-water gutters or troughs. It was further correlated with experimental data obtained by J. W. Ellms[9,10] at the Cincinnati filtration plant and has proved its adaptability on numerous occasions.

Wash-water troughs should be leveled in place so that the wash water overflows into every section uniformly. It is convenient to design filter troughs with an adjustable weir which permits of easy alignments. When such construction is provided, the weir should be made of brass, plastic, or other noncorrodible material. It is of the utmost importance that the troughs be made large enough to carry the maximum amount of wash water required for adequate expansion of the filter material.

SURFACE WASHING EQUIPMENT

Various schemes have been used to keep the surface filter media clean and free from mud balls and similar accumulations. However, the most satisfactory control for maintaining the surface of a filter clean is to agitate the bed by water or mechanically operated devices. The type of cleaning apparatus most extensively used in recent years is a Palmer surface washer.[11] The equipment is designed to scrub the surface media in the filters by jets of water under fairly high pressure as the pipes carrying the water are rotated around the surface. The rotation is effected by propulsion of the pipes by the jets of water impinging on the filter media. Such equipment is applicable to both gravity and pressure filters and, when properly designed and operated, will keep the surface areas of filters free from deposits.

Figure 14 illustrates the arrangement of the washers in a gravity filter. The figure illustrates the location of the rotating washers with respect to the sand bed and the wash-water gutters. Similar types of rotating washers are used in pressure filters and zeolite units.

During operation, the surface washers should be started after the surface of the water has been lowered to the edge of the wash-water trough. The regular backwash water is subsequently started; and as the bed expands, the rotating washers will be submerged in the bed for a few inches, and the combined surface, rotating washer, and the backwashing by the upward flow of water assure removal of deposits, which are carried away to the sewer.

MULTIPORT VALVES

Single-control valves by which the entire operation of the filters may be controlled are now used extensively. There are several different

* In J. W. Ellms, "Water Purification," Bibliography, p. 571, McGraw-Hill Book Company, Inc., New York, 1928.

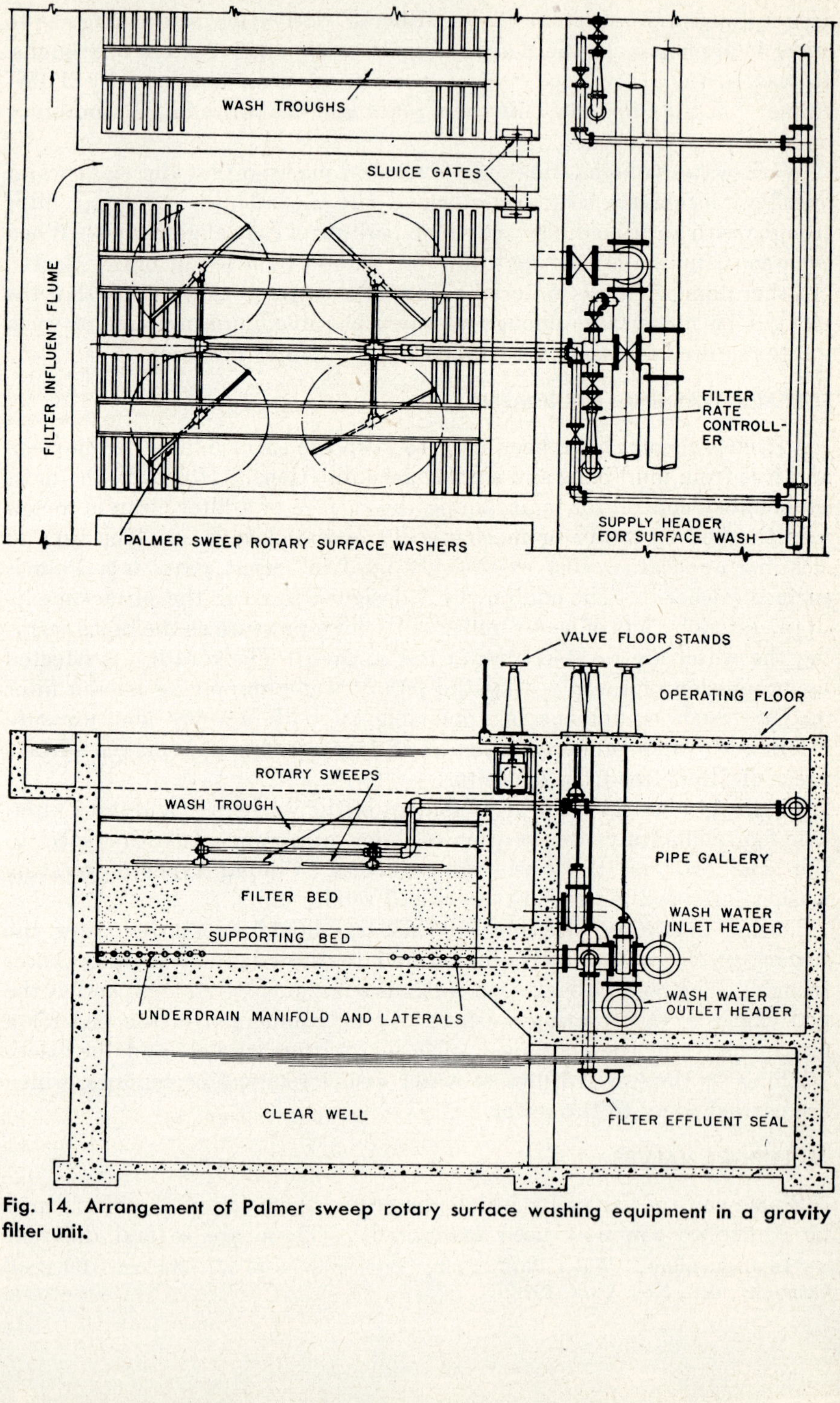

Fig. 14. Arrangement of Palmer sweep rotary surface washing equipment in a gravity filter unit.

designs offered by manufacturers. One of these designs, which reduces the chances of improper operation to a minimum, is shown in Fig. 15. The positions of the valves are clearly indicated on a dial by a pointer, and the operating mechanism is rotated in one direction only. Valves of this type are often motorized and are applicable to fully automatic or semiautomatic operation of both filters and zeolite units.

Fig. 15. Multiport single-control valve. (*Courtesy of The Permutit Company.*)

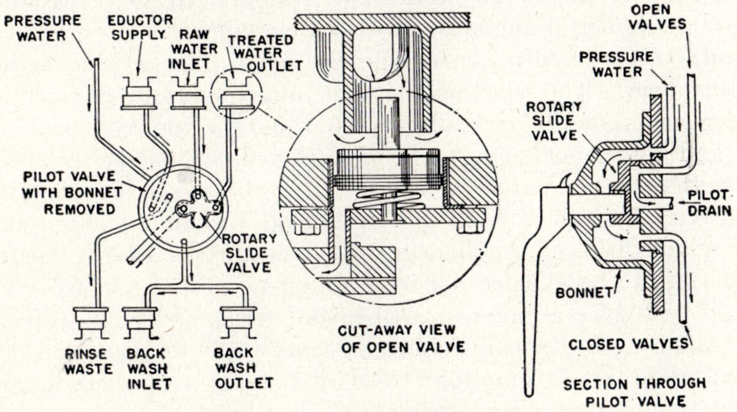

Fig. 16. Cochrane hydromatic valve.

Several rotor multiport valves are available for similar service. In one type, it is necessary to lift the valve member off the port plate to turn it, but the duration of this operation is so short that leakage is minor. When the pipe size of a rotary valve is large, the water pressure impeding its operation may be excessive, and in this case it is necessary to close a gate valve in the supply line before operation.

Figure 16 is a diagram of a multiport valve in which the ports are opened or closed by hydraulically operated pistons in response to flow through a rotary pilot valve. This eliminates leaks due to scoring, which have occurred in some types of rotary valves. Since the ports open slowly, hydraulic shock to the bed is avoided. Although the diagram shows a valve designed for a zeolite softener, it is also furnished for pressure filters.

RATE CONTROLLERS

A fairly uniform flow of water through filter beds without sudden or extreme fluctuation of rates is necessary for good filter performance. The sudden opening of filter effluent valves may cause a filter to "break through," permitting solids to penetrate the bed. Rapid closing of valves may produce a similar condition by liberation of entrained air previously released from the water and trapped in the bed. Many appliances for maintaining uniform rates of filtration have been employed in the past. Such equipment ranges from simple rate control, by regulating the height of water over a weir by means of a float, to elaborate hydraulically, pneumatically, or electrically manipulated, recording control devices. Relatively satisfactory filter operation may be obtained by a simple butterfly- or a ball-float-control-valve control holding the level of the effluent at a fairly constant rate over a weir or submerged orifice. However, if filter requirements demand a more accurate control device, there are a number of appliances which are available.

Manufacturers of filtration equipment can furnish a wide variety of rate controllers. The selection from the numerous types of stock equipment depends upon the particular use to which the apparatus will be subjected and the supervision to be maintained. Some types are more sensitive than others, but it is generally preferable to sacrifice some degree of sensitivity for rugged construction and positive functioning under all operating conditions. Appliances which require constant repair and attention are to be avoided. Direct-acting venturi types of controllers have met with general favor in filter installations, and such devices contribute infinitely to the convenience of semiautomatic operation. They are more adaptable to large than to small installations, where there is no regular attendant or where other activities limit the attention of an operator to occasional visits.

For controlling and limiting wash rates, devices similar to filter-rate controllers may be used in large filtration plants. For boiler-feedwater treatment, it will generally suffice to interpose in the wash line an orifice calculated to deliver at available pressure the maximum amount of wash water which will be required.

Recently, filter-rate controllers have been designed to operate pneumatically, and many large and small installations have adopted this type of regulating equipment, since they have some advantages over hydraulic controllers. A desirable feature of this type of equipment is that it eliminates all cables and wires which often cause high maintenance.

METERING DEVICES

Rate-of-flow Gauges. Efficient plant operation demands the use of rate-of-flow gauges or recorders at several points in the system. The

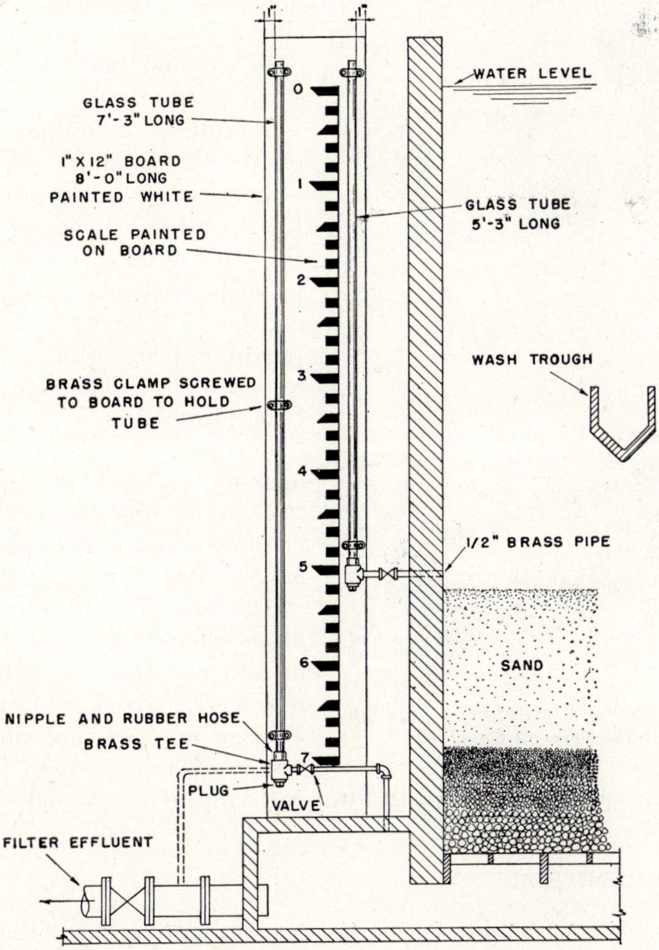

Fig. 17. Simple design of a loss-of-head gauge.

exact number of, and places for, such apparatus will depend upon local conditions, but the installation of such equipment is desirable on the following pipelines: (1) the raw water, (2) each filter effluent, and (3) wash-water supply.

Orifices, venturi tubes, pitot tubes, and rotating displacement meters all have their places as means of measuring water flow. Weir readings yield sufficiently accurate results and are applicable to the measurement of constant flow in open channels. A choice of these devices depends upon the accuracy desired, the convenience and cost of installation, and, to some extent, upon individual preference. Weirs and certain types of orifices are easily designed and may be installed at small cost.

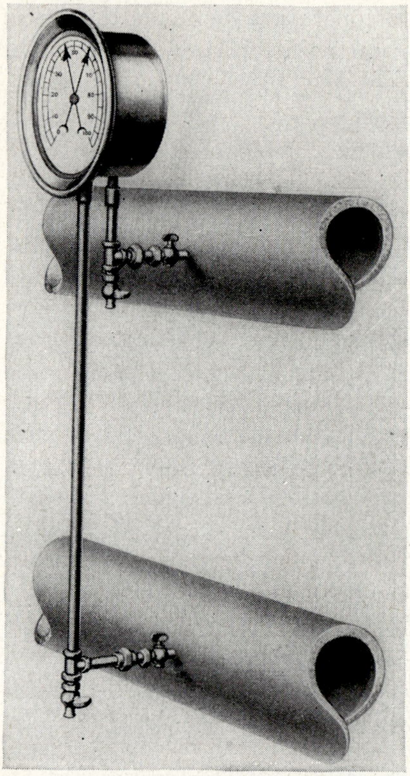

Fig. 18. Duplex loss-of-head gauge. (*Courtesy of The Permutit Company.*)

Loss-of-head Gauges. Apparatus for determining the loss of head is essential on all types of filters. Loss of head develops as a result of the filter medium becoming fouled by suspended solids or by air binding due to the release of gas within the filter bed. Simple designs of loss-of-head gauges are shown in Figs. 17 and 18.

For pressure filters, differential mercury manometers interposed between the influent and effluent connections or ordinary pressure gauges may be used.

The loss of head through clean filters should not exceed 18 to 20 in. and may be as low as 6 in. where relatively coarse sand is used as the filtering medium and where filtration rates are relatively low. The maximum head loss for a dirty filter will vary generally between 5 and 8 ft.

FILTER EFFLUENT SEAL

Under extreme conditions, particularly in cases where filter washing has been delayed, the head loss may exceed 8 ft. There may even be a negative pressure in the gravel layer, underdrain system, and filter effluent piping. To insure against the entrance of air into the filter, the end of the effluent piping must be kept submerged or "water-sealed." It is inevitable that some air may become entrained within a filter if an appreciable negative head is present any distance below the surface of the sand, since gases in equilibrium at a given pressure will be liberated when the pressure is reduced. Porous concrete in the lower sections of the filter has been known to give rise to premature air binding and short filter runs. In such cases, the filters must be washed more frequently than would otherwise be required and frequently before the head differential exceeds 6 to 8 ft.

FILTER TO WASTE

It is advisable to provide means by which filtered water may be diverted directly to waste. A filter-to-waste valve arrangement (sometimes termed "rewash") is utilized only so long as may be required for the effluent to become clear just after a filter has been washed. By allowing the filter to stand idle for a very short while and then opening the effluent valve slowly, a filter which has just been washed can usually be made to deliver a fairly clear effluent without running water to waste. Where there is a limited quantity of water, this type of operation can be used to avoid water wastage. The facility for filtering to waste can usually be easily and inexpensively provided by interposing a gate valve in a connection between the effluent line and sewer.

OPERATION REQUIRED TO ENSURE SATISFACTORY FILTER EFFICIENCY

The efficiency of any water-purification system is directly proportional to the care exercised by the operator in the control of the process. The functioning of filters is a relatively simple procedure, but the operator should possess some knowledge of the principles upon which the process is based.

For many years, it was common practice to design filters used for domestic service at a filter rate of 2 gpm per sq ft of surface area. For industrial uses, these rates were stepped up somewhat, and rates of from 2.5 to 3.5 gpm per sq ft of filter area were employed. Recently, developments in coagulation, especially when activated silica is employed, have demonstrated that, for gravity filtration, the rates can be stepped up to as high as 5 to 6 gpm per sq ft of filter area and still obtain satisfactory results. Such high rates, however, should not be employed with pressure filters, and maximum rates of about 3 gpm per sq ft should be the upper limit.

The primary object of filtering water is to remove suspended solids. To accomplish this successfully requires:

1. Maintaining the filter medium in good condition
2. Keeping the filter in a good state of repair
3. Intelligently controlling the application of chemicals
4. Regular and thorough cleaning of the filter bed
5. Keeping daily records of operation

The filter medium is the most important part of the filter and should be inspected at regular intervals to determine its condition. Where sand or similar material is employed, there is a tendency for certain areas on the surface to become coated with accumulations which, if not removed when the units are backwashed, can cause a serious operating problem (Fig. 19). Conditions of this kind occur frequently when muddy waters are filtered but may also take place in filtering flocculated lime or magnesium salts in softening plants. The condition generally results when filters are washed inefficiently but may occur also in well-designed and -operated plants.

When the surface of the filter becomes clogged with deposit, the filter may have to be cut out of service and drained and the deposits scraped from the surface by hand skimming. If deposits are not removed by

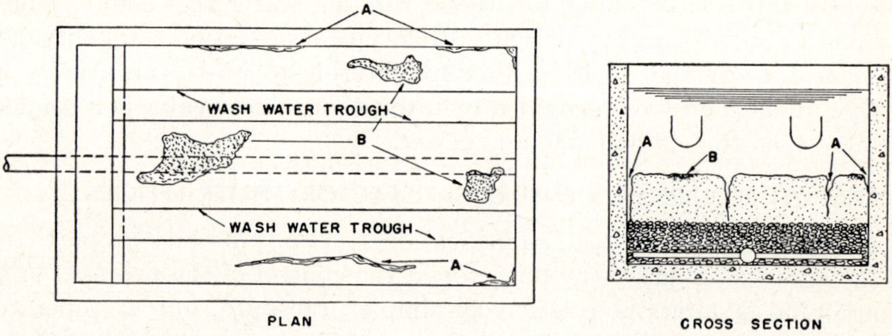

Fig. 19. Filter operating difficulties. ***A*, fissures in bed and along side; *B*, mud balls.**

washing, the portion of the bed where they occur operates at a lower rate, and this results in uneven filtration through different areas of the filter. The deposit can penetrate into the filter bed and so foul it as to reduce the rates of flow and prevent good clarification.

Uneven distribution of wash water is indicated by unevenness of the surface of the filter bed. Where there is evidence that the gravel bed has been disturbed, the filter medium should be removed and the underdraining system inspected and repaired if necessary. When the gravel bed or other material used for supporting the sand or anthracite coal is dislodged, there is danger of the filter media penetrating into the filtered-water collecting system. This condition is particularly objectionable in pressure filters which are installed on pressure pipelines. Filter media passing through a filter will be discharged into the water lines, causing serious trouble in pumps and other appurtenances.

When a filter has been properly designed, there should be a uniform distribution of wash water, which will remove practically all of the foreign matter that collects in the filter during operation. This should be accomplished without any material loss of filter material. Care should be exercised by the operator not to exceed the maximum designed rate of flow of wash water or agitate the filter bed too violently when air or mechanical agitation is employed. No predetermined washing period may be established, since the length of time a filter should be operated

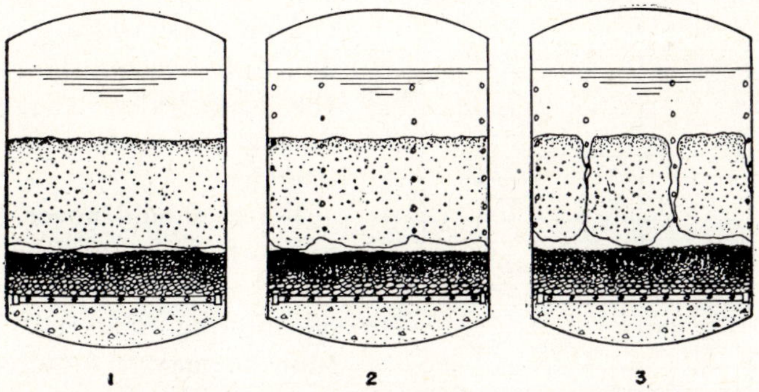

Fig. 20. Sectional views of a vertical unit, showing action of entrained air in filter beds.

between washings is controlled by the amount and character of the suspended material in the raw water, the temperature of the water, particularly changes in temperature of the water passing through the system, and a number of other factors. Figure 20 shows the effect of air entrainment due to too infrequent backwash of filters.

Upsetting the gravel beds in filters results in poor backwashing and

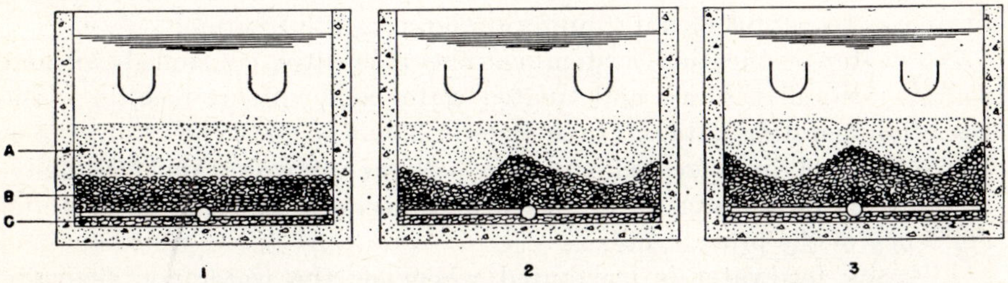

Fig. 21. Filter operating difficulties (upset gravel). *A*, sand; *B*, gravel, showing development of hills and valleys; *C*, underdrains.

poor filtration and causes many other operating difficulties. Because of this problem, every effort should be made to maintain the layers of gravel or other supporting material uniform, since if this is not done, good quality water cannot be obtained. Displacement of gravel layers is shown diagrammatically in Fig. 21.

Filters require backwashing usually every 6 to 24 hr. Much longer or shorter periods may be desirable or necessary. Wash water should always be admitted to filters slowly, since a sudden rush of water may dislodge the strainers or gravel bed. The amount of wash water used in comparison with the volume filtered will range from 0.5 to 5 per cent. In well-operated plants, 2 to 3 per cent of wash water is necessary. The amount of wash water necessary is influenced to a marked degree by the design of the filter. An illustration of this is given in Fig. 22. This curve shows the reduction in the percentage of wash water necessary for efficient backwashing of a filter due to increasing the capacity of the wash-water trough and reducing the spacing between the troughs.

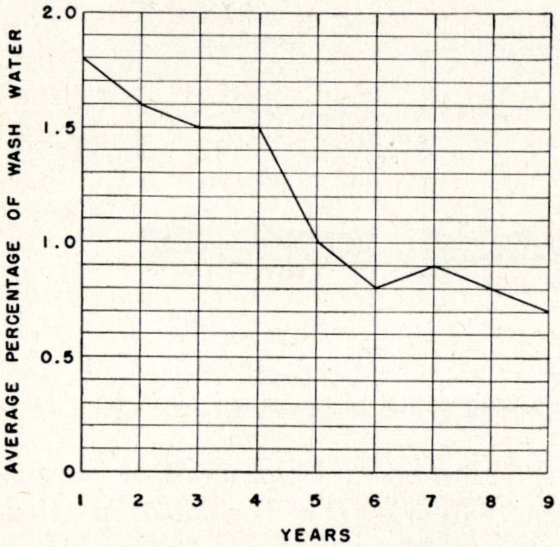

Fig. 22. Reduction in wash water effected by increased wash-water trough capacity.

Water should never be permitted to fall directly from the wash troughs onto the exposed surface of filtering material. At such times as may be required to draw the water level below the surface of the filter, for inspection or other purposes, the filter should be refilled to a height near the lips of the wash troughs by applying water through the wash piping and underdrain system.

The application of chemicals employed for the purpose of either softening or clarification should be controlled by a dosing apparatus so calibrated as to permit of fairly uniform and accurate treatment.

No water-purification system can be efficiently controlled without routine tests of the raw and treated water and without records of the essential operating data. These records should include chemical tests of the untreated and treated supplies, volume of water purified daily, amount of wash water used, length of time between cleaning of units, efficiency of the process, etc.

Filtering feedwater is warranted whenever the water contains any appreciable amount of suspended solids. The return on the capital investment of the apparatus will be proportional to the amount and character of the suspended solids and the degree of efficiency of operation. Finely divided suspended solids in feedwater are an important factor in the foaming and priming of boiler water, increase blowdowns, and are frequently the cause of burned tubes. Grease, sewage, and industrial-waste products reduce the value of feedwater greatly. These substances may be eliminated or reduced in the water by well-chosen and efficiently operated filter systems.

The cost of filtration, including the capital and operating charges, fluctuates widely for different systems and also for varying operating conditions.

BACKWASHING

The cleaning of filters is termed, generally, "backwashing." There are two methods of backwashing: (1) passing filtered water upward through the filter at medium or high velocities (15 to 30 gpm per sq ft of filter area) without other means of agitating the sand bed and (2) passing filtered water upward through the filter at a rate of 7.5 to 12 gpm per sq ft of filter area while the bed is agitated by means of revolving power-driven rakes or water jets or by the use of compressed air or steam. These methods of washing are termed, respectively, high-velocity wash and low-velocity wash. The use of either method depends upon the construction of the filters. Installations designed for functioning one way may not be operated otherwise without radical changes in construction details. Mechanical agitation of sand by rakes has not been employed to any great extent in recent years. Air and steam agitation are not now favored by engineers except for special requirements. Air agitation is effective for removal of deposits adhering to grains of filter media in manganese and iron removal plant filters. The use of steam as an aid for filter backwashing is practiced in chemical softening systems and for oil-removal filters.

Since the efficiency of a filtration plant depends largely upon the maintenance of a proper condition of the sand in the filters and since satisfactory washing is an absolutely necessary prerequisite to the fulfillment of this requirement, wash rates are now considerably higher than those which were formerly used. The earlier mechanically driven revolving rakes were efficient for the elimination of clogged areas but were rendered impractical with the advent of filters of rectangular section. In the absence of auxiliary agitation, high wash rates are essential. Other things being equal, effective filter washing depends upon:

1. Rate at which wash water is applied
2. Uniformity of wash-water distribution
3. Length of time of washing
4. Kind of filtering material used
5. Size and gradation of the filter media
6. Temperature of the water
7. Depth of the filter bed
8. Amount of freeboard
9. Design and spacing of wash-water troughs

For satisfactory washing under most conditions a sand expansion of from 50 to 100 per cent should be provided.

AUTOMATIC BACKWASHING

This type of operation has been practiced for some years for control of pressure filters operated either hot or cold, and similar backwashing and regeneration controls have been incorporated into the design of pressure ion-exchange softeners. It is only within recent times that the automatic backwashing cycle has been adapted to gravity rapid sand filters. A number of large installations are now equipped for either fully or semiautomatic (push-button) operation. Such units are designed to operate either electrically, hydraulically, or pneumatically.

RELATIVE MERITS OF PRESSURE AND GRAVITY FILTERS

In general, open gravity filters are preferable to pressure filters where it is desirable to minimize operating and maintenance cost and to insure the best quality of filtered water at all times. However, pressure filters are satisfactory for some service and are very extensively used for treatment of water for various industrial requirements. The relative advantages and disadvantages of this type of filter are summarized below:

Advantages of Using Pressure Filters

1. The units may be interposed in the water mains under pressure, thereby eliminating low lift pumping.
2. They require less space than gravity filters of the same capacity.
3. They can be constructed rapidly.
4. They may be operated at a high loss of head.
5. Additional units may be added without materially affecting the general design of the installation.
6. When they are used in the purification of water from deep wells, there is less chance for outside bacterial contamination.
7. When they are equipped with multiport valves, their operation is simpler than that required for gravity units.

Disadvantages

1. The filter sand is not easily inspected, and the efficiency of the backwashing cannot be observed.
2. Application of the coagulant is usually not so well controlled as it is in gravity units.
3. Undecomposed alum or other chemicals which are used may pass through the filters because of the short period of reaction between the point of chemical application and the passage of water through the filter bed.

4. The passage of sand into the filtered-water system is possible when the underdraining system fails, since no clear-water storage basin is provided in most installations.

5. It is more difficult to maintain constant rates of filtration with pressure filters than with gravity filters.

In pressure filters, considerable penetration of suspended matter occurs throughout the period of operation. Break-through of solids is especially likely to occur when the filter run between backwashing periods is prolonged. This is also true of gravity filters, especially those employing coarse filtering media.

Filter-operation technique varies widely at different times of the year, particularly in the treatment of surface supplies, because the character of the solids varies with the seasons. Generally, coagulated matter passes filter beds more rapidly in winter than in summer. Such occurrences are due, in part, to temperature, entrained gases, and other factors. Loss of head, on the other hand, may build up more rapidly in warmer periods than in cold weather.

Bibliography

1. Baker, M. N.: "The Quest for Pure Water," American Water Works Association, New York, 1948.
2. Baker, M. N.: "The Quest for Pure Water," American Water Works Association, New York, 1948.
3. Turner, Homer G., and G. S. Scott: Loss of Head and Backwashing Velocities for Anthracite Coal, *Water Works & Sewerage,* **80**: 66 (1933).
4. Turner, Homer G., and G. S. Scott: Clarification of Water Supplies by Filtration through Anthracite, *Water Works & Sewerage,* **80**: 135 (1933).
5. Turner, Homer G., and G. H. Young: Attrition Loss of Anthracite in Rapid Filters, *Water Works & Sewerage,* **80**: 246 (1933).
6. Turner, Homer G., and G. S. Scott: Accounting for Anthracite's Superiority to Sand As a Filter Medium, *Water Works & Sewerage,* **80**: 330 (1933).
7. Nordell, Eskel: "Water Treatment for Industrial and Other Uses," p. 309, Reinhold Publishing Corporation, New York, 1951.
8. Private communication from P. B. Streander (formerly chief engineer, Roberts Filter Manufacturing Company).
9. Ellms, J. W.: "Water Purification," McGraw-Hill Book Company, Inc., New York, 1928.
10. Hazen, Allen: "The Filtration of Public Water Supplies," John Wiley & Sons, Inc., New York, 1910.
11. Rehler, J. E.: Experiences with the Palmer Filter Agitator at Olean, *J. AWWA,* **32**: 1381 (1940).

6 COLD-PROCESS SOFTENING BY CHEMICALS

INTRODUCTION

The softening of water by chemicals depends upon the conversion of the soluble compounds in the water which are responsible for hardness into insoluble compounds, which can then be removed by mechanical separation.

The process involves the following steps:

1. Calculation of the amounts of chemicals required based on chemical analysis of the raw water.
2. Application of the reagents in the predetermined amounts.
3. Rapid mixing of the chemicals with the water to effect thorough and uniform admixture, followed by flocculation preferably by stirring of the water by mechanical means.
4. Settling for the required period of time, either in a quiescent state or at flows sufficiently slow to permit separation of the precipitated solids.
5. Removal of the clarified water for use, either with or without subsequent filtration.

HISTORICAL

The process of softening water by chemicals has been known for nearly 200 years. Softening water by the addition of lime is credited to Cavendish, who in 1766 noted that, when lime was added to hard water, it precipitated lime and magnesium in the form of carbonates. Later, Thomas Henry in England, about 1800, contributed additional information on the subject, but commercial advancement of the art was due to the classical work of Dr. Thomas Clarke,[1] of Aberdeen, Scotland, who in 1841 patented a process of softening water for municipal use. The Clarke process was subsequently improved by Porter in 1876, and for many years this type of treatment was known generally as the Porter-Clarke process. The improvements of Porter consisted of mechanically mixing lime with the water to be softened and removing the precipitated sludge by means of cloth filter presses. The Porter-Clarke process was limited to the removal of carbonate hardness only, although a process to remove noncarbonate hardness by the addition of soda ash and sodium silicate was patented as early as 1856. A similar patent was issued in 1856 based on removal of both carbonate and noncarbonate hardness by the combined use of sodium salts and lime.

A complete account of the history of chemical softening would include descriptions of many ingenious types of apparatus which are now obsolete.[2] Some of these will be discussed briefly in introducing the subject of equipment for cold- and hot-process softening.

HARD WATERS: THEIR CAUSE AND GEOGRAPHICAL DISTRIBUTION

The amount of hardness present in natural water varies widely throughout the world and depends on the geological formations in contact with the water, either on the surface or underground. The extent and variation of the average hardness of surface and underground water supplies in the United States are shown graphically on the maps prepared by the U.S. Geological Survey and reorganized by the National Natural Resources Board (Figs. 1 and 2).

Hardness is due to the presence of calcium and magnesium salts: bicarbonates, carbonates, sulfates, chlorides, and nitrates. Iron, aluminum, and manganese also cause water to be hard, but these substances are not ordinarily present in appreciable quantities. Normal carbonates are found only occasionally, in highly alkaline waters. Nitrates are usually present in small amounts, and, on the average, sulfates are likely to exceed chlorides. There are, of course, many water supplies to which these generalizations do not apply.

It has been customary in the past to classify hardness as temporary or permanent, both commonly but not always being present. Preferable terms are "carbonate" (more precisely, "bicarbonate") and "noncarbonate" hardness, which are more accurately descriptive. Waters containing temporary, or carbonate, hardness are partially softened by boiling, while permanent hardness is not reduced by this action.

The reason for this behavior is obvious from the chemistry of the two types of hardness. Although calcium carbonate (limestone) is only slightly soluble, it dissolves in carbonic acid to form soluble calcium bicarbonate:

$$\underset{\substack{\text{Calcium}\\\text{carbonate}}}{CaCO_3} + \underset{\substack{\text{Carbon}\\\text{dioxide}}}{CO_2} + \underset{\text{Water}}{H_2O} \rightarrow \underset{\substack{\text{Calcium}\\\text{bicarbonate}}}{Ca(HCO_3)_2}$$

The carbon dioxide which combines with calcium carbonate to form calcium bicarbonate is often referred to as "half-bound" carbon dioxide.

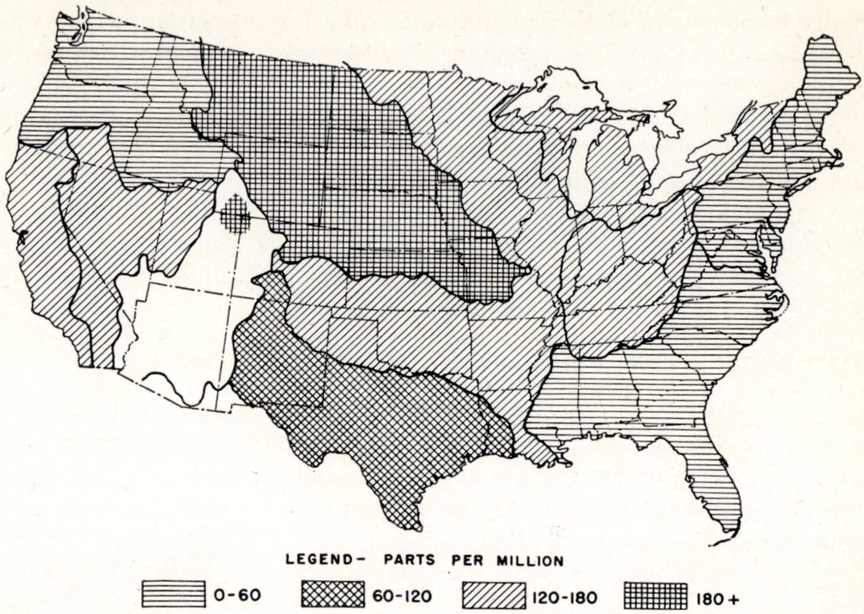

Fig. 1. Map showing average hardness of surface water weighted against population served in major drainage areas.

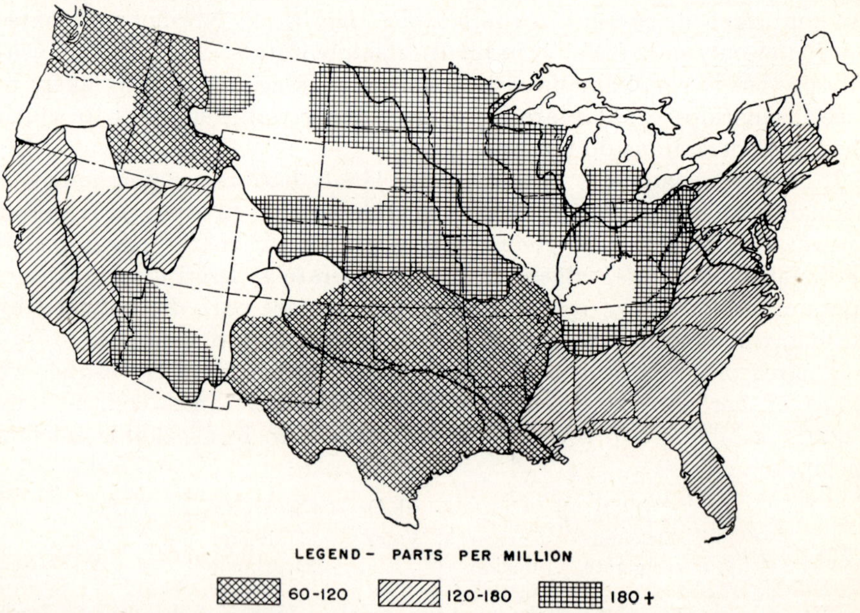

Fig. 2. Map showing average hardness of ground water weighted against population served in major drainage areas.

Heat reverses the reaction by driving off the half-bound carbon dioxide, thus restoring the normal calcium carbonate, and the water is softened by precipitation of this compound.

Noncarbonate hardness is due to calcium sulfate or chloride, which remain soluble when the water is heated and must be converted to carbonates by soda ash in order to be removed. For example,

$$\underset{\text{Calcium sulfate}}{CaSO_4} + \underset{\text{Soda ash}}{Na_2CO_3} \rightarrow \underset{\text{Calcium carbonate}}{CaCO_3} + \underset{\text{Sodium sulfate}}{Na_2SO_4}$$

Chemical requirements and softening results differ for carbonate and noncarbonate hardness and also for calcium and magnesium hardness.

TABLE 1. TYPES OF HARDNESS

Classification	Carbonate hardness	Noncarbonate hardness*
Calcium hardness	Calcium bicarbonate Calcium carbonate	Calcium sulfate Calcium chloride
Magnesium hardness	Magnesium bicarbonate Magnesium carbonate	Magnesium sulfate Magnesium chloride

* Noncarbonate hardness includes nitrates, but these are usually minor in quantity and are omitted to simplify the discussion.

Both bases of classification must, therefore, be used to interpret a water analysis and calculate results of treatment. These are shown in Table 1.

TERMINOLOGY OF CHEMICAL SOFTENING

It is common practice to identify water-softening procedures by the chemicals used in the process. A classification of these several treatments and a description of the operation, purpose, and adaptability for various services are given briefly in the following text.

CHEMISTRY OF LIME-SODA SOFTENING

Removal of Carbon Dioxide. Before softening can occur, lime will be consumed by any free carbon dioxide present in the water:

$$\underset{\text{Carbon dioxide}}{2CO_2} + \underset{\text{Calcium hydroxide (hydrated lime)}}{Ca(OH)_2} \rightarrow \underset{\text{Calcium bicarbonate}}{Ca(HCO_3)_2}$$

This first softening reaction produces calcium bicarbonate, adding to the amount of this type of hardness originally present in the water.

Reduction of Calcium Carbonate Hardness with Lime. The second and most important reaction in water softening by chemicals is the conversion of soluble calcium bicarbonate to the relatively insoluble normal carbonate, in which form it is removed by settling and filtration. This is accomplished by adding lime in amounts sufficient to react with the free CO_2, after which further addition of lime precipitates the carbonate according to the following reaction:

$$\underset{\text{Calcium bicarbonate}}{Ca(HCO_3)_2} + \underset{\text{Calcium hydroxide (hydrated lime)}}{Ca(OH)_2} \rightarrow \underset{\text{Calcium carbonate}}{2CaCO_3} + \underset{\text{Water}}{2H_2O}$$

Partial softening may be accomplished by adding enough lime to convert only part of the calcium bicarbonate to carbonate. Of the calcium carbonate thus formed, some will remain in solution (according to its solubility), while the remainder will precipitate. This practice is followed in treatment of public water supplies where complete softening is not justified. However, in softening boiler feedwater and many other supplies, all the bicarbonates, including those of magnesium and sodium, are converted to normal carbonates by lime to obtain the maximum degree of softening that can be accomplished by this chemical.

In case magnesium or sodium bicarbonate is present, the lime consumption for complete treatment will be increased by the following reactions:

$$\underset{\text{Magnesium bicarbonate}}{Mg(HCO_3)_2} + \underset{\text{Calcium hydroxide (hydrated lime)}}{Ca(OH)_2} \rightarrow \underset{\text{Magnesium carbonate}}{MgCO_3} + \underset{\text{Calcium carbonate}}{CaCO_3} + \underset{\text{Water}}{2H_2O}$$

$$\underset{\text{Sodium bicarbonate}}{2NaHCO_3} + \underset{\text{Calcium hydroxide* (hydrated lime)}}{Ca(OH)_2} \rightarrow \underset{\text{Sodium carbonate}}{Na_2CO_3} + \underset{\text{Calcium carbonate}}{CaCO_3} + \underset{\text{Water}}{2H_2O}$$

Sodium and magnesium carbonates formed by the foregoing reaction are soluble and are not precipitated. Therefore, the additional lime used does not directly remove an equivalent amount of hardness. However, the soluble carbonates which are formed improve the performance of the process because of the excess of these soluble carbonates which is produced.

Reduction of Magnesium Hardness with Lime. In the third reaction in chemical softening, lime precipitates magnesium hydroxide:

$$\underset{\text{Magnesium carbonate}}{MgCO_3} + \underset{\text{Lime}}{Ca(OH)_2} \rightarrow \underset{\text{Magnesium hydroxide}}{Mg(OH)_2} + \underset{\text{Calcium carbonate}}{CaCO_3}$$

$$\underset{\text{Magnesium sulfate}}{MgSO_4} + \underset{\text{Lime}}{Ca(OH)_2} \rightarrow \underset{\text{Magnesium hydroxide}}{Mg(OH)_2} + \underset{\text{Calcium sulfate}}{CaSO_4}$$

$$\underset{\text{Magnesium chloride}}{MgCl_2} + \underset{\text{Lime}}{Ca(OH)_2} \rightarrow \underset{\text{Magnesium hydroxide}}{Mg(OH)_2} + \underset{\text{Calcium chloride}}{CaCl_2}$$

* It is customary in water-softening practice to refer to calcium hydroxide or hydrated lime merely as lime, and where this term is used elsewhere in the text, it refers to calcium hydroxide or hydrated lime.

The soluble magnesium carbonate has already used one equivalent of lime in its formation from magnesium bicarbonate as shown in the preceding equation, and it requires a second equivalent for its conversion to the hydroxide. It should be noted that magnesium sulfate or chloride, when precipitated with lime, produces a corresponding calcium salt which adds to the calcium noncarbonate hardness of the water.

In general, magnesium hardness can be left in solution or removed by adding more lime, according to the requirements for the finished water. However, when lime is fed for the purpose of removing calcium bicarbonate hardness, some magnesium precipitates, and, as it consumes equivalent lime, the dosage must be increased to meet this demand. Such "incidental" magnesium removal may vary from 10 per cent of the initial magnesium to much higher amounts where the total lime dosage is heavy.

Reduction of Noncarbonate Hardness with Soda Ash. The fourth fundamental reaction in lime-soda softening is the conversion by soda ash of soluble calcium sulfate and chloride to insoluble calcium carbonate as follows:

$$\underset{\text{Calcium sulfate}}{CaSO_4} + \underset{\text{Soda ash}}{Na_2CO_3} \rightarrow \underset{\text{Calcium carbonate}}{CaCO_3} + \underset{\text{Sodium sulfate}}{Na_2SO_4}$$

$$\underset{\text{Calcium chloride}}{CaCl_2} + \underset{\text{Soda ash}}{Na_2CO_3} \rightarrow \underset{\text{Calcium carbonate}}{CaCO_3} + \underset{\text{Sodium chloride}}{2NaCl}$$

The calcium carbonate thus formed precipitates along with that which is produced from calcium bicarbonate.

The soda ash added not only reduces the calcium noncarbonate hardness but is required, also, to remove magnesium noncarbonate hardness. This occurs because magnesium sulfate and chloride, upon addition of lime, form an equivalent amount of calcium noncarbonate hardness which must then be precipitated with soda ash. In other words, the soda ash dosage must be based on the total amount of noncarbonate hardness, and no distinction need be made between calcium and magnesium salts in estimating treatment requirements.

Four fundamental processes are thus involved in softening with lime and soda ash, resulting in the formation of calcium carbonate and magnesium hydroxide. The final hardness of the water will be governed by the solubility of these compounds under the existing conditions of temperature and other substances present.

SOLUBILITY OF CALCIUM CARBONATE

The solubility of calcium carbonate is fundamental to water softening and control of scale and corrosion. Familiarity with these data is necessary to predict results of many of the softening reactions occurring in this process. Extensive research work has been done on the solubility of calcium carbonate.*

If pure calcium carbonate is added to distilled water not in contact with atmospheric carbon dioxide and allowed to dissolve until saturation is reached, the solubility thus determined would be of the order of only 15 to 20 ppm.† However, in the operation of a cold-process softening plant,

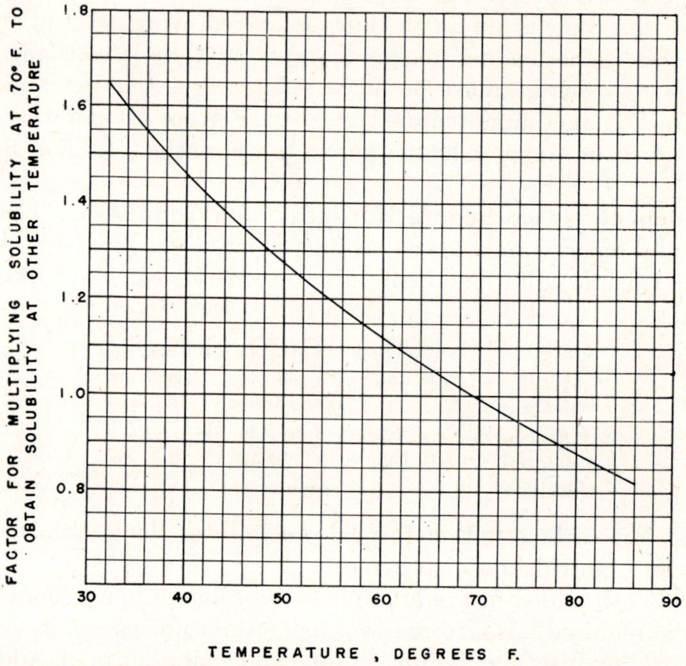

Fig. 3. Effect of temperature on the solubility of calcium carbonate.

this low level is never reached mainly because it is impractical to allow sufficient time. Formerly, a final calcium carbonate hardness of 50 to 70 ppm was considered acceptable in cold lime-softened water. With modern equipment providing for improved sludge contact, the calcium carbonate hardness can be reduced to about 35 ppm without use of excess chemicals.

Effect of Temperature. Figure 3 shows the effect of temperature on the solubility of calcium carbonate over the range encountered in cold-

* Seidell's "Solubilities of Inorganic and Metal Organic Compounds"[3] presents abstracts and data from 9 papers on the solubility of calcium carbonate in water with varying amounts of CO_2 in the atmosphere and 21 papers concerning its solubility in the presence of other salts.

† Based on solubility-product calculations.

process softening.[3] Referring to the curve, if the solubility is known at the base temperature of 70°F, the solubility at 60°F can be obtained by multiplying by 1.12 or, at 80°F, by 0.88, which is a change of 12 per cent for a 10°F change in temperature.

More practical evidence is the markedly greater efficiency of hot-process softening than of cold-process softening. The curves for calcium carbonate hardness vs. other excess soluble carbonates at 70°F and at

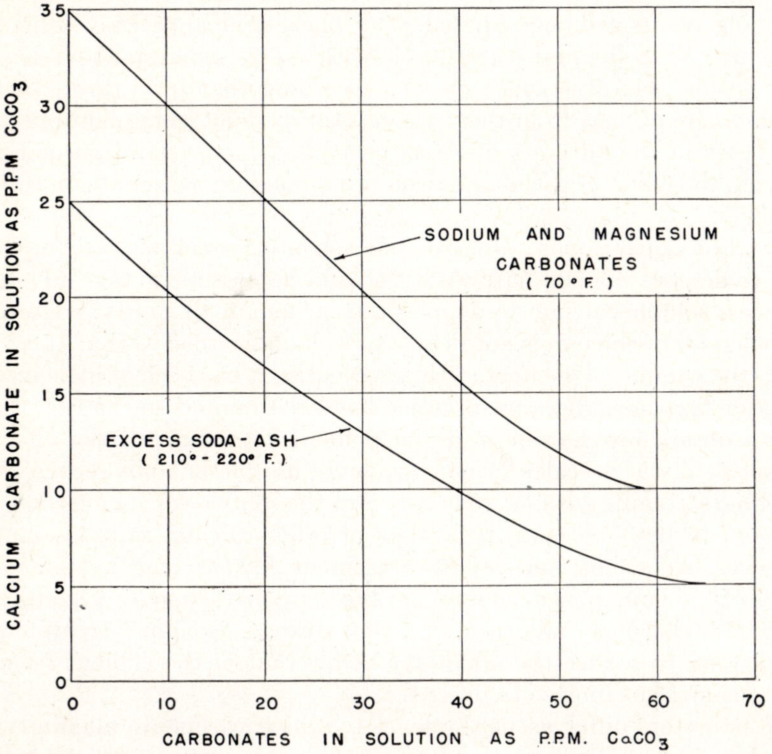

Fig. 4. Effect of soluble carbonates in reducing calcium carbonate hardness in cold- and hot-process softening.

220°F, Fig. 4, indicate the advantage of hot-process over cold-process softening.

The Effect of Carbonate Ions on Softening Reactions. When water is treated with lime to convert *all* the calcium bicarbonate to the normal carbonate, the calcium carbonate hardness can be reduced to about 35 ppm at 60 to 70°F. However, if in addition to calcium bicarbonate the water contains magnesium bicarbonate, which is fairly common, then soluble magnesium carbonate will be produced by lime treatment. In the presence of this soluble carbonate, the solubility of calcium carbonate will be depressed *below* 35 ppm. The same condition occurs if the water contains sodium bicarbonate, because this compound reacts with the hydrated lime added as the softening chemical to form sodium carbonate (soda ash).

This mechanism of the "common-ion" effect is fundamental to lime-soda softening practice. Graphs of residual hardness vs. excess soda ash have been published by several authors and by manufacturers of water-softening equipment. From these curves the expected performance of softening plants can be predicted.[4,5,6] There is no appreciable difference of opinion as to the limits attainable with excess soda ash above 17 ppm, but considerable divergence exists concerning that level down to zero. A composite of data from these various sources and from plant operating log sheets in the author's files has been used for preparation of Fig. 4, showing the effect of excess soda ash or magnesium carbonate on residual hardness.

Effect of Calcium Ions. Just as soda ash or other soluble carbonates are used to depress calcium hardness, it should be possible to take advantage of excess soluble calcium to depress the carbonate alkalinity. However, the behavior of calcium is not nearly so well understood as that of carbonate in the common-ion effect. Until recently, it has been of little interest in water treatment because practice has been limited to the use of soda ash to obtain low hardness in the effluent. In general, the effect on carbonate of excess calcium noncarbonate hardness is not recognized in cold lime softening without soda ash, and the soluble calcium carbonate is assumed to be 35± ppm, regardless of other calcium salts in solution. However, since hot lime-zeolite treatment has become available, the effect of calcium noncarbonate hardness upon carbonate alkalinity at 220°F has become important.[7,8] The "excess-calcium" treatment[9] is relied upon to reduce the alkalinity below that of the effluent from the conventional hot lime-soda process.

Investigators differ as to the level to which carbonate alkalinity can be reduced by excess calcium. Curves showing residual calcium carbonate vs. excess calcium have been published[6,10] in which the carbonate approaches 15 and 18 ppm, respectively. Others consider that calculations should be based on higher carbonate alkalinity, up to 30 to 32 ppm. Figure 5 is a compromise based on these and additional data from the author's files. It is tentatively presented as indicative of the low alkalinity obtainable by hot lime-zeolite softening.

A companion curve for 70°F is not necessary since, as indicated, the calcium carbonate alkalinity will be assumed to be 35± ppm as soon as equivalent lime is added.

Effect of pH. Calcium passes through a zone of minimum solubility of calcium carbonate at a pH of about 9.4, rising thereafter owing to the formation of the more soluble calcium hydroxide. On the other hand, the solubility of magnesium hydroxide drops steeply as the pH rises above 9.5 and approaches zero a little below 10.5.[11] This is shown graphically in Fig. 6.

Effect of Inhibitors. One of the most important causes for the differences between the commercial performance of water-softening plants and the theoretical solubility data is the presence of contaminants in the water which retard precipitation or distort crystal formation. In their presence, the apparent solubility of calcium carbonate may be increased.

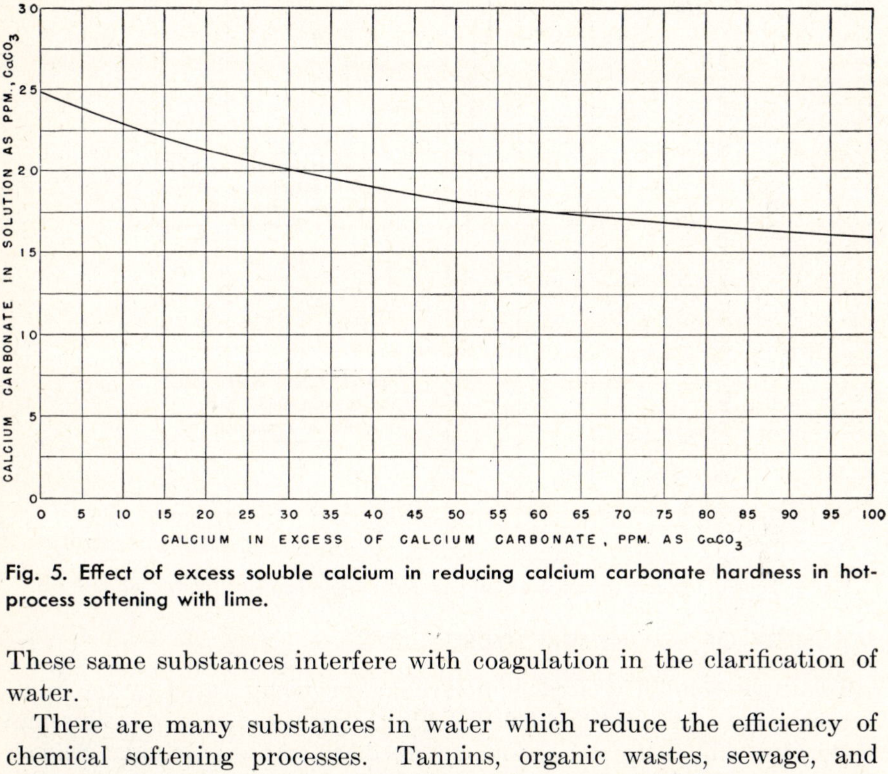

Fig. 5. Effect of excess soluble calcium in reducing calcium carbonate hardness in hot-process softening with lime.

These same substances interfere with coagulation in the clarification of water.

There are many substances in water which reduce the efficiency of chemical softening processes. Tannins, organic wastes, sewage, and other contaminants are objectionable in this respect. If they are present in a water supply, it is difficult to predict accurately the results of softening by chemicals without making laboratory jar tests and interpreting these data to determine adjustment of treatment necessary in full-scale operation.

Chlorination will often remove the effect of these contaminants by oxidation of the organic compounds. Such pretreatment is essential in many cases to insure good softening results. The failure of many softeners to produce the desired residual hardness in the treated water is traceable to failure to remove organic contaminants before adding the softening reagents.

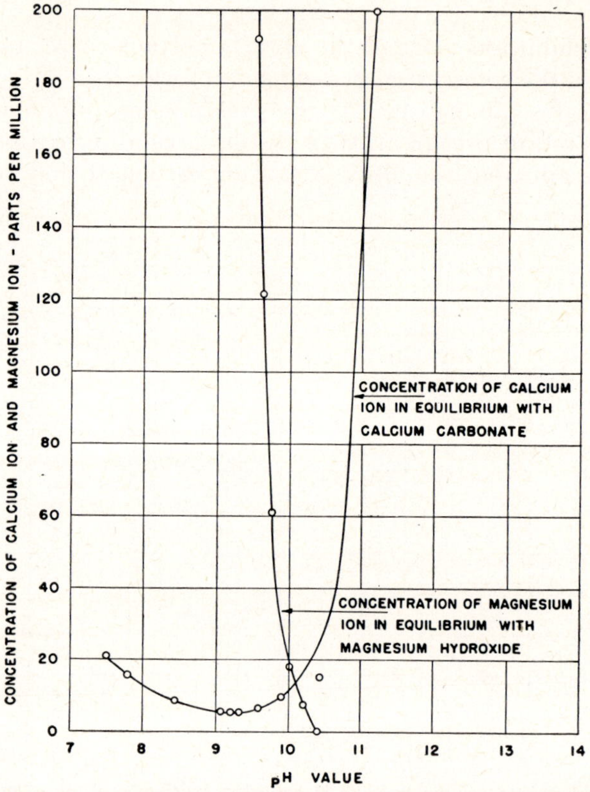

Fig. 6. Equilibrium concentration of calcium and magnesium ions plotted against pH.

SOLUBILITY OF MAGNESIUM HYDROXIDE

While the solubility of calcium carbonate is complicated by the interaction of carbonic acid and calcium bicarbonate, the case of magnesium hydroxide is much simpler and has received less extensive study. Seidell[3] has published a number of measurements varying from about 10 to 4 ppm over a temperature range of 65 to 212°F. The solubility curves shown in Fig. 7 graphically illustrate these conditions.

When lime is added to water containing magnesium hardness, the free carbon dioxide and the bicarbonate alkalinity first react with it. When these demands are satisfied, some magnesium hydroxide will begin to precipitate to the extent of about 10 per cent of the magnesium hardness. Thereafter, the addition of more lime brings down an exactly equivalent amount of magnesium hardness until it is reduced to a little less than

35 ppm (when treating cold water). Further reduction requires not only lime to precipitate the magnesium hydroxide but also an excess to furnish soluble hydroxide, as shown by the lower curve in Fig. 7. This shows a common-ion effect similar to that described for the solubility of calcium carbonate. Sodium hydroxide, which may be added directly or formed

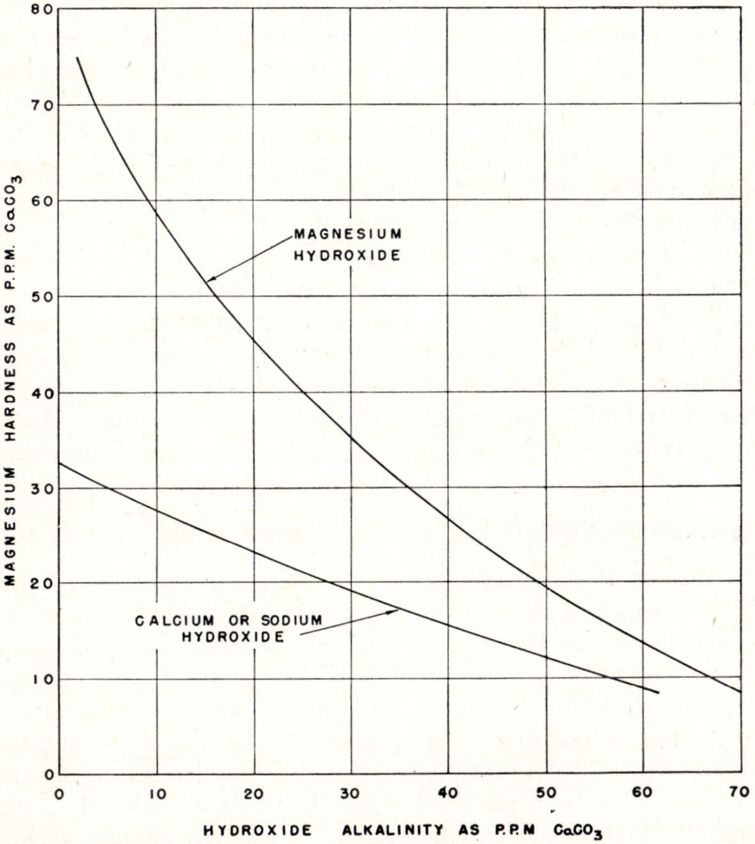

Fig. 7. Effect of excess calcium or sodium hydroxide in reducing magnesium hardness in cold-process softening.

by lime and soda ash, will furnish hydroxyl ions and have the same effect. By such an operation in cold-process treatment, the magnesium hardness can be depressed to about 8 ppm when using a heavy excess of another soluble hydroxide.

The factors which were discussed in detail as influencing the solubility of calcium carbonate also affect magnesium hydroxide. Since a high pH implies a high excess of hydroxyl, it is the most important factor in depressing the solubility of magnesium hydroxide. Like that of calcium carbonate, the effective solubility of magnesium hydroxide is modified by inhibitors of various types.

No curve for 220°F is available or required for Fig. 7. In hot lime-soda softening with an excess hydroxide of 0 to 10 ppm, it has been customary in practical operation to expect the magnesium hardness to vary between 2 and 6 ppm. In such cases, the pH of the treated water is likely to be 10.5 or higher, not only because of the sodium hydroxide but also because of the excess sodium carbonate. By contrast, in hot lime softening without soda ash, even though the excess hydroxide may be 0 to 10 ppm, the pH may be only 10.0±. The solubility of magnesium hydroxide will rise appreciably. To obtain a low magnesium hardness in hot-process softening, it is apparently necessary to assure the presence of enough sodium hydroxide or carbonate to raise the pH well above 10.0.

Treatment of magnesium bicarbonate with lime (essential for maximum calcium bicarbonate reduction) first produces beneficial soluble carbonate and then reduces total hardness by precipitating magnesium hydroxide within the above solubility limits. However, treatment of magnesium sulfate and chloride merely replaces them with an equivalent amount of calcium noncarbonate hardness, accomplishing no softening. As will be shown in an example, there is a limit beyond which magnesium removal is not profitable, depending on the water analysis.

CALCULATING TREATMENT RESULTS AND CHEMICAL DOSAGE REQUIRED

The compounds responsible for hardness in water are present in relatively dilute solution and are in ionic form, namely, as magnesium, calcium, bicarbonate, sulfate, and chloride ions. Lime and soda ash added for softening dissolve and dissociate to form calcium, hydroxide, sodium, and carbonate ions. Ions are now present which combine to form slightly soluble salts so that the excess over saturation precipitates and softening occurs. Before estimating chemical requirements, the analysis of the water before and after softening should be calculated, guided by the solubility principles discussed above. From the change in composition, the chemical dosage required for softening can be calculated.

Type of Water Analysis Required for Calculations. Table 2 lists the data for the raw and softened water which are used in calculating treatment requirements and results and shows how they are obtained from simple water-testing procedures.

Calculations are greatly simplified by the standard procedure of expressing all types of hardness as calcium carbonate, $CaCO_3$, which has a molecular weight of 100. Similarly, bicarbonate, carbonate, and hydroxide alkalinity are reported as calcium carbonate, as well as sulfate, chloride, and any of the other ions mentioned above. As examples will show, this practice of reporting an analysis in units of equal weight results in an exact balance which is a check on accuracy.

Table 3 shows the molecular weights of salts and ions commonly found in water and used in treatment and convenient factors to convert them from any of these expressions to parts per million of calcium carbonate.

Correction for Effect of Alum or Other Coagulants. Alum or other coagulants are often used in cold-process softening for the formation of floc which will aid settling or help the efficiency of sludge-contact treatment tanks. Alum acts as an equivalent amount of sulfuric acid.

TABLE 2. ANALYTICAL DATA REQUIRED FOR SOFTENING CALCULATIONS

1. Free carbon dioxide	ppm $CO_2 \times 2.27$.
2. Bicarbonate	M (methyl orange) alkalinity.*
3. Calcium bicarbonate	The bicarbonate is paired with calcium, magnesium, and sodium in the order given.
4. Magnesium bicarbonate	Any magnesium required to match an excess of bicarbonate over calcium.
5. Sodium bicarbonate	Any sodium required to match an excess of bicarbonate over calcium and magnesium.
6. Excess of bicarbonate over calcium	This is obviously item 4 or 5 or their sum. After lime treatment, it is converted to soluble carbonate, which affects the final hardness.
7. Noncarbonate hardness	The excess of total hardness over M alkalinity, if present.
8. Calcium noncarbonate hardness	The calcium in excess of the calcium bicarbonate. On addition of lime it is increased by any magnesium noncarbonate hardness which is removed, yielding an equal amount of calcium noncarbonate hardness.

* Rarely, normal carbonates are present in raw water, in which case there will be phenolphthalein alkalinity. The carbonate equals $2 \times P$ alkalinity; the bicarbonate is the M alkalinity minus the carbonate.

Hence, the alkalinity will be decreased and the noncarbonate hardness increased, and the chemical dosage requirements will change. Therefore, if it is anticipated that alum will be used, the raw water analysis should be corrected by a simple transfer from bicarbonate to sulfate. One ppm of alum will neutralize 0.45 ppm of bicarbonate and increase the sulfate an equal amount, both expressed as calcium carbonate. In the analysis shown in Column B of Table 7, such corrections have been made.

If prechlorination is applied, it is necessary to increase the chloride and decrease the bicarbonate. Table 4 shows the effect of various treatment chemicals on the composition of the water and gives factors to be used in correcting the raw water analysis.

Sodium aluminate is used with, or instead of, alum. It is an alkaline source of the aluminum needed for coagulation. Instead of increasing the noncarbonate hardness of the water, it furnishes an equivalent amount of caustic soda, which is useful for softening. For each part per million of sodium aluminate used, credit should be taken for 0.45 ppm of caustic soda, as calcium carbonate.

Chemicals Required. The equations given earlier for the four basic softening reactions show that one equivalent weight of lime is required for

TABLE 3. CONVERSION FACTORS FOR CALCULATING PARTS PER MILLION EXPRESSED AS $CaCO_3$, FROM PARTS PER MILLION OF IONS AND CHEMICALS

			For ppm, as $CaCO_3$	
			Divide* by—	Multiply by—
Cations:				
Calcium	Ca		40.1	2.495
Magnesium	Mg		24.3	4.112
Sodium	Na		46	2.174
Potassium	K		78.2	1.278
Anions:				
Chloride	Cl		70.9	1.414
Sulfate	SO_4		96	1.041
Nitrate	NO_3		124	0.806
Phosphate	PO_4		63.33	1.579
Bicarbonate	HCO_3		122	0.819
Carbonate	CO_3		60	1.666
Hydroxide	OH		34	2.939
Compounds:†				
Hydrated lime	$Ca(OH)_2$	92.5%	80	1.250
Quicklime	CaO	90%	62	1.613
Soda ash	Na_2CO_3	99%	108	0.926
Sulfuric acid	H_2SO_4	66°Bé	105	0.952
Hydrochloric acid	HCl	20°Bé	232	0.431
Sodium chloride	NaCl		117	0.855

* Multiply the quotient by 100. These divisors have been rounded off, where permissible, for speedy calculation.

† These values represent the weight of chemical of commercial purity equivalent to 100 ppm of $CaCO_3$; for example, 105 lb 66° H_2SO_4 will neutralize 100 ppm of alkalinity; 100 lb will neutralize 95.2 ppm of alkalinity.

each equivalent of carbon dioxide, alkalinity, and magnesium converted and one equivalent of soda ash for each equivalent of noncarbonate hardness removed. Obviously, these terms overlap, magnesium alkalinity requiring two equivalents of lime and magnesium noncarbonate hardness requiring both lime and soda ash. Conventional rules for calculating chemical dosage recognize these dual requirements. However, confusion will be avoided by considering separately each of the following to the

TABLE 4. CHARACTERISTICS OF ACIDIC CHEMICALS COMMONLY USED IN PRIMARY WATER TREATMENT AND THE EFFECT ON WATER ANALYSES OF EACH PART PER MILLION OF THE COMMERCIAL CHEMICALS ADDED

Chemical	Specification	Formula	Formula weight	Purity	Change in cations and anions, as ppm $CaCO_3$: Increase		Alkalinity decrease*
Aluminum sulfate (filter alum)	15.3% Al_2O_3	$Al_2(SO_4)_3 \cdot 18H_2O$	666	95–99%	Sulfate	0.45	0.45
Ammonium alum	....	$Al_2(SO_4)_3 \cdot (NH_4)_2SO_4 \cdot 24H_2O$	906	....	Sulfate	0.44	0.33
					Ammonium	0.11	
Potassium alum	....	$Al_2(SO_4)_3 \cdot K_2SO_4 \cdot 24H_2O$	949	....	Sulfate	0.421	0.316
					Potassium	0.105	
Ferrisul	25.2% Fe^{3+}	$Fe_2(SO_4)_3$	400	94.4%	Sulfate	0.71	0.71
		$FeSO_4$	152	0.4%			
Ferri-floc	20.5% Fe^{3+}, 0.42% H_2SO_4	$Fe_2(SO_4)_3$	400	73.4%	Sulfate	0.57	0.57
		$FeSO_4$	152	2.04%			
		H_2SO_4	98	0.42%			
Copperas	19.68% Fe	$FeSO_4 \cdot 7H_2O$	278	98%	Sulfate	0.35	0.35
Chlorine	99.6% Cl_2	Cl_2	71	100%	Chloride	1.41	1.41
Chlorinated copperas	....	8 $FeSO_4 \cdot 7H_2O$ +1 Cl_2 by wt	...	....	Sulfate	0.31	0.47
					Chloride	0.15	
Ferric chloride	60% $FeCl_3$	$FeCl_3$	162	60%	Chloride	0.56	0.56

* Usually the bicarbonate ion is decreased, but carbonate and hydroxide may have to be considered if the water has previously been softened. In this case, write out the reaction with the acid involved.

extent that they are to be removed: carbon dioxide, bicarbonate (alkalinity), magnesium, and noncarbonate hardness.

To calculate the chemical dosage required, it is only necessary to know the weights of one equivalent of lime and of soda ash which will combine with 100 parts of any of the substances listed above expressed as calcium carbonate. The equivalent weights of chemical lime, hydrated lime, and soda ash are 56, 74, and 106, respectively. The commercial products vary in purity. To correct for this, the following dosages are used for each 100 ppm of carbon dioxide, alkalinity, magnesium, or noncarbonate hardness:

	Ppm
Chemical lime, CaO	62
Hydrated lime, $Ca(OH)_2$	80
Soda ash, Na_2CO_3	108

Concentrations in chemical analyses are expressed in parts per million, and dosage may be calculated in the same units and also as pounds of

TABLE 5. FACTORS FOR CALCULATING CHEMICAL DOSAGE REQUIRED FOR LIME-SODA SOFTENING, FROM ANALYSES OF RAW AND TREATED WATER

To find the LIME dosage required for softening:

ADD the following (all expressed as ppm of $CaCO_3$):	MULTIPLY the sum by—
1. Free CO_2*	For 93% hydrated lime:
2. Bicarbonate to be converted to carbonate†	0.8 ppm, or 0.0066 lb per 1,000 gal
3. Magnesium hardness to be removed	For 90% chemical lime (CaO):
4. Excess hydroxide required	0.62 ppm, or 0.00516 lb per 1,000 gal

To find the SODA-ASH dosage required for softening:

ADD the following (all expressed as ppm of $CaCO_3$):	MULTIPLY the sum by—
1. The noncarbonate hardness to be removed	For 98% soda ash: 1.08 ppm, or 0.009 lb per 1,000 gal
2. The excess lime used above	
3. The excess soda ash required, if any	

* Omit the carbon dioxide from calculations for hot-process softening.

† This must include the calcium carbonate remaining in solution in addition to the calcium carbonate removed.

chemicals per 1,000 gal of water treated. The latter unit is generally used for estimating treatment costs. Table 5 shows the factors for calculating chemical dosages in either unit.

CONTROL OF SOFTENING TREATMENT BY WATER-TESTING RESULTS

It has been shown that varying degrees of softening may be obtained by adjusting the dose of lime or soda ash or both in accordance with simple rules based on solubility data. In practical operation, the adjustment of the treatment must be guided by testing the treated water to determine whether the desired objectives have been attained. Simple tests will show in which direction the control has deviated and what adjustments must be made to bring about the proper balance.

TABLE 6. CHEMICAL CONTROL DATA FOR LIME-SODA SOFTENING

Symbol	Name and synonyms	How measured or calculated
H	Total hardness	Titration with Versenate or soap
P	Phenolphthalein alkalinity ("partial" alkalinity, "causticity")	Titration of treated water sample to phenolphthalein end point
M	Methyl orange alkalinity ("total" alkalinity)	Continue *P* titration to methyl orange end point
A	Hydroxide, if positive (excess lime) bicarbonate, if negative (Undertreatment with lime)	$2P - M$
B	Carbonate alkalinity (total carbonate)	$M - A$, using the absolute value of *A*, disregarding the minus sign if *A* is negative
C	Excess soda ash, if positive, deficiency of soda ash, if negative	$B - H$

NOTE: 1. Other procedures give carbonate as $B = 2P$ when *A* is positive, $B = 2(M - P)$ when *A* is negative. The above simplified calculations are not in conflict with these procedures.

2. Precise measurement of bicarbonate, carbonate, and hydroxide requires consideration of pH and other factors, but these calculations are satisfactory for water-softening control.

TIME	HARDNESS			ALK.		TREATMENT RESULTS				FEED SET		TANK LEVEL READINGS AND CHEMICAL CHARGES			
	Ca	Mg	TOTAL	P	M	2P-M A	M-A B	B-H C	pH	NO. 1	NO. 2	LIME		PHOS.	SUL.
7 AM															
8 AM															
11 AM															
2 PM															
5 PM															
8 PM															
11 PM															
2 AM															
5 AM															
7 AM															
	RAW WATER			NC		TOTAL INCHES USED									
						TOTAL POUNDS USED									
						TREATMENT P.P.M.									

Fig. 8. Water-softening and chemical-treatment log.

Samples of treated water are tested for hardness and for phenolphthalein and methyl orange alkalinity, and from these data are calculated the excess or deficiency of lime and soda ash. Table 6 shows the symbols that are used, their meaning, and how the quantities they represent are measured analytically or calculated. Figure 8 is a water-treatment log sheet for a softening plant, which provides for the tabulation of these data. Samples of treated water are tested in a routine manner at regular intervals during the day. The necessary frequency of testing may be as little as once a day in the case of stable water supplies but may be as often as one to four times per shift when the quality of the raw water, the flow and other factors are highly variable.

For examples described in the next section, the treated water analyses are shown in Table 7. This table includes the chemical control data defined in Table 6.

Thus, in the partially softened water shown in Column C, the lime is deficient, and the value for soda ash is, of course, negative because none has been added. The analysis for "complete" softening indicates zero for lime, showing that the bicarbonate has been exactly neutralized by the lime dosage. Such precise control would be difficult to maintain continuously in practice. Column E, for railway lime-soda treatment, shows 70 ppm excess lime and 54 ppm excess soda ash. The chemical control data of Table 7 show that the excess or deficiency of lime and soda ash are exactly in accordance with the treatment planned. Had they been otherwise, the operator would adjust the chemical feeders accordingly.

EXAMPLES OF COLD-PROCESS SOFTENING WITH LIME AND SODA ASH

In the examples given in the following sections, enough technical data will be included to illustrate the results that can be obtained by virtue of the solubilities of calcium carbonate and magnesium hydroxide under various conditions.

Softening Lake Michigan Water. *Partial Softening with Lime.* Suppose that it is desired to soften Lake Michigan water of the analysis shown in Table 7, Column A, to a residual hardness of 85 ppm and that 10 ppm of alum will be added as a coagulant. The analysis of Column B, modified to show the effect of the alum, will be used to calculate chemical dosage.

The required hardness reduction is

$$130 - 85 = 45 \text{ ppm}$$

Assuming that 5 ppm of magnesium will precipitate, it is necessary to remove 40 ppm of calcium. Column C is calculated as follows:

1. Enter the reduced values for calcium and magnesium.
2. Assume calcium carbonate to be 35 ppm and enter carbonate.
3. Enter sodium and mineral acids, without change, from Column B.
4. A simple balance shows that bicarbonate must be 30.5 ppm. The reduction of 80 ppm checks against the sum of carbonate removed and residual, plus magnesium bicarbonate removed.

TABLE 7. RESULTS OF COLD-PROCESS SOFTENING OF LAKE MICHIGAN WATER
ALL DATA IN PARTS PER MILLION

	Raw water	Plus alum	Partial softening	Complete softening	Railway lime-soda softening
	A	B	C	D	E
As calcium carbonate:					
Calcium hardness	90	90	50	27.5	8
Magnesium hardness	40	40	35	35	8
Total hardness	130	130	85	62.5	16
Sodium and potassium	15	15	15	15	158.5
Total cations	145	145	100	77.5	174.5
Chloride	10	10	10	10	10
Sulfate	20	24.5	24.5	24.5	24.5
Mineral-acid ions	30	34.5	34.5	34.5	34.5
Bicarbonate	115	110.5	30.5	0	
Carbonate	...	.....	35	43	70
Hydroxide	...	.....	0	0	70
Total anions	145	145.0	100.0	77.5	174.5
Free CO_2	12	12	0	0	0
Noncarbonate hardness:					
Calcium	0	0	5	5	0
Magnesium	15	19.5	14.5	14.5	0
Total	15	19.5	19.5	19.5	0
Chemical control data:					
H = hardness	...	.....	85	62.5	16
P alkalinity	...	.....	17.5	21.5	105
M alkalinity	...	.....	65.5	43	140
$A = 2P - M$ (lime +)	...	.....	−30.5	0	+70
$B = M - A$ (carbonate)	...	.....	35	43	70
$C = B - H$ (excess soda)	...	.....	−50	−19.5	+54
Chemicals:					
Alum	...	.....	10	10	10
Lime	...	.....	78	102	170
Soda ash	...	.....	....	.....	155
Sludge produced:					
Calcium carbonate	...	.....	137	190	295
Magnesium hydroxide	...	.....	3	3	19
Total	...	.....	140	193	314

The lime dosage is, therefore, equivalent to the free carbon dioxide, the magnesium removed, and bicarbonate converted—a total of 97 ppm, as calcium carbonate, or 78 ppm of hydrated lime. Column C shows the effect of these changes, a finished water of the desired hardness.

Maximum Removal of Calcium Bicarbonate. This treatment is desirable to utilize to the greatest possible extent the effect of lime, usually the cheapest softening chemical, and to reduce alkalinity, calcium hardness, and total solids by precipitation. It often precedes final zeolite softening. In this case, the lime dosage for the water shown in Table 7 would be increased to convert all the 110.5 ppm of bicarbonate to carbonate instead of only 80 ppm. There is an excess of 15.5 ppm of soluble carbonate over the calcium, and, according to Fig. 4, this will depress the hardness to about 27.5 ppm. Column D of Table 7 shows the final results. The lime dosage is 127.5 ppm, as calcium carbonate, or 102 ppm of hydrated lime. Increasing the lime feed by 24 ppm has reduced the hardness and alkalinity by 22.5 ppm. This would be warranted for most industrial water supplies.

Further reduction of calcium hardness would require the addition of about two parts of soda ash for one part of calcium removed. The specifications for the final effluent will determine whether this treatment is justified.

Magnesium Removal. If it were desired to soften the water shown by Column D of Table 7 by removing more magnesium than the assumed 10 per cent, lime would be added to convert the remaining 35 ppm to magnesium hydroxide. However, examination of Fig. 7 shows that only a few parts per million will precipitate, the balance remaining in solution. This is due to the formation of complex basic magnesium carbonates, which are soluble.[12] Further magnesium removal can be effected only by adding an excess of lime. As the curves show, this requires a disproportionately high lime dosage and final alkalinity. For services where high alkalinity can be tolerated but magnesium must be reduced below 32 to 33 ppm, soda ash may be added to convert the excess lime to caustic soda.

Railway Lime-Soda Treatment. This process is used for wayside treatment of locomotive-boiler feedwater, and its aim is to produce the minimum hardness attainable by lime-soda softening. An excess of 62 ppm of lime is used to reduce the magnesium hardness to 8 ppm. Soda ash is added to convert the excess lime to caustic soda plus an additional 62 ppm to depress the calcium hardness to 8 ppm. The effluent contains 16 ppm of hardness, 70 ppm of carbonate, and 70 ppm of hydroxide—a total alkalinity of 140 ppm.

Split Treatment. "Split treatment" is used for some public water supplies to correct the undesirable qualities caused by excess lime. Water treated and softened in the first basin to the maximum extent by the use of excess lime is mixed with unsoftened water in a second settling tank, and the excess hydroxide of the first supply is neutralized by the bicarbonate alkalinity of the second. Since in this process calcium hydroxide precipitates calcium carbonate, the final effluent is a softer water than could have been produced by single treatment.

Recarbonation. Water treated by partial softening or for maximum calcium carbonate removal should have a pH of 9.4 to 9.6, since calcium is least soluble in this range. If the best modern equipment is used under careful control, the water may be "stable," that is, no further calcium carbonate will precipitate until the temperature is raised appreciably. This favorable alkalinity balance is not always attainable and is, of course, impossible where excess-lime treatment is used. In such cases, recarbonation may be employed to produce a stable and palatable water supply.

The reactions in recarbonation are as follows:

$$\underset{\text{Carbon dioxide}}{CO_2} + \underset{\text{Calcium carbonate}}{CaCO_3} + H_2O \rightleftharpoons \underset{\text{Calcium bicarbonate}}{Ca(HCO_3)_2}$$

$$\underset{\text{Carbon dioxide}}{2CO_2} + \underset{\text{Calcium hydroxide}}{Ca(OH)_2} \rightleftharpoons \underset{\text{Calcium bicarbonate}}{Ca(HCO_3)_2}$$

As shown by the equations, the carbonate and hydroxide alkalinity is converted to bicarbonates, lowering the pH to a level best suited to prevent both scale and corrosion.[13,14]

Carbon dioxide gas is produced by burning oil, gas, or coal, is scrubbed and cooled by water in a counterflow tower, and is blown through submerged diffusers or a pipe grid in a recarbonation chamber. Alternately, recarbonation may be effected by submerged combustion of natural gas if there are no objectionable products.

The dosage of carbon dioxide can be controlled by varying the amount of the combustion air admitted to the producing equipment. This has been adapted to automatic pH control, but in most instances, simpler manually operated apparatus is used.

Acid Treatment. Reduction of caustic and carbonate alkalinity can also be accomplished by feeding acid, and in most cases, sulfuric acid is used. The reactions are as follows:

$$\underset{\text{Sulfuric acid}}{H_2SO_4} + \underset{\text{Calcium carbonate}}{2CaCO_3} \rightarrow \underset{\text{Calcium bicarbonate}}{Ca(HCO_3)_2} + \underset{\text{Calcium sulfate}}{CaSO_4}$$

$$\underset{\text{Sulfuric acid}}{H_2SO_4} + \underset{\text{Calcium hydroxide}}{Ca(OH)_2} \rightarrow \underset{\text{Calcium sulfate}}{CaSO_4} + \underset{\text{Water}}{2H_2O}$$

This treatment is used to prevent formation of calcium carbonate films on zeolites and other ion-exchange materials. It is simpler and requires less constant supervision than recarbonation and is, therefore, well adapted to small public water-supply systems. Its main disadvantage is the hazard of handling sulfuric acid.

Cold Lime-Barium Treatment. This process is restricted to industrial uses and should not be employed for the treatment of drinking water because barium is a poison, even in small amounts, when taken internally. The treatment has very limited application. It is uneconomical because of the relatively high cost of barium salts, which is several times the cost of soda ash.

Barium is an efficient reagent for removal of noncarbonate hardness in the form of calcium and magnesium sulfates. The reactions which occur when barium carbonate is added to water containing such noncarbonate hardness are as follows:

$$\underset{\substack{\text{Calcium sulfate}\\ \text{(soluble)}}}{CaSO_4} + \underset{\substack{\text{Barium carbonate}\\ \text{(soluble)}}}{BaCO_3} \rightarrow \underset{\substack{\text{Barium sulfate}\\ \text{(insoluble)}}}{BaSO_4} + \underset{\substack{\text{Calcium carbonate}\\ \text{(insoluble)}}}{CaCO_3}$$

$$\underset{\substack{\text{Magnesium sulfate}\\ \text{(soluble)}}}{MgSO_4} + \underset{\substack{\text{Hydrated lime}\\ \text{(soluble)}}}{Ca(OH)_2} + \underset{\substack{\text{Barium carbonate}\\ \text{(soluble)}}}{BaCO_3} \rightarrow$$

$$\underset{\substack{\text{Barium sulfate}\\ \text{(insoluble)}}}{BaSO_4} + \underset{\substack{\text{Calcium carbonate}\\ \text{(insoluble)}}}{CaCO_3} + \underset{\substack{\text{Magnesium hydrate}\\ \text{(insoluble)}}}{Mg(OH)_2}$$

For each part per million of sulfate noncarbonate hardness, expressed as calcium carbonate, there is required 1.97 ppm of barium carbonate, or 0.017 lb per 1,000 gal of water treated. This dosage does not include a slight excess required to insure complete reaction.

This process is not practical for removing sodium sulfate, since an equivalent amount of soda ash remains in the water as a result of the reaction:

$$\underset{\substack{\text{Sodium}\\ \text{sulfate}}}{Na_2SO_4} + \underset{\substack{\text{Barium}\\ \text{carbonate}}}{BaCO_3} \rightarrow \underset{\substack{\text{Barium}\\ \text{sulfate}}}{BaSO_4} + \underset{\substack{\text{Sodium}\\ \text{carbonate}}}{Na_2CO_3}$$

By this reaction, the alkalinity of the water and also the total dissolved solids are increased.

Cold Lime-Zeolite Softening. This type of softening has a number of modifications and has been fairly extensively employed in recent years for conditioning boiler feedwater. The process consists in the reduction of the carbonate hardness by the addition of hydrated lime and, usually, a coagulant to speed up settling of insoluble calcium carbonate formed.

The effluent is then discharged to zeolite softeners, where complete softening takes place. It is customary to filter the pretreated water before it is applied to the zeolite units, but in some cases filters are omitted. The omission of filters is practical if the residual calcium carbonate floc carried by the water is dissolved by recarbonation effected by bubbling carbon dioxide gas through the water or by the addition of sulfuric or hydrochloric acid. Sulfur dioxide, generated by burning sulfur, can also be used for this purpose. Where this gas is employed, it is preferable to put it into solution in water, thereby forming sulfurous acid. The reaction is:

$$\underset{\text{Sulfur dioxide}}{SO_2} + \underset{\text{Water}}{H_2O} \rightarrow \underset{\text{Sulfurous acid}}{H_2SO_3}$$

SILICA REMOVAL

Silica can be removed from water by precipitation methods which may accompany clarification or cold- or hot-process softening, the latter being more efficient.* Most of the precipitates formed in the softening or clarification of water will adsorb some silica, especially compounds of alkaline-earth and metallic elements. Of these, calcium carbonate is least effective, and freshly formed hydroxides of iron, magnesium, and aluminum are most effective. Only ferric and magnesium compounds have been employed commercially to an appreciable extent. One investigator used ammonium zincate and sodium hydroxide in laboratory studies[15] but concluded that the ammonia made the process unsuitable for boiler-feedwater purification. Aluminum hydroxide, precipitated from sodium aluminate at a pH of 8.3 to 8.7, was found by Lindsay and Ryznar[16] to reduce silica from over 6 ppm to less than 2 ppm, but the dosage required was excessive. Subsequently, the Alhydro process of preparing aluminum hydroxide electrolytically from an aluminum anode was proposed for silica-removal treatment.[17] Behrman and Gustafson[18] experimented with hydrogen zeolite treated water, prepared in such a manner as to minimize sulfuric and hydrochloric acid while emphasizing carbonic acid. This water was passed through a magnetite bed and then softened with lime and soda ash, which reduced the silica to 2 to 4 ppm from a range of 20 to 64 ppm. Two other early papers discussing the fundamentals of the magnesium process are those by Tiger[19] and by Betz, Noll, and Maguire.[20]

Processes for removing silica by ferric and magnesium oxides have certain similarities and contrasts that should be noted before discussing the two methods separately:

* Silica removal by ion exchange is discussed in the chapter on water softening by zeolites.

1. As silica is reduced in concentration, the efficiency of the process falls off sharply, and increasingly larger quantities of ferric or magnesium hydroxide are required for each additional part per million of silica removed.

2. There is an optimum pH range for each process, and temperature affects their efficiency.

3. For a series of experiments carried out at constant temperature, if the silica removed per unit of reagent is plotted against residual silica on a double logarithmic scale, the graph is a straight line which conforms to a Freundlich adsorption isotherm. This shows that removal occurs by adsorption.*

4. The degree of hydration of the oxide used for silica removal is a controlling factor in its effectiveness for silica removal. Ferric hydroxide forms hydrous ferric oxide and then dehydrates in the absence of water:

$$\underset{\text{Ferric hydroxide}}{2Fe(OH)_3} \rightleftharpoons \underset{\text{Hydrous ferric oxide}}{Fe_2O_3 \cdot xH_2O} \rightleftharpoons \underset{\text{Anhydrous ferric oxide}}{Fe_2O_3}$$

Magnesium oxide combines with water to form magnesium hydroxide:

$$\underset{\text{Magnesium oxide}}{MgO} + \underset{\text{Water}}{H_2O} \xrightleftharpoons{\text{slow}} \underset{\text{Magnesium hydroxide}}{Mg(OH)_2}$$

In contrast to calcium oxide, which hydrates rapidly with evolution of heat, magnesium oxide hydrates very slowly. In all silica-removal investigations so far published, it has been found that the adsorption occurs on the hydrated form of the removal agent and that hydroxides precipitated *in situ* are far more effective than externally prepared oxides. However, prolonged contact with sludge which is not saturated with silica may effect some further removal.

5. The adsorption reactions are rapid and occur in much less time than the detention periods usually provided in water softening.

Silica Removal by Ferric Hydroxide. The commercial types of ferric sulfate marketed as coagulants are generally used in the process and precipitated with caustic soda or lime. The latter produces calcium sulfate, increasing the noncarbonate hardness of the water. The optimum pH value is about 9.0. For waters containing less than 10 ppm of silica, the dosage of ferric sulfate required is 15 to 20 ppm for each part per million of silica removed.[22] At higher silica concentrations, the dosage of ferric

* Nordell[21] considers that magnesium removes silica as magnesium silicate, although conceding conformity of the data to the Freundlich isotherm.

sulfate for each part per million of silica removed may be as low as 10 ppm. The lower limit attainable is about 2 ppm of silica, but this varies with different waters. In contrast to adsorption by magnesium hydroxide, efficiency decreases as the temperature rises; hence, this method is best suited to cold-process softening. The treatment is well adapted to waters which require clarification and color removal and where final zeolite softening is used. In one plant, color is removed in a first-stage clarification tank at a low pH, and silica is then removed in a second tank in which the pH is raised by means of caustic soda. The ferric hydroxide process for silica removal has the disadvantage of raising the sulfate content of the water, but, nevertheless, there are specific applications in which it has advantages over the magnesium process.

Silica Removal by Magnesium Compounds. The removal of silica by magnesium is much less efficient and flexible in cold-process softening than in hot-process softening, which is discussed in Chap. 7. At 50°F, the removal of each part per million of silica in excess of 15 ppm requires the precipitation of about 30 ppm of magnesium hardness, and this low efficiency falls off rapidly as the silica is reduced below this level. At 70°F, the ratio is 1 ppm silica removal for about 6 or 7 ppm magnesium hardness removal, and the efficiency increases rapidly as the temperature is raised further. It is apparent that cold-process treatment should not be relied on as a method of silica reduction unless temperate conditions can be assured. Sometimes this can be done by first using the water for cooling to pick up waste heat.

In a few cases, the natural magnesium hardness of the raw water may produce sufficient silica removal when excess lime treatment is applied, but often additional magnesium is needed. This can be supplied from several sources: soluble magnesium sulfate, dolomitic lime, or magnesium oxide.

Magnesium sulfate is of doubtful merit in the relatively inefficient cold-process for removing silica because of cost and the increase in the sulfate content of the water.

Magnesium oxide is available as a compound prepared for water treatment or as an ingredient of hydrated dolomitic lime, whose composition is generally in a ratio of 32 per cent MgO to 62 per cent $Ca(OH)_2$. If magnesium oxide is added to cold water in an environment of high pH, it will not dissolve. Since the calcium hydroxide of dolomitic lime provides a high pH rapidly, the associated magnesium oxide remains inert.

To bring magnesium oxide into solution, it is necessary that the pH be lower than that which prevails for the precipitation of magnesium hydroxide, and this requires a separate dissolving tank. The bicarbonate alkalinity or the free carbon dioxide of the raw water will dissolve magnesium oxide according to the following reactions:

$$\underset{\text{Magnesium oxide}}{MgO} + \underset{\text{Calcium bicarbonate}}{Ca(HCO_3)_2} \rightarrow \underset{\text{Magnesium carbonate}}{MgCO_3} + \underset{\text{Calcium carbonate}}{CaCO_3} + \underset{\text{Water}}{H_2O}$$

$$\underset{\text{Magnesium oxide}}{MgO} + \underset{\text{Carbonic acid}}{H_2CO_3} \rightarrow \underset{\text{Magnesium carbonate}}{MgCO_3} + \underset{\text{Water}}{H_2O}$$

If the natural water will not dissolve sufficient magnesium to give the necessary silica reduction, carbon dioxide must be added from a generator or available stack gases. The dissolver should preferably provide about 1 hr of detention.

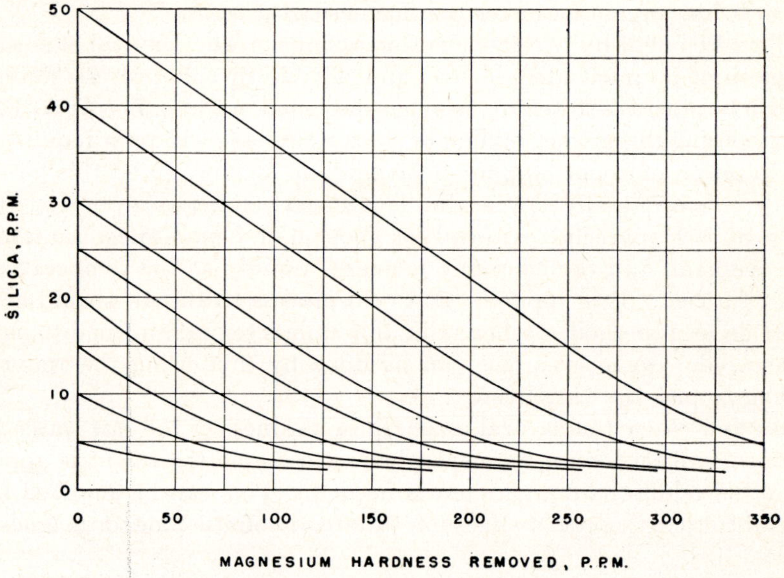

Fig. 9. Silica removed by precipitation of magnesium at 70°F.

Magnesium oxide prepared from calcined magnesite or other special sources can be added directly to raw water in the dissolver and will react in 20 or 30 min. Dolomitic lime cannot be so applied. It is added directly to the softening tank, where it serves to enrich the magnesium content of the sludge that settles in the concentrator zone. Part of this sludge is then recirculated to the dissolver, and the remainder is diverted to waste to keep removing as much silica as comes in with the raw water.

Tiger[19] observed that silica-adsorption powers were exhibited by undissolved magnesium oxide and magnesium-containing sludge, although to a lesser extent than freshly precipitated ionic magnesium. In practical operation, the efficacy of all these forms is utilized, even when no magnesium-dissolving compartment is provided. This is accomplished by utilizing a rapid-flow treatment tank, designed for retention and recirculation of a considerable amount of sludge. Detention time of at least 90 min is specified to take maximum advantage of the silica-removal power of the precipitates.

It is impossible to make precise predictions of the amount of ionic magnesium which must be precipitated from any water supply to produce a specific silica residual. However, experiments by the investigators cited and observations in many plants are reflected in the curves of Fig. 9 for ionic magnesium. For instance, if the silica concentration is 30 ppm, the precipitation of 150 ppm of magnesium hardness will reduce silica in direct proportion down to about 8 ppm. Beyond that, the curve flattens out and has been terminated at 2 ppm because results in that range may vary widely, depending on recirculation and other factors. For design purposes, these curves may be used to evaluate the adequacy of the magnesium hardness of the raw water and the order of magnitude of necessary magnesium additions.

EQUIPMENT FOR COLD-PROCESS SOFTENING

Cold-process softening plants may be broadly classified as *intermittent* or *continuous*. Current practice is dominated by the latter type.

Intermittent Softening Systems. The first precipitation softeners used in this country were of this type and were installed by the Southern Pacific Company at Port Los Angeles about December, 1896. This plant was designed by Howard Stillman. Subsequently, such equipment found wide favor particularly with the railroads for wayside water treatment. Batch softening systems have given good service for years in many boiler plants during the development of modern boiler-feedwater purification methods.

Intermittent softening systems include two or more reaction and settling tanks with chemical feeders, usually gravity filters, and softened water storage. The settled water is withdrawn through a swing-pipe skimmer supported by a float. The units are operated on a definite fill-and-draw cycle. The dosage of lime, soda ash, and coagulant is calculated and weighed out in advance. These chemicals are added to each tank as it is filled with water, and the initial excess of chemicals during the filling is favorable to low hardness. The water is mixed by motor-driven paddles, and the insoluble salts are then allowed to precipitate while the water remains quiescent. During all this period, softened water is drawn from another tank in which the cycle of filling, treatment, and settling has previously been completed. The system must include enough tanks to permit an adequate settling period, usually 4 to 8 hr. After each tank is emptied, the precipitated sludge is flushed out, and the cycle is repeated.

The effectiveness of these fill-and-draw softeners in producing an effluent of low hardness and turbidity depends on the excess of chemicals used, temperature, and the detention time which is allowed for reaction and settling. The temperature is a major factor, and often the raw water is first passed through surface condensers or other station equipment to elevate the temperatures of the water, thereby accelerating the reaction.

Early Types of Continuous-flow Cold-process Softeners. Continuous softeners were developed to furnish an uninterrupted supply of feedwater for stationary boiler plants. Many ingenious treatment tanks and associated chemical feeders and filters were developed for this purpose. The softeners usually consisted of a reaction tank with internal baffles, means for agitation, and compartments for the chemical reaction and settling of the precipitates. Chemical treatment was applied in proportion to the volume of hardness of water as it entered the unit. Thus, the early designs have many features in common with modern softeners.

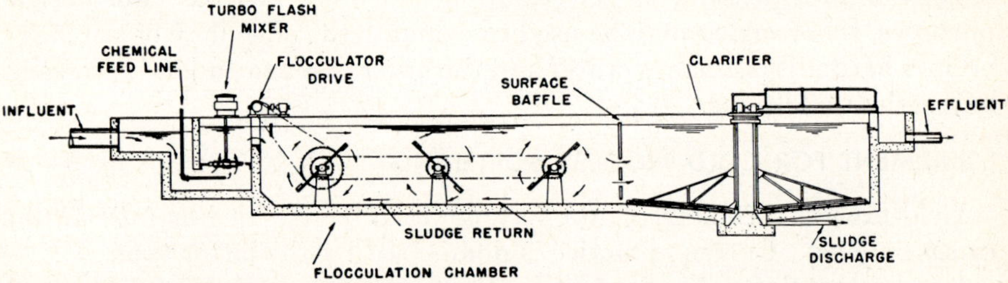

Fig. 10. A conventional horizontal-flow lime-soda softening system.

Modern Long-detention Cold-process Softening Plants. This type of equipment includes some industrial and most public water-supply softening systems installed prior to 1940. The basic elements are those employed in coagulation for clarification and color removal, described in Chap. 4. The water will pass through (1) a flash mixer, (2) a flocculating basin or clarifier or both, (3) a settling basin, and (4) filters. Recirculation of sludge to the zone of chemical reaction is often employed to "seed" the softening reaction and promote large-particle formation. Recarbonation and acid treatment are widely employed before or after filtration or both to prevent growth of calcium carbonate on the filter medium, to produce a stable effluent which will cause neither scale nor corrosion, and to meet public-health requirements for the absence of caustic alkalinity.

Although the basic treatment units in cold-process lime softening systems are similar to those used in clarification, the weight and volume of the precipitate is so great that mechanical facilities must be provided for continuously removing it. Typical units are shown in Fig. 10. The bulk of the sludge is often best removed in a basin with rakes which scrape the material to a central drainage sump rotating around the vertical axis. Horizontal-flow settling basins used for softening should also be equipped with flight scrapers or rakes traveling longitudinally toward a sump at the inlet end.

Rapid-flow Treatment Tanks for Cold-process Softening. The Spaulding Precipitator was developed for the Springfield, Illinois, water softening plant.[23] One of the main objectives was to produce a stable water by bringing the already formed sludge into contact with the water and the chemicals in order to accelerate precipitation and cause the reaction to go to completion. Behrman and Greene stressed the same advantages in the first published information on the Accelator.[24] Illustrations and descriptions of these designs will be found in Chap. 4. At the present time, a number of manufacturers offer equipment designed for cold-process softening which take advantage of prolonged contact with the precipitated calcium carbonate and magnesium hydroxide.

By using equipment of this type, it is possible to reduce carbonate hardness to 35± ppm with a detention time of not over 1 hr. These tanks are advantageous for silica removal by adsorption on precipitated magnesium hydroxide, but when this effect is important, the detention is usually increased to 1½ hr. A further advantage is a continuous withdrawal of sludge as fast as it is formed, facilitating disposal or recovery.

The great majority of cold-process softening plants built within recent years have utilized one of the types of rapid-flow treatment tanks.

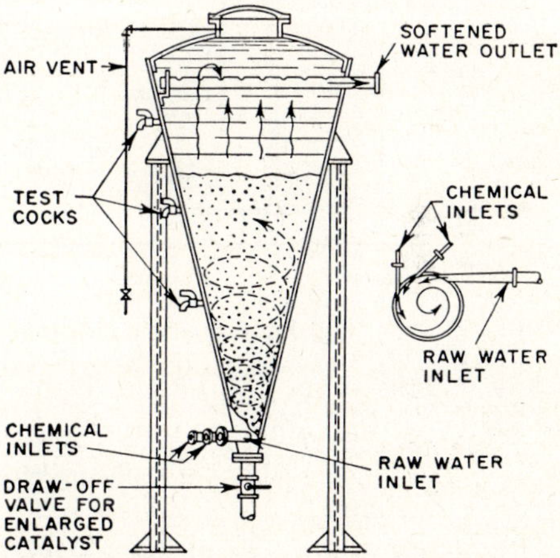

Fig. 11. Cross-sectional view of Permutit Spiractor, showing flow. Inset shows detail of arrangement of raw-water and chemical inlets.

"Catalytic" Removal of Calcium Hardness. An ingenious design of softening unit known as the Spiractor is shown in Fig. 11. This is an adaptation of a cold lime-soda system which was originally developed by Dr. E. T. Zentner of Czechoslovakia.* The tank in which the reaction takes place is an inverted cone, which is about half full of finely divided granules or beads of calcium carbonate. The raw water and softening reagents enter, in a tangential direction, the bottom of the tank where intimate mixture of the water and chemical takes place and the body of water is subjected to a whirling, rotating motion as it rises through the tank. The precipitated

*Marketed by The Permutit Company, New York.

calcium salts come into direct contact with the grains of calcium carbonate, deposit on the surface, and increase their size. The velocity of the water passing through the unit is high enough to maintain the grains of calcium carbonate in suspension but is insufficient to carry the granules out of the tank with the effluent. The suspended solids (mainly magnesium hydroxide) not adhering to the floating granules are removed by pressure filters.

The process of treatment is particularly suitable for softening hard water containing calcium salts but only a limited amount of magnesium.

When the grains of calcium carbonate build up to a certain diameter, they are removed by blowing down the tank, and additional fine granules of the catalyst are added to the bed.

AUXILIARY TREATMENT EQUIPMENT

Chapter 4 describes chemical feeders used in connection with coagulation and settling processes for clarification, color removal, and other pur-

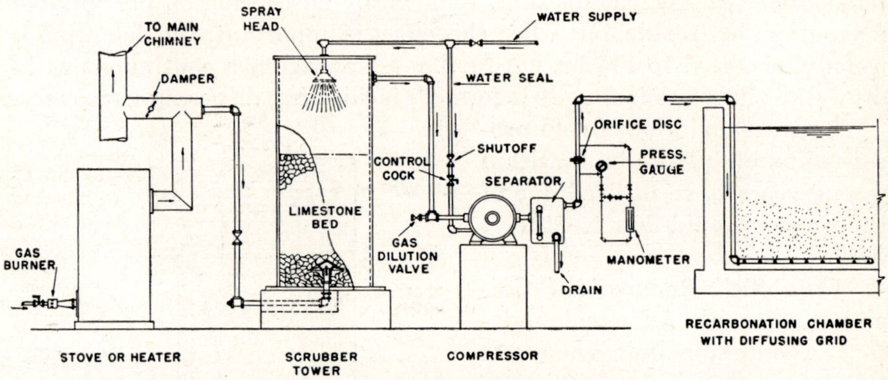

Fig. 12. Schematic diagram of a recarbonation system.

poses. The same equipment is used for cold-process softening. The present section adds equipment for some auxiliary treatments which are specific to lime-soda softening: recarbonation, acid feed, and decarbonation.

Recarbonation. Equipment for recarbonation generally includes: a burner for gas, oil, or solid fuels, a counterflow water scrubber for cooling and purifying the gas, a drier and compressor, and a distribution grid or diffusers submerged in a recarbonation tank. The combustion process may be carried out either in a simple firebox or in a water heater where there is a use for the hot water, which may often be employed advantageously for slaking lime, building heating, or sanitary fixtures. Figure 12 is a schematic flow diagram of a simple system furnishing carbon dioxide at a constant, manually adjustable rate of feed. This is suitable where the water-treating plant is operated on a constant-flow, on-and-off basis. Proportioning recarbonation systems and those involving pH control are more complicated.

Figure 13 shows a carbon dioxide generator developed especially for recarbonation service. Gas or fuel oil is burned under forced-draft conditions at a high temperature to produce carbon dioxide of high purity which does not need scrubbing. Fuel requirements are based on about 23 lb of carbon dioxide per gallon of oil or 11.5 lb per 100 cu ft of natural

Fig. 13. Forced-draft carbon dioxide generator. *(Courtesy of Walker Process Equipment, Inc.)*

gas. The combustion gases contain 14.5 per cent of carbon dioxide when oil is used or 10 per cent when gas is used. Chemical and combustion efficiency are higher than for natural draft burners. The blower delivers the gas at any pressure up to 7 psi for delivery to the distributor.

The distributor may be a simple perforated grid of brass pipe or a more complicated design, with which higher absorption efficiency can be obtained. The manufacturer of the package unit shown in Fig. 13 furnishes a distributing grid in which each of the gas nozzles is combined with a high-velocity water jet which impinges on it and produces very fine bubbles, promoting high absorption efficiency.

Acid Treatment. Equipment for feeding sulfuric acid to lime-soda softened water may be preferred to recarbonation equipment for industrial or small public water-supply plants because it is easier to operate, especially for proportioning or pH-controlled feeding. In general, dilute acid (5 per cent) is prepared in a lead- or rubber-lined tank and applied by any of the liquid feeders described for coagulants in Chap. 4. It may be satisfactory to control the acid by proportioning the feed to the flow of raw water entering the system in parallel with the coagulant-proportioning equipment. This is applicable where only enough acid is used to remove any caustic alkalinity and assure prevention of calcium carbonate

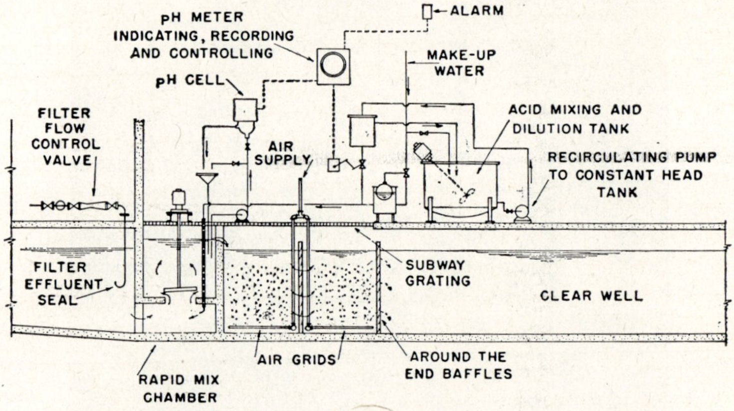

Fig. 14. Equipment for acid feeding and decarbonation.

deposition in filter or zeolite beds, always leaving a relatively large fraction of the alkalinity. However, if it is desired to produce boiler feedwater of relatively low alkalinity, the acid feed should be actuated by a pH recorder-controller to avoid the danger of overtreatment.

Equipment for this purpose is described in Chap. 12, Fig. 6, covering the application of chemicals for secondary treatment of feedwater and boiler water. The basic requirements are the same when the treatment is to be used after cold-process softening. Figure 14 shows a typical application.

Decarbonation. Acid treatment of lime-soda softened water will convert hydroxides to sulfates and normal carbonates to bicarbonates and sulfates, in neither case releasing any free carbon dioxide. However, for each part per million of *bicarbonate* alkalinity removed, 0.88 ppm of carbon dioxide will be released. If the resultant concentration of carbon dioxide is high in relation to the residual alkalinity, the pH will be low and the water will be corrosive; therefore, the gas is frequently removed immediately after acid treatment.

There are three basic methods for removing carbon dioxide from water: thermal degasification, vacuum degasification, and aeration. Deaerating heaters may reduce the carbon dioxide concentration to nearly zero, but if bicarbonate alkalinity is low or absent, the last trace cannot be removed.[25] If the bicarbonate alkalinity is 10 ppm, the free carbon dioxide can be reduced to zero; while if the alkalinity is zero, the free carbon dioxide may remain as high as 5 ppm.

Vacuum deaerators designed for removing oxygen from cold water to control corrosion operate at temperatures where the solubility of carbon dioxide and the stability of carbonic acid are less favorable to carbon dioxide removal, but the residual may be as low as 3 to 5 ppm.

The residual carbon dioxide in water treated by aeration processes approaches, as a limit, the solubility of carbon dioxide in contact with the partial pressure of this gas in the air at the operating temperature. At 70°F and normal atmospheric composition, this is of the order of 1 ppm. Actual concentrations after aeration will depend on the time allowed and the surface of the air-water interface. In simple spray or cascade aerators, both of these factors are at a minimum, and the final carbon dioxide may be of the order of 20 to 30 ppm. The efficiency of a decarbonation basin with submerged air diffusers, as shown in Fig. 14, is improved by producing the maximum number of small air bubbles, thus increasing the total surface. Porous Alundum, carbon, or special screen diffusers are used for this purpose. However, frequently a grid of perforated brass pipe may be adequate for the removal required. Detention should be 60 min and air 1 cu ft per min per gal for a final carbon dioxide as low as 10 ppm if the initial carbon dioxide is 100 ppm. More liberal design is required for the removal of higher concentrations. The forced-draft degasifier is one of the most efficient aerators for carbon dioxide removal. This apparatus consists of a tower filled with wood slats or other packing in which the water flows downward counterflow to air from a blower. Column heights range from 14 to 20 ft, depending on the initial carbon dioxide and the required residual of 5 to 10 ppm. Flow is of the order of 20 gpm per sq ft, and air requirements 80 cu ft per min per sq ft.

DISPOSAL OF WATER-SOFTENING SLUDGES

Quantities Involved. Precipitates formed in water softening are transferred from the treatment tank by draining under gravity head or pumping from a selected zone of highest concentration designed to accumulate settled solids. Thickeners, clarifiers, and some conventional horizontal-flow basins and rapid-flow treatment tanks are equipped with scrapers or rakes which bring the solids to a drainage sump. Other designs employ expanded settling surfaces in sludge concentrators from which perforated collector pipes withdraw the sludge to waste. It is desirable to remove the sludge at the highest possible solids concentration to minimize loss of treated water and to keep down the volume requiring disposal. In practice, the sludge as drawn from the plant may contain 5 or, rarely, 10 per cent of solids on a "bone-dry" basis, while if performance is poor, the solids may drop to a fraction of 1 per cent.

The basic reactions in lime-soda softening produce calcium carbonate equivalent to the calcium hardness removed plus all of the lime added and magnesium hydroxide equivalent to the magnesium removed. Since the molecular weight of calcium carbonate is 100 and that of magnesium hydroxide is 58, the weight of the sludge on a bone-dry basis is readily calculated when estimating the results of water treatment. The quantities of sludge produced are included in Table 7, showing the results of softening Lake Michigan water.

TABLE 8. SLUDGE PRODUCED BY PARTIAL SOFTENING OF ONE MILLION GALLONS OF LAKE MICHIGAN WATER WITH LIME

Moisture content,* %	Remarks	Weight, incl. H_2O, lb	Volume, cu ft	Sp gr†
95	As blown down from treatment system	23,333	366	1.01–1.02
85	After thickening	7,777	118	1.04–1.06
75	Dewatered by skimming and evaporation in drying beds or lagoons	4,666	69	1.07–1.10
50	Dewatered on vacuum filter	2,333	32	1.12–1.23
35	Dewatered by centrifuge	1,795	23	1.20–1.32
0	Precipitate on bone-dry basis	1,166‡	11	1.33–1.60

* The moisture content will vary considerably, but those given are typical or values cited in the literature referred to.

† Calculated according to the formula $S = 100s/[100s - \% (S - 1)]$, where S = sp gr of sludge, % = per cent solids on dry basis, and s = sp gr of dry solids in finely divided state. The range of specific gravities given corresponds to variations in s for calcium carbonate between 1.60 and 1.33 for 50- to 200-mesh particles.

‡ The solids contain 2 per cent magnesium hydroxide but, to simplify calculations, have been assumed to be 100 per cent calcium carbonate.

Table 8 shows the sludge produced by partial softening of 1 million gal of this water (Column C of Table 7) and its weight and volume at various stages in dewatering. If the sludge is 95 per cent moisture, then for every 10 mgd of plant capacity, facilities must be provided for the daily removal of over 3,600 cu ft of sludge. If it can be thickened to 85 per cent, the volume which must be disposed of drops to 1,180 cu ft per day, and so on.

For other cases, calculate the pounds of dry solids and the pounds of associated water as defined by the per cent moisture content, *e.g.*, 15 lb of 15 per cent sludge carries with it 85 lb of water. The volume of solids estimated at 100 lb per cu ft plus the volume of the water at 62.5 lb per cu ft will give a close approximation of the total sludge volume.

The example discussed above covers the removal of a relatively small amount of hardness from a water supply that is initially fairly soft. However, numerous public and private water supplies will produce many times as much sludge per million gallons when softened to an acceptable degree. Since water softening sludges must be removed continuously as fast as they are formed, in some cases this presents a serious problem.

Methods of Disposal. The methods that have been used include disposal in rivers or other surface waters, lagooning, dewatering and trucking to waste land, drying and marketing as calcium carbonate, and calcining to quicklime, reusing what is required and marketing the excess.

Nonrecovery Methods. *Disposal in a Receiving Body of Water.* The traditional procedure, which was ready-made for early designers, has been to locate the softening plant on a river or lake which provides dilution volume and flow velocity adequate to disperse the sludge solids. There are still some watercourses that are essentially industrial sewers in some sections and which have capacity for diluting and dissolving sludge from water-softening plants. Obviously, small industrial installations can be accommodated where large public-supply plants might encounter difficulty. However, in many locations where lime-soda softening plants are needed, this disposal method is not available either because there is no adequate receiving body of water or because of pollution-abatement laws.

Lagooning. This method of disposal is widely used not only for water-softening sludges but for the liquid and solid wastes from many industrial processes, such as the manufacture of acetylene from calcium carbide, the causticizing of soda ash, etc. Lagoons are constructed by moving earth from a shallow excavation of the area selected to form a surrounding dike. Depth is limited by the permissible hydraulic thrust against the dike, which, in section, is a broad-based triangle. Sluice gates must be provided for overflow to avoid exceeding this safe depth and for skimming off the clear supernatant so that the lagoon will retain storage capacity for solids.

The dewatering characteristics of different sludges vary widely, and this has an important influence on the requirements for disposal facilities. A 5 per cent slurry of calcium carbonate and magnesium hydroxide will gradually settle and dewater to a solids content of 20 to 35 per cent. The time required depends on particle size and other factors and may be excessively long. In the above example of Lake Michigan water, the volume of the daily thin sludge (5 per cent solids) per 10 mgd is 3,666 cu ft, but if it will compact to 25 per cent solids, the volume decreases to 690 cu ft. Under the latter condition, a year's storage in a lagoon 5 ft deep will require 1.10 acres. For final solids contents of 50 and 65 per cent, the annual storage area required would be 0.54 and 0.38 acres, respectively. Retention volume must be added for settling the thin sludge to the final density.

Disposal of sludge by lagooning requires large areas of land that can be abandoned to this purpose. Such areas are not always adjacent to the sites of water-treatment plants. The useful life of a sludge lagoon has a definite limit; adjacent ground must also be available for future needs, or

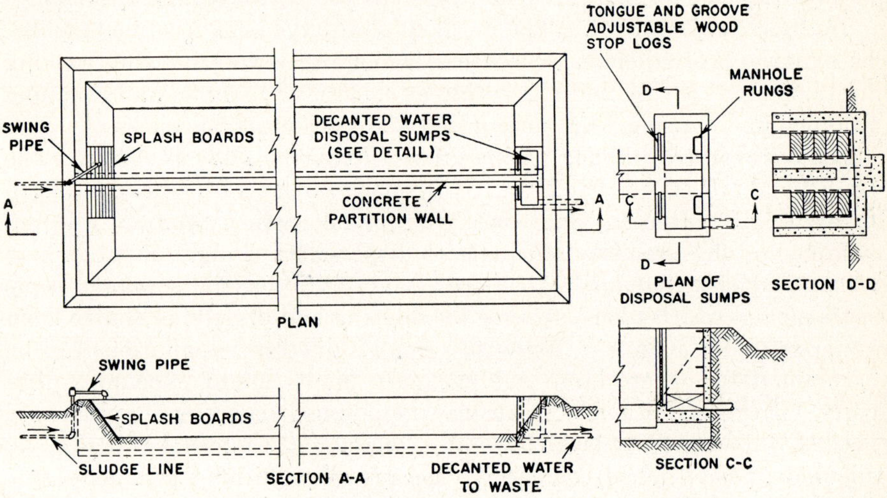

Fig. 15. Settling basin for sludge from lime-softening plant.

the compacted sludge must be removed by dredging and hauling to waste land where it can be dumped.

Dewatering for Hauling and Dumping. Where hauling away and dumping on waste land or for fill is the accepted method of sludge disposal, dewatering can be accomplished in a smaller area than that required for long-term lagooning. Figure 15 shows a settling basin designed for this purpose. The basin is divided longitudinally into two sections so that one can be used while the other is being dredged. The thin sludge enters through a delivery pipe which may be swung over either section. A weir at the end of a long, narrow basin skims off the clear supernatant. The design may be based on a settling time of one or more days (depending on the characteristics of the sludge) plus two to four weeks' storage of settled sludge at the density attainable in the time permitted. Settled solids will pile up at the inlet end and slope off toward the outlet weir. Before the volume is reduced enough to impair settling efficiency, the first basin is taken out of service and dredged while the other is used.

From the data of Table 8 for sludge at 75 per cent moisture content, it will be seen that the daily sludge accumulation to be handled in this manner for a 10-mgd plant softening Lake Michigan water would be of the order of 690 cu ft, but it might vary ±30 per cent, depending on the compacting properties of the sludge.

Mechanical dewatering by centrifugal or vacuum filters would further reduce the volume of this sludge to 200 to 300 cu ft per day. Such treatment is indicated where disposal area is limited and recovery methods cannot be justified.

Recovery Methods. *Quantity of Product.* Calcium carbonate is a product of commerce as agricultural limestone, as a filler for various products, and as an industrial raw material for various uses. If the calcium carbonate produced by water softening is dried and is of satisfactory chemical and physical quality, it can be sold in competition with other supplies on the market. Alternately, it can be calcined to yield quicklime according to the following reactions:

$$\underset{\text{Calcium carbonate}}{CaCO_3} \rightleftharpoons \underset{\text{Calcium oxide}}{CaO} + \underset{\text{Carbon dioxide}}{CO_2}$$

The carbon dioxide produced can be compressed and used for decarbonation where this treatment is used.

The yield of calcium carbonate or oxide varies with the quality of the raw water and the degree of softening. In the simplest case, when only calcium bicarbonate hardness is removed with lime, the calcium carbonate precipitate is equivalent to *all* the lime added *plus* the decrease in calcium hardness in the raw water. Table 7 shows that, in the partial softening of Lake Michigan water (Column C), the calcium carbonate precipitated, if calcined to lime, would obviously exceed the lime used, continuously producing a surplus. Calcium carbonate from carbon dioxide neutralization and noncarbonate-hardness removal with soda ash would swell the excess. However, losses of solids in dewatering and in other steps in the recovery will reduce the yield, as found in the several plants mentioned below.

Experience with Recovery Systems. Although drying and calcining of calcium carbonate sludge from industrial processes is common, recovery systems for water-softening plants are few in number. In England, the process was installed at the Southend Water Works Company at Langford and was described in a symposium published in 1933.[26,27] In 1939, Aultman[28] noted that there were no recovery plants in the United States. However, by 1944, operating data were reported for plants recovering dried calcium carbonate at the Wright Aeronautical Company, Cincinnati,[29] the Kankakee, Illinois, Water Company,[30] and plants for burning it to lime at Marshalltown, Iowa,[31] and the Columbia Steel Works, Provo, Utah.[32]

The 1949 report of the AWWA Committee on Disposal of Softening Plant Wastes[33] summarized the status of recovery systems as of that date when considerable progress had been made since the early work on the plants listed above. The 80-ton-per-day recovery plant at Miami, Florida, was the largest one reported.[34] This installation treats not only the sludge produced by current operation but is recalcining accumulations of 8 million cu ft of sludge in a 20-acre lagoon.

Equipment Used. Illustrations and performance data for the dewatering, drying, and calcining equipment are given in detail in the articles cited.

If the sludge, as drawn from the softening plant, contains more than 85 per cent moisture, it should be dewatered to that in a thickener. It is further dewatered to 50 ± per cent moisture on a vacuum filter or to 35 ± per cent moisture by means of a solid-bowl centrifugal filter. The "centrate" from the latter is the water removed plus some of the solids, while the cake contains the preponderance of the solids and the balance of the water, both phases being controllable within limits.

The dewatered cake is ultimately flash-dried in suspension to produce calcium carbonate or calcined to calcium oxide. Three types of apparatus have been used for the latter operation: rotary drum kilns, multiple-hearth furnaces, and flash-drying furnaces which produce either dried calcium carbonate or calcium oxide. Figure 16 is a flow sheet of the latter process.

Control of Quality. In addition to calcium carbonate, water-softening sludges contain magnesium hydroxide, silica, iron, and other contaminants if the raw water is a surface supply. If the sludge is calcined to lime and recycled, the magnesium and other inert materials will build up in the recovered lime until its softening value is too low for satisfactory operation. To overcome this, the magnesium must be purged from the system by displacing part of the reburnt lime by new lime. Where marketable calcium carbonate is the product recovered, iron and other sources of color may be objectionable.

Two general types of corrective measures are available for controlling the quality of recovery products from water-softening sludges. In the Lykken-Estabrook or Hoover processes,[28] split treatment is used: Calcium is precipitated selectively in one clarifier at a low pH and magnesium in another at a high pH, the high magnesium sludge being rejected. Numerous modifications of this basic scheme can be devised to suit individual water supplies. The second procedure[29] is dewatering by means of a centrifuge and splitting the solids between the cake and the centrate so that the latter contains a higher proportion of magnesium, iron, etc.

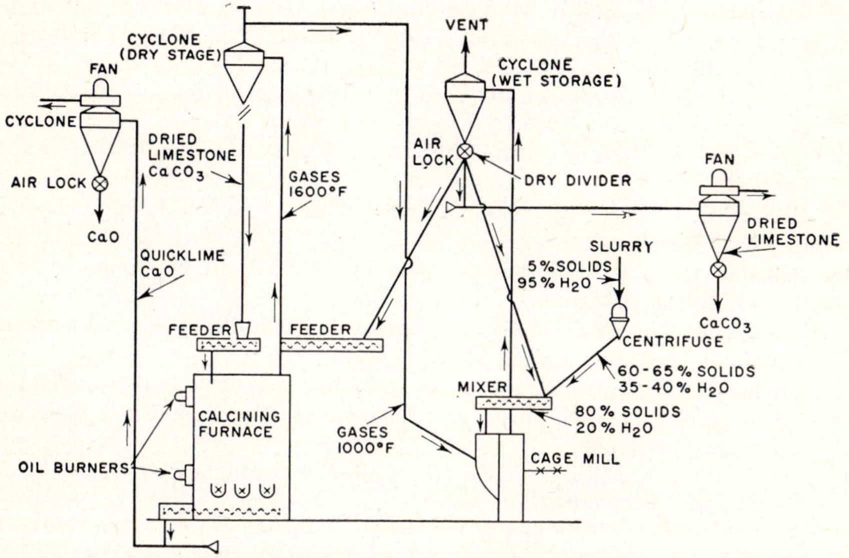

Fig. 16. Process for drying and calcining sludge from a water-softening plant.

Recovery processes for water-softening sludges are now considered to be feasible for a wide variety of water supplies.

Bibliography

1. Scott, W. B.: "Water and Water Supplies," 1884.
2. Archbutt, Leonard: The Softening of Water by the Archbutt-Deeley Process, *Eng. News*, **40**: 403 (1898).
3. Seidell, A.: "Solubilities of Inorganic and Metal Organic Compounds," 3d ed., D. Van Nostrand Company, Inc., New York, 1940.
4. Nordell, Eskel: "Water Treatment for Industrial and Other Uses," Reinhold Publishing Corporation, New York, 1951.
5. Joos, C. E., and Frank Kemmer: Operation of Hot-process Softener at 50 Pound Gauge Improves Performance and Saves Chemicals, *Combustion*, **14** (10): 45–47 (April, 1943).
6. Calise, V. J.: Hot Zeolite Softening, *Power Eng.*, **54** (9): 54 (September, 1950).
7. Wirth, L. F., and W. S. Butler: "Power Plant Operation with Hot Lime-Zeolite Process," ASME Paper 50-A-77, abstracted in *Mech. Eng.*, vol. 28, January, 1951.
8. Butler, W. S.: An Inspection Report after 18 Months with Hot Lime Zeolite, *Proc. 12th Ann. Water Conf. Engrs. Soc. West. Penna.*, 1951.

9. Kemmer, F. N.: Conditioning Make-up by High-temperature Sodium Ion-exchange, Excess Calcium, Hot-lime, Zeolite Process, *Combustion,* **21** (10): 59 (April, 1950).
10. Applebaum, S. B.: Discussion of Butler's paper, Ref. 8.
11. Greenfield, R. E., and A. M. Buswell: Investigation by Means of the Hydrogen Electrode of the Chemical Reactions Involved in Water Purification, *J. ACS,* **44**: 1435 (1922).
12. Baylis, John R.: Hydrogen-ion Control in Water-softening, *Ind. Eng. Chem.,* **20**: 1191 (November, 1928).
13. Hoover, C. P.: Recarbonation of Softened Water, *Ind. Eng. Chem.,* **19**: 784 (1927).
14. Hoover, C. P.: Stabilizing Lime-softened Water, *J. AWWA,* **34**: 1425 (September, 1942).
15. Reimers, Hans A.: Solution of Silica by Water, and Its Removal, *Vom Wasser,* **7**: 175 (1933).
16. Lindsay, F. K., and J. W. Ryznar: Removal of Silica from Water by Sodium Aluminate, *Ind. Eng. Chem.,* **31**: 859 (1939).
17. Stephan, Paul G., technical director, Alhydro, Inc.: Private communication.
18. Behrman, A. S., and H. Gustafson: Removal of Silica from Water, *Ind. Eng. Chem.,* **32**: 468 (1940).
19. Tiger, Howard: Silica Removal by an Improved Magnesia Process, *Trans. ASME,* **54** (I): 49 (1942).
20. Betz, L. D., C. A. Noll, and J. J. Maguire: Removal of Silica from Water by Cold-process, *Ind. Eng. Chem.,* **34**: 1320 (1940); Removal of Silica from Water by Hot-process, *Ind. Eng. Chem.,* **32**: 1323 (1940). See also *Trans. ASME,* **63**: 713 (1941).
21. Powell, Sheppard T., Lewis V. Carpenter, and John J. Coates: Complex Silicate Scales in High Pressure Boilers—Their Occurrence and Correction, *Combustion,* **11**: 18 (1939).
22. Schwartz, M. D.: Removal of Silica from Water for Boiler Feed Purposes: The Ferric Sulphate and Hydrous Ferric Oxide Process, *J. AWWA,* **30**: 551 (1938).
23. Spauling, C. H.: Conditioning of Water Softening Precipitates, *J. AWWA,* **29**: 1697 (November, 1937).
24. Behrman, A. S., and W. H. Greene: Accelerated Lime-Soda Water Softening, *Ind. Eng. Chem.,* **31**: 128 (February, 1939).
25. Kittredge, Arthur E.: The Removal of Carbon Dioxide from Alkaline Water, Midwest Power Conference, 1947.
26. The Chelmer and Blackwater Supplies of the Southend Waterworks Company—A Symposium, *Water and Water Eng.,* **35**: 321 (1933).
27. Lime Recovery from a Water Treatment Plant, *Ind. Chemist,* **9**: 229 (1933).
28. Aultman, W. W.: Reclamation and Re-use of Lime in Water Softening, *J. AWWA,* **31**: 640 (April, 1939).
29. Sheen, Robert T., and Herbert B. Lammers: Recovery of Calcium Carbonate or Lime from Water Softening Sludges, *J. AWWA,* **36**: 1145 (November, 1944).

30. Nelson, Frederick G.: Recalcination of Water Softening Sludge, *J. AWWA*, **36**: 1178 (November, 1944).
31. Pedersen, H. V.: Calcining Sludge from a Softening Plant, *J. AWWA*, **36**: 1170 (November, 1944).
32. Black, A. P.: Disposal of Softening Plant Wastes, Committee Report, *J. AWWA*, **41**: 819 (September, 1949).
33. Black, A. P., C. F. Wertz, C. R. Henry: Recalcining Softening Sludge at Miami, Florida, *Water Works Eng.*, **104**: 214 (March, 1951).
34. Gordon, Charles W.: Calcining Sludge from a Water Softening Plant, *J. AWWA*, **36**: 1176 (November, 1944).

7 HOT-PROCESS SOFTENING BY CHEMICALS

Hot-process softening extends the principles and reactions of cold-process softening to higher temperatures, where solubilities are favorable to much greater hardness reduction, and permits variations in treatment that would not be economical in the cold. In hot-process softening, the carbon dioxide is removed by flashing and needs no lime. The treatment is almost always carried to the maximum extent by using excesses of chemicals, and the partial softening which is commonly practiced in cold process is not given consideration. Hot-process softening is one of the most versatile methods of treating water, simultaneously heating, deaerating, and removing hardness, silica, excessive alkalinity, turbidity, and other impurities. The water produced by hot-process softening is almost always used for boiler feed make-up, although there are a few other applications, such as sulfur mining. The standard treatment which predominates in this process is softening with lime and soda ash. Some of the modifications have been mentioned in introducing the concept of lime-soda softening in Chap. 6. Variations include the use of magnesium oxide or soluble magnesium salts to remove silica and gypsum to treat sodium bicarbonate waters. This process is used to treat waters high in sulfate by barium and by second-stage softening with phosphates or phosphoric acid. Aftertreatment with acid to reduce pH and alkalinity and secondary zeolite softening are employed as combined treatments.

HISTORICAL

By 1900, there were numerous devices in use for heating and purifying boiler feedwater. The technical literature on this subject published early in the century[1] shows a number of complicated vessels and tanks in which sludge was produced by thermal softening as the water was heated before injection into boilers. The next step was the addition to an oversized open heater of the soda ash used for boiler-water conditioning, thereby keeping the softening products outside the boiler. However, there usually remained considerable bicarbonate hardness unconverted by the heat. About 1910, the practice of feeding milk of lime along with soda ash was adopted. Cold-process softening was being developed for various uses at the same time, and knowledge of the lower solubility and more rapid separation at high temperature led to improvement in the hot-process treatment equipment. To permit better removal of the sludge, the treatment tank was fitted with a conical hopper bottom and a central uptake, and the open heater was installed on top. Thus, the softeners of the second decade of the century were roughly similar in shape to current designs, although many improvements have been made and new processes added.

CONVENTIONAL HOT-PROCESS SOFTENERS

Figure 1 is a phantom view of a conventional hot lime-soda softening system and illustrates the function of the component parts. The raw water flows through a metering orifice and float-controlled inlet valve actuated by the level in the treated-water section of the sedimentation tank. The water then passes through a vent condenser and spray valves mounted in the rosette of a preheater. Chemicals are fed in proportion to flow, controlled by the meter transmitter. The chemical pump injects the lime and soda into the softening tank in the path of the sprayed water. The sprays produce fine droplets offering a large surface, and the exhaust steam which is supplied to the upper section of the tank heats the water to within one or two degrees of steam temperature, causing the release of oxygen and carbon dioxide. In tanks with well-designed primary heaters, the oxygen content of the water will be reduced to 0.1 or 0.2 ppm at this point. In the upper zone of the sedimentation tank there should be sufficient agitation by the sprayed water to cause thorough mixing of the water with the chemicals. Beneath this zone, precipitation and sedimentation take place rapidly. It is usual to provide volume between the surface of the water and the uptake funnel to allow 1 hr of detention.

The water flows around the rim of the inverted uptake funnel, where the density of the precipitate and the change in direction give effective separation. The solids settle to the cone bottom and are drawn off by blowing down. The clarified water then passes upward to a secondary deaerating heater, in which the steam entering the tank atomizes the water into fine drops and removes final traces of noncondensible gases. The treated, deaerated water then passes to a battery of filters and thence to the boiler feed pumps. Water is drawn from the sedimentation tank above the uptake funnel and used for backwashing the filters and returned to the upper section of the sedimentation tank.

The assembly shown in Fig. 1 is designed for softening and deaerating make-up only. Where appreciable condensate is available, three general arrangements may be used:

1. The deaerator may be omitted from the sedimentation tank, and both make-up and condensate may be discharged to a separate deaerating heater. This has some advantages of flexibility and makes it possible

Fig. 1. Hot lime-soda softening system. (*Courtesy of Cochrane Corporation.*)

always to furnish heated boiler feedwater in the event the softener must be taken out of service and raw water used.

2. If the percentage of condensate is moderate, a separate condensate deaerator with its own separate storage compartment is incorporated in the sedimentation tank, and its outlet is connected to the line from the filters to the boiler feed pump suction. Head conditions should be arranged to favor preferential use of the condensate.

3. If the condensate is a major portion of the feedwater, it is often desirable to store the mixed condensate and make-up after deaeration.

In this case, the settled water is discharged to the filters and returned to a deaeration compartment. The condensate enters a primary spray-heater section which discharges to the same deaerator. This arrangement permits storage of a relatively large amount of feedwater. Additional tank height is required because the deaeration compartment must be far enough below the water level in the main tank to offset loss of head through the filters. There are numerous variations in the design of such systems, especially with respect to modification of internal arrangements of compartments.

DESIGN FEATURES OF REACTION AND SEDIMENTATION TANKS

These tanks differ hydraulically as to upflow or downflow design and functionally as to the omission or inclusion of deaerating heaters, backwash-water compartments, etc.

Figure 2 shows simplified sectional drawings of downflow, hot-process sedimentation tanks of the simple nondeaerating type offered by five manufacturers.

The first requirement for good operation is efficient atomization of the water in the primary heater, *H*. The pressure drop across the spring-loaded valves aids in immediate flashing of gases out of solution, and the approach of the water temperature to steam temperature depends on the fineness of the spray. Partial failure of the sprays may cause the water temperature to drop as much as 15° below saturation temperatures, and this will, in turn, affect the velocity and completeness of the softening reaction and the clarity of the effluent. Chemicals which enter the point *C* should be thoroughly mixed by the violence of the spray action or, in the case of tanks with central downcomers (Design 2), by turbulence and high velocity.

One hour's retention in the tank between the water level and the effluent point *E* is generally considered necessary, although it is known that the softening reactions, at the temperature usually employed, go to completion in less than a third of this time. The most important function after mixing is separation of the solids from the water. Opinions vary as to the proper relation between the downward velocity *V* of the water passing from the reaction zone to the upward velocity *W* in the zone of solids separation. In Design 3, advantages have been claimed for equal velocities throughout, whereas in the other units shown, the upward velocity in the zone of separation is from one-half to one-fourth the downward velocity. More important is the actual rising rate in the separation zone. This has been called the "upward settling rate."[2] Obviously, the downflow tank with a central offtake cone does not permit so large a rising area or so low a rising velocity as does the upflow tank with a central downcomer. In general, in either type of tank, if a very low effluent turbidity is desired, the rising rate should not be over 3 gpm per sq ft of tank area.

Lack of symmetry in the offtake facilities can cause serious short-circuiting of water, resulting in reduction of detention to as little as ¼ the

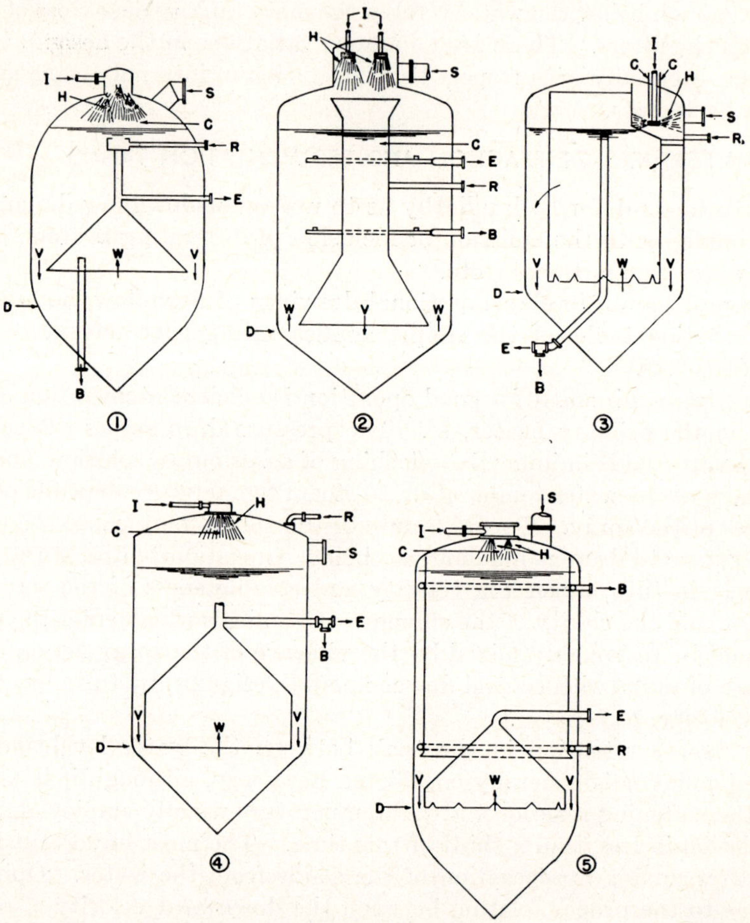

Fig. 2. Nondeaerating hot-process-softening sedimentation tanks. [*Courtesy of* (1) *Cochrane Corporation,* (2) *Graver Water Conditioning Corporation,* (3) *Infilco Incorporated,* (4) *The Permutit Company,* (5) *Worthington Corporation.*]

nominal design. This has occurred where the effluent pipe is located inside the offtake cone and thence drops vertically through the cone bottom. Short-circuiting has occurred where unsymmetrical internal deposits build up in the tank or when the primary sprays on one side were out of service. When these conditions occur, the effluent may be turbid and the reaction incomplete, resulting in after precipitation in the piping and filters.

With one exception (Design 5), provision is made to withdraw water for backwashing filters from the settling zone or offtake and to return it to the upper section or mixing zone. This practice causes greatly increased velocities in the zone between these two points and has at times been responsible in some installations for poor reaction and separation efficiency. A preferred method of operation is to provide storage for filtered water in a separate compartment which drains back slowly to the main tank. In some cases, the author has recommended the use of a separate storage tank with the installation of a bulkhead to make compartments for clean and dirty water; in other cases, the returned water is settled in a cone-bottomed tank, the solids are blown down, and the supernatant returned. Either of these arrangements requires a transfer pump to return the dirty backwash water to the reaction and sedimentation tank. Such refinements cannot always be justified, particularly in small installations where the increase of the equipment for the improved design cannot be justified economically.

The unit shown as Design 5 has an annular compartment from which backwash water is drawn through a high offtake and to which it is simultaneously returned through a lower ring and settled for reuse. This avoids interference of the backwash operation with settling and separation of solids and does not increase velocities in the main portion of the tank.

Variations in Internal Arrangements of Sedimentation Tanks. As stated above, the hot-process sedimentation tank may include steam atomizing or scrubbing deaerators for make-up or condensate or both. A further variation is the occasional use of tray-type deaerators for either purpose on top of the tank or in an internal compartment. The quantity of condensate and make-up, temperature, pressure, head requirements, and space availability may call for many different modifications of arrangement.

The schematic drawing (Fig. 3) shows the internal compartments for backwash-water storage and return. The figure also shows a design with a compartment for secondary phosphate softening within the main lime-soda sedimentation tank. While most two-stage lime-soda-phosphate systems utilize a secondary tank of conventional design like those described above, consideration of space and economy may dictate the two-compartment arrangement. Since the formation and precipitation of calcium and magnesium phosphates is very rapid, as little as 30 to 45 min may be allowed for this portion of the treatment. Separation of the bulk of the solids should occur in the primary lime-soda softener, and the emphasis on clarification is not so great for the phosphate softening equipment which precedes the filters.

a meter register and transmitter. The chemical solution is pumped into a pressure chamber leading to a chemical orifice. This discharges into a receiving bowl and thence to the feed pump which delivers it to the softener. The pressure in the chamber and, thus, behind the orifice is varied by a valve which allows some of the liquid to escape and recirculate to the main solution tank. This valve, in turn, is actuated by a piston in a cylinder, whose upper and lower compartments are connected to the orifice taps in the raw-water line. As the raw-water flow varies, the piston rises and falls and alters the pressure of the liquid behind the chemical orifice. In this manner, the chemical feed is proportioned to

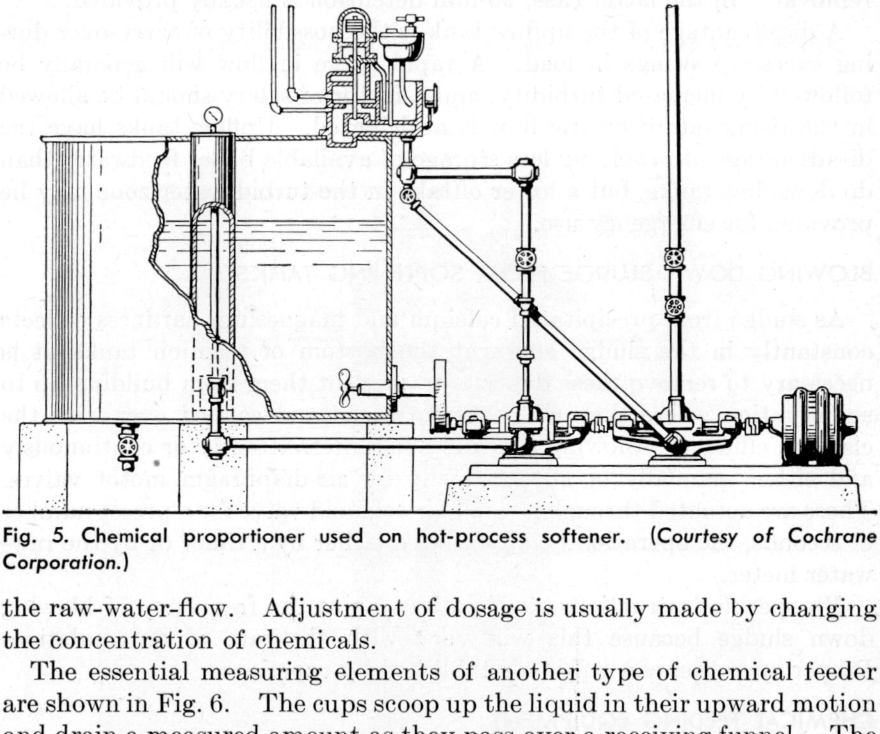

Fig. 5. Chemical proportioner used on hot-process softener. (*Courtesy of Cochrane Corporation.*)

the raw-water-flow. Adjustment of dosage is usually made by changing the concentration of chemicals.

The essential measuring elements of another type of chemical feeder are shown in Fig. 6. The cups scoop up the liquid in their upward motion and drain a measured amount as they pass over a receiving funnel. The width of funnel opening governs the quantity which is collected for treatment purposes, and cutoff shields over the funnel are the means of adjustment.

In early installations, it was customary to feed both lime and soda ash solution from a single tank. This is often fairly satisfactory if the chemical composition of the hard raw water is uniform, as in the case of well water. Where the quality of the water varies appreciably, the practice of feeding two chemicals from one tank imposes serious operating difficulties and leads to poor softener performance. For these reasons, it is always preferable to use two solution tanks so that changes in dosage can be made easily and quickly to comply with the needs of changing water quality.

One of the most troublesome operating problems encountered in hot lime-soda softening is keeping chemical pipelines free from deposits and incrustations. This may be minimized by returning a small amount of sludge from the reaction tank to the suction of the lime pumps. Hard water used for dilution will cause rapid build-up of scale; one remedy is to

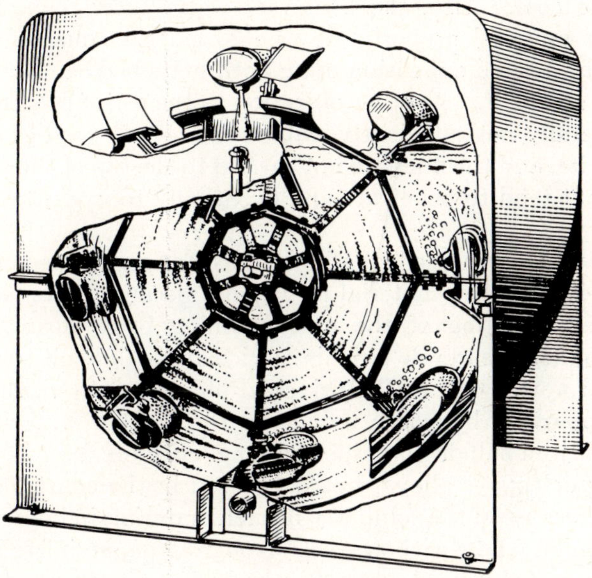

Fig. 6. Chemical mixer and feeder. *(Courtesy of Infilco Incorporated.)*

eliminate dilution make-up water and recirculate to the pump part of the concentrated slurry. In another case, continuous boiler blowdown water cooled in a heat exchanger was used as dilution water at the chemical pump suction.

To simplify cleaning deposits from lines, they should be so constructed as to be readily dismantled, and all lines carrying lime solution should be installed in duplicate. Cleanout crosses and tees should be provided to simplify removal of the accumulated deposits by blowing out by air, steam, or water or by cleaning with inhibited hydrochloric acid.

FILTER DESIGN AND OPERATION

Since the treated water from the reaction tank is at or above boiling temperatures, only pressure filters can be used to completely remove suspended solids from the treated water.* Their design is essentially the same as filters for cold-water service, with the exception of the backwash rate that must be used and the limitation on filter material and underdrains due to the high temperature and high pH of the softened water.

No sand or siliceous material can be used as the filter medium because of the high solubility of silica when in contact with hot alkaline water. Calcite (calcium carbonate) was generally employed in early installations and some filters of this kind are still used. The main objection to this material is that incrustations build up more rapidly on the grains than on anthracite coal. Also, the backwash-water rates required to clean filters of this material properly may be 20 gal per square foot of filter area, while the desired expansion of an anthracite bed can be attained by a 15-gal rate at 220°F. Another objection to the use of calcite is that the beds are generally not sufficiently expanded to effect good grading of the grains of calcite, and the fines are not carried to the top of the bed, as with anthracite, but tend to penetrate and eventually to work through to the underdrains.

Magnetite† has been used to a limited extent and has some favorable characteristics as a filter material principally because of its high density, which prevents shifting of the underdraining beds during backwash. Grades of magnetite suitable for this service are so difficult to obtain that it is not used now. One of the principal objections to magnetite is the danger of imparting hydrogen sulfide to the treated water due to the breakdown of iron sulfide contained in magnetite ore.‡

No copper or copper alloys should be used in the construction of the underdrains because of the solubility of such materials in hot water having a high pH value. Low-carbon steel, wrought iron, stainless steel, or, in special cases, monel is used.

So much difficulty has been experienced in keeping filters clean of heavy deposits that a number of schemes have been tried to insure better operation and less maintenance. The use of surface washing devices has proved quite satisfactory, and there is now a trend toward the use of such auxiliary appurtenances. Rotating or stationary surface washing devices are now used successfully for this purpose. The rotating type of washer is similar to those shown in Fig. 2, Chap. 5.

* For economy, filters are sometimes omitted and the partially clarified water from the settling tank is delivered directly to the boiler-feedwater supply or to hot zeolite units. Generally, this is not good practice, and filters should always be used except where some suspended solids can be tolerated.

† Iron oxide, Fe_3O_4, an important ore, sometimes possessing polarity and then called "lodestone."

‡ The author experienced two serious cases of hydrogen sulfide production in steam due to the use of magnetite. In one case, heavy losses were incurred by a manufacturer who used the steam for processing of a food product.

CALCULATING CHEMICAL DOSAGE AND PREDICTING RESULTS

Instructions for calculating chemical dosage to effect cold-water softening are given in Chap. 6, but a discussion of modifications in dosage required by the elevation of the water temperature is necessary for the guidance of operators.

It is common practice in hot-process softening to add soda ash in excess of that required to precipitate noncarbonate hardness, and when the excess dosage is about 40 ppm the calcium hardness in the treated water can be reduced to about 10 ppm, reported as calcium carbonate. By maintaining higher excess, the hardness can be still further reduced. The degree of reduction obtainable can be determined by use of Fig. 4, Chap. 6.

Magnesium can be reduced to 2 to 6 ppm when excess of lime (hydroxide alkalinity) is in the range of zero to 5 ppm. To secure this low reduction in the magnesium, excess soda ash is also necessary.

If a hot-process unit is to be operated by adding lime only, the residual hardness will not be reduced to as low a concentration as that when soda is used because the noncarbonate hardness will remain in the treated water. The total residual hardness in the treated water when the system is so operated can be determined by reference to Fig. 5 to estimate the calcium carbonate, to which must be added the noncarbonate hardness. When such an operation is carried out, there will be no sodium alkalinity present in the treated water, unless it is present in the raw water. The residual magnesium in the effluent from the reaction tank when such operation is practiced will not be materially reduced unless the pH value of the water is raised to 10.0 or higher.

The results which can be obtained when treating several different types of water are shown in Table 1. The softening reagents used have been determined as specified in Table 5, Chap. 6. No lime is required to remove free carbon dioxide, since the gas is practically all expelled when the water is heated to 212°F or higher. From Table 5, it will be seen that the lime dosage required for the raw water being treated will be 204 ppm of hydrated lime. This is the lime required for reaction of 164 ppm of bicarbonate, 86 ppm of magnesium, and an excess of 5 ppm, all constituents being calculated as calcium carbonate.

For reaction with the noncarbonate hardness, 142* ppm of soda ash is necessary. This amount includes the soda ash for removal of the 87 ppm of noncarbonate hardness and 5 ppm of the excess lime which has been added plus the excess dose of 40 ppm required to depress the solubility of the residual hardness to the desired value to assure the lowest

* Soda ash, 142 ppm, expressed as Na_2CO_3, is equivalent to 132 ppm, expressed as $CaCO_3$.

TABLE 1. RESULTS OF HOT-PROCESS SOFTENING OF ILLINOIS RIVER WATER
ALL DATA IN PARTS PER MILLION

	Raw water	Lime and soda	Corrected for dilution	Lime alone	Corrected for dilution
	A	B	C	D	E
As calcium carbonate:					
Calcium hardness	165	10	10	76	68
Magnesium hardness	86	2	2	36	32
Total hardness	251	12	12	112	100
Sodium and potassium	21	153	138	21	18
Total cations	272	165	150	133	118
Chloride	23	23	20	23	20
Sulfate	85	85	73	85	73
Mineral-acid ions	108	108	93	108	93
Bicarbonate	164	0			
Carbonate	...	52	52	20	20
Hydroxide	...	5	5	5	5
Total anions	272	165	150	133	118
Free CO_2	10	0	...	0	
Noncarbonate hardness:					
Calcium	1				
Magnesium	86				
Total	87				
Chemical control data:					
H = hardness	...	12	12	112	100
P alkalinity	...	31	31	15	15
M alkalinity	...	57	57	25	25
$A = 2P - M$ (lime ±)	...	+5	+5	+5	+5
$B = M - A$ (carbonate)	...	52	52	20	20
$C = B - H$ (excess soda)	...	+40	40	−92	−80
Chemicals, ppm:					
Lime	...	204	...	167	
Soda ash	...	142	...	0	

residual in the treated water commensurate with economy of treatment and reasonably low total dissolved solids.

Following the rules referred to in Chap. 6, all calcium noncarbonate hardness will be removed and calcium carbonate reduced to 10 ppm. The magnesium hydroxide will be 2 ppm. Sodium will be increased by the soda ash added, expressed in terms of $CaCO_3$. Sulfates and chlorides remain unchanged. The result is shown in Column B of Table 1.

A further change in the composition of the water occurs because of dilution of soluble salts by the steam which is condensed in heating the water. Figure 7 shows the percentage of the total feedwater which is accounted for by condensed steam, assuming the raw water to be at 60°F and the pressure of the steam used for heating to vary between zero and 30 psig. For instance, when water is heated with 5-psig steam in a hot-process softener, only 85.8 per cent of the final effluent is raw water, and the balance is condensed steam. Therefore, the sodium sulfate and

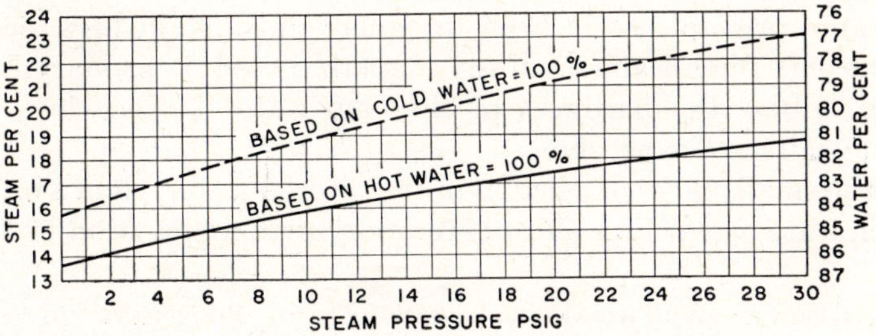

Fig. 7. Steam required to heat boiler feedwater.

chloride in the softened water will be only 85.8 per cent of their concentration in the raw water. Column C of Table 1 shows the effect of this dilution.

HOT-PROCESS SOFTENING WITH LIME ONLY

Since base-exchange resins suitable for softening hot water have been available, hot-process softening with lime alone has become attractive. Three pounds of salt will do the work of one pound of soda ash, and the alkalinity of the effluent will be lower because no sodium carbonate will be present. Softening with salt is often cheaper than with soda ash, but the alkali market and freight to the site may reverse this situation. Softening with lime will reduce the calcium carbonate concentration to 15 to 30 ppm, depending on the excess of calcium in the water, as shown in Fig. 5, Chap. 6. The magnesium hardness will remain fairly high, unless sodium alkalinity is present. Column D in Table 1 shows results that might be expected in hot-process softening using lime alone, and Column E shows the additional reduction in noncarbonate hardness and sodium salts resulting from dilution by condensed steam.

USE OF CHEMICALS OTHER THAN LIME AND SODA ASH

Second-stage Softening with Phosphates. Since phosphates are commonly added to boiler water to complete the external softening process, a logical development was the use of an external softening tank to keep the phosphate precipitates outside the boiler. In 1935, the author recommended this treatment for one of the first large central stations operating modern relatively high-pressure boilers with a high percentage of treated make-up.[3] In this case, the primary softened water was available from an intermittent fill-and-draw lime-soda softening plant that was furnishing make-up for older low-pressure boilers. The analysis of the water is shown in Table 2. Caustic soda and disodium phosphate are added in a hot-process softener, where the hardness is reduced to zero by the soap test, with an excess of only a few parts per million of phosphate.

Since the successful operation of this plant, numerous others operating on the same principle have been designed and installed.[4,5] Secondary phosphate softening is almost always superimposed on primary lime-soda softening, unless the hardness of the raw water is very low, because phosphate salts are expensive.

Even the small traces of calcium in the filtered water, shown in Table 2, will precipitate as calcium phosphate upon further elevation of tempera-

TABLE 2. RESULTS OF TWO-STAGE LIME-SODA AND PHOSPHATE SOFTENING OF GENESEE RIVER WATER
ALL DATA IN PARTS PER MILLION

	Raw water	Lime-soda softened	Phosphate softened	Acid treated
	A	B	C	D
As calcium carbonate:				
Calcium hardness	118	10	Tr.	Tr.
Magnesium hardness	42	5	Tr.	Tr.
Total hardness	160	15	Tr.	Tr.
Sodium and potassium	99	132	239	239
Total Cations	259	247	239	239
Chloride	80	80	80	80
Sulfate	63	75	75	92
Phosphate	...	...	6	6
Mineral-acid ions	143	155	161	178
Bicarbonate	116	...	...	11
Carbonate	...	72	72	50
Hydroxide	...	6	6	
Total anions	259	233	239	239
P alkalinity	...	42	39	8
M alkalinity	116	78	72	
pH	7.4	9.7	9.9	8.9

ture. The size of the crystals thus formed is different from that which occurs in a boiler, and the product is a very hard scale. Therefore, where the system includes boiler economizers or stage heaters in which the temperature will be raised and cause further precipitation, it is necessary to lower the pH to a level at which tricalcium phosphate will not form. This is generally done by the addition of sulfuric acid by a feeder operated by a pH recorder-controller. Equipment for this purpose is shown in Fig. 12, Chap. 6.

Sodium Aluminate. Sodium aluminate acts as a coagulant when used in hot-process softening and frequently lowers the final residual hardness considerably. The chemical is available in solution or solid form and is commonly applied at a dosage of 5 to 10 ppm with the lime or soda ash used in hot-process softening. The commercial grades vary in sodium aluminate content from 72 to 92 per cent and in water of hydration and free alkali. In hot-process softening, the total sodium content of the aluminate has the same effect as an equivalent amount of caustic soda. Calculations should take this into account. For one grade commonly used in feedwater softening, each 10 ppm is equivalent to 3.75 ppm of caustic soda.

Use of Gypsum or Calcium Chloride to Remove Sodium Bicarbonate. There is a large area, extending along the southern Atlantic coast and west to southern Louisiana and Texas, in which underground water is likely to contain high concentrations of sodium bicarbonate and sometimes relatively little hardness. The analysis of the well water shown in Table 3 is an example of such water. Its use for boiler feedwater without drastic alkalinity reduction would be impossible. Water of this kind can be treated in a hot-process softener with lime, which converts all of the bicarbonates to normal carbonates, and with gypsum, to precipitate the calcium carbonate and leave the sodium in the form of sodium sulfate. The reactions are shown below:

	$2NaHCO_3$ +	$Ca(OH)_2$ →	Na_2CO_3 +	$CaCO_3$ +	$2H_2O$
	Sodium bicarbonate	Lime	Sodium carbonate	Calcium carbonate	Water
Combining weights:	100 (as $CaCO_3$)	74	106	100	

	Na_2CO_3 +	$CaSO_4$ →	Na_2SO_4 +	$CaCO_3$
	Sodium carbonate	Gypsum	Sodium sulfate	Calcium carbonate
Combining weights:	106	136	142	100

Calcium chloride can be substituted for gypsum if high sulfate concentrations are objectionable. Where a high ratio of sulfate to alkalinity

TABLE 3. RESULTS OF LIME AND GYPSUM SOFTENING OF A WELL WATER CONTAINING SODIUM BICARBONATE
ALL DATA IN PARTS PER MILLION

	Raw water	Plus lime	Plus gypsum
	A	B	C
As calcium carbonate:			
Calcium	82	10	10
Magnesium	38	2	2
Total hardness	120	12	12
Sodium and potassium	518	518	518
Total cations	638	530	530
Sulfate	8	8	308
Chloride	170	170	170
Mineral-acid ions	178	178	478
Bicarbonate	460		
Carbonate	...	350	50
Hydroxide	...	2	2
Total anions	638	530	530
Chemicals, ppm:			
Lime	...	406	
Gypsum	...	...	408

is definitely sought to control caustic embrittlement, gypsum is preferred.

The deliberate introduction of excess gypsum to supply soluble calcium which will depress the carbonate alkalinity is sometimes considered for hot lime-zeolite softening systems. This is effective, as shown in Fig. 5, Chap. 6, but fairly large excesses of calcium are needed for relatively moderate alkalinity reduction. Since it must later be replaced with sodium by hot-process zeolite softening or by the use of phosphates, the "excess-calcium" treatment has limitations.

SILICA REMOVAL BY HOT-PROCESS SOFTENING

In Chap. 6, the fundamental principles of silica removal by the magnesium process were discussed in detail. It was pointed out that magnesium hydroxide, freshly precipitated from magnesium hardness (ionic magnesium), is the most effective removal agent. Magnesium oxides from several sources, if they can be hydrated, are next in effectiveness. There are several light, porous forms of the oxide on the market that are said to be "activated" because of the relatively greater ease of hydration.* Finally, dolomitic lime was rated as a source of magnesium oxide which could not be hydrated in the presence of the associated calcium hydroxide.

As the temperature rises above about 180°F, the rate of hydration of various magnesium oxides increases greatly, and in hot-process softening at 220°F, their effectiveness as silica-removal reagents does not differ greatly. Betz, Noll, and Maguire (reference 20, Chap. 6) published data

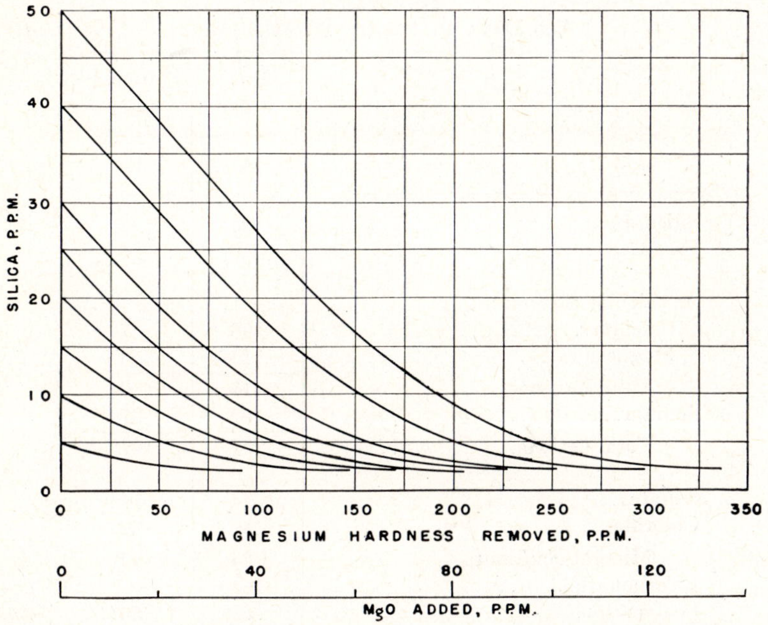

Fig. 8. Silica removed by precipitation of magnesium and addition of MgO in hot-process softening.

for comparison experiments on nine different types of magnesium compounds.

The curves of Fig. 8 may be used to determine the amount of magnesium hardness that must be precipitated to reduce silica from any given value down to a desired residual. In practice, the effect of the ionic magnesium present in the water should first be allowed for, and the deficiency of magnesium hardness calculated. The hydrated lime dosage for softening the water is calculated, and this requirement may be met by using dolomitic lime on the basis of 62 per cent $Ca(OH)_2$ and 32 per cent MgO. If there is still a deficit, magnesium oxide may be added.

* These include Remosil, marketed by W. H. & L. D. Betz, and Silimite, marketed by Allis-Chalmers. Activated magnesium oxide is also a by-product from certain metallurgical and chemical manufacturing operations.

Prolonged contact with the silica-adsorbing sludge is necessary for best results. This may be accomplished by recirculating sludge from the bottom of the tank to the upper mixing zone or by utilizing a sludge-blanket tank, or both. The latter may be desirable if the magnesium oxide is dense and tends to settle out without sufficient contact with the water.

TABLE 4. EXAMPLE OF HOT-PROCESS SOFTENING WITH DOLOMITIC LIME FOR SILICA REMOVAL

ALL DATA IN PARTS PER MILLION

	Raw water	Softened water
	A	B
Silica, ppm	25	1–2
Free CO_2	10	0
As calcium carbonate:		
Calcium	96	7
Magnesium	14	3
Total hardness	110	10
Sodium	97	97
Total cations	207	107
Sulfate	14	14
Chloride	30	30
Mineral-acid ions	44	44
Bicarbonate	163	
Carbonate	...	53
Hydroxide	...	10
Total anions	207	107
Chemicals, ppm:		
Dolomitic lime	...	187

Table 4 illustrates the treatment of a well water in Louisiana for hot-process softening and silica removal. The raw water as shown in Column A contains more alkalinity than hardness; hence, sodium bicarbonate is present. According to the rules previously used, the excess hydroxide, bicarbonate, and magnesium will require 126 ppm of hydrated lime. This can be provided by 187 ppm of dolomitic lime which will carry with it 60 ppm of magnesium oxide, equivalent to 150 ppm of magnesium hardness. The curves of Fig. 8 show that this amount of magnesium plus the small amount in the raw water should reduce the silica from 25 ppm to about 4 ppm. Actually, it was found that the rate of sludge recirculation could be adjusted to obtain any desired final silica down to 1 ppm. The sodium bicarbonate will be converted by lime to sodium carbonate sufficient to depress the calcium carbonate to about 7 ppm. Column B shows the final effluent, which furnishes a satisfactory boiler-feedwater make-up.

CHEMICAL CONTROL OF HOT-PROCESS SOFTENING

In Chap. 6, instructions were given for controlling cold lime-soda softening by testing the water for hardness and phenolphthalein and methyl orange alkalinity. Table 6 of that chapter shows how the excess or deficiency of lime and soda ash can be calculated from these data.

The same procedures are applicable to hot-process softening with slight differences. Except when zeolite softening follows, soda ash is always used, and an excess is maintained. In general, narrower limits are imposed for lime and soda excess or deficiency than are necessary in cold-process softening.

Other methods of controlling the treatment have been published by various authorities and in equipment manufacturers' manuals. Some of these are graphs on which the limits for hardness and alkalinity are plotted with instructions for change of chemical feed whenever the tests fall outside the defined zones. Another method omits complete calculation of excess lime and soda ash. The author prefers the procedure described in Chap. 6, which is applicable regardless of the limits set up for lime and with excess of either soda ash or calcium.

Control tests and calculations for hot-process softening should be entered on a log sheet such as is shown in Fig. 8, Chap. 6. The separate reporting of calcium and magnesium hardness is generally omitted in lime-soda softening but is desirable in softening with lime alone where the excess of calcium over carbonate reduces the alkalinity. For second-stage phosphate softening, additional tests for hardness, phosphate, and pH are required for the final effluent.

Bibliography

1. De La Coux, H.: "Industrial Uses of Water," D. Van Nostrand Company, Inc., New York, 1903.
2. Calise, V. J.: Factors in Hot-process Softener Design and Operation, *Combustion*, **22**: 49 (August, 1950).
3. Powell, S. T., Irving G. McChesney, and Frank Henry: Boiler Feedwater Treatment at a High Pressure Station, *Ind. Eng. Chem.*, **30**: 400 (April, 1938).
4. Joos, C. E.: Primary Treatment of Feedwater by Phosphate, *Power Plant Eng.*, **40**: 244 (April, 1936).
5. Joos, C. E.: Phosphoric Acid—A New Reagent for Use with Hot-process Water Softeners, *Combustion*, **14**. 45 (September, 1942).

8 SOFTENING AND DEMINERALIZATION BY ION-EXCHANGE MATERIALS

HISTORY OF ION EXCHANGE

Although the phenomenon of ion exchange was noted and vaguely reported early in the history of man, the observations were recorded without knowledge of the fundamental chemical laws governing such phenomena or of the practical and commercial application of them. Credit for initiation of practical use of ion-exchange material is given to H. S. Thompson,[1] who, observed and reported in 1845, that, in the passage of a liquid containing ammonia through soil, the ammonia salt was absorbed and that, when the soil was subsequently treated with lime, the ammonia was released. Thompson's original study was continued by Thomas Way[2] during the years from 1850 to 1854. Way reported his findings in a comprehensive report before the Royal Agricultural Society of London. Subsequent research and development in recent years have confirmed the accuracy of some of his observations and conclusions. An abbreviated summary of these is as follows:

1. Thompson's findings, that calcium and magnesium ions can be removed by passage through some kinds of soil, have been confirmed.
2. Certain ions are more easily removed than others, and the ions are exchanged in equivalent amounts.
3. Silicates of aluminum are the active materials responsible for the exchange of ions.
4. The rate of exchange is less, as influenced by the temperature coefficient, than that required by true chemical reaction.
5. The ion-exchange properties are destroyed by heating the material.
6. The exchange of ions is not due to physical adsorption.
7. Ion-exchange material can be produced synthetically by combining soluble silicates and aluminum compounds.

Following Way's classical experiments, a large group of investigators conducted research in this field. Most of these investigations were related to agricultural phases and soil chemistry. Most of Way's findings have been largely verified, but the development of new materials, particularly resinous compounds, has greatly expanded the application and usefulness of ion-exchange processes. These synthetic materials possess many chemical properties differing widely from the ion-exchange soils studied by Way and others down to the present time.

Way failed to recognize the full value of the reversible action of the double silicates, an all-important function in the regeneration of the base-exchange bed which makes this process practical. Credit for this discovery is due to Eichorn,[3] whose investigation brought forth and established the essential theory of base exchange upon which zeolite water softening as now practiced is based.

It was not until 1905 that ion exchange for water conditioning was placed on a firm and practical basis. This development was due to Robert Gans,[4] a German chemist, who produced synthesized ion-exchange minerals by fusing clay, sand, and sodium carbonate (soda ash). Gans called these products "permutits," from the Latin *permuto*, which means interchange. The name "zeolite" has been much more widely used to describe such water-softening minerals, which are either processed natural clays or synthesized material. Gans followed the principle discovered by Thompson, Way, and others using sodium salts, usually sodium chloride, for regeneration to replace calcium and magnesium ions with sodium removed during softening. The process, therefore, was shown to be completely practical, since the hardness of water due to calcium and magnesium ions could be removed; and after the bed had been exhausted, it could be regenerated and was capable of further softening of water. The wide variety of applications of this process resulted in the installation of many softening systems.

The zeolite made originally by Gans had some limitations and later was replaced largely by processed greensands and gel-type materials prepared by precipitation. The double silicates prepared by fusion were used to some extent for special applications. In recent years, siliceous zeolites have been less widely used because of the development of different cation- and anion-exchange materials of higher efficiency.

During the early history of ion exchange and prior to about 1938, the process was used only for removal of calcium and magnesium for softening and, to a lesser extent, iron and manganese. The accepted name for the treatment was "zeolite softening." The term zeolite was coined from the Greek words *zein* and *lithos* and, literally translated, means "boiling stone." The Swedish geologist Cronstedt used the term to describe certain silicates which dehydrated when heated. Many naturally occurring or synthetic compounds are known which contain silicon dioxide and aluminum trioxide as a base with exchangeable sodium, potassium, calcium, magnesium, barium, strontium, and other alkaline ions. The term zeolite is accurately descriptive of these materials but is not truly applicable to a wide variety of ion exchangers which now dominate the field and are extensively employed, not only for water conditioning, but also for processing many solutions, for concentration and recovery of metals, and for analytical procedures.

ION-EXCHANGE PROCESSES USED IN WATER PURIFICATION

These processes include *sodium-ion exchange*, generally referred to as sodium zeolite softening; *hydrogen-ion exchange*, or hydrogen zeolite softening, usually employed in combination with sodium zeolite softening; and *anion exchange*, generally following hydrogen-ion exchange to effect demineralization or de-ionization but also applied to the removal of specific anions, as in dealkalization.

Sodium zeolite softening of water for process, boiler-feed, and domestic use is the form in which ion exchange has been longest familiar. This treatment serves well to introduce the fundamental concepts and phenomena of ion exchange. The sodium and hydrogen zeolite softening process and equipment will be discussed in detail before proceeding to the classification and properties of modern ion-exchange materials.

SODIUM ZEOLITE SOFTENING

The Basic Process. Zeolite softening consists of passing water through a bed of a material that possesses the property of removing calcium and magnesium from the water and replacing these ions with sodium or potassium. The exchange takes place rapidly, so that hard water in passing through a zeolite bed can be almost completely softened, regardless of variations in hardness. Calcium and magnesium will be removed from any of their salts in solution, bicarbonates, sulfates, chlorides, etc. Considering only chlorides as an example, for the sake of simplicity, the softening reactions are:

$$\underset{\substack{\text{Sodium}\\\text{zeolite}}}{Na_2Z} + \underset{\substack{\text{Calcium}\\\text{chloride}}}{CaCl_2} \rightleftharpoons \underset{\substack{\text{Calcium}\\\text{zeolite}}}{CaZ} + \underset{\substack{\text{Sodium}\\\text{chloride}}}{2NaCl}$$

$$\underset{\substack{\text{Sodium}\\\text{zeolite}}}{Na_2Z} + \underset{\substack{\text{Magnesium}\\\text{chloride}}}{MgCl_2} \rightleftharpoons \underset{\substack{\text{Magnesium}\\\text{zeolite}}}{MgZ} + \underset{\substack{\text{Sodium}\\\text{chloride}}}{2NaCl}$$

Figure 1 shows diagrammatically the changes in the zeolite bed as the softening progresses. At (*a*), calcium and magnesium chloride pass through the surface of a fresh bed of sodium zeolite and are immediately converted to sodium chloride, changing the surface layer of the bed to calcium and magnesium zeolite. At (*b*), half of the bed has been so changed, or "exhausted." The zeolite material will thus continue to soften water until a high percentage of the sodium ions have been exchanged. However, at (*c*), the layer of sodium zeolite has become so thin that some calcium and magnesium salts pass through without exchange, and hardness appears in the effluent. After this "break-through" point, the hardness rises rapidly. The softening run is ended at some predetermined effluent hardness, as measured by testing samples. The total capacity of the mineral to exchange sodium for calcium and magnesium has not been exhausted, but the "Operating Exchange Value" has been utilized.

In many cases, it is permissible to continue beyond the break-through point until the effluent contains as much as 10 ppm of hardness, since

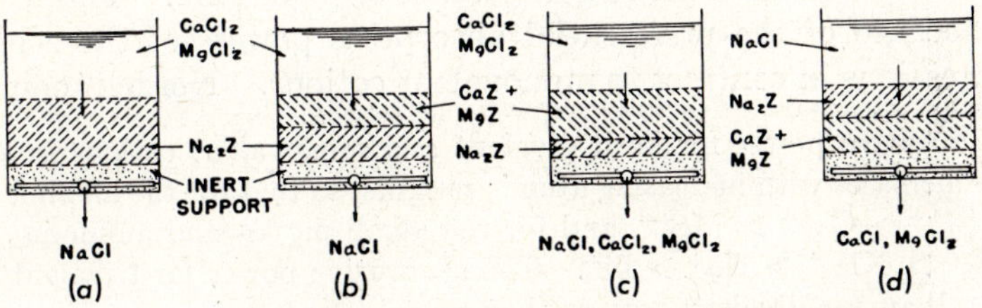

Fig. 1. Diagrammatic sketch showing changes in zeolite bed during softening and regeneration.

this water is a small fraction of the product of the entire run. Economy in use of salt is thus effected. Commercial ratings of Operating Exchange Value are based on this practice, which corresponds to regenerating as soon as a "five-drop" soap test cannot be obtained.*

When the Operating Exchange Value is exhausted, the softener must be regenerated; that is, the sodium ions removed during the softening process must again be replaced. Regeneration is effected by treating the bed with a saturated solution of sodium chloride (common salt). The reverse chemical reaction takes place, and the calcium and magnesium ions extracted from the water and fixed in the grains of ion-exchange material are released and pass into the brine solution. In this manner they are removed from the zeolite and washed to waste. The sodium ions are taken up by the zeolite, and the unit is again capable of softening water.

The chemical reactions which occur during regeneration may be expressed as follows:

* This test is made by adding five 0.05-ml drops of special soap solution to 40 ml of effluent and shaking to produce a lather. A titration test with a color indicator is preferred by many operators. See Chap. 24.

$$\underset{\text{Calcium zeolite}}{CaZ} + \underset{\text{Sodium chloride}}{2NaCl} \rightleftharpoons \underset{\text{Sodium zeolite}}{Na_2Z} + \underset{\text{Calcium chloride}}{CaCl_2}$$

$$\underset{\text{Magnesium zeolite}}{MgZ} + \underset{\text{Sodium chloride}}{2NaCl} \rightleftharpoons \underset{\text{Sodium zeolite}}{Na_2Z} + \underset{\text{Magnesium chloride}}{MgCl_2}$$

The regeneration process is shown schematically at (*d*) in Fig. 1. Actually, the zeolite is not 100 per cent regenerated, traces of calcium and magnesium remaining combined with the granules. Similarly, in commercial softening practice, it is not possible to extract all traces of the sodium. This does not interfere with the practical utility of the process.

In softening, the divalent cations, such as calcium, magnesium, manganese, and iron, are removed from the water and replaced by equivalent amounts of the monovalent ions of sodium or potassium. This exchange occurs because of the preferential absorptive powers of the zeolite for divalent cations in contrast to monovalent cations. Nordell[6] points out:

> The absorptive power for both divalent and monovalent cations in aqueous solutions increases with increasing atomic weights, so that for the alkaline earths, $Ba^{++} > Sr^{++} > Ca^{++} > Mg^{++}$, and for the alkali metals and ammonia, $Cs^+ > Rb^+ > K^+ > NH_4^+ > Na^+ > Li^+$. The absorptive power for trivalent cations is greater than for divalent cations, but no trivalent cations are concerned in softening water by the zeolite (sodium-cation-exchange) process. (Soluble iron or manganese is always present in the divalent form in waters which contain bicarbonate alkalinity.) Obviously, also, as far as zeolite water softening and regeneration of the zeolite are concerned, the principal ions involved are Ca^{++}, Mg^{++}, and Na^+; to a lesser degree, K^+, NH_4^{++}, Fe^{++}, and Mn^{++} may also be of interest.

The mechanism by which ion exchange occurs, when an electrolyte is brought into contact with organic and inorganic compounds possessing ion-exchange properties, is complex, and a number of theories have been proposed to explain this action. Kunin and Myers[5] discuss three theories, namely, (1) the crystal-lattice exchange theory, (2) the double-layer theory, and (3) the Donnan membrane theory. It is not necessary to master the physical chemistry and mathematics of these concepts to comprehend the practical aspects of ion exchange, and for a more advanced technical dissertation on the theoretical aspects of ion exchange, reference is made to these authors and others.[7]

Exchange Value: Compensated Hardness, Low-hardness Waters, and Conservative Effluent Standards for Boiler Feedwater. The exchange capacity of siliceous zeolites and other cation exchangers is commonly expressed in terms of kilograins* of hardness removed per

* One kilograin (kgr) = 1,000 grains; 7,000 grains = 1 lb.

cubic foot of softening material. The exchange values vary with the type of exchanger, the quantity of regenerant used, the total hardness, the sodium salt concentration, and other factors. In designing softeners, the true or compensated hardness is calculated to take account of excessive hardness and sodium present in the water. When treating water containing more than 100 ppm of sodium ions and more than 400 ppm of hardness, the compensated hardness may be calculated by the following formula*

$$H_c = H_a \times \frac{9{,}000}{9{,}000 - \text{total cations}}$$

where H_c is compensated hardness, H_a the actual hardness due to calcium and magnesium, all expressed as parts per million of calcium carbonate.

To illustrate, if a water contains 350 ppm (20.5 grains per gal) of calcium and magnesium hardness and 200 ppm of sodium salts, the total cations are 550 ppm. Then the calculated compensated hardness would be

$$350 \times \frac{9{,}000}{8{,}450} = 372 \text{ ppm or 21.8, say 22, grains per gal.}$$

A further correction to the Operating Exchange Value of carbonaceous zeolites is applied when the softening run is terminated as soon as the effluent does not show zero hardness by test, which the author recommends for boiler-feedwater treatment.† If this practice is contemplated, a further reduction of 10 per cent should be applied to the capacity after correcting for high sodium salts or hardness.

Regeneration Requirements. Since softening by zeolites depends upon replacement of sodium or potassium ions for calcium and magnesium, regeneration can be effected by any of the sodium or potassium salts. Sodium chloride, however, is most commonly used because it is low in cost and is readily available commercially in most areas. The quantity of salt required will depend, obviously, upon the quantity of sodium ions replacing the calcium and magnesium removed from the water which has been treated. The actual consumption of salt for regeneration of the exchange material is greatly in excess of the theoretical requirement of the 0.167 lb per 1,000 grains (1 kgr) of hardness removed, calculated as calcium carbonate. This excess is needed because the greater affinity of the zeolite for the calcium and magnesium than for sodium in regenerating zeolite softeners requires "mass action" to overcome it.

* This formula is used by one manufacturer for siliceous, carbonaceous, and resinous sodium cation exchangers. Graphic data on various ion-exchange resins are available.

† This corresponds to operating on a "three-drop" test or a zero-hardness test by color-indicator titration.

The regeneration efficiency of zeolites and other ion-exchange resins is defined as the pounds of regenerant per kilograin of ions removed.* In using natural greensand and synthetic zeolites for softening water, it was common practice for many years to employ 0.5 lb of salt per kilograin of hardness removed. Better materials and processes led to an improvement in efficiency to as low as 0.3 lb per kgr in some cases. The regeneration level, that is, the pounds of regenerant per cubic foot of zeolite, varies over a wide range from the greensand zeolites to the high-capacity resins which are used in the sodium cycle. The items which influence the salt consumed during regeneration depend upon:

1. The type of exchange material.
2. Degree of hardness, very hard water requiring higher salt consumption for regeneration than do waters of low or medium hardness.
3. Effectiveness of distribution of the brine through the bed.
4. Salt-saving procedures applicable to some types of zeolite materials.
5. Incomplete softening and regeneration, known as "salt starvation."†

In very small-capacity units, the salt brine solution is prepared, as required, by dissolving the necessary amount of salt in a solution tank and pumping or injecting the entire quantity of solution into the unit. Ordinarily, however, a saturated brine solution is made by charging a storage tank with a large quantity of salt and flooding with water, so that the tank at all times contains a saturated solution of brine. Such salt storage tanks may have a capacity ranging from a few hundred pounds to one or more carloads of salt. A typical wet storage tank, having a capacity of one and one-half carloads of salt, is shown in Fig. 2. Where the softening system has sufficiently high capacity to justify such operation, large-capacity wet storage tanks should be provided, since they are more economical to operate and minimize the labor of daily preparation and handling of the brine.

Saturated brine may flow by gravity or under pump pressure to a brine-measuring tank from which the quantity required for regeneration of a softener is drawn. Alternately, the required amount may be pumped through a meter directly to the softening unit.

Regeneration by Natural Brines and Sea Water. Although it is the usual practice to regenerate sodium zeolite softeners with a concentrated solution of common salt, sea water, which contains only about 2.7 per cent of sodium chloride and other soluble salts, is sometimes used in

* Salt consumption, in pounds per million pounds of water, is calculated by multiplying these factors by 7,000/1,000, and the hardness in parts per million. Thus, 0.45 lb per kgr is equivalent to 3.15 $\times$ parts per million of hardness. The theoretical dosage is 1.17; hence, the excess to be rinsed out is 1.98 ppm.

† This operation is impractical for boiler-feedwater conditioning and many industrial processes but has an application to municipal softening where the completely softened water can be sacrificed for economy.

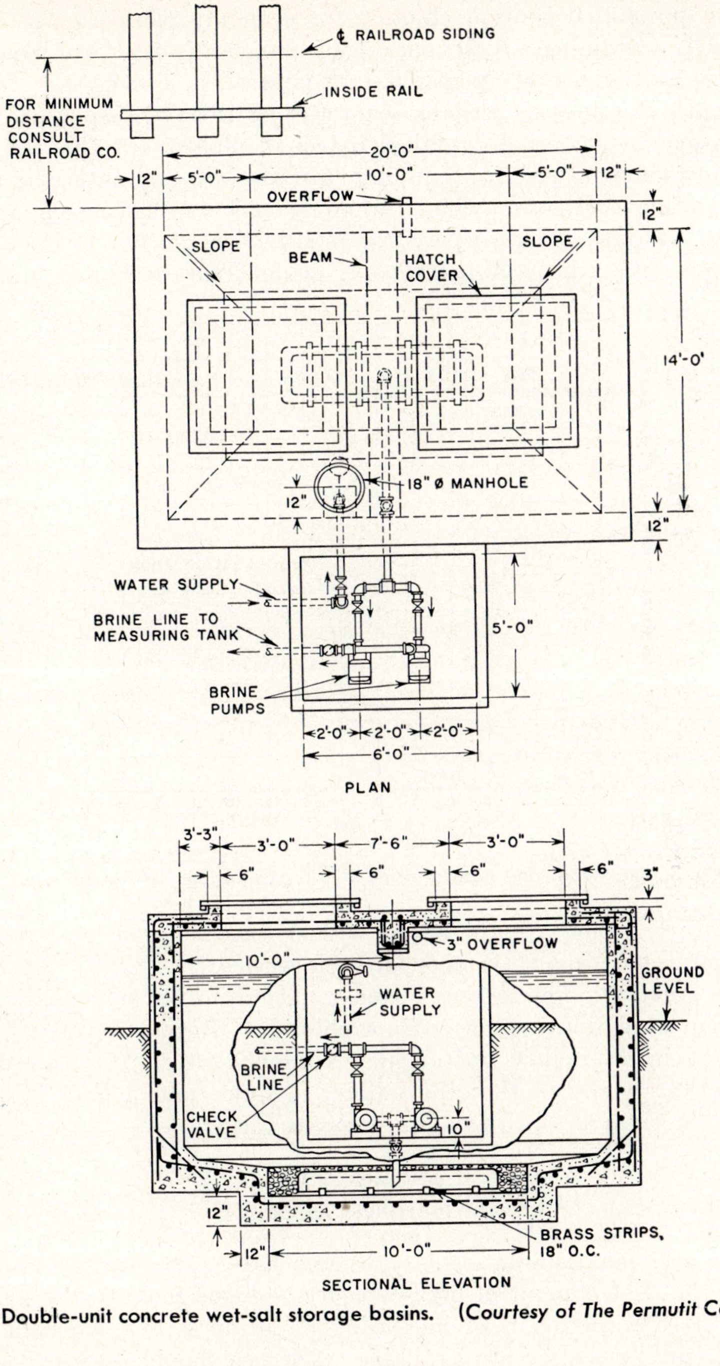

Fig. 2. Double-unit concrete wet-salt storage basins. *(Courtesy of The Permutit Company.)*

place of purchased sodium chloride for economy of operation. Brine from wells containing a high concentration of salt is also used where it is available at the site of operation. In general, it is necessary to filter these natural brines as well as sea water to remove suspended solids. Such regeneration is not practical for small plants, but the use of sea water and some natural brines in large installations can materially reduce the operating cost, even when the additional clarification charges are included. Brines containing hydrogen sulfide should not be used unless it is first removed, because this gas attacks the exchange materials.

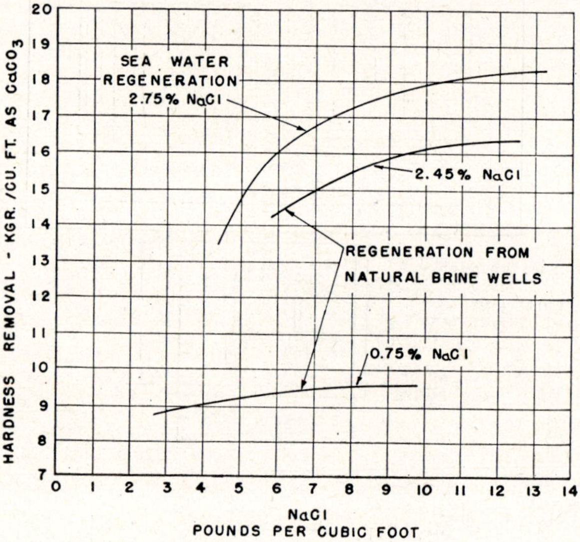

Fig. 3. Sea-water and natural-brine-well regeneration. (*After Lindsay.*)

Also, natural brines contaminated with sewage, organic material, and industrial waste are not suitable for such service.

Much greater volumes of sea water or natural brines are required than of concentrated solutions of sodium chloride. Regeneration is usually performed in an upflow operation. Nordell[6] states: "In general, the specifications would be that the salinity is 3 per cent or more; that the ratio of chlorides as Cl to total hardness as $CaCO_3$

$$\left(\frac{\text{Chlorides as Cl}}{\text{Total Hardness as CaCO}_3}\right)$$

should not be less than 3. . . ."

With the development of high-capacity resinous zeolites, the value of sea water regeneration is of increasing importance, and where natural brines from any source are available, their use should be investigated.

Lindsay[8] has prepared a family of curves showing the hardness removal by sea water and natural brines. These are shown in Fig. 3.

DESIGN OF EQUIPMENT FOR CATION-EXCHANGE SOFTENING

Zeolite softening equipment is designed to operate either by gravity flow or under pressure. Gravity softeners are constructed in rectangular concrete tanks or cylindrical steel or wood-stave tanks, similar in most respects to gravity filters, and pressure softeners are similar to pressure filters. In all cases, a brine piping assembly must be provided to distribute the salt solution. Gravity softeners are used where large capacity and, therefore, large bed area are needed; such softeners have the advantage of the ready accessibility for inspection and maintenance which characterizes gravity filters. However, the great majority of zeolite softener installations are of the pressure type in horizontal or vertical cylindrical steel shells.

The sizes of the shells vary, depending on the volume of water to be treated and its hardness. Permissible softening rates vary from 3 gpm per sq ft of bed area to 8 gpm or more for some of the newer resins. Bed depths vary from 30 to 72 in., and the cubic feet of zeolite required to soften the desired volume of water between regenerations will also affect the selection of shell diameter. The diameters of the shells vary in size from about 10 in. for household units to 10 or 12 ft or greater for large industrial plants.

Most softeners are constructed as vertical units, but some large softeners, especially if head room is limited, are installed in horizontal cylindrical shells. The latter type may be as much as 25 ft long and 8 to 12 ft in diameter. Because of the curvature of the sides and heads of the tanks, maximum cross-sectional area will not be available for passage of the water down through the bed vertically, and some of the bed will not be useful for softening. It is customary, therefore, when using horizontal shells, to derate the bed capacity from 10 to 15 per cent by comparison with the same volume for vertical units in which the entire bed is available for softening. Horizontal units are not widely used, are more difficult to maintain in satisfactory condition, and impose some operating problems.

A cutaway view of a typical installation and arrangement of an automatic single zeolite softener, regeneration apparatus, and other auxiliaries is shown in Fig. 4.

Underdraining System. The design and construction of the underdraining system is governed by the same requirement as for filters, namely, uniform collection of water through the bed and uniform backwashing. The principal difference between filter- and zeolite-underdrain construction and operation is in the selection of material and volume of backwash water required. In zeolite softeners, the collection system should be constructed of red brass or other corrosion-resisting materials because of the corrosive action of the brine and the zero hardness water.

The underdrains are covered with graded gravel* or anthracite coal, which serves as support for the ion-exchange bed and to distribute backwash water and promote uniform collection of the treated water. The

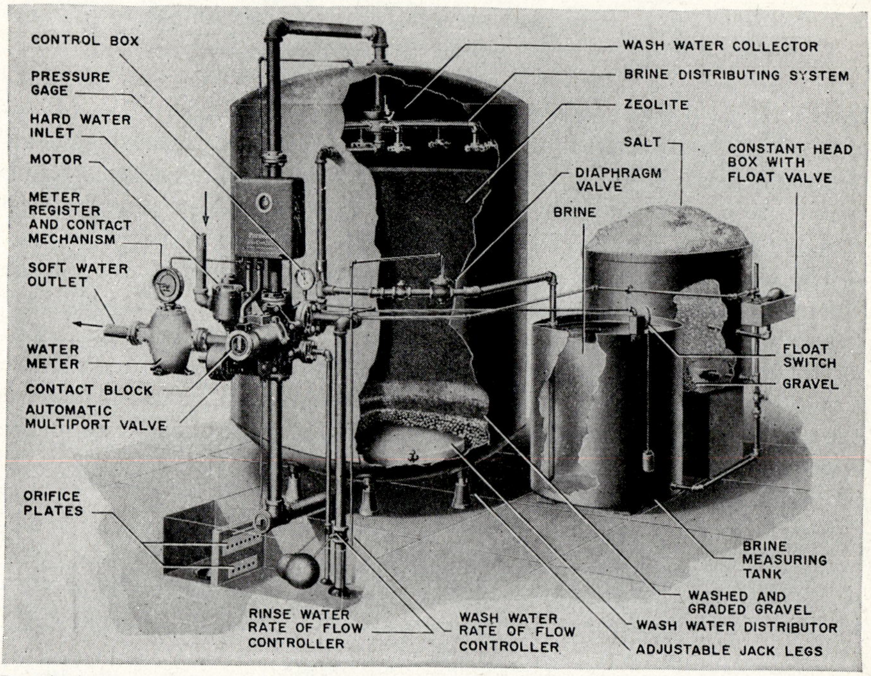

Fig. 4. Automatic pressure zeolite softener, showing brine distributor, mineral bed, and underdraining system. *(Courtesy of The Permutit Company.)*

depth of the supporting bed varies with the design of the tanks, size of the units, and other factors. The gravel or coal layers will range from 12 to 24 in. in thickness and should be properly graded so that each successive layer will support and not mix with the next layer above it and will not permit penetration of the zeolite into the supporting bed. Number 1 Anthrafilt (effective size 0.65 to 0.75 mm) is placed immediately under the zeolite.

When a unit is backwashed, the water should be distributed uniformly and passed upward through all portions of the bed. If the underdraining system is poorly designed or is not maintained in good repair defec-

* Gravel or other siliceous material should not be used when the water is hot because of the solution of silica under such conditions.

tive softener operation may occur owing to channeling of the bed, to upset gravel layers resulting in hill-and-valley packing of the softening material, and to other objectionable operating conditions.*

Strainer heads or nozzles, specially designed deflector plates, and perforated manifolds and laterals similar to those of filter collection systems are employed, since their purpose and function are the same in both types of equipment.

Backwashing, Regeneration, and Rinsing. Before a unit is regenerated, it is taken off the line and backwashed by discharging water up through the bed. This is done principally to loosen the bed, which becomes packed during operation. Backwashing also removes accumulated deposits from the surfaces of the zeolites, although, if the raw water has been properly clarified, few or no deposits should be present. After the backwashing has been completed, brine solution is discharged through the bed at controlled rates. Brining of the bed is followed by rinsing to remove all salt solutions from the equipment. When rinsing has been completed, the unit is again cut into service and is capable of softening. The rate of backwashing and rinsing is regulated by rate controllers, which may be float-operated butterfly valves located in the sewer lines, and the predetermined flow of wash and rinse water is governed by flow over a weir or through orifices. In some installations, especially larger units, the flows are regulated by venturi-type rate controllers or equivalent devices.

The rate at which the backwash water flows through the units is less than that used in backwashing filters, because the specific gravity of ion-exchange materials is lower than that of sand and coal. The rates vary from 7 to 8 gpm per sq ft of bed area for greensand to as low as 4 to 5 gal for carbonaceous or resinous products (higher for hot-process zeolites, up to 15 gpm per sq ft).

An important consideration in designing zeolite pressure softeners is to provide the proper freeboard, or rising space, between the normal surface of the bed and the backwash-water collector. If it is too high, surface deposits may not be removed with the wash water; if the rising space is too shallow, exchange material may be carried out of the unit and lost. The rising rate selected will depend upon the type of zeolite used, greensands requiring a higher wash-water rate and less freeboard than do carbonaceous or resinous exchangers. Ordinarily, a rising space of 50 to 75 per cent of the bed depth is provided.

Brine Distributors. The distributors for admitting the brine to the unit during the regeneration operation should be constructed of noncorrodible metal, preferably red brass or other alloys, and should be so designed as to insure uniform distribution of the brine over the entire

* See Figs. 19–21, Chap. 5, which illustrate these conditions.

bed. Where the nonferrous metal of which the distributors are constructed passes through the shell, these dissimilar metals should be kept from direct contact with each other by insulating joints. If this is not done, accelerated corrosion of the steel will occur. Similar protection should be provided elsewhere in the system at locations where dissimilar metals occur. Figure 5 shows the construction (insulating flanges) used for a brine-distribution assembly.

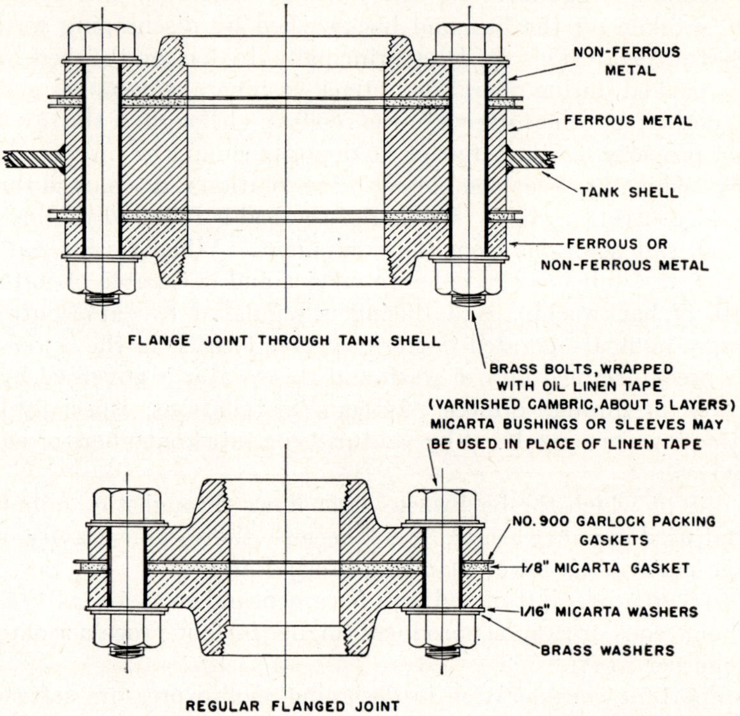

Fig. 5. Insulating flanged pipe joints to prevent contact of dissimilar metals.

Control Valves. Regulation of softeners is performed by manually operated, automatic, or semiautomatic valves. Many small installations are equipped with standard gate valves, nested as a central valve assembly for control of the raw and softened water and of the backwash water, brine, and the backwash and rinse discharged to the sewer. In many installations, practically all the operational functions are controlled by a single multiport valve. Such valves avoid the manual operation of a number of gate valves and minimize the hazard of incorrectly opening or closing valves out of sequence.

The complete cycle of removing a unit from service, backwashing, brining, rinsing, and returning to service can be accomplished automatically. This is done by hydraulically or electrically operated valves actuated by timing devices or by volumetric meters. The complete regeneration cycle can be accomplished by a single valve or a series of valves. Figure 4 shows an automatic assembly which is operated electrically by means of a single multiport valve.

The semiautomatic regeneration is similar to fully automatic operation, except that the operation is not on a metered cycle but must be started by push-button control. The justification for either of these controls depends on the size of the plant and variation in the quality of the raw water. The selection of either a fully or semiautomatic system should be made after a careful evaluation of predicted operating conditions.

APPLICATIONS OF SODIUM CATION EXCHANGE

The basic process and equipment described above may be used directly for sodium zeolite softening of any water supply that contains no interfering substances such as turbidity, organic matter, etc. In many cases, the influent is from a public supply of potable quality. Otherwise, pretreatment by coagulation, filtration, and other processes may be required.

Sodium Zeolite Softening without Pretreatment. Table 1 shows the changes in composition accomplished by sodium zeolite softening of a municipal water supply. To convert the influent water analysis to the softened water analysis, it is necessary to show only a simple subtraction from calcium and magnesium and equivalent addition to sodium. In recognition of leakage of trace quantities of hardness, calcium and magnesium are each shown as 1 ppm, although tests of the effluent usually show zero hardness.

It should be noted that the chloride, sulfate, and bicarbonate go through the process unchanged. The *ionic analysis* is included to show that the soluble solids (sum of the ions) increase by the exchange of sodium for calcium and magnesium.

In Table 1, salt consumption is based on 0.45 lb per kgr of hardness removed, which has been shown above to equal 3.15 ppm per ppm of hardness.

Cold Lime Softening Followed by Zeolite Softening. Preliminary cold-process softening by lime, followed by complete treatment in cation exchangers, is especially adapted to waters having a high carbonate hardness. The effluent from the cold-process softener is filtered and finally completely softened by cation exchange. Before organic exchange materials were available, it was necessary to reduce the pH value of the lime-softened water by acid to avoid destruction of greensand or synthetically prepared siliceous zeolites, because such materials are readily attacked by water having a pH above 8.0 to 8.5. Carbonaceous and resinous

TABLE 1. ZEOLITE SOFTENING OF BALTIMORE CITY WATER
ALL DATA IN PARTS PER MILLION

Constituents		Raw	Softened
As $CaCO_3$:			
Calcium		34	1
Magnesium		13	1
Total hardness		47	2
Sodium and potassium		13	58
Total cations		60	60
Chloride		9	9
Sulfate		11	11
Mineral-acid ions		20	20
Bicarbonate		40	40
Total anions		60	60
Ionic analysis:			
Calcium	Ca	13.6	0.4
Magnesium	Mg	3.2	0.2
Sodium and potassium	Na	6.5	26.7
Chloride	Cl	6.4	6.4
Sulfate	SO_4	10.6	10.6
Bicarbonate	HCO_3	50.0	50.0
Silica	SiO_2	6.5	6.5
Sum of ions		96.8	100.8
Salt ($H \times 3.15$)		....	148.
Lb/1,000 gal (ppm $\div$ 120)		....	1.23

exchange materials are not detrimentally affected by water having a high or low pH value. Although acid treatment is not required to protect the cation-softening materials, it may be provided to prevent the coating of the granules by suspended matter or by afterprecipitation of calcium carbonate. At times, it is desirable to omit filters, and this can be safely done if the flocculent calcium carbonate contained in the effluent from the reaction tank is dissolved by acid.

Figure 6 is a schematic flow diagram showing equipment arrangement for this type of system. Table 2 shows the results obtained by treating Mississippi River water. This type of combined treatment is well suited for treatment of water high in bicarbonate hardness and for silica removal to low values (1.0 to 3.0 ppm) without resorting to demineralization.

Hot Lime-Zeolite Softening. Until sulfonated styrene resins were made available, combined chemical and cation-exchange softening was limited to the treatment of cold water. These newly developed resins can stand water temperatures up to 225°F or higher and are not affected

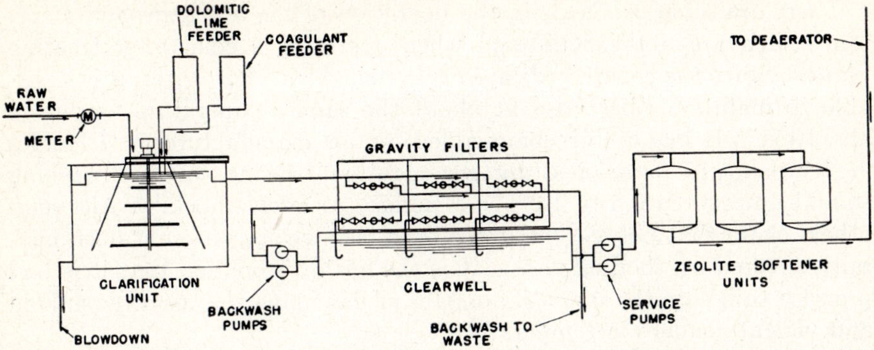

Fig. 6. Flow diagram of water-treatment system for cold lime softening followed by zeolite softening.

TABLE 2. COLD-PROCESS LIME SOFTENING, SILICA REMOVAL, AND ZEOLITE SOFTENING OF MISSISSIPPI RIVER WATER
ALL DATA IN PARTS PER MILLION

Constituents	Raw	Lime softened	Zeolite softened
As $CaCO_3$:			
Calcium	118	91	1
Magnesium	40	18	1
Total hardness	158	109	2
Sodium	84	84	191
Total cations	242	193	193
Sulfate	44	73	73
Chloride	96	96	96
Mineral-acid ions	140	169	169
Bicarbonate	102	0	0
Carbonate	.....	24	24
Total anions	242	193	193
Silica	5.7	3.5	3.5
pH	8.0	10.0	10.0

by high-pH water. Because of these properties, it is now possible to effect the desired reduction of carbonate hardness and silica in hot-process reaction tanks and to soften the effluent from the primary tank to zero hardness by these cation-exchange materials. This development has opened a new field of water conditioning, permitting the economical and efficient treatment of hot water from various sources. A schematic flow diagram of equipment of this type of system is shown in Fig. 7.

There are some problems in the operation of the hot lime-zeolite system which are not encountered when operating a cold-water treating plant. Since the zeolite beds are backwashed with hot water, it is desirable to maintain this water at about the same temperature as during operation. It has been recommended by one manufacturer:[9] "Changes in temperature occasioned by regeneration, rinsing, or backwashing should not exceed 0.5 to 1.0°F per minute to avoid shock." Although others claim that polyvinyl-styrene base resin zeolites are not detrimentally affected by thermal shock, it is the author's opinion that it is bad practice to ignore the potential danger of damaging the resin by sudden and violent changes in temperature.

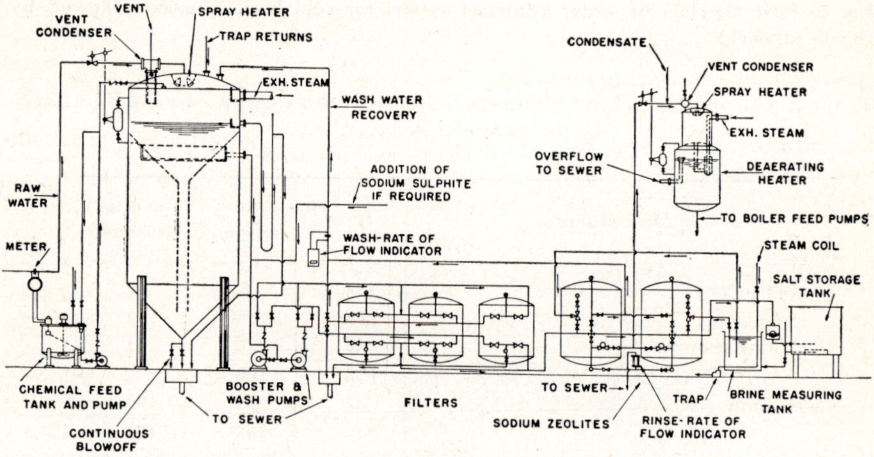

Fig. 7. Two-stage hot-process lime and zeolite softeners.

There is a difference of opinion among manufacturers as to the need for deaerating the brine solution used for regeneration of hot zeolites. The solubility of oxygen in cold saturated brine is about 2 ppm. Serious damage to equipment resulted from corrosion of uncoated tanks and underdrains at one plant where the brine was heated but not deaerated. Most of the damage may have been caused by the fact that complete deaeration of the water was not obtained in the primary heater of the reaction tank; therefore, secondary deaeration should precede final zeolite softening, otherwise it is necessary to add sulfite to the water leaving the tank before delivery to the zeolite units.

The softening capacity of the cation exchangers used in a hot cycle is 18,000 to 25,000 grains of hardness removed (as calcium carbonate) per cubic foot of resin, depending on the amount of salt used. Apparent reduction in zeolite unit capacity may occur if the loss of head through the resin bed is allowed to build up until channeling occurs. One or more backwashes may be required between regenerations to fluff up the bed.

The backwash rate should be high enough to remove excessively fine particles. Sludge from the lime softening process, even in amounts corresponding to turbidities below 10 ppm, cements the finer particles of resin into a dense, impervious mass. Therefore, filters should always precede the zeolite units and must be maintained in a condition which will produce a crystal-clear effluent.

Advantages of Lime-Zeolite Softening. Either cold- or hot-process softening with lime followed by zeolites has the marked advantage of reducing alkalinity to the closest practicable approach to the solubility of calcium carbonate. In the absence of excess sodium or magnesium carbonate, this is approximately 35 ppm in cold water and 20 to 25 ppm at 220°F.[10–12] By contrast to conventional lime-soda softening, practically all the hardness is removed without the use of an excess of soda ash, thereby reducing blowdown requirements when the water is used for boiler-feed purposes. The free carbon dioxide in the steam is also decreased. Silica can be removed in the lime softening process, and some further reduction occurs by adsorption on the magnesium-exhausted resin.[15] The economy of substituting salt for soda ash as a softening reagent has often been cited. Approximately 3 lb of salt will replace 1 lb of soda ash, but freight rates and variable alkali-market conditions may make one treatment as costly as the other in some localities. However, reduced phosphate consumption and boiler sludge formation are great advantages of the hot zeolite process.

TABLE 3. HOT-PROCESS LIME SOFTENING, SILICA REMOVAL, AND ZEOLITE SOFTENING OF LAKE MICHIGAN WATER
ALL DATA IN PARTS PER MILLION

Constituents	Raw	Lime softened	Zeolite softened
As $CaCO_3$:			
Calcium	90	25	1
Magnesium	40	10	1
Total hardness	130	35	2
Sodium	15	15	48
Total cations	145	50	50
Sulfate	20	20	20
Chloride	10	10	10
Mineral-acid ions	30	30	30
Bicarbonate	115	20	20
Carbonate			
Total anions	145	50	50
Silica	15	2	2

Table 3 shows results obtained by softening Lake Michigan water by the hot lime-soda and the hot lime-zeolite process. The data are from plant log sheets in the author's files.

HYDROGEN CATION EXCHANGE

The Basic Process. Softening water by cation exchange in the sodium cycle does not reduce the total dissolved solids but slightly increases them, since 46 parts of sodium replace 40 parts of calcium or 24 of magnesium. The anions, namely, sulfates, chlorides, nitrates, and bicarbonates, pass through the unit unchanged and remain in the treated water combined with sodium. With the development of organic cation-exchange materials of the sulfonated-coal type,[13] the removal of sodium was made possible by operating the zeolite beds in a hydrogen instead of a sodium cycle. To accomplish this, the material is regenerated by a mineral acid, usually sulfuric or hydrochloric acid, instead of sodium chloride. Subsequently, resinous ion exchangers have been adapted to hydrogen-cycle operation.

The chemical reactions which occur when operating and regenerating hydrogen cation exchangers are as follows, the hydrogen zeolite being expressed as H_2Z:

Softening.

$$Ca(HCO_3)_2 + H_2Z \rightarrow 2H_2CO_3 + CaZ$$
$$Mg(HCO_3)_2 + H_2Z \rightarrow 2H_2CO_3 + MgZ$$
$$2NaHCO_3 + H_2Z \rightarrow 2H_2CO_3 + Na_2Z$$
$$CaSO_4 + H_2Z \rightarrow H_2SO_4 + CaZ$$
$$MgSO_4 + H_2Z \rightarrow H_2SO_4 + MgZ$$
$$Na_2SO_4 + H_2Z \rightarrow H_2SO_4 + Na_2Z$$
$$CaCl_2 + H_2Z \rightarrow 2HCl + CaZ$$
$$MgCl_2 + H_2Z \rightarrow 2HCl + MgZ$$
$$2NaCl + H_2Z \rightarrow 2HCl + Na_2Z$$

Regeneration.

$$CaZ + H_2SO_4 \rightarrow H_2Z + CaSO_4$$
$$MgZ + H_2SO_4 \rightarrow H_2Z + MgSO_4$$
$$Na_2Z + H_2SO_4 \rightarrow H_2Z + Na_2SO_4$$
$$CaZ + 2HCl \rightarrow H_2Z + CaCl_2$$
$$MgZ + 2HCl \rightarrow H_2Z + MgCl_2$$
$$Na_2Z + 2HCl \rightarrow H_2Z + 2NaCl$$

From the foregoing reactions, it will be seen that the bicarbonate, chloride, and sulfate ions in the untreated water combine with hydrogen ions, while the calcium, magnesium, and sodium remain in the bed combined with the ion-exchange material. The treated water contains the corresponding acids, namely, carbonic acid, H_2CO_3, hydrochloric acid, HCl, or sulfuric acid, H_2SO_4.

Practically all the carbonic acid can be removed as free carbon dioxide by aerating the effluent or treating it by vacuum deaeration, while the hydrochloric and sulfuric acids remain in solution. It is obvious that free mineral acid cannot be tolerated in water for boiler feed or for most processes; the acid must be neutralized. This can be accomplished in two ways, either by adding caustic soda in amounts sufficient to neutralize the acid and raise the alkalinity to any desired concentration or by mixing the acid effluent with the effluent from a sodium cation unit, which contains sodium bicarbonate. By blending of the acid and alkaline-softened water in proper amounts, the desired alkalinity can be secured. In this case, aeration or deaeration follows blending to remove the additional carbon dioxide which is released.

APPLICATIONS OF HYDROGEN ZEOLITE SOFTENING

Hydrogen cation exchangers are used in water purification in four arrangements: (1) hydrogen-sodium zeolite softening systems, (2) hydrogen zeolite softening with caustic feed to neutralize the mineral acids in the effluent, (3) hydrogen zeolite softening with sufficient raw water bypassed to produce a neutral effluent, and (4) de-ionization. The several modifications of de-ionization will be considered after a discussion of cation and anion exchangers.

Hydrogen-Sodium Zeolite Softening. Figure 8 is a flow diagram of a hydrogen zeolite softening system. The flow is divided between hydrogen- and sodium-ion exchangers by means of a blending control, adjusted with the aid of rate-of-flow gauges measuring each stream. Since mineral acids are used in regeneration and produced by ion exchange in the hydrogen units, corrosion-resistant construction is necessary. Softener shells, valves, and piping are lined with rubber or plastic, and hard-rubber pipe is used for underdrains and other internal parts.

Table 4 shows the results of the treatment applied to a Lake Erie water supply. The treatment completely removes bicarbonates down to any desired level; a residual of, say, 10 to 20 ppm is usually considered the minimum compatible with safety to avoid the hazard of mineral acids in the effluent. The treatment is well adapted to reduction of total solids and alkalinity in boiler feedwater where the supply is high in bicarbonates.

The blended water contains carbonic acid produced in the hydrogen unit and by the reaction of the mineral acids in its effluent with the bicarbonate in the sodium zeolite softener. This is removed as free carbon dioxide in a forced-draft degasifier, a cold-water vacuum deaerator, or a deaerating heater with stainless-steel shell and internals. A counterflow, forced-draft degasifier is shown in Fig. 8. This is a tower of wood-stave or coated-steel construction, into the top of which the water is sprayed and allowed to trickle down over a packing which will be covered by a thin film of water offering a large surface favoring release of gases. Air is discharged upward through the tower by a low-head centrifugal blower. The water cascades down and flows through a sealed outlet into a receiving reservoir under the tower, from which a transfer pump discharges it to use. Design and performance vary with initial carbon dioxide, temperature, and bicarbonate alkalinity, but commercial standards usually

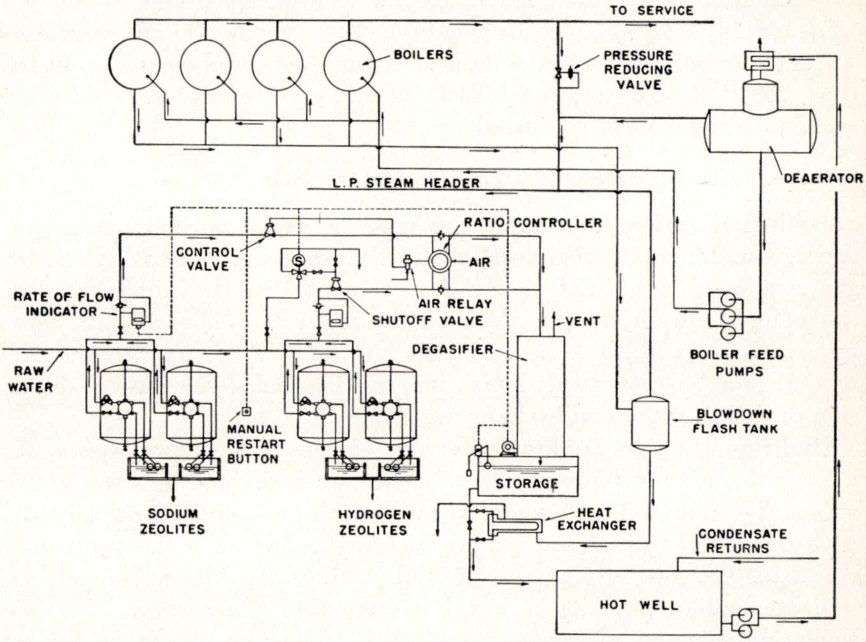

Fig. 8. Schematic flow diagram of feedwater system using hydrogen and sodium zeolites.

call for a maximum final carbon dioxide of 5 or 10 ppm. For initial carbon dioxide contents commonly encountered, the required slat-packed tower heights are 18 and 14 ft, respectively. The blower capacity required is 4 cu ft per min per 1 gpm of water. When Raschig rings are used, the tower height can be reduced 25 per cent or more.

Vacuum deaerators are discussed in Chap. 10. They reduce the corrosiveness of the hydrogen zeolite softened water by removing dissolved oxygen, whereas forced-draft degasification saturates the water with oxygen. Where the carbon dioxide produced by hydrogen zeolite softening is not excessive and the supply is to be used for boiler feedwater, direct discharge to a corrosion-resistant deaerating heater is recommended for economy and simplicity.

TABLE 4. SODIUM-HYDROGEN ZEOLITE SOFTENING OF LAKE ERIE WATER
ALL DATA IN PARTS PER MILLION

Constituents	Clarified raw water	Treated water
As $CaCO_3$:		
Calcium	98	1
Magnesium	31	1
Total hardness	129	2
Sodium	33	80
Total cations	162	82
Chloride	27	27
Sulfate	35	35
Mineral-acid ions	62	62
Bicarbonate	100	20
Total anions	162	82
Silica	2.6	2.6

Blending Controls. Several types of blending controls are available. That shown in Fig. 8, designed for variable flow, uses an air-operated control valve in the sodium zeolite effluent line and a ratio controller connected across orifices in both effluent lines. Since such controls lose accuracy at low flows, a solenoid valve in the control air line is connected through contacts in the rate-of-flow gauge to shut the system down at some minimum flow, usually 20 per cent of maximum.

A simpler constant-flow blending control is practical when a large storage tank is provided for the softened, degasified water. In this case, float switches start the system at low level and stop it at high level.

Calculations of Blend for Alkalinity Reduction. For a system in which sodium zeolite effluent or raw water is blended with hydrogen zeolite effluent, the per cent of the latter required is

$$\text{Per cent } H_2Z = \frac{M_1 - M_2}{M_1 + A}$$

where M_1 = initial methyl orange alkalinity of raw water
M_2 = desired final alkalinity
A = free mineral acid in effluent from hydrogen zeolite units

Alkalinity and acidity are expressed as parts per million of equivalent calcium carbonate. Acidity is the sum of the sulfates and chlorides, and if sodium is absent or low in amount, these may be taken as the difference between the total hardness and the alkalinity.

Mixed Hydrogen and Sodium Exchangers. When the ratio of alkalinity to mineral-acid ions in the influent water is reasonably constant, both hydrogen and sodium exchange can be effected in the same unit. Salt is used at a rate of 0.45 lb, and sulfuric acid at a rate of 0.18 lb, per kgr of total capacity. This "starvation" use of acid is for the purpose of avoiding mineral acidity in the effluent. Since capacity use is divided between hydrogen and sodium exchange, regenerant efficiency is less than when separate beds are used. This may be offset by simplicity and lower equipment cost.

Other Arrangements. There are many water supplies in which the mineral acids produced by hydrogen zeolite softening can be economically neutralized by feeding caustic soda solution proportionately to the degasified effluent, sometimes with pH control superimposed. Some waters initially low in hardness, but high in sodium bicarbonate, produce an acceptable supply when untreated water is bypassed and blended with the hydrogen zeolite effluent.

PROPERTIES OF ION-EXCHANGE MATERIALS

From the standpoint of utility for water purification, the three major groups of ion-exchange materials are (1) sodium cation exchangers, (2) hydrogen cation exchangers, and (3) anion exchangers. Essentially, ion exchangers are carriers which are insoluble and largely inert and to which are attached active polar groups that confer the power to exchange ions.

The carriers for the exchangeable cations and anions include inorganic and organic materials. However, although ion exchange is exhibited by phosphates and other precipitates, the only inorganic exchangers of commercial value are aluminosilicates, which were used exclusively from 1905 until 1935. Organic ion exchangers were introduced with sulfonated coal, a versatile and durable medium, but synthetic resins now dominate the fields of cation and anion exchange.

Development of Ion-exchange Resins. After sodium and hydrogen zeolite softening processes and equipment had been fully developed on the basis of siliceous zeolites and sulfonated coal, a few synthetic cation-exchange resins having higher capacities were offered commercially. However, the current prolific development of new ion-exchange resins was stimulated by successful de-ionization of water after the perfection of anion exchangers.

A knowledge of the structure and synthesis of resins is not necessary for applying them to water treatment, but descriptive terms by which they are commercially designated should be defined. Resins are polymers or copolymers* of organic compounds such as phenols, aldehydes,

* A polymer of a compound contains the same elements in the same proportions but has a higher molecular weight. A copolymer consists of multiples of two or more compounds.

styrene, and vinyl derivatives. A familiar example is the plastic Bakelite, a phenol-formaldehyde resin. Representing the unit organic molecule by R, polymerization of n groups is represented as follows:

$$nR \rightarrow (R_1—R_2—R_3— \cdot \cdot \cdot —R_n)$$

In addition, the chains are bound together by cross linkages. The extent of cross-linkage has an enormous influence on physical and chemical properties. By comparison with the original simple structural units, these polymers may have enormous molecular weight and size. Whereas the compound R is often a volatile liquid of low viscosity, soluble in many solvents, as the polymer grows its boiling point rises, and it becomes a viscous liquid or, ultimately, a noncrystalline solid. With the proper degree of cross linkage, the polymer becomes insoluble in solutions of practically all electrolytes, although the active groups hold ions accessible for exchange. Kunin and Myers[5] state: "The various cation and anion exchange resins may be considered as insoluble, high molecular weight polymeric electrolytes."

Published data on ion-exchange resins identify the types as phenolic, polystyrene-polyvinyl, etc., to describe the insoluble carrier. The identification is completed by specifying the active groups to which they owe their exchange capacity, such as sulfonic, carboxylic, phenolic hydroxyl ($—HSO_3$, —COOH, —OH), etc., for cation exchange, amino or quaternary nitrogen for anion exchange. Variations in polymer type and cross linkage affect the insolubility and life of the resin and the diffusibility of ions in exchange processes. The active groups as well as the resin type determine the exchange capacity for specific ions at various regeneration levels. Accordingly, the design of ion-exchange equipment for water treatment requires careful selection of resins adapted to the water analysis and effluent specifications.

Ion-exchange resins may be produced in the form of small spherical particles, described as "beads," or in larger masses, which are crushed and screened. For water-purification purposes, the effective size is usually 0.40 to 0.45 mm. Head loss and backwash characteristics, but not exchange capacity, are affected by size. Considerable bed expansion occurs during regeneration, and the freeboard provided for backwashing must take this into account.

Sodium Cation Exchangers. Table 5 summarizes the properties of sodium exchangers, including siliceous zeolites and sulfonated coal.

Siliceous Exchangers, or True Zeolites (*Groups* 1 *and* 2). Natural greensand zeolite (glauconite) is found on the Atlantic coastal plain, notably in shallow deposits in New Jersey, and elsewhere in the country. It is mined by open-pit methods. The material is washed, screened, and treated with caustic to remove clay and with alum, silicates, and other salts to produce a stable medium which is durable and inert except for its ion-exchange properties. Its exchange capacity is 2.8 kgr per cu ft at a regeneration efficiency of 0.45 lb per kgr of hardness. Capacities up to 3.5 or 5.0 kgr per cu ft may be obtained with special treatment to increase porosity but at the expense of durability and long life. Siliceous zeolites may be described in terms of oxides; for example, $x\mathrm{Na_2O}\cdot y\mathrm{Al_2O_3}\cdot z\mathrm{SiO_2}\cdot a\mathrm{H_2O}$, etc. Also present may be FeO, MgO, CaO, and other compounds.

Greensand zeolite is generally rugged but is resistant over a pH range of only 7.2 to 8.0. Ferrous and manganous salts are removed simultaneously with hardness. Ten parts per million of soluble iron will be removed for each 342 ppm of hardness.

Synthetic sodium aluminosilicates of the gel type have exchange capacities of 8 to 10 kg per cu ft if 0.4 to 0.5 lb of salt per kilograin is used for regeneration. The gels are especially suitable for hard waters where higher capacity is needed to keep down equipment size. They may be used in a pH range from 6.9 to 8.3, but their tolerance and capacity for iron is lower than greensand, usually not over 3 ppm per 342 ppm of hardness or 5 ppm with harder waters.

Sulfonated Coal (*Group* 3). Following the development and widespread use of siliceous zeolites, it was discovered that capacity for ion exchange could be conferred upon a wide variety of organic materials by sulfonating them, that is, by attaching sulfonic acid,—HSO_3, groups to the inert organic residue through treatment with sulfuric acid, sulfur dioxide, and other reagents. Beginning about 1935, the literature contains a large number of patents covering materials made from coal, peat, and a variety of waste fibers or other carbonaceous substances. The commercially available product is prepared from a selected grade of bituminous coal and is a rugged, durable ion exchanger that has proved very satisfactory. This type of exchange material does not contribute silica to the softened water as do siliceous zeolites and will tolerate a pH as low as dilute mineral acids and as high as lime-softened water, provided the latter is stable and does not precipitate calcium carbonate. These so-called carbonaceous zeolites will remove iron in a manner similar to greensand. Operating in the sodium cycle, the exchange capacity of this material is 7 kg per cu ft when regenerated with 0.45 lb of salt per kilograin of hardness removed. High concentrations of sodium salts or high hardness reduce the unit-exchange capacity of the softening material.

The development of carbonaceous zeolites provided a material with more than double the capacity of the natural zeolites and affords considerable economy in equipment size. Further, the sulfonated-coal ion exchangers can be operated at higher flow rates than can siliceous zeolites. It is satisfactory to design on the basis of 6 gpm per sq ft of softener bed when one unit is out of service for regeneration.

TABLE 5. CHARACTERISTICS OF CATION EXCHANGERS

A. Physical and chemical data

Group	Description	Ex-changer	Trade name*	Polar exchange groups	Recommended operating limits			
					Flow rate, gpm/sq ft, normal and max†	pH‡	Temp., °F	Chlorine
1	Natural greensand zeolite (sodium alumino-silicate)	(1) (2)	Zeo-Dur Inversand	....	3–4	6.8–8.0	110	No limit
2	Synthetic gel zeolite (sodium alumino-silicate)	(3) (4) (5)	Decalso Super Nalcolite Aridzone	....	3–4	6.0–8.3	140	No limit
3	Carbonaceous zeolite (sulfonated coal)	(6) (6) (7) (8)	Zeo-Karb Cochranex CCA Catex-55 (Na) Catex-12 (H)	Methylene sulfonic, carboxylic, phenolic OH	4–6	No limit	120	Zero
4	Phenolic resin	(9)	Duolite C-3	Nuclear sulfonic	5–8	No limit	120	Limited
5	Styrene, medium capacity (styrene polyvinyl resin)	(10) (11) (12)	Amberlite IR-112 Chempro C-25 Nalcite MCR	Nuclear sulfonic	5–8	No limit	§	No limit
6	Styrene, high capacity (styrene polyvinyl resin)	(13) (14) (15) (16)	Amberlite IR-120 Chempro C-20 Nalcite HCR Permutit Q	Nuclear sulfonic	5–8	No limit	250	No limit

TABLE 5. CHARACTERISTICS OF CATION EXCHANGERS.—(*Continued*)

		B. Capacity data—Sodium cycle				C. Capacity data—Hydrogen cycle‖, ¶			
Group	Description	Exchanger	Regenerant, lb/cu ft	Capacity, kgr/cu ft	Efficiency, lb/kgr	Acid	Regenerant, lb/cu ft	Capacity, kgr/cu ft	Efficiency, lb/kgr
1	Natural greensand zeolite	(1), (2)	1.25	2.8	0.45				
2	Synthetic gel zeolite	(3)	3.2	8	0.40				
		(3)	4	9	0.45				
		(4), (5)	4	12	0.33				
		(3)	5	10	0.50				
		(4)	5	14	0.36				
		(4), (5)	6	15	0.40				
		(5)	9	17	0.53				
3	Sulfonated coal	(6), (7), (8)	3.15	7	0.45	H_2SO_4	2	8	0.25
4	Phenolic resin	(9)	.....	....	....	H_2SO_4	4	8.2	0.49
		....	.....	....	....		6	10.0	0.60
		....	.....	....	....		8	11.1	0.75
		....	.....	....	....		10	12.0	0.80
		....	.....	....	....	HCl	4.5	10.5	0.43
		....	.....	....	....		8.0	12.5	0.64
		....	.....	....	....		12.5	14.5	0.86
		....	.....	....	....		17.0	16.0	1.06
		....	.....	....	....		21.0	17.0	1.23

5	Styrene, medium capacity	(10), (11), (12)	3	12	0.25	H_2SO_4	2.5	9.5	0.26
		....	4	13.8	0.29		4.0	11.8	0.34
		....	6	16.4	0.37		5.0	13.5	0.37
		....	8	18.2	0.44		6.0	14.2	0.42
		....	10	21.5	0.47		8.0	16.2	0.49
6	Styrene, high capacity	(13), (14)	5	18	0.28		2.5	9.0	0.28
		(15), (16)	6	24	0.25		4.0	10.5	0.38
		....	6.6	22	0.30		5.0	11.0	0.45
		....	7.5	22	0.34		6.0	12.0	0.50
		....	8.4	24	0.35		7.5	13.5	0.55
		....	10.0	25	0.40		8.0	15.0	0.53
		....	13.5	27	0.50		10.0	25.0	0.40
		....	15	30	0.50				
						HCl	10.0	19.0	0.53
							15.0	21.6	0.70
							20.0	25.0	0.80

* Trade names are identified with the following manufacturers: Arizona Mineral Corp. (5); Chemical Process Co. (9), (11); Dow Chemical Co. (Nalco Inc. distributors) (12), (15); Nalco Inc. (4); Infilco Inc. (7), (8); The Permutit Co. (1), (3), (6), (16); Rohm & Haas Inc. (10), (13).

† Based on minimum bed depth of 30 in. Deeper beds may permit higher rates.

‡ Where no upper limit is given for pH it is assumed that the water is nonscaling.

§ Group 6 exchangers are preferred for hot water service.

‖ Capacity values for cation exchangers in the hydrogen cycle in the higher regeneration levels are median values selected from manufacturers' data, based on treating a water in which alkalinity is 50 per cent of the anions and sodium is 50 per cent of the cations. It is assumed that leakage must not exceed 2 per cent. Better capacity and efficiency are attainable when the sodium is lower or the specifications less rigorous.

¶ Weights of acid based on 66°Bé H_2SO_4 and 30 per cent HCl.

One of the disadvantages of carbonaceous zeolite is its vulnerability to oxidation by chlorine, in the same manner as activated carbon. It was observed during its early use that residual chlorine in excess of 0.5 ppm caused the production of fines, which were lost during backwashing, resulting in an excessive annual loss of zeolite. Subsequently, dechlorination with sulfite has been used ahead of carbonaceous zeolite softeners, or exchangers of the resinous type have been substituted where residual chlorine is present in the water and cannot readily be eliminated.

Sodium-exchange Resins (*Groups* 4, 5, *and* 6). Table 5 lists several distinct types of sodium-exchange resins which have been developed. The first of these (group 4) were phenolic polymers having a higher capacity than any of the then-available materials. They were followed by the group 6 copolymers of styrene, cross-linked with vinyl derivatives and containing strongly acidic nuclear-sulfonic* exchange groups. These have still higher capacity, resistance to free chlorine, freedom from "color throw," lower raw-material cost, and other advantages. Production of phenolic ion-exchange resins for water treatment was greatly reduced in favor of these materials. Subsequently, the need for intermediate-capacity exchangers led to the improvement of phenolic exchangers and of the styrene resins containing a smaller number of nuclear-sulfonic groups per unit volume.

Ion-exchange resins vary in exchange capacity in proportion to the regeneration level over a much wider range than siliceous or carbonaceous zeolites. Table 5 shows conservative ratings for typical water supplies rather than total capacities that can be realized under the most favorable conditions. The grouping of products in this table implies functional similarities but not identical performances, and capacity data for design purposes should be obtained from manufacturers to fit the specific water supply.

Hydrogen Cation Exchangers. Of the cation exchangers whose characteristics in the sodium cycle have been described above, all except the silicates can be regenerated with acids and operated in the hydrogen cycle. Sulfuric acid is the regenerant most commonly used. When sulfuric acid is applied at high concentrations to regenerate base exchangers exhausted by treating waters containing a high percentage of calcium, the removed calcium ions may precipitate in the bed as calcium sulfate. During the rinsing, this precipitate dissolves and exhausts part of the capacity of the material, leaving calcium in the bed but still producing a mineral-acid effluent. For this reason, if calcium exceeds 20 per cent of total cations, the sulfuric acid is used in dilute solution (2 per cent) and at high flow rate during the first part of regeneration, and the concentration

* Attachment of the strongly acid —SO_3H group directly to the polymer chain, as distinguished from the weaker methylene sulfonic group, —CH_2SO_3H.

is increased stepwise to the optimum level. The use of hydrochloric acid eliminates this difficulty but is more costly and requires different corrosion-resistant materials.

Leakage of sodium is due to the mineral acids formed in the upper part of the bed percolating through and replacing sodium left in the lower layers by incomplete regeneration. Hence, it is important to use the regenerant acid in quantity, concentration, and contact time which will *completely* regenerate the bed and reduce leakage to the level dictated by specifications for the finished water. Low leakage is frequently obtainable only by using relatively high amounts of acid at high final concentration. This practice generally involves less-efficient regeneration levels, requiring stepwise dilution.

Sulfonated Coal. This material, the first hydrogen zeolite available, has long been accepted as having an exchange capacity of 8 kg of cations per cubic foot at a regeneration level of 2 lb of 66°Bé sulfuric acid per kilograin, an efficiency of 0.25 lb per kg. However, the total cation content of the water (and, hence, exchange capacity and efficiency) is compensated by using factors varying from 0.88 to 2.5 for different combinations of alkalinity and mineral acidity. For design purposes, this material was initially considered capable of removing all the cations associated with the bicarbonate ion and 88 per cent of those associated with sulfate and chloride ions—the "theoretical mineral acidity." Calcium and magnesium in the effluent are negligible, and the 12 per cent leakage, according to this estimate, is practically all sodium. In practice, however, the sodium leakage has been found to be considerably less. The weakly acid carboxyl and phenolic hydroxyl groups, which supplement the sulfonic acid groups as the active exchangers in sulfonated coal, favor complete regeneration with sulfuric acid as dilute as 2 per cent; hence, sodium leakage is low, difficulties with calcium sulfate precipitation are avoided, and regeneration efficiency is high.

Hydrogen-exchange Resins. The performance of cation-exchange resins in the hydrogen cycle is shown in Table 5. Exchange capacities vary over a wide range, depending on influent analysis and regeneration level. The data given are based on de-ionization of boiler feedwater to 5 ppm or less of ionizable solids, requiring conservative capacity ratings and regeneration levels to yield low sodium leakage.

The capacity of a cation-exchange resin in the hydrogen cycle is increased, and leakage is decreased, by a high percentage of bicarbonate in the influent anions. Capacity is also greater for waters high in sodium, but leakage increases rapidly as sodium increases above 50 per cent. The high-capacity nuclear-sulfonic polyvinyl-styrene resins in group 6 may require progressive or stepwise increase in acid concentration up to 8 per cent for the most complete and efficient regeneration.

TABLE 6. CHARACTERISTICS OF ANION EXCHANGERS

A. Physical and chemical data	Group 1	Group 2	Group 3
Anion exchanger	Weakly basic	Strongly basic, type 1	Strongly basic, type 2
Trade names* and resin base	(1) Deacidite — Aliphatic (polyethylene) (2) Nalcite WBR — Styrene-polyvinyl (3) Amberlite IR-45 — Phenolic (4) Duolite A-2 — Phenolic (5) Duolite A-7 — Phenolic	(6) Nalcite SBR (7) Permutit S-1 (8) Amberlite IRA-400 (9) Duolite A-42 (6)–(8): Styrene-divinyl benzene	(10) Nalcite SAR (11) Permutit S-2 (12) Amberlite IRA-410 (13) Duolite A-40 (10)–(12): Styrene-divinyl benzene
Polar exchange groups	Primary amino	Quaternary nitrogen	Quaternary nitrogen
Tolerance:			
Temperature, °F, max	105	140–150	105
Recommended flow rate:			
Gpm/sq ft, normal and max	5–7.5	5–7.5	5–7.5
Gpm/cu ft	2–3	2–3	2–3
Regeneration requirements:			
Regenerant and concentration, %	Caustic, soda ash, ammonia	Caustic, 4	Caustic, 2 or 4
Temperature, °F	Cold	95–130	95 for best efficiency
Contact time, min	20–40	12/lb/cu ft	12/lb/cu ft
Special properties	The phenolic resins are sensitive to chlorine.	Preferred for weak acids 30%, or more, of total exchangeable anions. Less efficient for strong acids. Higher basicity than type 2. Retains capacity better than type 2.	Preferred for weak acids under 30% of total exchangeable anions. More efficient for strong acids. Lower basicity than type 1. Slowly loses capacity during many cycles. Preferred for chloride-bicarbonate exchange.

B. Capacity† data	Group 1 Ex-changer	Lb/cu ft	Kgr/cu ft	Lb/kgr	Group 2 Ex-changer	Lb/cu ft	Kgr/cu ft	Lb/kgr	Group 3 Ex-changer	Lb/cu ft	Kgr/cu ft	Lb/kgr
Caustic, 100%	(4), (5)	3.2	25.00	0.13	(6)	3.5	11.00	0.32				
	(3)	3.5	25.00	0.14	(9)	4.0	9.00	0.44	(10)	4.0	12.50	0.33
	......	...	.....	....	...	...	.....	....	(13)	4.0	15.00	0.27
	......	...	.....	....	(7)	4.5	11.00	0.41	(11)	4.5	14.00	0.32
	......	...	....	....	(6)	5.0	11.75	0.42	(10)	5.0	13.50	0.37
	......	...	.....	....	(8)	6.0	10.00	0.60	(12)	6.0	16.20	0.37
	......	...	.....	....	(9)	6.0	10.00	0.60	(13)	6.0	16.00	0.37
	......	...	.....	....	(6)	7.5	13.00	0.57	(10)	7.5	14.50	0.52
	......	...	.....	....	(8)	8.0	10.90	0.73	(12)	8.0	16.90	0.47
	......	...	.....	....	(9)	8.0	11.00	0.72	(13)	8.0	17.00	0.46
Soda ash, 100%	(1)	4.2	19.00	0.22								
	(4), (5)	4.2	17.00	0.25								
	(3)	6.0	27.00	0.22								
Ammonium hydroxide, 100%	(3)	3.0	25.00	0.12								
	(4), (5)	3.0	22.00	0.13								

* Trade names are identified with the following manufacturers: Chemical Process Co., (4), (5), (9), (13); Dow Chemical Co. (Nalco, Inc., distributors), (2), (6), (10); The Permutit Co., (1), (7), (11); Rohm & Haas Co., (3), (8), (12).

† The capacities of strongly basic anion exchangers have been selected for silica-removal applications in which the total silica is of the order of 10 ppm and is 30 per cent of the total exchangeable anions. Higher capacities and efficiencies can be realized when silica removal to a low residual is not required. Since actual capacities vary with the influent-water analysis, the above typical figures in boiler-feedwater practice should not be used for design.

Also, for high-sodium waters, the regeneration level, commonly 2 to 4 lb of 66°Bé sulfuric acid per cubic foot, may have to be several times as high if effluent specifications require low total solids.

The medium-capacity phenolic resins described under group 4 contains weakly acidic exchange groups similar to sulfonated coal, and the medium-capacity nuclear-sulphonic styrene-base resins in group 5 are essentially diluted counterparts of the high-capacity exchangers. Both groups can be completely regenerated with 2 per cent sulfuric acid; stepwise regeneration is, therefore, unnecessary but may be used to obtain better performance. Because of high efficiency and low leakage, these resins are frequently advantageous for de-ionization.

Carboxylic Exchangers for Bicarbonate Removal. In addition to sulfonated coal and the resins described above, a hydrogen cation exchanger of the carboxylic type is available for removing bicarbonates of calcium, magnesium, and sodium without affecting the sulfates and chlorides.[14] This material is regenerated with sulfuric acid and is especially suitable for reducing excessive alkalinity. Enough raw water is bypassed and mixed with the effluent to furnish sufficient bicarbonates to avoid corrosion. Of course, hardness is not reduced to zero unless sodium zeolite softening is also used.

Anion Exchangers. The characteristics of anion exchangers are summarized in Table 6. They are essentially inert, polymeric compounds containing active nitrogen groups that react like ammonia and ammonium hydroxide. Ammonia, NH_3, and the primary amine, RNH_2 (in which R is an organic radical), will both absorb inorganic acids such as hydrochloric by addition reactions:

$$NH_3 + HCl \rightarrow NH_4Cl$$
$$RNH_2 + HCl \rightarrow RNH_3Cl$$

An alkali will reverse the reaction:

$$NH_4Cl + NaOH \rightarrow NH_3 + NaCl + H_2O$$
$$RNH_3Cl + NaOH \rightarrow RNH_2 + NaCl + H_2O$$

These equations typify the reactions and regeneration of some weakly basic anion exchangers which absorb strong acids only. Soda ash, ammonia, or other weak bases serve as well as caustic for regeneration.

Ammonium hydroxide and strongly basic anion exchangers are *quaternary* nitrogen compounds capable of holding hydroxide ions which neutralize weak as well as strong acids to form corresponding salts:

$$2NH_4OH + H_2CO_3 \rightarrow (NH_4)_2CO_3 + 2H_2O$$
$$2RNH_3OH + H_2CO_3 \rightarrow (RNH_3)_2CO_3 + 2H_2O$$

Regeneration, in this case, requires a strong base such as caustic soda:

$$(NH_4)_2CO_3 + 2NaOH \rightarrow 2NH_4OH + Na_2CO_3$$
$$(RNH_3)_2CO_3 + 2NaOH \rightarrow 2RNH_3OH + Na_2CO_3$$

Strongly basic exchangers have aptly been called "solid insoluble caustic."

It would be an oversimplification to explain all anion exchange in terms of one or the other of the above reactions, which may not be mutually exclusive, although dominant in the behavior of specific exchangers. However, the performance of resins used in water treatment is conveniently described by analogy to these addition and neutralization reactions.

Anion-exchange resins must be considered expendable items, since they do not have chemical stability of the same order as cation exchangers. Generalizations or categorical ratings of the useful life of exchangers listed in Table 6 would be unwarranted and shortly rendered obsolete by new developments in resin manufacture. It is known that, during repeated cycles, there is a slow leaching of amines from the matrix, and that the highly basic quaternary groups revert to tertiary groups which will only absorb strong acids.

Weakly Basic Anion Exchangers. The first anion exchangers used commercially, about 1937, were aliphatic (polyethylene) amine resins capable of removing strong acids, such as hydrochloric and sulfuric. They will not remove silicic, carbonic, or other weak acids. Since this development, polymers of phenolic and other organic groups have been adapted as carriers for the active amino radicles. These materials are used in conjunction with the various available hydrogen exchangers for the de-ionization of water for boiler-feed and process use, where it is unnecessary to remove silica. Their capacity is less for chlorides than for sulfates; the over-all capacity varies with the quantity of regenerant used over a practical range of 15 to 27 kg per cu ft. The total solids in the effluent depend on the sodium leakage through the preceding cation exchanger and, under favorable conditions, may be quite low, 5 ppm or less of ionizable solids. The weakly basic exchangers have the advantage of high efficiency, using for regeneration soda ash or ammonia instead of caustic, and the resins are less costly than the more basic materials. It is often worthwhile to introduce a weakly basic anion-exchange step into a cycle terminating in silica removal, in order to carry the burden of strong-acid removal more economically.

Strongly Basic Anion-exchange Resins. Several types of resins containing quaternary nitrogen are available for anion exchange, as shown in Table 6. Each of these has properties which are advantageous for specific conditions, leading to several subclassifications.

Two types of quaternary exchangers consisting of styrene-polyvinyl resins are produced by several manufacturers.[16] One type has alkyl (methyl) groups attached to the nitrogen atom; the other type, alkanol (ethanol) groups. These structural differences are accompanied by marked contrasts in chemical and physical properties.

Type-1 exchanger is more highly basic but has lower exchange capacity than type 2 when operated at the same regeneration level* and, hence, is less efficient for strong-acid absorption. Type 1 and type 2 differ in selectivity, type 1 being best for removing silica and weak acids. Type 2 loses basicity on repeated cycling more rapidly than type 1 and has less stability for high temperatures.

Resin structures based on other raw materials, but containing quaternary nitrogen groups, are available. They possess still different specific selectivities, capacities, efficiencies, and stabilities, paralleling to a certain extent types 1 and 2 described above but overlapping enough to make classification difficult. These are listed in Table 6 along with their more important characteristics.

The very weakly acidic silica will break through an anion-exchanger bed before strong-acid ions, since these displace the silica downward as the run progresses. Hence, the capacity of the resins, based on the silica-break through end point, will obviously depend on the ratio of silica to total anions. Capacity data given in Table 6 are for waters in which silica is equal to 30 per cent of total anions and does not exceed 10 ppm. For other conditions, manufacturers' data should be consulted.

Strongly basic exchangers should be regenerated with 4 per cent caustic warmed to 95 to 100°F, with a contact time of 60 to 90 min.

Salt-splitting Anion Exchangers. The chloride form of certain of the type-2 resins will remove bicarbonates, sulfates, nitrates, sulfides, or fluorides from water. The resin is then regenerated with sodium chloride. For purposes of water purification, the most important application appears to be the reduction of bicarbonate alkalinity in boiler feedwater. The capacity of the resin is relatively low in this function, 1 to 8 kg of alkalinity per cubic foot, as the bicarbonate in the raw water varies from 20 to more than 90 per cent. The corresponding regeneration efficiencies are also low, 0.5 to 5.0 lb of salt per kilograin of alkalinity removed. Use of caustic soda as well as salt for regeneration and of sodium cation exchanger with the anion exchanger are two modifications that have been suggested. When caustic is used, some silica removal is accomplished.

As in the case of the other ion-exchange materials whose characteristics have been considered, anion exchangers are being developed with

* This is analogous to the comparison of a 5 per cent solution of sodium hydroxide with a 10 per cent solution of ammonium hydroxide: the sodium hydroxide is more basic; the ammonium hydroxide contains more alkalinity.

emphasis on specific properties which are desired in certain applications. It is therefore necessary to consult the most recent technical data published by manufacturers to select the materials best suited for any water-purification system.[16,17]

DE-IONIZATION

For many years, the only means available for essentially complete removal of soluble salts from water was distillation. The various types of equipment used for this purpose are described in Chap. 9. Since the development of the anion-exchange resins described above, it has been possible to produce water of the highest purity by a two-step ion-exchange process. This has been called "demineralization" and "de-ionization" by various authors, but since some of the ions exchanged are not necessarily of mineral origin, the term de-ionization appears more inclusive.

The first step in the process is hydrogen cation-exchange treatment, as described in a preceding section. The effluent from this step, containing acids of the anions originally present in the raw water, is then passed through an anion-exchanger bed in which these are removed.

EQUIPMENT AND APPLICATIONS

Figure 9 shows various arrangements of equipment that may be used in de-ionization. The selection will depend on the quality of the influent, specifications for the effluent, and economy of operation. From the foregoing discussion of characteristics of cation and anion exchangers, the following considerations governing the flow cycle and ion-exchanger selection will be obvious:

1. Cation exchangers vary with respect to capacity, sodium leakage, economy of regeneration, durability, and unit cost.

2. Anion exchangers have similar variables, and, in addition, the weakly basic resins cost less and can be regenerated more economically by soda ash or ammonia, while the more expensive strongly basic resins require caustic.

3. When strongly basic anion exchangers are used, it is often best to remove carbonic acid as carbon dioxide to reduce the burden on this more expensive ion-exchange process.

Removal of Sulfates and Chlorides by Ion Exchange. The first de-ionization systems followed the simple cycle shown as (1) in Fig. 9, utilizing sulfonated coal as the hydrogen exchanger, and phenolic or polyethylene amines as the anion exchanger. Many installations of this type have been made for treating water for boiler-feed or process use. Where a very low ionic content is required in the effluent, the same materials are used in multiple-stage de-ionization, as in cycle 2.

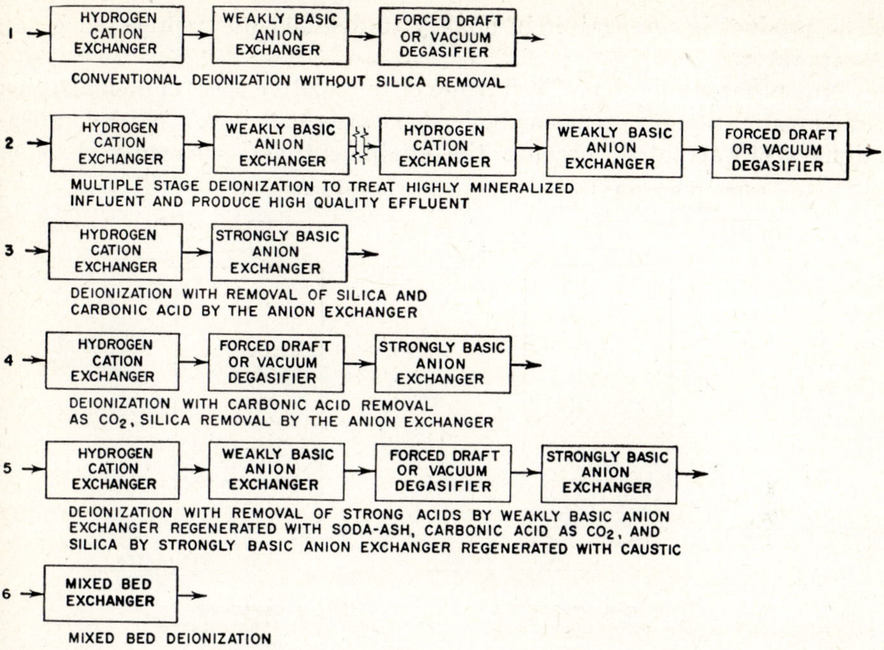

Fig. 9. Flow sheet showing arrangements of de-ionization cycles used for water purification.

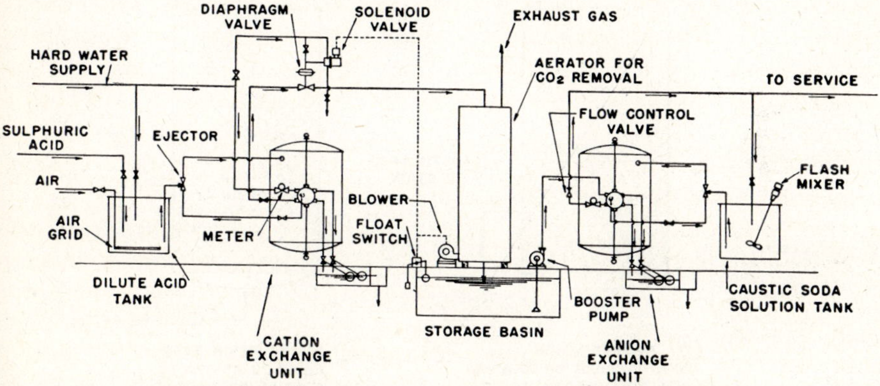

Fig. 10. Demineralizing system including silica-removal unit.

Silica Removal. Figure 10 shows a boiler-feedwater de-ionization system designed for silica removal, according to cycle 4. The influent to the system passes through the cation-exchange unit into a forced-draft degasifier where the carbonic acid is removed as carbon dioxide. The degasified water, containing hydrochloric, sulfuric, silicic, and other acids is pumped through the highly basic anion-exchange unit in which the silica is reduced to 0.05 to 0.2 ppm and the other acids are neutralized.

The product is comparable in quality to distilled water prepared by an evaporator.

Numerous studies have been made of the relative costs of de-ionization and evaporation for preparing boiler-feedwater make-up. No generalizations are warranted, because the equipment and operating costs of

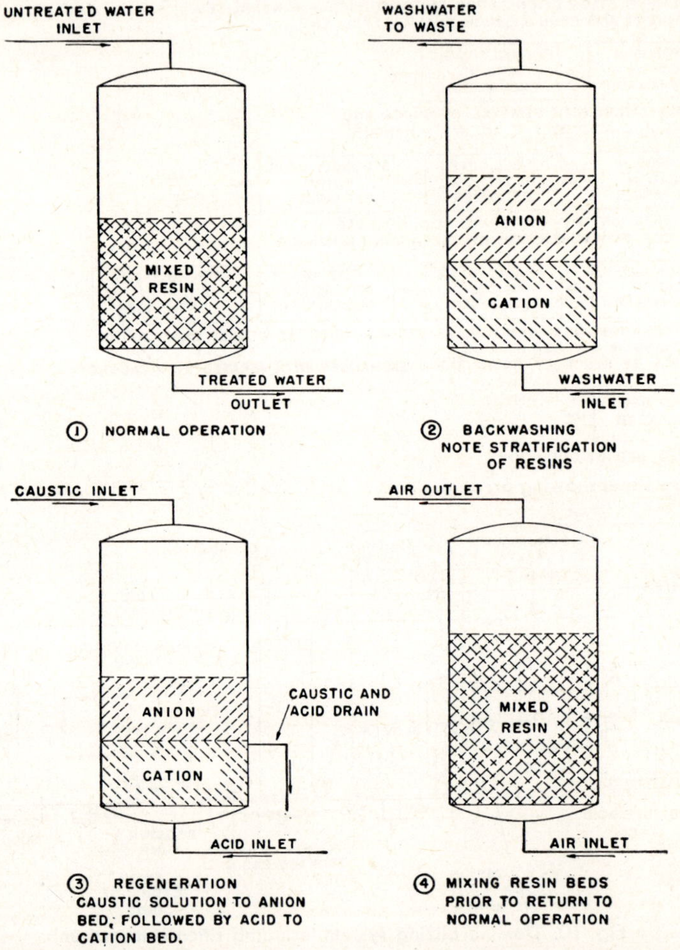

Fig. 11. Diagrammatic sketch of a mixed-bed de-ionizer during various stages of operation, backwashing, and regeneration.

demineralization vary in proportion to the ionic concentration in the water supply. For waters that are moderately mineralized, a case can often be made for saving by de-ionization in comparison with evaporation.[18,19]

One method for silica removal not shown in these diagrams is of historical interest but has been superseded by the use of strongly basic resins. In this process, sodium fluoride was added to the raw water before its entry into the hydrogen exchangers, and this resulted in the production of fluosilicic acid, H_2SiF_6. This is strong enough to be removed effectively by the weakly basic anion exchangers. The theoretical quantity of sodium fluoride required is 4.2 ppm for each 1.0 ppm of silica, expressed as calcium carbonate.* A number of systems of this type were installed and gave fairly good results, but the process is no longer comparative.

Mixed-bed De-ionization. This process is illustrated in Fig. 11. Water is passed through a bed which is a mixture of hydrogen- and anion-exchange resins, which have the effect of multiple-cycle de-ionization, each pair of contrasting resin particles functioning as a stage in the treatment. This produces an effluent of very low ionic concentration, far superior to that available from the conventional two-bed operation.[23]

When the resins are exhausted, the bed is backwashed, and the difference in density causes a sharp separation in the two layers with a relatively thin stratum in which some mixture occurs. The rubber-lined steel shells are equipped with transparent windows through which the operator can observe this phenomenon. After hydraulic separation, regenerants are introduced, as shown in the drawing. The bed is then rinsed, and the resins are mixed by injecting air into the bottom of the shell.

INTERFERENCE BY INORGANIC COATINGS AND BACTERIAL SLIMES

Defective operation of cation and anion exchangers is often caused by inorganic materials which coat the individual grains of the softening material. These coatings may consist of calcium and magnesium compounds, turbidity, aluminum hydroxide floc, and miscellaneous material not completely removed from the water prior to reaching the softening beds. As an example, the condition of grains of greensand zeolite coated by deposits of alum floc is shown in Fig. 12. The coatings shown were revealed by a study made by means of a *camera lucida.*† Similar coatings adhering to the zeolite grains are deposited by miscellaneous organic materials and by delayed precipitation of calcium, magnesium, and iron compounds. The treatment required for removal of these accumulations in order to restore the efficiency of a softener bed depends upon the type of deposit and on the cation- or anion-exchanger material. Chlorine and compounds of chlorine can be used to remove organic and bacterial slimes

* To convert SiO_2 to $CaCO_3$, multiply by 0.833.

† This is an apparatus used for reproducing objects magnified under the microscope by projecting the image in the microscopic field to a piece of paper located beside the apparatus. The image may then be traced so as to give an exact reproduction of the material viewed in the microscope under low-power magnification.

from material not attacked by chlorine. These include the resinous zeolites and the double silicates, either natural or synthetic. Chlorine treatment should not be used routinely on sulfonated coal or anion-exchange resins, because free chlorine attacks this type of zeolite. Frequently, acid is used to destroy films on carbonaceous and resinous material because they can withstand low pH values. A 5 per cent solution of caustic containing 1 per cent sodium sulfite and a wetting agent, circulated at 110°F, effectively cleans cation exchangers.

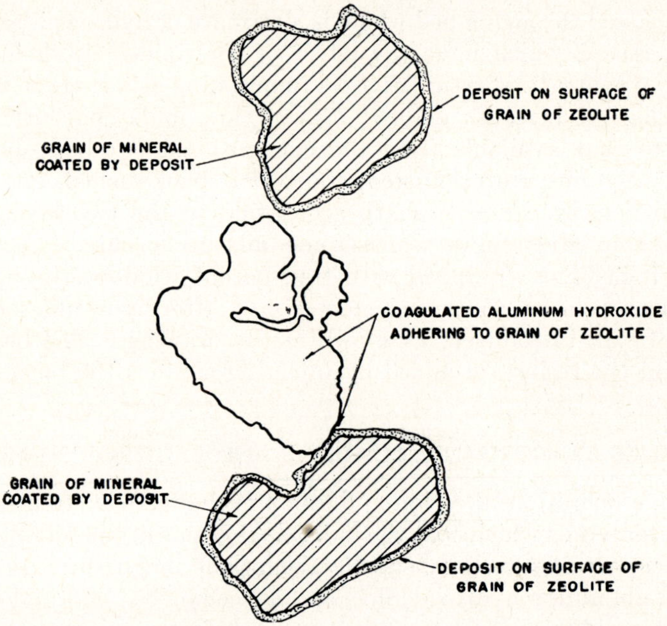

Fig. 12. Camera lucida drawing of zeolite grains with organic matter on surface of grains and flocculant material (aluminum hydroxide) adhering to grains of zeolite showing how coagulating material can cement grains of zeolite resulting from inefficient washing.

If deposits are permitted to accumulate, they can seriously depreciate the exchange capacity of the ion-exchange beds by coating the surfaces of the grains or by retarding uniform flow through the unit. Some relief from fouling can be obtained by providing for surface washing of beds during backwashing or for subsurface washers to break up the compact top layer of the beds. De-ionization systems are especially sensitive to suspended matter in the influent, and even a filtered water supply of low turbidity may cause trouble.[20] For this reason, it may be necessary to install filters ahead of the cation exchanger.

The formation of bacterial groups resulting in the development of slimes is now recognized as a troublesome operating problem requiring control. Several authorities[21] have confirmed the fact that active growth of organisms occurs in all types of exchange material, and control of such growths by chlorine, where suitable, or other sterilizing agent is necessary to insure satisfactory softening.

DESIGN OF SOFTENING AND ION-EXCHANGE SYSTEMS

In order to establish dimension and capacity specifications for an ion-exchange system, it is necessary to know the flow Q, in gallons per minute, the influent-water analysis, the effluent-quality specifications, the period between unit regenerations, and the number of units. Ion-exchange materials are selected on this basis, and their performance characteristics also enter into the calculations. An example will make this clear.

Sodium Zeolite Softening System. Suppose it is desired to design a sodium zeolite softening system to furnish 100,000 lb per hr of boiler feed-water make-up, using Youngstown, Ohio, city water as the influent. Treatment of the above data is as follows:

1. $$Q = \frac{100,000}{60 \times 8.33} = 200 \text{ gpm}$$

2. Express the water analysis as parts per million, in terms of calcium carbonate, and tabulate:

Cations	ppm	Anions	ppm
Calcium	66	Chlorides	7
Magnesium	20	Sulfates	79
Hardness	86	Mineral acids	86
Sodium	42	Carbonate	28
		Hydroxide	14
Total	128	Total	128

Dividing by 17.1, hardness = 5 gpg

3. The effluent quality should be zero hardness. This means zero by the Versenate test* or a three-drop test by the B & B* soap solution. Calcium and magnesium in the effluent are usually shown as 1 ppm each in recognition of leakage in trace quantities, but zeolite capacity and salt consumption are based on removal of *all* the hardness.

4. Assume a 24-hr regeneration cycle for each softener unit. Except in the case of very hard waters, shorter cycles are inadvisable because of the extra labor required—a fact often ignored in an effort to secure a low equipment price.

* See Chap. 24.

5. The system must obviously consist of at least two units, one capable of handling the entire load. When Q is great enough, a larger number of units is desirable to reduce the percentage of system outage during regeneration. In this example, 200 gpm warrant two units, but three would be questionable.

6. As a first approximation, size the units for a sulfonated-coal exchange material having a capacity of 7 kgr per cu ft at a regeneration efficiency of 0.45 lb salt per kilograin, and a permissible peak flow of 8 gpm per sq ft.

TABLE 7. TOTAL FLOW THROUGH VERTICAL CYLINDRICAL FILTERS AND SOFTENERS

Dimensions		Flow, gpm/sq ft of bed area					
Diameter, in.	Area, sq ft	3	4	5	6	8	10
18	1.77	5	7	9	11	14	18
24	3.14	9	13	16	19	25	31
30	4.91	15	20	24	30	39	49
36	7.07	21	28	35	42	56	71
42	9.62	29	38	48	58	77	96
48	12.57	38	50	63	75	100	126
54	15.9	48	64	80	95	127	159
60	19.6	59	78	98	118	157	196
66	23.8	71	95	119	143	190	238
72	28.3	85	113	142	170	226	283
78	33.2	100	133	166	199	266	332
84	38.5	116	154	193	231	308	385
90	44.2	133	177	221	265	354	442
96	50.3	151	201	252	302	402	503
108	63.6	191	254	318	382	509	636
120	78.5	236	314	392	471	628	785
132	95.0	285	380	475	570	760	950
144	113.1	339	452	566	679	905	1131

EXAMPLES:

1. Suppose the flow Q through a two-unit zeolite system is 240 gpm and the maximum permissible rate when one unit is out of service is 6 gpm per sq ft, then

$$\text{Maximum flow per unit} = 240 \text{ gpm}$$

In the 6 gpm per sq ft column, the smallest number which exceeds 240 is 265, requiring a 90-in. diameter softener.

2. For a three-unit system, $Q = 850$ gpm, rate limit $= 8$ gpm per sq ft.

$$\text{Maximum flow} = 850/2 = 425 \text{ gpm}$$

In the 8-gpm column, minimum = 509 gpm = 108-in. diameter.

7. The hardness removed per 24 hr is

$$5.0 \times 200 \times 1{,}440 \times 1/1{,}000 = 1{,}440 \text{ kgr}$$

equivalent to 720 kgr per unit, or

$$\frac{720 \text{ kgr}}{7 \text{ kgr/cu ft}} = 103 \text{ cu ft of zeolite per unit}$$

8. The *shell diameter* should be selected by reference to Table 7, showing the flow rates in gallons per minute per square foot against flow for various standard shell sizes. It is necessary to use the *maximum* flow per unit, in this case, 200 gpm. In Table 5 a 72-in. shell is the minimum size that will give 200 gpm at a rate of 8 gpm per sq ft.

9. The *bed depth* should be checked to see whether it is acceptable:

$$\frac{103 \text{ cu ft} \times 12}{28.3 \text{ sq ft}} = 43.7 \text{ in. deep}$$

Suitable bed depths are maximum, 72 in., minimum, 30 in. If the calculation gives *less* than the minimum, a 30-in. bed is nevertheless specified, and the run will be proportionately *longer* than 24 hr. For example, if a high-capacity resin rated at 20 kgr per cu ft were used in the above, the exhaustion per 24 hr would be 720/20 = 36 cu ft, or

$$(36 \times 12)/28.3, = 15.25 \text{ in. of depth.}$$

The length of run for a 30-in. minimum bed would then be

$$24 \times 30/15.25 = 47 \text{ hr.}$$

Manufacturers of zeolite softening systems will solve the foregoing problem by referring to tables of standard assemblies, and their resulting proposals may differ considerably in the intrinsic value of the material offered. They can be kept on a competitive basis by stipulating the following in the specifications:

Hardness of water to be softened	5 gpg
Hardness of effluent by Versenate test (or not requiring more than 3 drops of B & B soap solution to produce lather stable 5 min in 40-ml sample of effluent)	Zero
Quantity of water softened per regeneration of entire system	288,000 gal
Flow	200 gpm
Salt required per kilograin hardness removed, not over	0.45 lb
Number of softener units	2
Diameter of shells	72 in.
Zeolite bed depth: not more than	72 in.
Zeolite bed depth: not less than	30 in.

Other details of the specification include shell design pressure, red brass or bronze shell internals and brine piping, insulating couplings, linings for shells and salt-storage and brine-measuring tanks, furnishing of back-wash- and rinse-rate controllers, single multiport or individual gate valves, etc. The bidders should be asked to identify the ion-exchange material by the *manufacturer's designation*, to state the quantity furnished, the rated capacity, and the guaranteed maximum capacity loss per year, and to fill in a questionnaire covering make, catalogue number, size of accessories (all dimensions not specified), metal thickness, etc. When the resulting proposals are compared, a selection based on price and intrinsic worth will be obvious.

Hydrogen-Sodium Zeolite System. Consider a system to furnish 500,000 lb per hr of feedwater to be heated in a deaerator at 5 psig, with no condensate returns, using filtered Lake Erie water as the influent. It is desired to reduce the bicarbonate alkalinity to 10 ppm.

1. Q = 500,000 lb per hr = 1,000 gpm.

The 5-psig steam condensed in heating the feedwater will amount to nearly 15 per cent of the total flow, or 150 gpm (Fig. 7, Chap. 7), and cold water will be 850 gpm.

2. The maximum water analysis is

Cations	ppm	Anions	ppm
Calcium	98	Chloride	27
Magnesium	31	Sulfate	35
Hardness	129	Mineral acids	62
Sodium	33	Bicarbonate	100
Total	162	Total	162

Total hardness, 129 ppm = 7.5 gpg; total cations, 162 ppm = 9.5 gpg. Sodium = 21 per cent of total cations; alkalinity = 62 per cent of total anions.

3. The effluent quality should be zero hardness and 10 ppm bicarbonate. Assume that a high-capacity resin will be used for hydrogen exchange and that sodium leakage with this low-sodium, high-bicarbonate water will be practically zero—in other words, the mineral acids in the hydrogen-exchanger effluent will be 62 ppm.

The per cent of hydrogen exchanger effluent required is

$$\frac{100 - 10}{100 + 62} = 55.5 \text{ per cent}$$

This maximum may be taken as 56 per cent × 850 gpm = 475 gpm. The corresponding sodium-exchanger effluent would be 850 − 475 = 375 gpm, but to take care of possible *increase* in the mineral-acid ions of the raw water, provide for 50 per cent $\times$ 850 = 450 gpm.

4. Assume a 24-hr cycle for all units.

5. Because of the large flow, try a three-hydrogen-, three-sodium-unit system. Then the flows per unit are as follows: hydrogen, average, 158 gpm, maximum, 238 gpm; sodium, average, 150 gpm, maximum, 225 gpm.

6. From manufacturer's data for ion-exchange resin, use the following characteristics:

Acid cycle, 12 kgr per cu ft, using 6 lb 66°Be H_2SO_4 per cubic foot. Efficiency = 0.5 lbs per kgr.

Sodium cycle, 20 kgr per cu ft, using 9 lb salt per cubic foot. Efficiency = 0.45 lb per kgr.

Peak softening rates, 8 gpm per sq ft.

7. The cations removed per 24 hr will be

$$\text{H} = 9.5 \times 475 \times 1{,}440 \times 1/1{,}000 = 6{,}500 \text{ kgr}$$

equivalent to 2,166 kgr per unit, or

$$\frac{2{,}166 \text{ kgr}}{12 \text{ kgr/cu ft}} = 180 \text{ cu ft per unit}$$

$$\text{Na} = 7.5 \times 450 \times 1{,}440 \times 1/1{,}000 = 4{,}860 \text{ kgr}$$

equivalent to 1,620 kgr per unit, or

$$\frac{1{,}620 \text{ kgr}}{20 \text{ kgr/cu ft}} = 81 \text{ cu ft per unit}$$

8. The shell diameters from Table 7, required for the maximum unit flows, are

$$\text{H}: d = 78 \text{ in.}, A = 33.2 \text{ sq ft}$$
$$\text{Na}: d = 78 \text{ in.}, A = 33.2 \text{ sq ft}$$

9. Bed depth

$$\text{H}: \frac{180 \times 12}{33.2} = 65 \text{ in.}$$

$$\text{Na}: \frac{81 \times 12}{33.2} = 29.2 \text{ in. (specify minimum, 30 in.)}$$

The data developed above should be incorporated in specifications following the pattern given for sodium zeolite units but should cover rubber-lined shells, piping and valves, a blending-control system, and a degasifier to remove the carbon dioxide.

DESIGN OF DE-IONIZATION SYSTEMS

The basic principles governing ion-exchange softening systems also apply to de-ionization, but the design is complicated by additional decisions that must be made. It is necessary, first, to determine the specifications for ionizable solids in the effluent, the tolerance for any particular ion (*e.g.*, sodium), and whether silica is to be removed. On this basis ion-exchange resins and one of the cycles shown in Fig. 9 are selected which will give the best assurance of effluent quality and economy of operation. Usually, a number of possible combinations must be evaluated.

A simple example is the demineralization of make-up amounting to a maximum of 2 per cent of 1,000,000 lb per hr of feedwater for boilers operating at 950 psig, using well water of the following analysis (part per million as calcium carbonate):

Cations	ppm	%	Anions	ppm	%
Calcium	50	31	Chloride	21	12.5
Magnesium	26	16	Sulfate	25	15
Hardness	76	47	Mineral acids	46	27.5
Sodium	86	53	Bicarbonate	116	69
Total	162	100	Total	162	96.5
			Silica	6	3.5
			Total exchangeable	168	100
			Grains per gallon	9.8	
			Excluding bicarbonate	3.1	

In view of the boiler pressure and the low percentage of make-up, the extremes of purity attainable only by multiple-cycle or mixed-bed systems cannot be justified. Soluble solids should be reduced to about 3 to 4 ppm, corresponding to a specific conductance of about 6 to 10 micromhos. Silica should be as low as possible. Carbonic acid can be removed as carbon dioxide gas or by a strongly basic anion exchanger. Since the bicarbonate is 69 per cent of the total anions and the resin and regenerant are expensive, it would appear that removal of carbonic acid as carbon dioxide is preferable.* Therefore, the system will consist of a hydrogen exchanger followed by a degasifier and an anion exchanger which will remove silica as silicic acid, and the flow sheet will resemble cycle 4 in Fig. 9. A cation exchanger and regeneration procedure which will give low sodium leakage must be used so that the total solids and silica in the

* For small flows, it is often better to burden the anion-exchange units with carbonic acid removal to gain simplification and to avoid secondary pumping by omitting a degasifier from the cycle.

effluent will meet specifications. It will be found that one of the high-capacity resins in group 6 of Table 5 may be used at a regeneration level of 8 lb of sulfuric acid per cubic foot, capacity, 14 kgr per cu ft—an efficiency of about 0.6 lb of acid per kgr. A Type-1 strongly basic anion-exchange resin should be used (see Table 6) at a regeneration level of 5 lb of caustic per cubic foot, capacity, 11 kgr—efficiency, 0.45 lb of caustic per kgr. Silica will not exceed 0.1 ppm within this capacity range.

From here on, the calculations parallel zeolite softener design. $Q = 20$ gpm—net yield of make-up. Assume that a two-train plant will be designed for safety; then the flow per train is, average, 10 gpm, maximum, 20 gpm.

Anions removed per 24 hr by each train are 44.6 kgr. Units of 24-in. diameter are required (Table 7), and the required minimum bed depth of 30 in. (7.9 cu ft) will provide a regeneration cycle of about 50 hr. Each anion unit will require 70 gal per cu ft of rinse water from the hydrogen units, or about 600 gal per cycle, at a flow rate of 10 gpm. The hydrogen units must, therefore, be designed for a peak flow of 30 gpm.

The hydrogen units should also have 50-hr cycles, delivering 30,600 gal, including rinse water. This requires a capacity of 290 kgr, equivalent to 20.7 cu ft. The peak flow of 30 gpm can be handled by 30-in.-diameter units, requiring a bed depth of about 51 in.

From the foregoing, regenerant quantities per unit regeneration are 166 lb of 66°Be acid and 40 lb of 100 per cent caustic. Facilities should be provided for diluting the acid for stepwise regeneration, beginning with 2 per cent, since calcium is 40 per cent of the cations. Batch preparation of caustic at the final dilution of 4 per cent and temperature of 95°F is safer than blending and steam injection and gives better assurance of complete regeneration after every cycle.

Specifications should include the flow, the influent- and effluent-water analysis, and other data similar to those in the first example.

The design of mixed-bed de-ionization equipment requires similar calculations and study of a number of factors peculiar to each problem.[23]

Bibliography

1. Thompson, H. S.: Absorbent Power of Soils, *J. Roy. Agr. Soc. Engl.*, Vol. 11, 1850.
2. Way, J. T.: Power of Soils to Absorb Manure, *J. Roy. Agr. Soc. Engl.*, **11**: 313 (1850), **13**: 123 (1852).
3. Behrman, A. S.: Early History of Zeolites, *J. AWWA*, **13**: 223 (1925).
4. Gans, Robert: German Patent 197,111, 1906; U.S. Patents 914,405, Mar. 9, 1909; 943,535, Dec. 14, 1909; 1,131,503, Mar. 9, 1915.
5. Kunin, R., and R. J. Myers: "Ion Exchange Resins," John Wiley & Sons, Inc., New York, 1951.

6. Nordell, E.: "Water Treatment for Industrial and Other Uses," Reinhold Publishing Corporation, New York, 1951.
7. Nachod, F. C.: "Ion Exchange," Academic Press, Inc., New York, 1949.
8. Lindsay, F. K.: High-capacity Cation Exchangers, *J. AWWA*, **42**: 75 (1950).
9. Calise, V. J.: Hot Zeolite Softening, *Power Eng.*, **54**: 54 (September, 1950).
10. Kemmer, Frank N.: Conditioning Makeup by High-temperature, Sodium Ion-exchange, *Combustion*, **21**: 59 (April, 1950).
11. Butler, William S.: An Inspection Report after Eighteen Months with Hot Lime Zeolite, *Proc. 12th Ann. Water Conf., Engs. Soc. West. Penna.*, October, 1951. See also ASME Paper 50-A-77, December, 1950.
12. Applebaum, S. B.: Discussion of above paper. *Proc. 12th Ann. Water Conf., Engs. Soc. West. Penna.*, October, 1951.
13. Tiger, Howard L.: Carbonaceous Zeolites—An Advance in Boiler-feedwater Conditioning, *Trans. ASME*, **60**: 315 (May, 1938).
14. Research Products Corp., Madison, Wis.: Bulletins on Alkalex.
15. Gowdy, G. H., and S. B. Applebaum: "Conversion of a Two-stage Hot Process Water Softener from Hot Lime Soda Phosphate to Hot Lime Zeolite," Reprint 81, The American Society of Mechanical Engineers, New York, Nov. 30–Dec. 5, 1952.
16. Wheaton, R. M., and W. C. Bauman: Properties of Strongly Basic Anion Exchange Resins, *Ind. Eng. Chem.*, **43**: 1088 (May, 1951).
17. Thompson, Joseph, and Francis X. McGarvey: Factors Influencing the Operation of Anion Exchange Resins, *Proc. Am. Power Conf.*, **15**: 597 (1953).
18. Yoder, J. D.: Demineralization as Compared with Evaporation of Makeup for Surface Condensing Power Plants, *Proc. Midwest Power Conf.*, 1950. See also *Trans. ASME*, **72**: 491 (1950).
19. Calise, V. J.: High Pressure Boiler Feedwater Treatment: Evaporation vs. Demineralizing and Silica Removal, *Proc. 11th Ann. Water Conf., Engs. Soc. West. Penna.*, Oct. 17, 1950.
20. Brines, M. E.: Dow's Demineralizing Experience, *Power Eng.*, **56**: 82 (1952). See also *Power*, **96**: 94 (1952).
21. Panel Discussion: Control of Bacterial Reproduction in Cation-exchange Layers, *J. AWWA*, **41**: 933 (1949). See also *Combustion*, **25**: 49 (1953).
22. Lindsay, F. K.: Acid Regeneration of Cation Exchangers, *Proc. Am. Power Conf.*, **15**: 588 (1953).
23. Thompson, J., F. X. McGarvey, John F. Wantz, S. F. Alling, M. E. Gilwood, and David R. Babb: Mixed Bed Ion Exchange: AIChE Symposium, *Chem. Eng. Progr.*, **49**: 341 (1953).

9 EVAPORATORS: THEIR DESIGN AND OPERATION

Distilling equipment or evaporators have been employed for many years in marine service to supply pure make-up water to steam-generating equipment and for other services. Evaporators were used only to a limited extent in stationary steam plants until about 1920.[1–5] Since that time, the trend toward higher boiler ratings and larger boilers has led to the adoption of evaporators for preparation of high-quality feedwater make-up. Where evaporator operation fits in with the heat balance of the plant, this type of equipment often is preferred, especially when extraction turbines are used and the make-up water requirements are low.*

There are many types of evaporators and several factors govern their selection. These include ease of operation, control of scale formation, accessability for cleaning, and purity of vapor. Consideration must be given to the specific heat balance and space requirements of each installation before an intelligent evaluation and selection of the equipment can be made.

For many years there were only minor changes in evaporator design and the quality of vapor obtained was not always satisfactory. However, as the number of high-pressure boiler installations increased, difficulties in operation demanded improved design of evaporators to minimize carry-over from these units, thereby insuring a higher-quality condensate. These improvements included better facilities for preventing carry-over, correction of foaming, and more effective descaling.

The early types of evaporators for stationary power plants were similar to those used in marine service. No pretreatment of the evaporator feedwater was provided, and scale deposited in the units. Scaling occurred to some extent despite blowing down concentrated salines and

* The preparation of make-up water of a quality equal to that of water from evaporators is now possible by means of chemical demineralization by ion-exchange minerals. For a discussion of this type of treatment, the reader is referred to Chap. 8.

cracking down the coils to remove loose scale and accumulations of deposits. To avoid these difficulties, pretreatment of the feedwater supplying the evaporator was initiated. Such treatment is now common practice and is generally justified because of improved evaporator performance. Many forms of treatment are used for such services. These include miscellaneous chemical conditioning, chemical softening, and ion-exchange processes.

THEORY OF EVAPORATION

By vaporization is meant a change from a liquid to a vapor state.[6–9] This may be accomplished by (1) ebullition, in which the vaporization

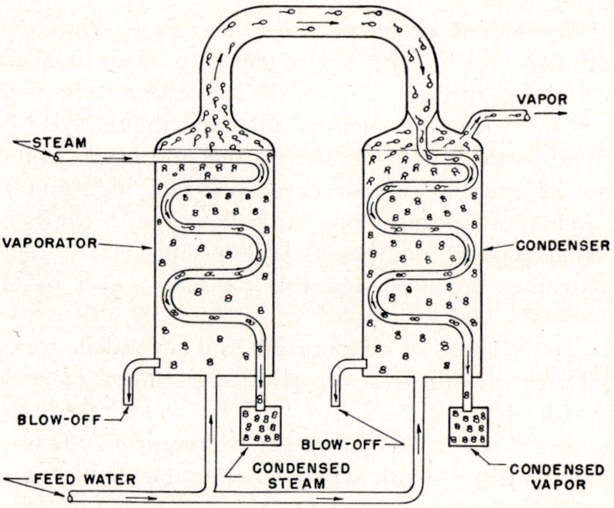

Fig. 1. Evaporator system.

takes place within the liquid and passes through the liquor, discharging at the surface; and (2) evaporation, or vaporization, which takes place from free surfaces only.

The generally accepted theory of evaporation assumes that above 0°K all molecules of a substance are in a constant state of motion and that the individual molecules collide with one another constantly. Because of this action, some of the molecules attain a velocity relatively greater than the average velocity of the molecules of the entire mass. When this action takes place at a free surface, the velocity of some molecules becomes so violent that they are hurled into space above the surface of the substance and then exist as vapor. The molecules in vapor phases occupy the spaces existing between the air or other gaseous molecules. Similarly, molecules in this state continually return to the liquid state, and at equilibrium, the rate of transfer across the liquid boundary is equal in the two directions. To effect evaporation

1. The necessary heat must be supplied to the liquid.

2. The vapor evolved must be continuously removed from above the liquid surface to prevent its accumulation.

Since evaporation requires that heat be transferred to the liquid to be evaporated, it follows:

1. Evaporation depends on the temperature difference between the heat source and the liquid to be evaporated.

2. Evaporation depends on the exposed surface area available for transfer of heat.

3. Evaporation depends on the coefficient of heat transfer which is affected by the viscosity of the solution, the rate of flow of the liquid, the noncondensible gas present, and the cleanliness of the surface.

The design and operation of all industrial evaporators, whether they are for use in the distillation of water or of other liquids, are based upon control of these fundamental factors. Figure 1 shows graphically, on an exaggerated scale, how evaporation of water occurs and how the vapors are condensed to form a liquid.

PRINCIPLE OF OPERATION

In its simplest form, an evaporator system consists of two essential parts:

1. The distilling chamber or evaporator, in which the water is heated and converted into vapor.

2. The condenser, in which the vapor is converted into a liquid.

The source of heat utilized to vaporize the water in steam-generating plants is either high- or low-pressure steam, which, upon passing through the heating coils, condenses, giving up its latent heat to the raw water to be evaporated. Thus, in an evaporator there are two sources of distilled water. One is the condensate from the steam which has been employed to heat the water and which replaces the steam used by the evaporator and cannot, therefore, be considered as "make-up." The other is the condensed vapor resulting from evaporation of the make-up water. When water is converted into vapor and then condensed, solids in suspension or dissolved in the water remain in the distilling chamber unless they are carried over mechanically with the vapor or pass over as gases.

In large-scale evaporator operations, the distillate obtained from evaporators generally contains a small amount of soluble or suspended solids and some gases. Solids are present because droplets of water are carried from the evaporator with the vapor. Well-designed and -operated evaporators can produce condensate having a solids content as low as 1 to 2 ppm, or lower. The amount of solids present in distillate is influenced by the rate at which the equipment is operated, the design of the evaporator, the vapor-purification equipment installed, and the solids present in the feed and concentrated water.

TYPES OF EVAPORATORS

Several types of evaporators are available for power-plant service. They may be classified as (1) submerged tube, (2) film, (3) flash, or (4) compression distillation. They may be designated, also, as single-, double-, or multiple-effect.

Submerged-tube Evaporator. The submerged-tube[10] evaporator usually consists of a cast-iron or steel shell which serves as a reservoir. The water is heated by live steam (frequently at reduced pressure) or by exhaust steam, passing through submerged coils. The equipment may be designed to operate either as a submerged-tube type or as a film-type unit.

Film Evaporator. Film evaporators,[10,11] as the name indicates, are those in which raw water is filmed over a series of steam coils heated by live or exhaust steam. The coils are enclosed in cast-iron or steel shells. The water flashes into vapor as it is filmed over the surfaces of the steam-heated coils and is subsequently converted to the liquid phase in a suitable condenser.

Flash Evaporator. This type[12,13] is operated generally under low pressure, using exhaust steam as the source of heat. The evaporator consists of a tubular heater in which the make-up water is heated to a temperature high enough for it to flash into vapor upon entry into the flash chamber, which is maintained under a partial vacuum. The vapor from the flash tank is condensed in the main condenser or in a separate condenser.

Multiple-effect Evaporator. An evaporator that consists of one heating chamber only is termed a single-effect evaporator. For economy, it may be desirable to place a number of units in series, and when this is done, the units are called multiple-effect evaporators.[14,15] Where this arrangement is used, the vapor from one effect is utilized to heat and vaporize the water in the effect following it in the series; the vapor coils of each effect then serve as the condenser for the previous effect. By thus utilizing the vapor from each successive effect, the efficiency of the apparatus is increased with respect to the amount of primary steam required to evaporate a given amount of water.

EVAPORATOR ECONOMY

The poundage of distilled water produced by an evaporator per pound of steam used is termed the "evaporator economy." Single-effect evaporators will deliver about 0.9 lb of distillate per pound of steam. A double-effect evaporator, that is, two units in series and of the same size, will produce about double this amount per pound of steam entering the first effect, a triple-effect installation about three times, and so on as more units are added. Figure 2 is a simplified diagram showing how the evaporator economy increases in the multiple units. Assuming that a single-effect installation is fed with exhaust steam at atmospheric pressure and the corresponding temperature of 212°F, that the raw water

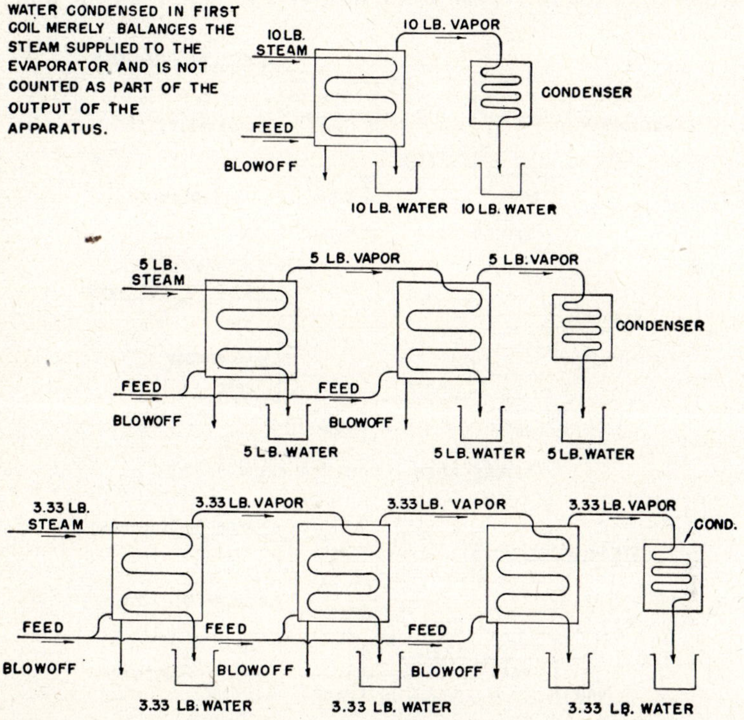

Fig. 2. Diagrams showing the theoretical increase in evaporator economy due to increasing the number of effects.

supplying the equipment is at the same temperature, that the vapor is under atmospheric pressure, and that no losses occur, then the equipment would function as indicated in the diagram.

Typical evaporator performance for multiple-effect units is about as follows:

	Ratio of Pounds of Steam Supplied to Pounds of Distillate Produced
Single-effect evaporator	1:0.9
Double-effect evaporator	1:1.65
Triple-effect evaporator	1:2.3
Quadruple-effect evaporator	1:2.8

Multiple-effect evaporators are termed "series feed" or "parallel feed." By series feed is indicated an arrangement in which the first effect receives its raw water from the evaporator pump but in which each succeeding effect is fed from the overflow of the preceding effect. Parallel feed of multiple-effect units is indicated where each unit receives its make-up supply directly from the evaporator-feedwater pump.

In some installations, multiple-effect units are classified still further with reference to the direction of the flow of water through the units and

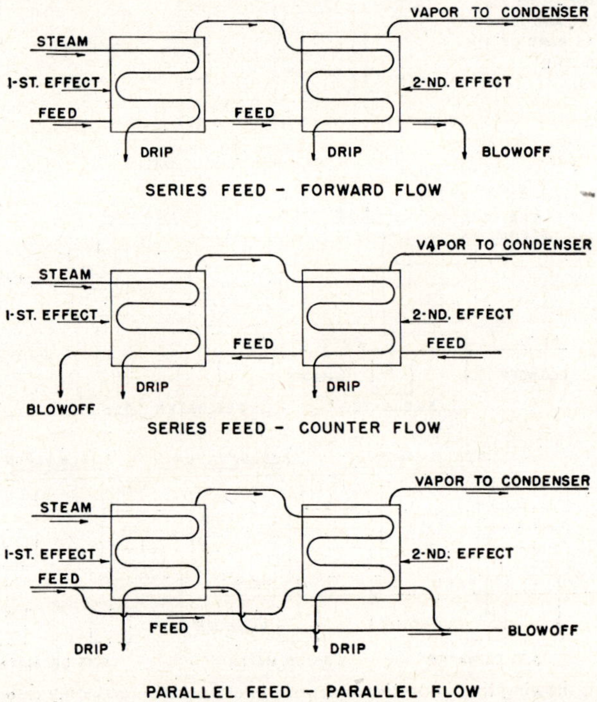

Fig. 3. Diagrammatic arrangements of steam and vapor flows with forward-, counter-, and parallel-feed flows.

the blowdown connection from each effect. These schematic classifications are termed "forward flow," "counterflow," and "parallel flow." The conditions for these terms are indicated in Fig. 3. In a double-effect installation where forward flow is used, the raw water passes from the first to the second effect, and the blowdown water is discharged from the second effect only. In the counterflow scheme, the raw water enters the cooler, or second effect, and is pumped from there to the first effect, and the blowdown is from the first effect only. By a parallel-flow scheme, each effect receives its raw-water supply directly from the evaporator pump, and each effect is blown down independently of the other effect or effects, in case there are more than two.

SPECIFIC DESIGN OF EVAPORATOR

Submerged-coil Unit. In Fig. 4 is shown a type of submerged-tube evaporator. The unit consists of a steel or cast-iron shell on top of which is a dome serving as a chamber for the collection of the vapor. The heating of the water is effected by steam which is passed through the

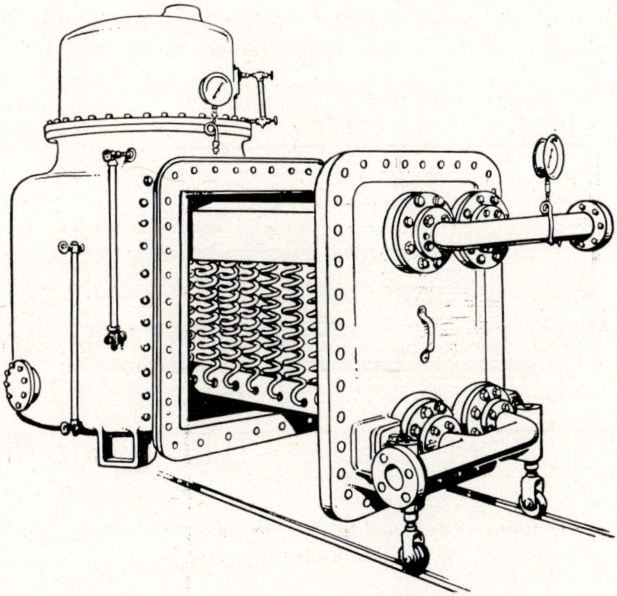

Fig. 4. Reilly submerged-tube single-effect evaporator showing coil nest removed from shell. *(Courtesy of The Griscom-Russell Company.)*

coils where the latent heat is given up to the raw water as the steam is condensed.

The coils are seamless brass tubes or other alloys, which in some cases are tinned. These coils, of which there are a number in each evaporator, are fastened into a top and a bottom header of cast iron. The manifolds are bolted to a door, so that the entire nest of coils may be removed readily from the shell for inspection or repairs. Baffles are placed above the coils to prevent water being carried over with the vapor. Scale formed upon the coils is removed at intervals by a sudden change of temperature in the coils. Owing to their construction in a spiral, relatively great expansion or contraction occurs during cracking down, which breaks the scale away from the tube surface. Two or more of such units are often installed in series and operated as a multiple-effect system.

Bent-tube Evaporator. This type of low-pressure apparatus (Fig. 5) is designed for vacuum and moderate-pressure operation. The distinguishing features in the design of this evaporator are the shape and arrangement of the evaporator tubes. The tube bundle is normally submerged.

The tubes of the evaporators are installed with an initial slight bow or curvature into fixed headers or tube sheets, so that all changes in length of the tubes must be taken up by increased bowing. Where no pretreatment is provided and scale formation is rapid, it may be cracked off through chilling the tube surface with cold water and then quickly admitting steam. Tubes are widely spaced to prevent bridging of the

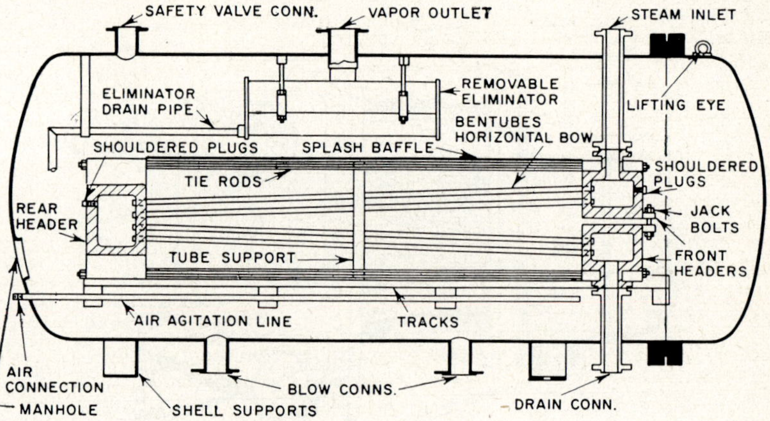

Fig. 5. Bent-tube evaporator designed to operate either as a submerged-tube or film-type evaporator. ***(Courtesy of The Griscom-Russell Company.)***

scale between them and to allow free dropping of cracked scale to the bottom of the shell.

The tubes are secured in the headers or tube sheets by expanding them into a number of annular grooves. For high-pressure or high-temperature conditions, a steel ferrule is expanded inside the tube end, further insuring permanent tightness. Tube holes in the headers or tube sheets are drilled at an angle to conform to the position of the bowed tubes at operating temperatures and are chamfered to prevent wear or cutting of the tube. The design permits replacement of single tubes without disturbing others, and access is provided to the exteriors of the tubes. The heating elements of the evaporators are readily removable without entering the shell.

A splash baffle above the topmost tubes prevents spouting of the water from violent ebullition. This baffle directs the liberated vapor towards the outward sides of the shell into the large vapor space, minimizing

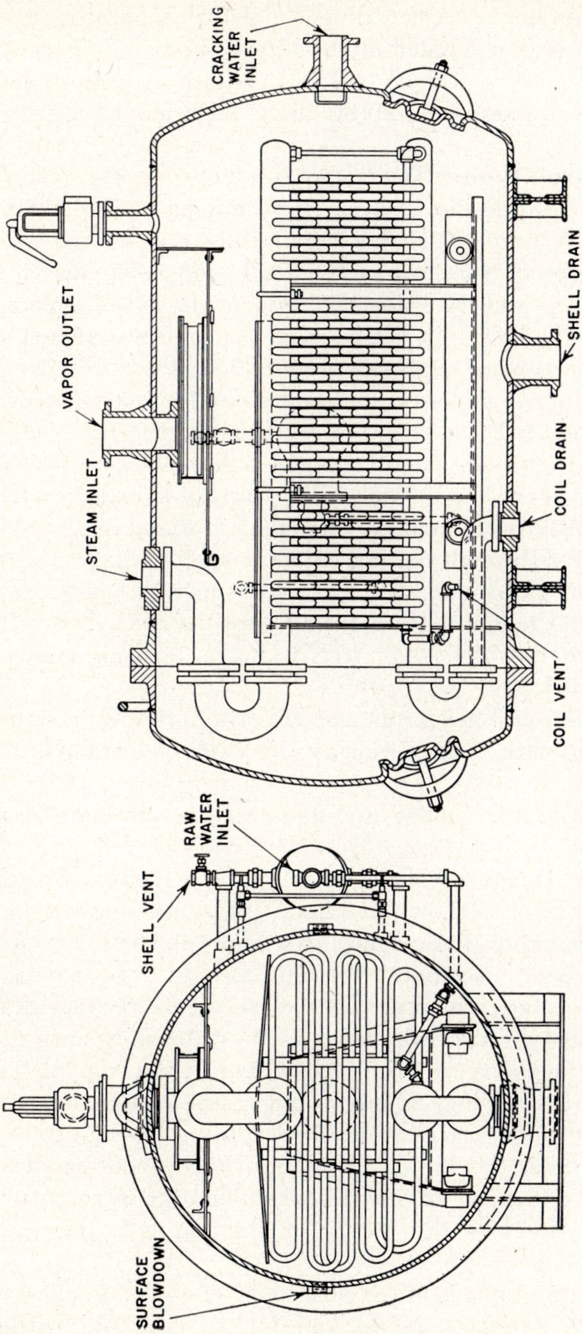

Fig. 6. Submerged-coil evaporator. *(Courtesy of Foster Wheeler Corporation.)*

moisture entrainment. A centrifugal moisture separator through which all vapor must pass is located at the top of the vapor space. The separator has cover plates at each end to facilitate inspection and cleaning, and the entire separator assembly may be removed by breaking one joint.

Submerged-coil Type. The evaporator shown in Fig. 6 is a submerged-coil unit of modern design, capable of producing high-quality water. It is applicable to large industrial installations and is used where process steam is generated from raw water, by condensing direct from main boilers in the coils and returning the condensate to the boilers.

As a heat-exchanger unit, the evaporator is fed by exhaust steam from a high-pressure turbine and can produce low-pressure steam for power or process use, or the vapor can be condensate for feedwater or other uses. The uncontaminated condensate is returned to the boiler circuit. This type of evaporator is designed for high- or low-pressure operation.

The shell cover for low pressures is assembled from a formed steel head and a high hub-type forged-steel flange, full diameter, welded together and arranged for through bolting with the shell flange.

The coils are flexible and arranged with sufficient bends per coil to permit descaling as required to maintain capacity. Coils are attached to horizontal headers which allow free drainage, insuring stable and satisfactory operation.

The entire coil and header assembly is supported on a structural steel frame equipped with flanged wheels, which roll on a track fastened to the inside of the shell.

Several evaporator baffles are installed to eliminate carry-over of solids.

Bowed-tube Design. The general arrangement of another high-capacity evaporator is shown in Fig. 7. The shells are proportioned to provide liberal vapor-disengaging area and tube bundles operate completely submerged. A square tube pitch is used to provide access to the exterior of the tubes, which are roller expanded into multigrooved holes in the tube sheets. The tube holes in the tube sheets are drilled at an angle so that the tubes are not pinched against the edge of the tube hole during thermal descaling, which would result in loosening of the tube joint. Tube-support plates are of the "floating" design so that they are free to move with the bundle assembly during thermal flexing of the tubes. Tubes are given an initial bow which, due to the restraint of the tube headers, provides the means of thermally flexing the tubes for descaling.

Internal steam piping is of the pipe-loop type to compensate for expansion. Internal joints to be broken for tube-bundle withdrawal are accessible from the cover end of the unit.

The incoming feed is admitted to the shell through an internal spray system by which it is distributed in a thin sheet above the surface of the main liquid body. The vapor, in rising from the liquid surface to the vapor separator, passes through the incoming feedwater and is washed of entrained droplets of the highly concentrated shell liquid. This contributes to the final vapor purity by reducing the impurity concentra-

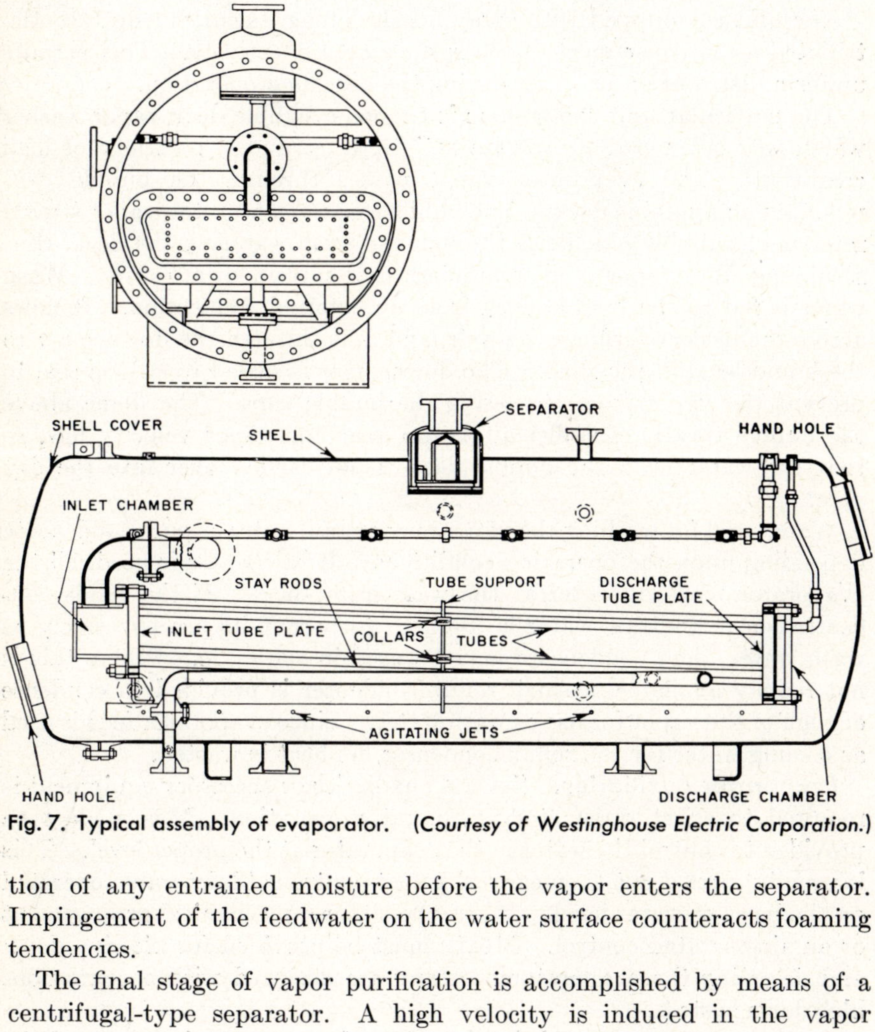

Fig. 7. Typical assembly of evaporator. *(Courtesy of Westinghouse Electric Corporation.)*

tion of any entrained moisture before the vapor enters the separator. Impingement of the feedwater on the water surface counteracts foaming tendencies.

The final stage of vapor purification is accomplished by means of a centrifugal-type separator. A high velocity is induced in the vapor stream because of the entrance of the vapor into the cylindrical separator body. By centrifugal force, moisture is removed to a "dead" space around the separator periphery formed between a perforated inner wall and the external separator wall.

Horizontal-tube Evaporator with Bubble-cap Vapor Separator. In this type of evaporator the horizontal U tubes are installed in tube sheets so that the entire tube bundle can be withdrawn on bundle tract supports.

The liquid level is maintained approximately at the top of the tube bundle. Referring to Fig. 8, it will be seen that the feedwater enters the shell through the inlet at the bottom of the shell, thus passing up through the tube bundles, which are supplied with steam of the desired pressure.

The unit is equipped with spray nozzles along its entire length so distributed as to cover most of the surface area effectively. This permits uniform distribution of water during the descaling operation.

The particular unit shown here is termed a bubble-deck vapor washer which was developed to operate satisfactorily under a variety of load conditions. The evaporator vapor passes through the bubble caps mounted on a bubble deck. The bubble caps break up the vapor stream into small bubbles which rise through the wash water on the deck, thus scrubbing the vapor and removing the entrained moisture. Wash water is fed to the bubble deck from an external connection. It flows across the deck, overflows the weir, and falls through the downcomer to the liquid level in the shell. The downcomer is sealed by a loop seal to prevent the vapor from bypassing the bubble caps. The dome above the bubble tray is of sufficient size to keep the vapor velocity low, so that the water from the bubble deck is not carried over into the dry pipe.

Water used for washing the vapor may be raw water or pure condensate depending upon the operating conditions. If raw water is used, all the evaporator feed can enter at the wash-water inlet. If the solids concentration in the raw water is too high to permit its use effectively as wash water, then condensate may be used in the cycle. Where this is not readily available, a small reflux condenser is provided to condense enough of the vapor for use as wash water. Since evaporator feed is used as cooling water for the reflux condenser, no heat is wasted.

Evaporator Auxiliaries. Several auxiliaries or accessory equipment is required to effect satisfactory evaporator operation. Means must be provided to control the water in the evaporator at the proper level. This is usually carried out by means of a float-controlled inlet valve operated on the basis of water level. It may be either a mechanical arrangement or an air-operated control. Means must be provided for blowing down and descaling the evaporators. Many are designed with a continuous blowdown system.

Controls must be provided to regulate the amount of steam and the pressure of the steam so as not to exceed certain preestablished pressures. In case high-temperature steam is used, it may have to be desuperheated before it can be used safely.

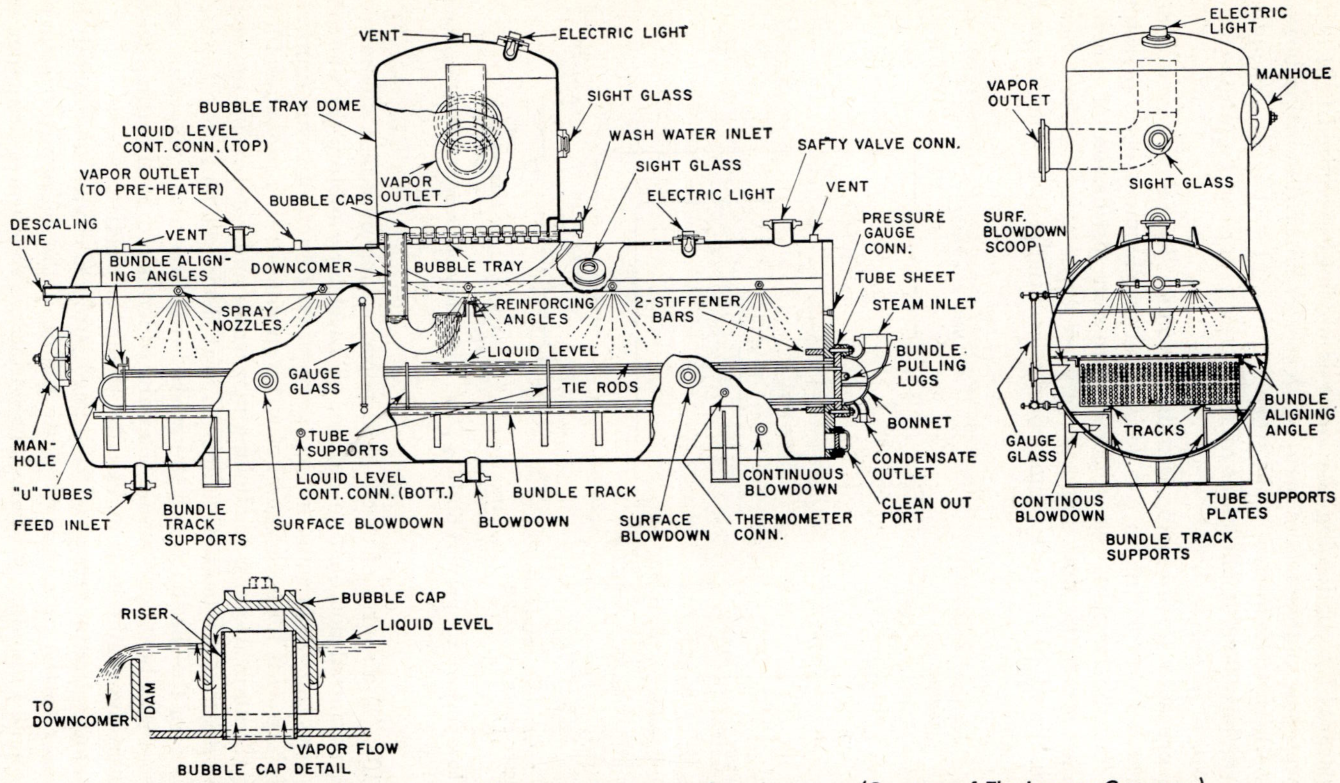

Fig. 8. Evaporator design for washing vapor by means of condensate. (*Courtesy of The Lummus Company.*)

Apparatus for condensing the vapor is obviously required. In some instances, this is done in the main condenser or in the deaerating heater, and in others, it is effected in the stage-heater heat exchanger, depending upon the heat level in the plant at which the vapor can be most readily used.

Low-pressure Compression Distilling Plants. Extensive research has been carried on relating to the design and development of evaporators based on the principle of vapor-compression distillation of water.[16] Simplified schematic flow diagrams of these types of system are shown in Figs. 9 and 10. The systems consist essentially of five major parts

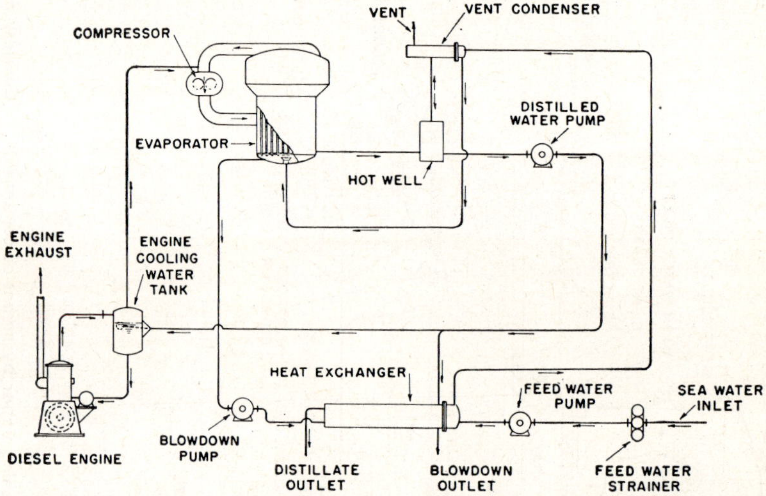

Fig. 9. Low-pressure compression distilling plant. (*Courtesy of Cleaver-Brooks Company.*)

(Fig. 10): (1) a source of mechanical power, which may be a steam turbine, an internal combustion engine, or an electric motor, (2) a steam compressor capable of compressing the required volume of steam (approximately 225 cu ft for each gallon of water produced) to a pressure of 3 to 5 psig, (3) an evaporator for boiling raw water and condensing the compressed steam, (4) a heat exchanger to heat the incoming raw-water feed by means of the distillate and the waste brine from the evaporator, and (5) a source of heat for starting the unit, which is ordinarily either waste heat from the engine jackets, exhaust steam, or electric heaters.

Auxiliaries that are required are a feed pump to supply and meter the raw water to the unit, a condensate pump, if the pure water is to be delivered at a distance or to an elevated tank, a steam-pressure gauge on the compressor discharge, a manometer on the evaporator, a feed-

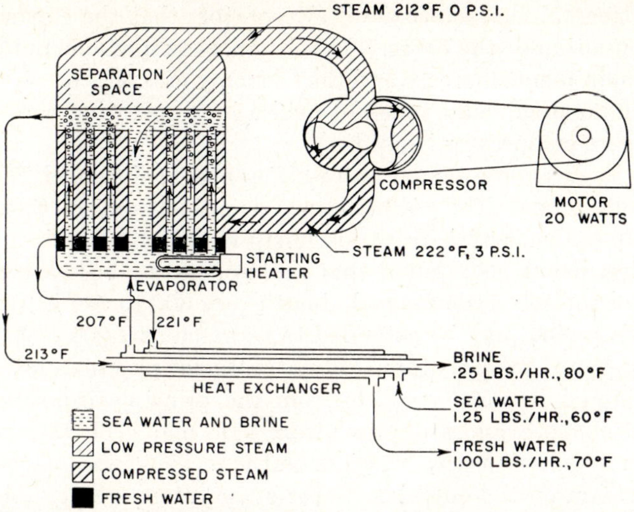

Fig. 10. Diagrammatic representation of compression distillation as applied to distillation of sea water. (*Courtesy of Badger Manufacturing Company.*)

Fig. 11. Vapor compression still. (*Courtesy of Badger Manufacturing Company.*)

regulating valve, a compressor steam bypass valve, drain and vent cocks, and equipment for cleaning the evaporator surfaces of scale. A typical installation is shown in Fig. 11.

It is a general characteristic of evaporators that the capacity falls off as scale builds up on the tubes. In the vapor-compression units, although power requirements increase and the economy falls off, the capacity of the units remains practically constant over a long operating period if a positive pressure compressor is used.

Owing to the lower temperature differences maintained in the vapor-compression process, the normal scale formation from sea water is a relatively soft sludge, which responds fairly readily to chemical treatment. In practice, it has been found that units operating on sea water require cleaning of the soft scale formed about every 300 to 700 hr. With fresh water, this period may be extended to as much as 2,000 or 3,000 hr.

Foreign-type Evaporators. Early foreign practice in evaporator design differed in some respects from the principles upon which most American-made evaporators function. In many of the British and European installations, the water is on the inside of the evaporator tubes and the steam on the outside. Some evaporators are designed, also, on the thermocompressor principle.

ADAPTABILITY OF DIFFERENT TYPES OF EVAPORATORS

The cost of operation of evaporators depends primarily upon the source of the heat utilized. The choice as to the type should be dictated by existing conditions in reference to the particular heat balance of the steam station.

Two specific conditions generally control: first, available means for utilizing the latent heat from the vapor discharged from the evaporator (from the last effect, where the installation is a multiple-effect evaporator); and, second, the possibility of utilization of steam which has performed a certain amount of useful work prior to discharge into the evaporator coils.

In considering small- and moderate-sized industrial plants in which the heat balance is simplest, evaporators may be installed for operation by either a vacuum system, low-heat-level system, or high-heat-level system. The distinction between these systems lies in the thermal level at which the evaporator operates. This is defined by the pressure at which the steam is supplied to the evaporators and the corresponding temperature at which the evaporator vapor is generated and condensed.

In a typical vacuum system, the raw-water supply for boiler-feed makeup is discharged directly to the evaporator, which receives steam at low pressure or under vacuum from an auxiliary exhaust or from a low-pressure stage of the turbine. The evaporator vapor is generated at less than atmospheric pressure and is condensed by the condensate from the main turbine in an evaporator condenser. The condensed vapor from the evaporator is absorbed by the turbine condensate and becomes part of the boiler-feedwater supply.

In a low-heat-level system, the raw evaporator feedwater may be preheated in an open heater using exhaust steam from the auxiliaries as the heating medium. The evaporator steam supply may be bled from one of the low-pressure stages of the turbine. The vapor from the evaporator is condensed in the boiler-feedwater open heater, thus eliminating the need for an evaporator condenser.

A high-heat-level system is one in which the evaporator feedwater is preheated as in the low-heat-level system, but the evaporator receives steam at high pressure. The high-temperature evaporator vapor is condensed in an evaporator condenser, which also serves as a high-pressure boiler-feedwater heater.

One important advantage of a high-heat-level system is that it leaves the boiler-feed open heater or deaerating heater free to condense more of the auxiliary exhaust steam, because the evaporator heat returns can be absorbed at the higher heat level. This type of system is adapted to plants in which there is a considerable differential between boiler temperature and the temperature at which the feedwater enters the boiler, such as a plant in which there is no economizer.

The determining factor in the choice of a high-heat-level evaporator is the layout of a steam station that will permit the economical use of the vapor from the last effect to raise the feedwater above 212°F by using the feedwater which has a temperature of 212°F or less for cooling water in the evaporator condenser. Since high-heat-level evaporators discharge the distillate from the apparatus at a temperature above 212°F, the operating conditions of the station should permit the utilization of this heat without loss of energy.

In the absence of economizers and high-pressure bleeder heating, these installations will deliver the feedwater at a higher temperature than could otherwise be accomplished, and this may be done without a greater loss than direct heating of the condensate by steam at boiler temperatures. Evaporators of this type are smaller and have relatively smaller steam passages than low-heat-level units, on account of the dense vapor of the steam produced. The coils and shells must be constructed to withstand high pressure, since the steam is admitted to the unit at full boiler pressure or at partly reduced pressure.

The high-heat-level plants are adaptable primarily to those stations where there is an insufficient amount of exhaust steam to operate the evaporator economically under varying load conditions and where it is inadvisable to use steam bled from the main turbine. These evaporators may be operated economically where economizers are used, provided the station auxiliaries are largely electrically driven. In this case, there may be an insufficient amount of exhaust steam for feedwater heating so that the high temperature of the vapor from the evaporator may be utilized to heat the feedwater. All the latent heat is absorbed, but the temperature of the water is raised to less than 212°F, so that there is no loss in passing through the economizers.

Where this condition exists, the adjustment of vapor available for feedwater heating to volume of make-up water to be heated may be regulated by the number of effects employed. Since the amount of vapor discharged from a single-effect is about twice as great as from a double-effect evaporator, increasing the effects will reduce the volume of vapor produced from the last effect and thus control the amount of vapor to be discharged from the equipment. This is an important factor in deciding the number of effects to be employed in some plants. It is necessary to balance the cost of the additional effects against the thermal efficiencies gained by this arrangement. The economic consideration of the number of effects to be employed is determined to a great extent by balancing the fixed and operating cost of the increasing number of effects against the utilization of the heat released in condensers in the final effect, under various load conditions of the station.

In some of the larger industrial plants and central power stations, evaporators often are arranged in an interstage manner so that the evaporator with its condenser is located between two consecutive stage heaters in the boiler-feedwater system.

The primary consideration in the design of an evaporator layout is to eliminate as far as possible heat losses due to the operation of the unit. Where considerable exhaust steam is available and the vapor may be employed economically to heat the boiler feedwater, a low-heat-level evaporator may be used. The unit may be a single-effect evaporator, provided all the heat from the vapor can be absorbed in the boiler-feedwater heater by the condensate from the main units. By this arrangement, all the heat from exhaust steam, except losses due to radiation and blowdown, will be absorbed and no losses will occur.

Should the vapor from a low-heat-level single-effect unit be greater than can be absorbed for useful work at various loads on the station, a multiple-effect evaporator is warranted. Since the cost of an evaporator installation is almost in direct proportion to the number of effects, it is desirable to use as few effects as will produce the desired amount of make-up water under full and low loads on the station. Further, the installation should meet these requirements at a minimum investment and operation cost.

OPERATION OF EVAPORATORS

As a rule, much more sludge will form in low-pressure evaporators than in high-pressure units when untreated raw water is used. The solids, particularly the carbonates of lime, will deposit as mud before they have an opportunity to bake upon the heated surfaces. Scale that forms upon the tubes of high-pressure evaporators is usually hard and dense and adheres tenaciously to the metal surfaces. Removal is effected by sudden changing of temperature of the metal of the coils. This cracks off the scale by the expansion or contraction of the tubes. Although this is effective in a general way, some scale will adhere to tubes after cracking down.

When the water level in an evaporator fluctuates and the coils are alternately under water and then exposed to the hot vapors in the shell, the scale on these coils is likely to be very hard. The cracking down of scale is more effective with hard, brittle scale than with soft scale and may be accomplished to the best advantage when the scale has attained a certain definite thickness. Where evaporation takes place at low temperatures, as it does in the last effects of multiple-effect evaporators, the scale formed is softer than in the hotter effects. Thus, it is necessary at times to remove the scale from the coils by mechanical or chemical means. It is possible to remove scale effectively from the tubes of film-type evaporators by shutting down the operation and putting the shell under a partial vacuum. When this is done, the moisture in the scale next to the tube flashes into steam, exerting sufficient pressure to blow the scale away from the metal surfaces.

Priming and foaming in evaporators are the results of the same conditions that cause these phenomena in boilers. It has been fairly well established that suspended solids have an important part in causing water to foam and prime. Therefore, it is essential to limit the concentration of both suspended and soluble solids in the water in the shell of the evaporator to a point sufficiently low that the tendency for the water to prime is eliminated. This is accomplished by blowing down the evaporator from time to time. The amount of blowdown required will depend upon the character of the raw water used in respect to the rate of concentration of soluble salts and the amount and character of the suspended solids.

Water containing sodium salts in conjunction with sewage and industrial wastes is more likely to prime at a given concentration of solids than water containing the same amount of solids in the absence of these constituents. On the average, under ordinary conditions, from 3 to 10 per cent of the evaporator feed must be blown down, although a much higher percentage of blowdown may be necessary. The failure to operate a unit satisfactorily with respect to blowing down sooner or later results in relatively large amounts of solids being carried over mechanically with the vapor and contaminating the condensate.

Many novel features in evaporator design have been employed to reduce the entrainment of solids in the moisture with the steam.

Difficulty on the water side of the evaporator coils is due to corrosion or scale formation, while on the steam side the heat transfer of the tubes may be reduced greatly by blanketing with noncondensible gases. A layer of air or other gas due to release from the vapor when it is condensed may form over the surface of the coils on the steam side. In installations where this action occurs, provision should be made to remove the gases. If this is not done, the effective heating surface may become materially reduced. Arrangement for venting the tubes to correct this difficulty is provided in most evaporators.

Corrosion of evaporator coils becomes a serious operating difficulty with some waters. In surface waters and many well supplies, the raw water contains dissolved oxygen, which is aggressive to ferrous materials. Water high in carbonate hardness may cause fairly rapid corrosion owing to the release of relatively large amounts of carbonic acid when the water is heated. The carbon dioxide generated may be carried over with the steam into the feedwater cycle, thus causing corrosion in the system. Corrosion may also occur because of contamination with sewage wastes, mineral or organic acids, or magnesium chloride. Where waters of this character are encountered, pretreatment of the evaporator feedwater is desirable. Many types of conditioning processes are available for such service.

PRETREATMENT OF RAW WATER TO EVAPORATORS

Raw water containing extremely high concentrations of solids can be fed directly to evaporators, but performance and maintenance can be improved by pretreatment of the evaporator feed.

As discussed in previous chapters, the hardness of water may be partially or totally removed by softening with chemicals or by ion-exchange processes. Both of these methods are applicable to the treatment of evaporator feedwater.

Lime-soda or zeolite softeners for use in conjunction with evaporators are warranted only when the expense resulting from the additional equipment and the operation of the equipment ensures greater economy in the operation of the evaporator, permits increased production from the unit, or results in improved quality of feedwater to the boiler. By the use of lime-soda softeners it is possible to reduce the hardness of the evaporator feed to about 3 grains per gal, if a cold-process treatment is used, or 1 grain if a hot process is employed. The use of zeolite softening will reduce the hardness of the incoming feedwater to less than 2 ppm. In some cases, presoftening by hydrogen or sodium zeolites can be justified, since this will reduce the carbon dioxide in the vapor and minimize corrosion of the condensate collector and return system.

Pretreatment by sodium zeolite softeners will eliminate practically all scale-forming solids from the evaporator. The total soluble solids in the zeolite-softened water will be somewhat greater after treatment than before. However, in recent years, it has been found desirable in many installations so to soften the make-up water to the evaporator in order to reduce outage period and increase the efficiency. Various combinations of treatments of evaporator feedwater are shown in Fig. 12. The need

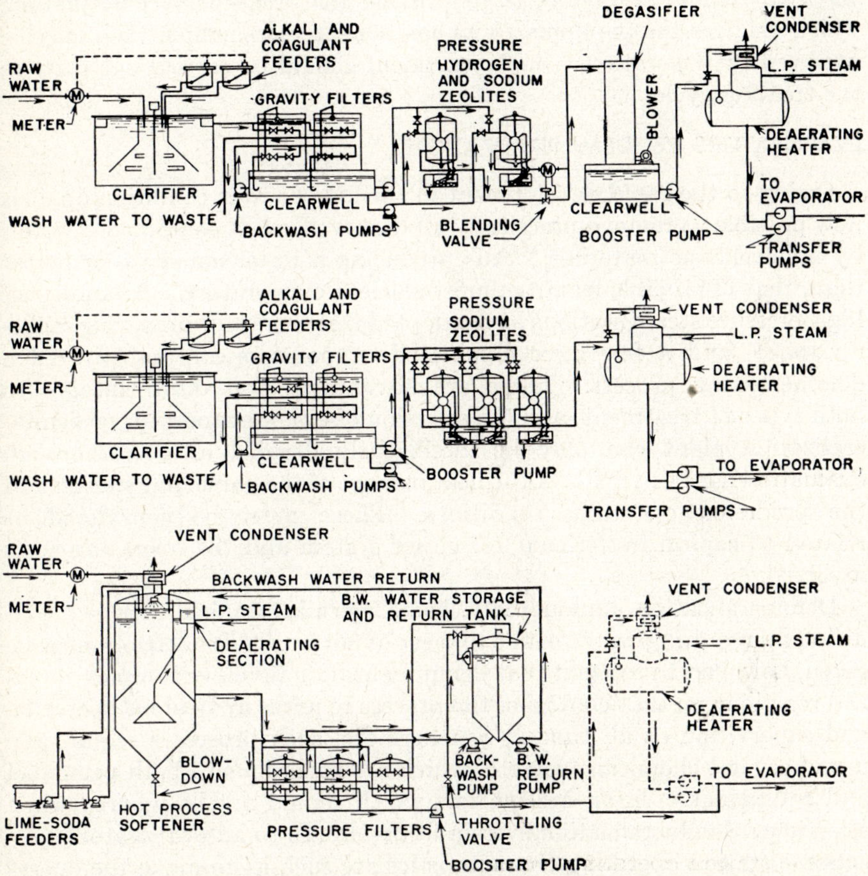

Fig. 12. Schemes for pretreatment of make-up water to evaporators.

and selection of any of these treatments will depend on the resulting economy effected.

DEAERATION OF FEEDWATER TO EVAPORATOR

Deaeration of feedwater to evaporators is desirable for reducing corrosion both in the evaporator and in condensate lines. In this way, oxygen can be eliminated completely, and free carbon dioxide can be removed to a low amount. Carbon dioxide carried over with vapor is dissolved when the vapor is condensed, and, because the condensed liquid contains few solids to buffer it, the pH is low, and the water may be aggressive.

When ammonia is contained in the evaporator feedwater, it cannot be completely removed by the deaerator; its removal must be effected at a early stage in the treatment. It has been found that one of the simplest ways to remove ammonia is to provide for break-point chlorination, which oxidizes the ammonia to its component gases, which then may be removed by deaeration. Such treatment should be carried out early in the treating cycle.

EVAPORATORS VS. DEMINERALIZATION

Owing to the rapid development in the art of water conditioning, it is now possible to remove practically all dissolved solids contained in water by ion-exchange materials,* thus producing a water equal to, or better than, that obtainable from the most efficient evaporators. A number of high-pressure steam stations have adopted this means for preparing makeup water, and a few have eliminated evaporators and substituted a demineralizing process. There are advantages and disadvantages to both types of treatment, and the selection of either evaporator or demineralization plant can only be intelligently decided after a thorough evaluation has been made of the heat balance of the particular station and the predicted operating conditions. These must be considered as related to capital investment for either system and the operating costs involved.

Demineralization, undoubtedly, has a place in the field of water conditioning requiring water of the highest quality. Consideration must be given, however, to the fact that demineralization involves complex chemical reactions, so that control of the process requires meticulous operation and supervision at all times. Furthermore, such processes require the use of several chemicals, some of which are hazardous to both personnel and equipment. These are matters which should not be ignored when selecting a demineralization system in preference to an evaporator.

Some station operating characteristics are such as to make the operation of evaporators uneconomical during low-load conditions, and in such cases, demineralizing plants have special merit. It is reasonable to predict that the installation of water-conditioning systems for the production of high-quality water by means of ion-exchange material will increase and replace evaporators in many plants. It is the author's considered opinion that such selection should be based on the merit of each type of treatment system. All factors favoring or penalizing each system should be evaluated in terms of their over-all effect on efficiency and

* This process is discussed in Chap. 8.

economy and not upon sales promotion advanced by vendors of either type of apparatus.

When properly chosen to meet the requirements of the specific power plant layout and so designed that they do not distort the heat balance, evaporators generally will reduce or eliminate feedwater difficulties, in so far as the boilers are affected. It should be borne in mind, however, that this method of water treatment has its limitation and that there are certain advantages and disadvantages in the use of such equipment.

Bibliography

1. Evaporators in the Stationary Power Plant, *Power*, Sept. 19, 1922.
2. Werner, Heinrichs: Distillation Apparatus for Producing Make-up Feed Water, *Arch. Wärmewirtsch.*, Vol. 3, 1922.
3. Distilled Water for Boiler Feed at River Rouge Plant, *Power*, Vol. 54, 1921.
4. Distilled Water for Boiler Feed, *Power Plant Eng.*, Vol. 26, 1922.
5. Klein, R.: Feed Water Requirements for High Pressure Boilers, *Wärme*, Vol. 48, Mar. 20, 1925.
6. Speller, F. N.: "Corrosion: Causes and Prevention," McGraw-Hill Book Company, Inc., New York, 1926.
7. Croft, Terrell: "Practical Heat," McGraw-Hill Book Company, Inc., New York, 1934.
8. Royds, R.: "Heat Transmission in Boilers, Condensers and Evaporators," D. Van Nostrand Company, Inc., New York, 1921.
9. Badger, W. L.: "Heat Transfer and Evaporation," Chemical Catalog Company, Inc., New York, 1926.
10. Price, J.: "Raw Water Distilling Plants for Producing Distilled Boiler Make-up Water," Griscom-Russell Company, 1922.
11. How a Lillie Evaporator Operates, *Power*, Nov. 21, 1922.
12. The Flash Evaporator, *Power*, Nov. 28, 1922.
13. Flash Evaporator Furnishes Pure Make-up Water to Boiler, *Power Plant Eng.*, Vol. 26, 1922.
14. Chapin, E. H.: Adapting Evaporators to Heat Balance, *Power Plant Eng.*, Vol. 28, June 1, 1924.
15. Bonilla, Charles F.: Design of Multiple-effect Evaporators for Minimum Area or Minimum Cost, *Trans. Am. Inst. Chem. Engrs.*, **41**: 529 (Oct. 25, 1945).
16. Latham, Allen, Jr.: Compression Distillation, *Mech. Eng.*, **68** (3): 221 (1946).

10 DEAERATION OF FEEDWATER

Certain gases, principally oxygen, O_2, and carbon dioxide, CO_2, dissolved in water accelerate its corrosive properties. The rate of corrosion is approximately in proportion to the quantity of gas dissolved. Moreover, higher temperatures accelerate the rate of attack. These rates are

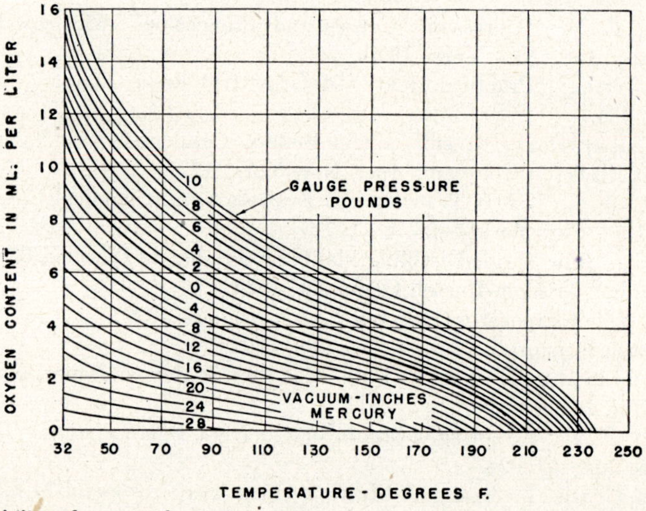

Fig. 1. Solubility of oxygen from air in water at various temperatures and total pressures. (*After Cochrane Corporation.*)

further influenced by changes in pH of the water which either assist in forming protective films on the metal surfaces or destroy such films, thereby exposing the surface to corrosive attack.*

THEORY OF DEGASIFICATION

Water dissolves oxygen, nitrogen, carbon dioxide, and other gases directly from the atmosphere up to the point of saturation at a specific pressure and temperature. The solution of air in water occurs in accord-

* The mechanism of corrosion is discussed in Chap. 16.

ance with Henry's and Dalton's laws, namely, that the solubility of a gas in a liquid is proportional to the partial pressure of the gas and inversely proportional to the temperature (Fig. 1). Each gas or vapor in a mixture exerts its specific pressure and is not influenced by the pressures exerted by the other gases.

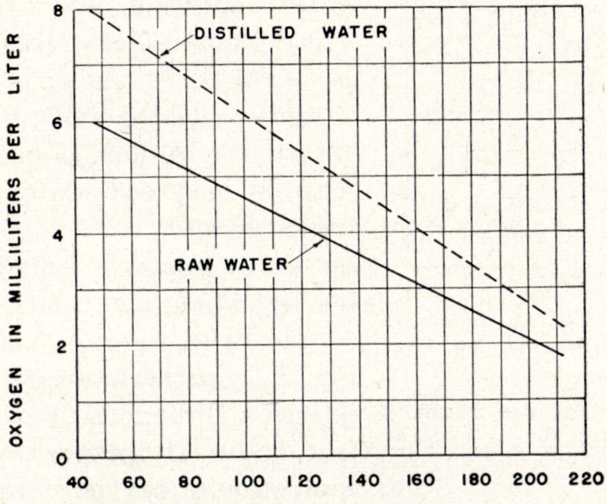

Fig. 2. Curves showing amount of oxygen present in raw and distilled waters at given temperatures.

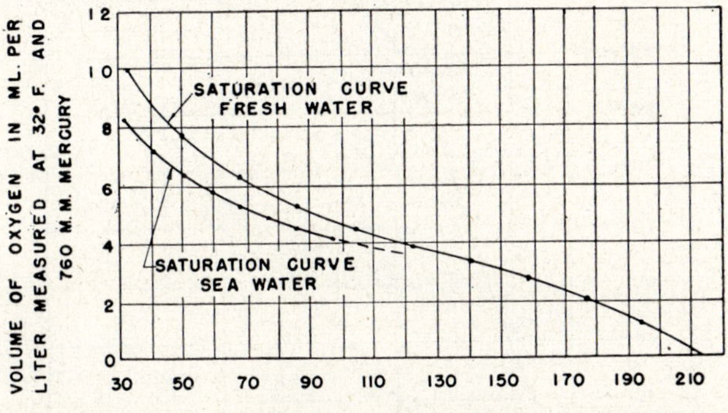

Fig. 3. Curves showing the absorption of oxygen in fresh and sea waters at various temperatures. (*After Speller.*)

The amount of air dissolved by the water is influenced by other constituents dissolved in it at that time. Distilled water absorbs more oxygen at a specific temperature and pressure than does water containing dissolved materials (Fig. 2). Hard water absorbs less oxygen than soft water. Similarly, sea water dissolves a smaller amount of oxygen at a given temperature and pressure than does fresh water under the same conditions (Fig. 3).

The rate at which gases dissolve in water is governed by the area of water exposed to the specific gas. The greater the exposed area of water in contact with oxygen, the greater will be the absorption in a given period of time; and the rate is altered by temperature, agitation, and other conditions. The volume of gas absorbed in a given period of time is shown in Fig. 4. Curve *A* represents the amount of oxygen absorbed for various periods up to 70 hr with an exposed water surface seven times greater than the surface represented in Curve *B*. A similar condition results when air comes into intimate contact with water by violent agitation, Fig. 5. Furthermore, the rate of oxygen diffusion through a film on a metal surface is directly proportional to the concentration of oxygen in the water.

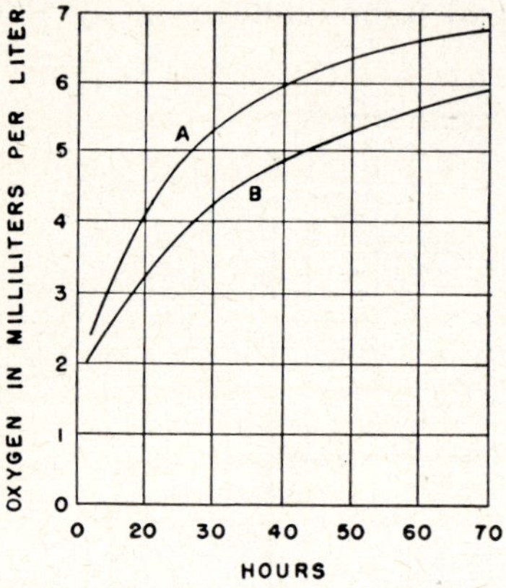

Fig. 4. Influence of surface on solubility. (*After Paris.*)

The problem of removing carbon dioxide from water is similar to that of removing oxygen but is considerably more complex and difficult.

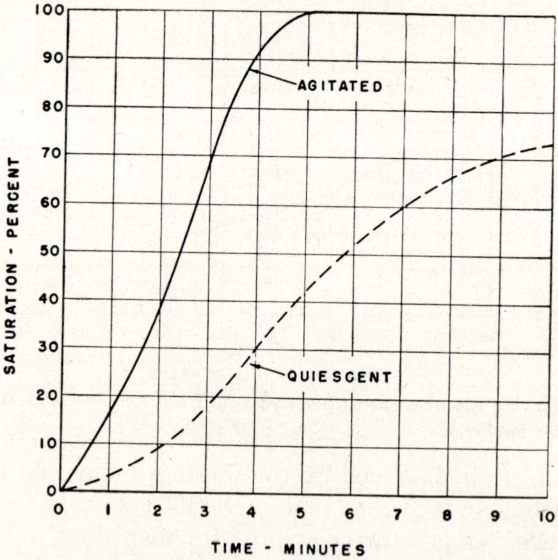

Fig. 5. Absorption of oxygen by water with and without agitation.

Carbon dioxide can exist in three forms in water, namely, as carbonic acid, H_2CO_3, as bicarbonates, HCO_3, and as normal carbonates, CO_3.

Carbonic acid is readily changed to free carbon dioxide and water ($H_2CO_3 \rightleftharpoons H_2O + CO_2$). In this form, it can be easily expelled by increasing the temperature of the water. The carbonates and bicarbonates, when chemically combined to form bases, however, are not readily removed by heat and are expelled as free carbon dioxide only to the extent to which they are dissociated. The degree of dissociation, in turn, depends upon the pH value of the water, the vapor pressure above the surface of the

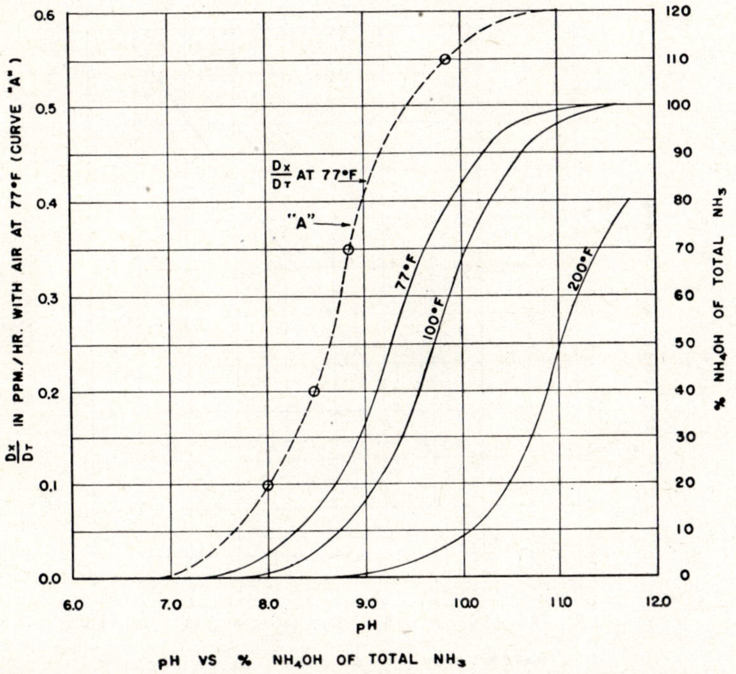

Fig. 6. The distribution of ammonia at various pH values and temperatures. *(After Kittredge.)*

liquid, and a number of other factors. For these reasons and because of its higher solubility, carbon dioxide gas cannot be reduced to as low a value as can oxygen when water is passed through any deaerating apparatus.

Gaseous ammonia, NH_3, dissolves readily in water, forming NH_4OH ($NH_3 + H_2O$), a much more soluble compound than either carbon dioxide or oxygen. Ammonium hydroxide is a strong base and does not readily attack ferrous metal except at very high concentrations. However, ammonium salts are aggressive to copper and copper alloys and, therefore, objectionable in feedwater and circulating cooling water when present in even relatively small amounts.

The degree of dissociation of salts of ammonia is influenced by a number of conditions, and they react differently than do carbonates and bicarbonates. The solubility of ammonia is greatly affected by the pH value of the water. Kittredge,[1] in discussing this problem, stated:

In the removal of ammonia from boiler feedwater, or any other gas that ionizes to an appreciable extent, it is necessary to take into consideration the pH value at which the equipment will be operated.

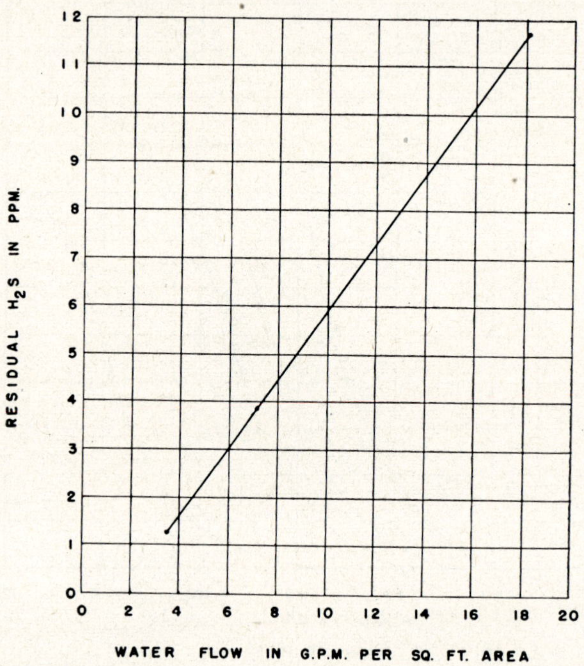

Fig. 7. Hydrogen sulfide removal by forced-draft aeration.

Figure 6 shows three curves at 77°F, 100°F and 200°F respectively, indicating the ratio of ammonia hydroxide to total ammonia in solution plotted against pH at these temperatures. The curves are calculated. Also plotted on the same chart is a test curve "A" indicating the relative rate of ammonia removal at different pH values as determined from a glass bead tower operating at 77°F.

Each individual test point indicated represents a separate test made with a buffered solution. The conclusion is indicated that the instantaneous rate of ammonia removal in a packed or tray type of equipment is reasonably proportional to the fraction of the total ammonia in the form of ammonia hydroxide—that is to say not ionized.

The author has observed similar conditions in studying the performance of many deaerating heaters and their ability to remove ammonia.

Sulfur dioxide, hydrogen sulfide, and other gases are present in some waters. These compounds are corrosive to nickel and copper alloys. The rate of attack of sulfur compounds is accelerated when carbon dioxide and oxygen are also present.

Where appreciable quantities of such compounds are present in water, it is preferable to remove them by aeration, either plain or with forced- or induced-draft fans, before the water is admitted to the boiler-feedwater

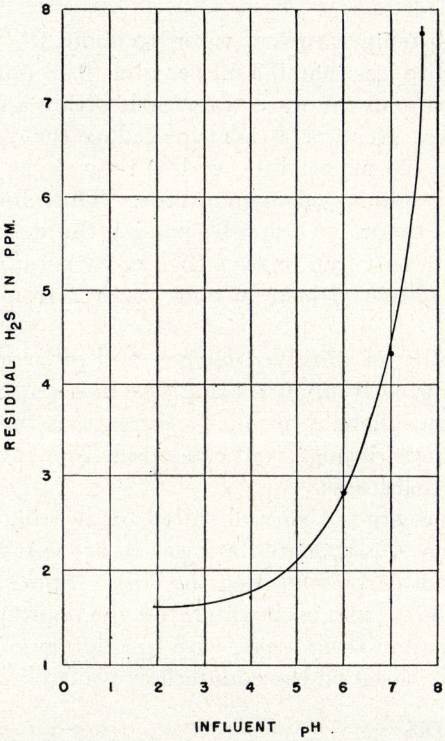

Fig. 8. Effect of pH on reduction of hydrogen sulfide through forced-draft aerators.

cycle. The author found that the efficiency of the aerating equipment for removal of some gases is greatly improved[2] by maintaining (1) low pH values in the water and (2) a sufficient air flow through the aerating equipment. These test data are shown in Figs. 7 and 8.

DEGASIFICATION PROCESSES

The removal of gases from water, as practiced in steam stations, is effected by mechanical and thermal treatment, by the addition of chemicals, or by combinations of these processes. The most widely practiced method is to heat the water in either open or contact heaters or in the more efficient types of heaters known as "deaerators."

All feedwater heaters, either simple open heaters or deaerating heaters, are distinguished by the manner in which the steam and water are routed through the units. The three designs used are (1) full counterflow, (2) cross flow–counterflow, and (3) parallel flow. Heaters are described, also, as tray- or spray-type units to identify them with respect to the means provided for heating and for gas removal.

OPEN HEATERS

By elevating the temperature of water to about 212°F, the oxygen in solution is reduced to less than 0.3 ml per liter (0.43 ppm). At the same time, the temperature of the water will reach within a few degrees of the steam temperature. Heaters of this type reduce the dissolved oxygen of the water to about .03 ml per liter, or less than .5 per cent of the initial oxygen, but are not sold on a firm guarantee. These heaters are designed to effect heating of the water either by passing the water downward over trays contained in a cast-iron or steel shell or by spraying the water into the heater. The source of heat in both cases is from steam at low or medium pressures.

Since the solubility of dissolved oxygen and other gases contained in the water reduces greatly upon heating, the gases are released and discharged to the atmosphere. In many installations the water vapor and the released gases pass through vent condensers to recover the condensate and minimize thermal losses.

Tray-type heaters are not so well suited for heating water containing high concentrations of carbonate hardness as are spray heaters, because precipitated calcium carbonates clog the tray. Spray heaters may also be clogged by deposits from such water, but the deposits can be removed with less difficulty than from trays, with less outage of equipment.

A large tray unit based on the counterflow principle is shown in Fig. 9.

DEAERATING HEATERS

Complete deaerating heaters, or deaerators as they are generally called, are similar to open heaters but are so designed as to insure practically complete removal of oxygen and the reduction of carbon dioxide and other gases contained in the feedwater to very low values. The greater efficiency of such units is obtained by providing sufficient tray areas or other arrangements, sufficient time of contact for release of the gases, and adequate means for removing the gases from the units without reentrainment. Because of these improvements it is now possible to purchase deaerators guaranteed to produce water of zero oxygen* content.

* The term "zero oxygen" indicates that the concentration of this gas does not exceed 0.005 ml per liter, which is the tolerance of accuracy of existing chemical testing procedures.

The deaerator in Fig. 10 is designed with a primary heating-tray section and a second bank of trays to insure complete removal of oxygen. The steam supply reaches the heating zone through a perforated steam baffle and flows down through all the tray sections, discharging finally to the vent condenser. The water enters the unit through perforated

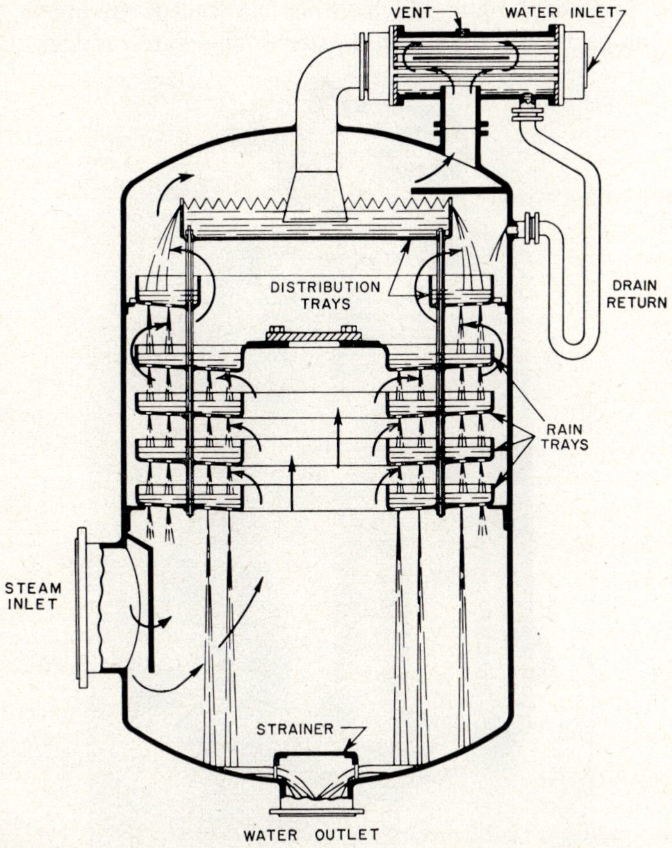

Fig. 9. Tray-type direct-contact heater with vent condenser utilizing the principle of reverse contact flow. *(Courtesy of Foster Wheeler Corporation.)*

manifolds. The lower portion of the heater is partitioned off for storage of deaerated water.

SPRAY DEAERATORS

A type of spray deaerator is shown in Fig. 11. The water enters the heater through the vent condenser to the spray valves, constructed of stainless steel or other corrosion-resistant alloy metals. These valves are spring-loaded to effect atomization of the water, which is then injected in thin sheets or droplets into the upper preheater compartment.

The water is heated to within 2 or 3°F of the steam temperature by the time it falls to the bottom of the preheater section, in which most of the oxygen and free carbon dioxide are removed. The preheated and partially degassed water then flows down through a narrow distributor to the steam scrubber. Additional steam enters the scrubber section through a perforated plate which forms one wall of the steam scrubber and extends partially around the heater. The water and steam mix at

Fig. 10. **Tray heater providing design for primary and secondary deaeration.** ***(Courtesy of Cochrane Corporation.)***

this point and rise through a concentric section in the heater to where the water overflows a weir into the storage compartment. In passing through a constricting port in this section, the water is scrubbed by the steam to remove traces of oxygen or carbon dioxide. Little condensation of steam occurs in this region, and the steam and noncondensible gases flow to the preheater, where the steam is condensed in heating the water from the spray valves. Noncondensible gases, collected in the preheater section, are vented to the atmosphere through the vent condenser. Excess steam reaching the vent condenser is recondensed and returned to the heater.

The vent condenser is a conventional type, and the tubes used are constructed of admiralty metal, arsenical copper, or other corrosion-resistant material, depending upon the service requirements. Both the vent condenser and the spray-valve water box are designed so that they can be easily opened for inspection and repairs.

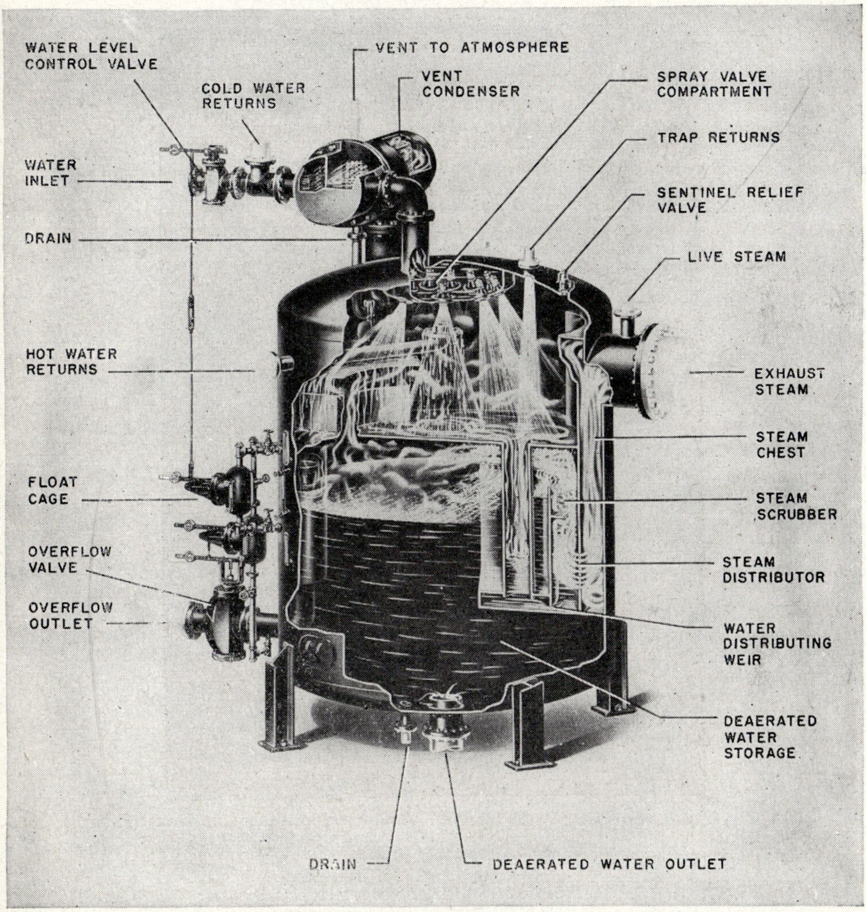

Fig. 11. Spray-type deaerator providing primary heating and secondary deaeration. (*Courtesy of The Permutit Company.*)

SPRAY HEATER WITHOUT VENT CONDENSER

Spray-type heaters have been developed recently in which external vent condensers have been eliminated, thus avoiding many operating and maintenance problems inherent in various types of equipment. In place of the vent condenser, an inlet spray unit with a vent collecting hood is provided in the upper section of the heater. The vapor is cooled by the incoming water and the gases pass through the water spray, thereby providing the equivalent of a direct-contact heat exchanger. For all practical purposes, this accomplishes the same results as the conventional vent condenser, and improved operation in some instances has been effected over the conventional design using vent condensers. A phantom cross section of a deaerator of this type is shown in Fig. 12.

Fig. 12. Spray-type deaerator with internal vent condenser. (*Courtesy of Elliott Company.*)

The water enters the heater through the inlet assembly and is sprayed across the opening of the vent collecting hood into the heating compartment. As the water travels across the spray assembly, it strikes against the inner head baffling and falls to the tray section of the heater, where the water is heated. This heat releases the gases, and they rise to the top of the heater and leave it through slots in the discharge vent.

HORIZONTAL DEAERATOR DESIGN

A horizontal heater in which the storage tank is an integral part of the design is shown in Fig. 13. It is designed with a cylindrical-shell storage compartment with the tray-type deaerating heaters located in the upper section of the shell. These are mounted either in the center or at the end of the storage chamber. The water flows through a vent condenser onto a distributing tray where it is filmed out so as to cascade from one tray to another. The trays have serrated edges to assure a definite amount of water flow from each tray. Steam is introduced from the side and flows around the trays, making contact with the film of water as it overflows each tray section. The units are constructed so that the tray area is enclosed. This protects the shell from contact with the corrosive gases released and permits nearly complete countercurrent flow of steam

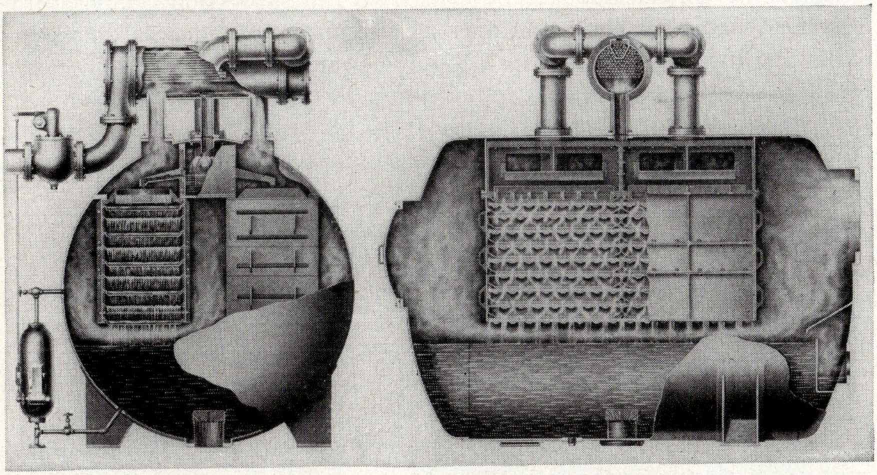

Fig. 13. Tray-type deaerator designed with storage compartment in bottom of heater. (Courtesy of Worthington Corporation.)

and water. The inlet water cools and condenses the excess steam, returning it to the deaerating section, and the noncondensible gases are discharged to atmosphere.

DEAERATING CONDENSERS

In many modern steam stations, the deaerating heaters have been eliminated from the cycle and the removal of noncondensible gases is effected in the hot well of the main condensers. Where the make-up is low and where secondary chemical deaeration is provided, this type of design has proved adequate. The omission of deaerating heaters from the steam-water cycle is a matter requiring special study in each case, and the decision to eliminate them should be governed by local station requirements.

A recent design providing a deaerating section in the hot well improves the efficiency of gas removal from the water in the hot well of the main condensers. In the latter equipment, condensate is collected in a trough having a staggered row of holes on each side. Double streams of condensate are sprayed from the trough through a curtain of steam. The source of steam is generally from the turbine exhaust. The steam is delivered to the hot well at full turbine exhaust pressure and temperature.

Economies can often be secured by this type of deaeration. This has been reported by Adams,[3] who indicated that the ultimate decision as to the most advantageous method of removing oxygen from feedwater may be a matter of individual judgment and arrangement of station facilities. The cycle using a deaerating heater shows slightly better heat economies because the condensate can be heated to full saturation temperature.

Deaeration in the condenser may require injection of deaerating chemicals into the feedwater to correct for contamination due to condenser leakage. In some systems, it is necessary to provide a small deaerating heater for make-up water. This is especially desirable if return drips are contaminated with carbon dioxide and ammonia, since corrosion of tubes may occur if such water is not corrected.

ACCESSORIES FOR MECHANICAL DEAERATORS

There are several accessories normally required for the satisfactory operation of deaerators. These appurtenances may be incorporated as an integral part of the deaerator or as auxiliary control devices.

Vent Condensers. Standard vent condensers are usually heat exchangers consisting of a shell and head with a removable bundle of U tubes or straight tubes expanded into the tube sheets. The incoming cold water circulates through the tubes, and the excess steam from the deaerator is condensed on the outside of the tubes. The U-tube construction is simple and eliminates strains and tube-sheet leaks caused by expansion and contraction. However, where heating of the raw water causes deposition of scale or suspended matter, it is preferable to use straight tubes rolled into tube sheets of the floating-head type of construction. This U-tube design permits the withdrawal of the entire tube bundle for cleaning or replacement of the tubes. The tube sheets are constructed normally of steel or muntz metal, or, where the water is corrosive, more resistant metals are required. Tube materials should be selected to insure maximum resistance to corrosion, thereby minimizing operating problems due to corroded or split tubes.

Water-level Control Valves. It is necessary to control the water level in the storage compartment of the deaerator. For this purpose, a float-operated type of valve is usually provided, which throttles the water flow to the deaerating section when the level in the storage compartment rises. When the load in the plant increases and the level falls, the raw-water valve opens. Small-type valves are usually directly float-operated; whereas, for larger installations, a diaphragm control valve may be operated by a pilot valve from air or water pressure. Where condensate is returned to the deaerator, it is often desirable to provide for free flow of the condensate, so that the float control valve operates only on the raw or treated make-up water.

Overflow Valves. It is necessary to protect against an emergency in which the amount of water delivered to the deaerator is in excess of the storage capacity of the unit. For this purpose, either an overflow trap or a loop seal may be used. When the pressures are low and the space is available, a loop seal is the simplest means of obtaining this protection. With higher pressures and a minimum of space available, overflow valves are provided which open on high water level in the storage section and discharge the excess water to waste or, preferably, to storage.

Relief Valves. The steam supply for deaerators is obtained, generally, from low-pressure sources to conserve economically the heat available from such steam. However, it is frequently necessary to use live steam in order to assure sufficient steam at all times. Where make-up steam is to be supplied from a high-pressure source, the pressure must be reduced to comply with the safe structural design of the equipment.

Relief valves similar to those used for other pressure vessels are also required. Various types of relief valve are available, including multiport valves in which several valve discs are installed in one multiport valve. These are adjusted by means of springs for a given pressure setting and have sufficient capacity to discharge excess steam to the atmosphere. Precautionary design features are specifically designated in the rules for pressure vessels in an American Society of Mechanical Engineers code.*

Oil Separators. Where the steam supply to the deaerator contains oil or grease, which occurs often in the exhaust from reciprocating pumps or engines, an oil separator must be included in the steam connection. If contamination is severe, more refined types of oil-removal appliances may be required.

Miscellaneous. All heaters, deaerators, and similar station equipment should obviously be equipped with gauge glasses, vacuum breakers, pressure gauges, and similar indicating or control devices. All gauge glasses should be protected by suitable guards to prevent injury to operators should the glass break.

EFFICIENCY OF DEAERATORS

It is difficult to define the efficiency of deaerators in normal terms of percentage of gas removed. Plain contact heaters are capable of removing a high percentage of gases from water, but the ultimate requirement

* "Rules for Construction of Unfired Pressure Vessels," 1950 ed., Sec. VIII, ASME Boiler Construction Code.

in a high-pressure station is that of as complete removal of all gases as possible. These critical requirements have stimulated the effort to improve heater efficiencies to the highest degree possible. Heater manufacturers can now furnish equipment guaranteeing practically complete elimination of oxygen, but absolute zero oxygen is not always possible. However, residual oxygen can be maintained in an amount of 0.005 ml per liter, with recorded values by refined testing methods as low as 0.001 or below.

CHEMICAL DEAERATORS

A type of deaerator was used at one time in this country to remove oxygen by oxidizing iron or steel. The unit, termed a "deactivator," consisted of a steel tank in which were a number of layers of steel laths or perforated sheets over which the water flowed. The Kestner deaerator, used to some extent in Europe, was based on a similar principle, although the design features were quite different.

EUROPEAN PRACTICE

Many different types of deaerators have been designed and operated in Europe for correcting the corrosive characteristics of feedwater. At least four different degassers were formerly used in Germany on a practical scale, and other designs have been employed in France and England. Among those which were known to function by mechanical separation of the gases are the Griescheim Elektron Lauta, the unit built by the Deutsch Sanitätswerke, and the Balcke deaerator. Of the chemical deaerators, the Kestner degasser and the degassers of Hulsmeyer and of Seiffert were best known at one time.

COLD-WATER VACUUM DEAERATION

Degasification of large volumes of cold water received little or no attention until about 30 years ago. The first large-scale system for such service was designed and installed at Coolgardie in Western Australia for the protection of a 30-in. steel pipeline 351½ miles in length. The pipeline was completed and put into service in 1902.[4]

The first large-capacity deaerator in this country was installed under the supervision of the author by the Freeport Sulphur Company at Port Sulphur, Louisiana,[5,6] for deaeration of Mississippi River water. This unit was built to deaerate approximately 4 million gal of water daily, which is pumped about 10 miles through a steel pipeline. The deaerator was placed in service in 1935 and has been in constant use since that time. A second unit was installed a few years ago at the same site. The design of the unit was less elaborate than that of the Coolgardie unit, since it was demonstrated from pilot plant data that equally satisfactory results could be secured by less complicated equipment.

A line drawing of the Freeport Sulphur Company deaerator is shown in Fig. 14. Cold water is delivered to the top of the deaerator and distributed over perforated plates. The water then falls vertically through a series of wooden-slat bundles to a storage reservoir at the bottom, in which the water level is regulated, and is delivered directly to the distribution line. Sodium sulfite is injected into the effluent from the deaerator to complete chemically the removal of dissolved oxygen. A vacuum line is connected at the top of the deaerator for the removal of the

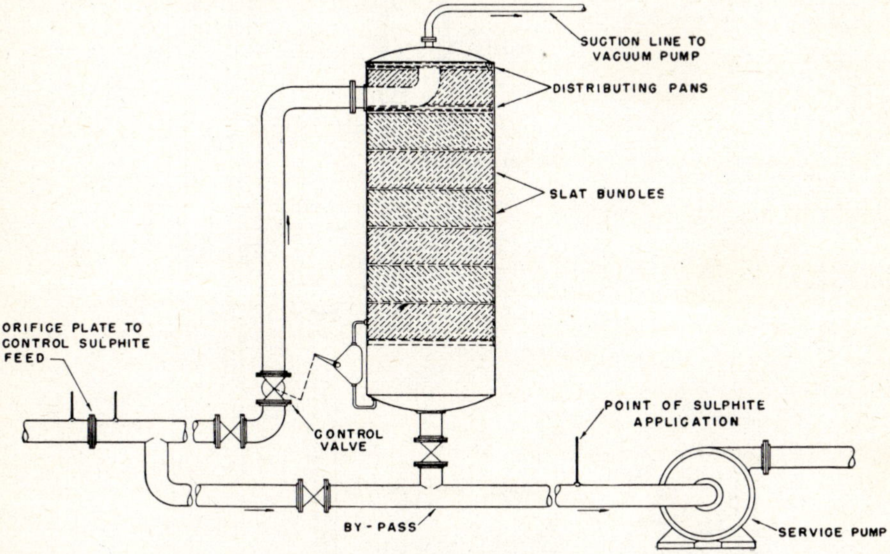

Fig. 14. Schematic arrangement of cold-water deaerator.

dissolved gases. It is of interest to note that, in this design, the water is drawn directly from the reservoir supplying the pipeline into the deaerator by the vacuum created. The vacuum is maintained at about 28 to 28½ in. Hg. This equipment has been in continuous and successful operation for more than 15 years.

Typical Design of Single-stage Unit. The selection and design of cold-water deaerators depend on the specified residual gas desired in the deaerated water, the physical characteristics of the collection and distribution system, and the economies involved for providing such facilities.

A typical unit for cold-water deaeration and subsequent shortage of the deaerated water is shown in Fig. 15. The cold water enters the deaerator near the top and is distributed uniformly over the cross-sectional area before falling vertically through a series of slat bundles. This arrangement facilitates diffusion of the gases by formation of thin water films. Gas vents are provided in the distributing tray to prevent interference with the removal of gas from the unit. Storage has been provided in the bottom of the unit with overlapping gauge glasses to indicate the water level. The vapor line leaving the top of the unit is baffled to minimize

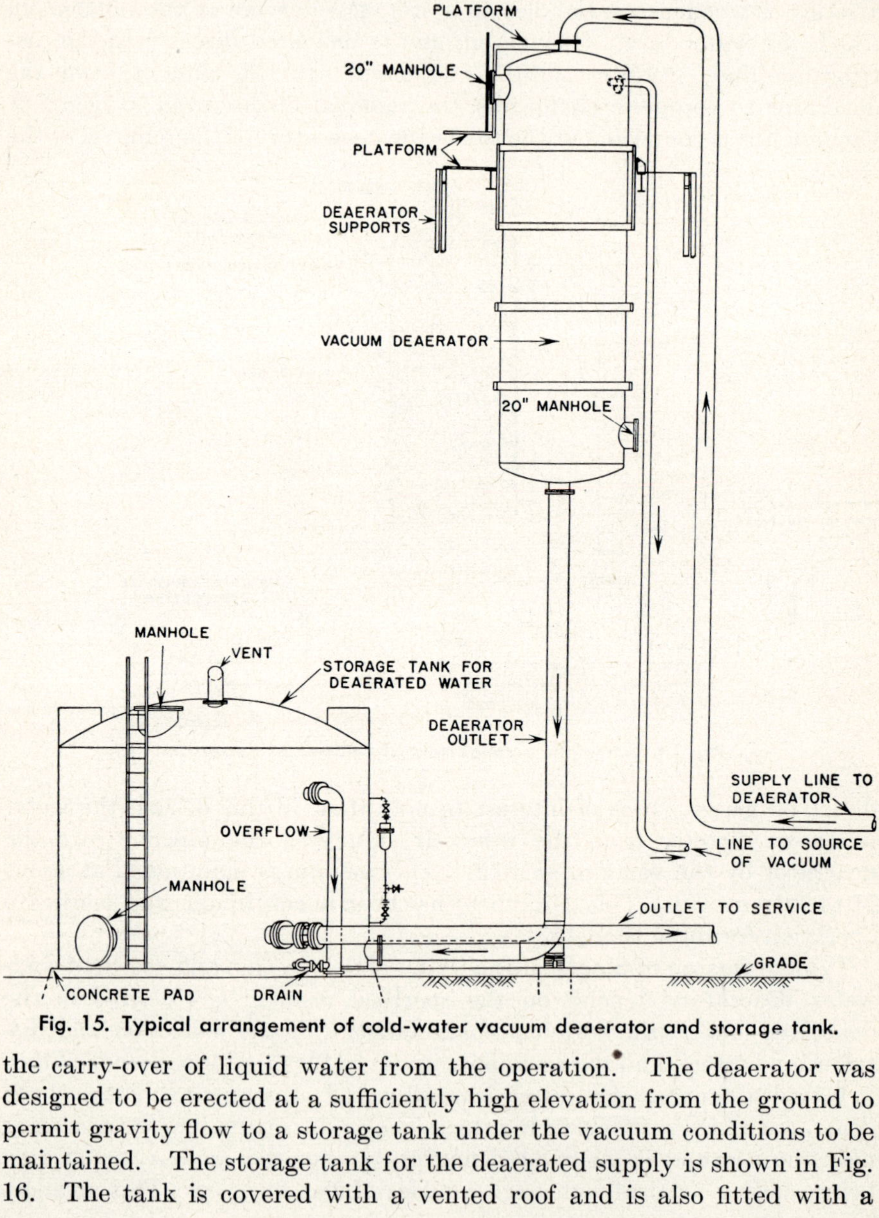

Fig. 15. Typical arrangement of cold-water vacuum deaerator and storage tank.

the carry-over of liquid water from the operation. The deaerator was designed to be erected at a sufficiently high elevation from the ground to permit gravity flow to a storage tank under the vacuum conditions to be maintained. The storage tank for the deaerated supply is shown in Fig. 16. The tank is covered with a vented roof and is also fitted with a

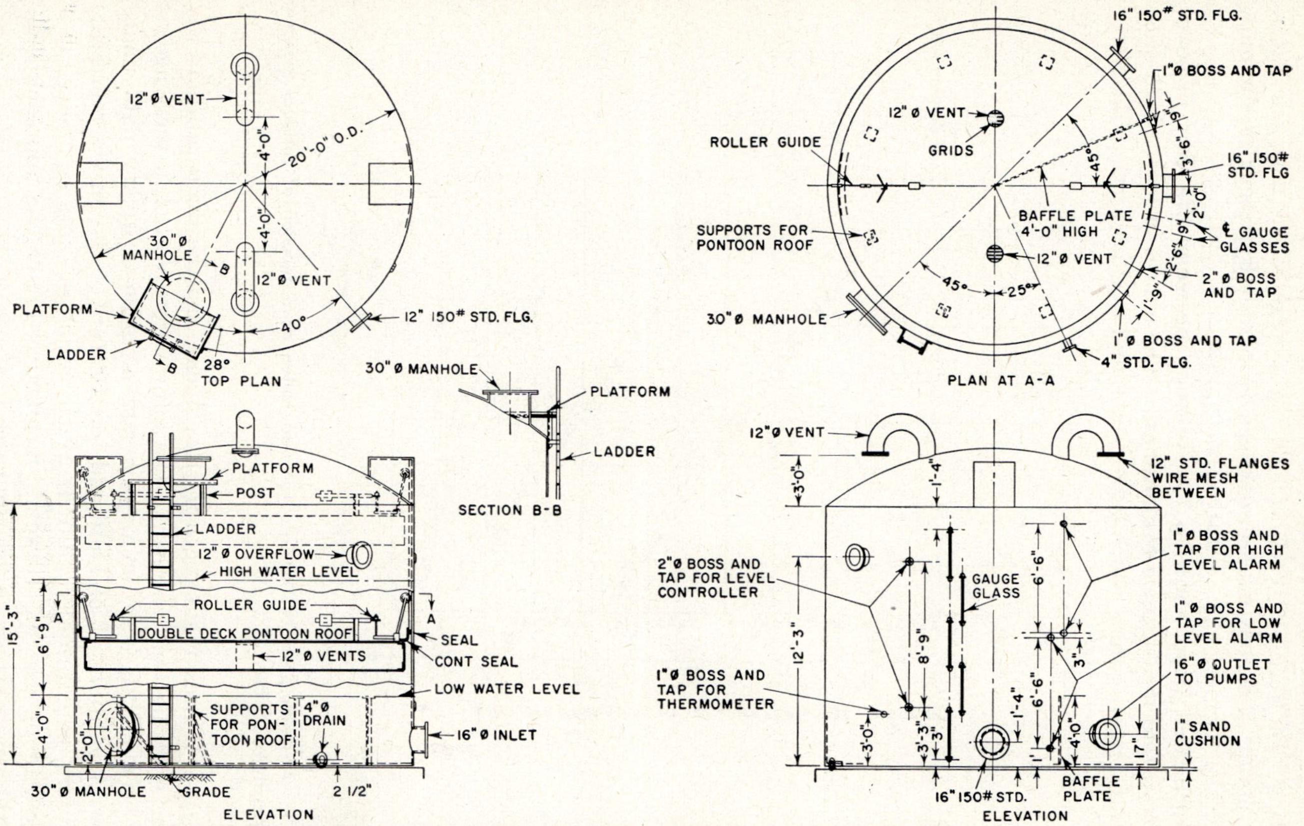

Fig. 16. Storage tank with floating cover to prevent reaeration of water after cold-water vacuum deaeration.

floating roof to avoid reaeration during storage of the degassed water. The piping and controls are arranged to permit bypassing of the unit, should such emergency operation be required.

Calculated Data Showing the Relation between the Gas Removal Effected and the Vacuum Maintained. The curves in Figs. 17–20 illustrate the relationships which have been found to exist between deaerator

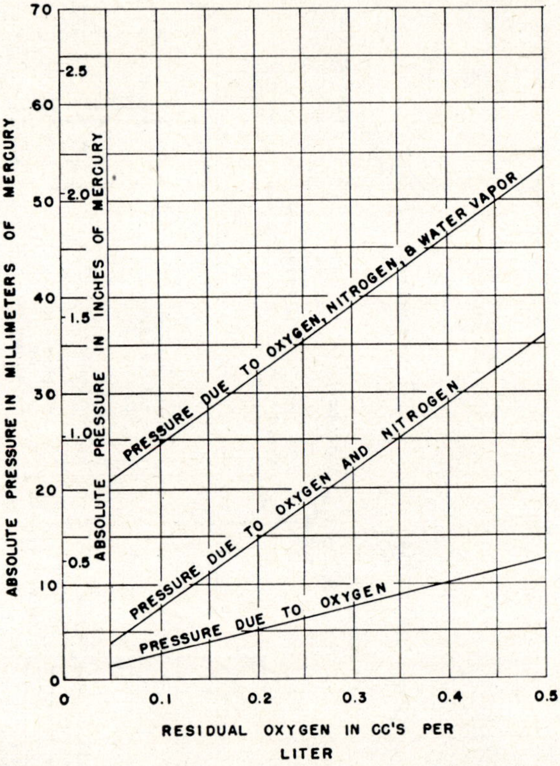

Fig. 17. Absolute pressure required for deaeration as a function of residual oxygen (temperature, 68°F).

pressure, the residual oxygen in the effluent, and the temperature of the water entering the deaerator. The information plotted in these figures has been calculated from oxygen and nitrogen solubility data under known conditions of pressure and temperature. The curves represent, therefore, equilibrium conditions which can never quite be realized in a large-scale operation. However, in practice, the results approach sufficiently close to these curves to justify their use.

Certain assumptions have been made in order to simplify the presentation of these data. It has been assumed that the water to be degasified is saturated with oxygen and nitrogen gases when entering the deaerator, since this condition imposes the greatest load on the equipment and is a limiting factor in the design of the apparatus. It has been assumed, also, that the temperature of the water is not changed during degasification. Actually, the temperature would decrease owing to

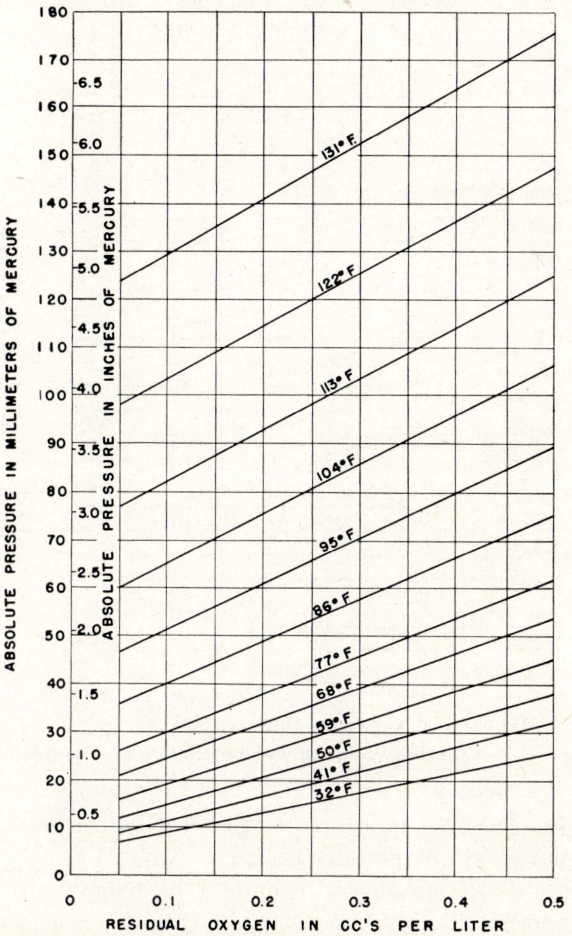

Fig. 18. Absolute pressure required for deaeration at various temperatures as a function of residual oxygen.

evaporation of the water resulting from evacuation of gases. It has been found that, when operating deaerators under normal outdoor conditions, the temperature of the water may be reduced from 3 to 5°. Obviously, the magnitude of the correction for this factor will depend upon the degree of saturation of the influent water with dissolved gases, the temperature of the fluid, and the vacuum maintained.

Other basic assumptions which must be made when designing such equipment and evaluating its potential efficiency are:

1. The flow of the liquid must be maintained sufficiently low, in terms of gallons of water per square foot of cross-sectional area, to prevent flooding of the packing.

2. The depth of the packing must be such as to provide enough time for equilibrium to be reached or closely approximated.

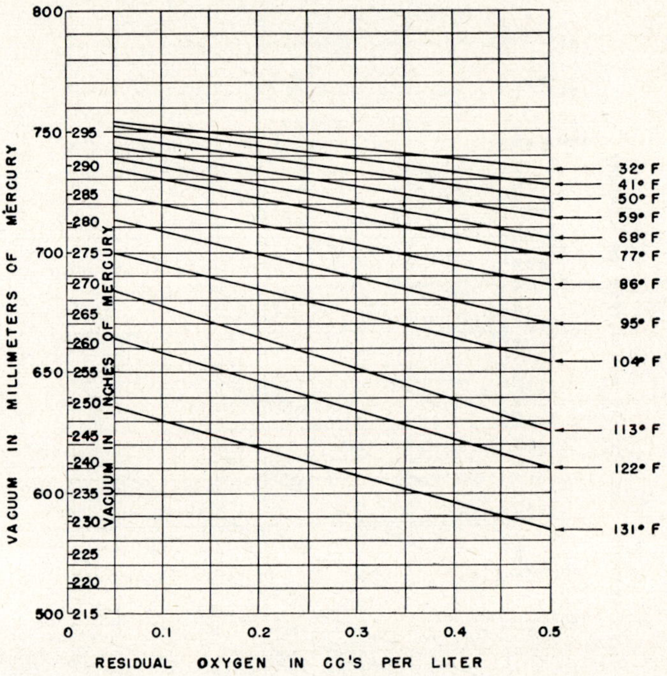

Fig. 19. Vacuum required for deaeration at various temperatures as a function of residual oxygen.

3. The packing should be properly spaced to effect good distribution and so designed as to prevent channeling.

4. Pipe lines of adequate size to remove the vapors with a minimum of pressure drop must be provided.

In evaluating these curves, it should be pointed out that they are not intended to serve completely as a basis for design, but are suitable to indicate performance and to serve as a guide to show the range of variables to be met in practice. Their proper use should aid in controlling the operation and in minimizing the number of control tests necessary to secure best operation. They also serve to define to the designer the limiting conditions which the equipment must meet for any particular installation.

Regarding the design of deaerators, it should be noted that most cold-water vacuum units have been constructed from empirical data collected principally on degasifiers in operation. The same applies to the selection of vacuum pumps, which must be designed large enough adequately to care for any leakage which might develop in the equipment. This statement is also applicable to steam jet ejectors or similar types of equipment.

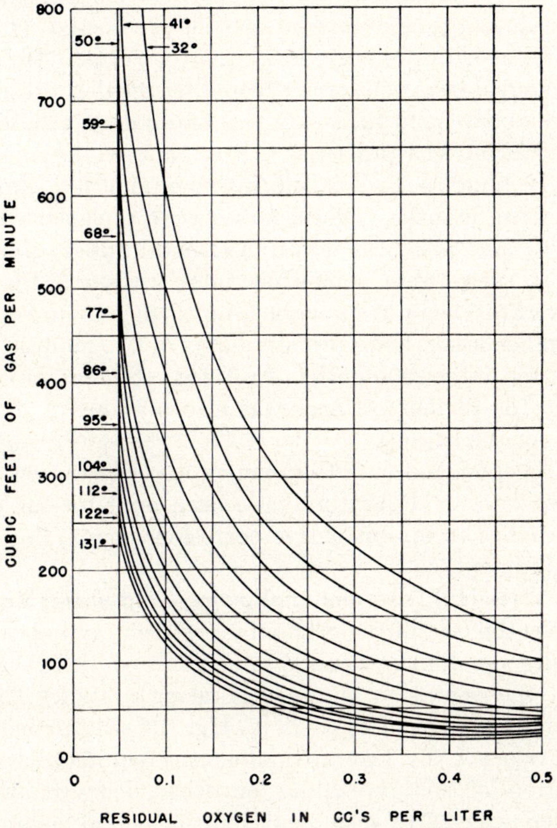

Fig. 20. Volume of saturated air to be removed under vacuum at various temperatures and a water flow of 1,000 gpm as a function of residual oxygen.

Applicability of Cold-water Vacuum Deaeration. Since the mechanism of corrosion is influenced by many factors, generalization as to the predicted effectiveness of deaeration may not be safely made without a comprehensive investigation of the local conditions. It must be apparent that degasification of a cold-water supply as a corrosion-prevention measure has limitations and is not a cure-all for all corrosion problems. The value of such processes is limited to correction of the corrosion-accelerating effect of certain dissolved gases, principally dissolved oxygen, carbon dioxide, hydrogen sulfide, and other gases present in the water which contribute to the aggressive attack on ferrous and nonferrous metals.

The effectiveness of cold-water vacuum deaeration* has been demonstrated in many plants and is now recognized as a practical means for corrosion control for pipelines and miscellaneous equipment.

W. H. Atwill[7] recorded the results obtained from a deaerating unit installed to reduce the corrosive properties of circulating water at the plant of the Texas Oil Company, Port Arthur, Texas. He reported that 98 per cent removal of oxygen was effected by the deaerator and that the residual oxygen content in the water was maintained at about 0.12 ppm.

Before deaeration was practiced, six of nine water-cooled admiralty condenser-tube bundles had failed in nine months, and two others within three more months. These tubes were replaced with red brass, copper, nickel, and arsenical admiralty metal tubes, since experience indicated that these were somewhat more resistant than admiralty. After the deaerator was put in service, no further failures occurred, and steel tube bundles which had normally failed in five months were in satisfactory condition after 26 months of service. Similar protection of the piping system for various services was also obtained as a result of the degasification of the supply.

Since the Freeport Sulphur Company's experiences were reported, a great many cold-water deaerators have been built for the deaeration of cold-water supplies to inhibit active corrosion in pipe lines and in miscellaneous equipment.

Secondary Chemical Treatment Following Cold-water Vacuum Degeneration. Since the cost of the original investment and operating charges increase rapidly as a higher vacuum is maintained, designers have sometimes limited the operation of the deaerating units by controlling the vacuum within the range of 25 to 28 in. Hg. If still further reduction of the residual oxygen in the deaerated effluent is required, it can be effected by the addition of sodium sulfite or other deoxygenating reagents. Operators have resorted to such treatment in several instances in preference to the installation and operation of the more costly mechanical design. The results obtained have been successful in further reducing the active corrosion due to the dissolved oxygen.

OXYGEN REMOVAL BY THE ADDITION OF CHEMICALS

There are many chemicals which can be added to water that will remove oxygen by direct chemical reaction resulting from oxidation of the material used. Sodium sulfite, ferrous iron, hydrazine, and various salts of

* Sheppard T. Powell and James A. Powell, "Apparatus for Degasifying Liquids," U.S. Patent 2,080,151, May 11, 1937.

hydrazine can be, and have been, used for this purpose. Of all these products, however, sodium sulfite is used most extensively as a deaerating chemical to remove traces of oxygen remaining in water after mechanical deaeration. This particular activity is often referred to as "scavenging."[8,9,10,11,12]*

Bibliography

1. Kittredge, Arthur E.: The Removal of Dissolved Gases from Boiler Feedwater, *Proc. 2d Ann. Water Conf., Engrs. Soc. West. Penna.*, Nov. 3–4, 1941.
2. Powell, S. T., and L. G. von Lossberg: Relation of Water Supply to Chemical Plant Location, *Chem. Eng. Prog.*, **45** (5): 289–300 (May, 1949).
3. Adams, C. A.: Two Ways to Oxygen-free Feedwater, *Power*, **93**: 88 (March, 1949).
4. Longley, Col. Francis F.: Coolgardie Pipeline in Australia and Measures Taken to Correct Corrosion, *J. New Engl. Water Works Assoc.*, 1925.
5. Powell, S. T., and H. S. Burns: Vacuum Deaeration Combats Cold-water Corrosion, *Chem. & Met. Eng.*, **43** (4): 180–184 (April, 1936).
6. Powell, S. T.: Vacuum Deaeration Rids Cold Water of Dissolved Gas: Saves Pipelines, *Power*, **91**: 99 (June, 1947).
7. Atwill, W. H.: Corrosion Prevention on Water Side of Refinery Equipment—Use of Deaerator, *Oil Gas J.*, May, 1941.
8. Alexander, R. C., and J. K. Rummel: The Quality of Steam Condensate as Related to Sodium Sulphite in the Boiler Water, *Trans. ASME*, **72**: 519 (1950).
9. Alexander, R. C., and J. K. Rummel: Feedwater Treatment During Early Operation of Steam-Electric Stations, *Trans. ASME*, **72**: 529 (1950).
10. Taff, W. O., H. F. Johnstone, and F. G. Straub: Decomposition of Sodium-sulphite Solutions at Elevated Temperatures, *Trans. ASME*, **60**: 261 (1938).
11. Hitchens, R. M., and J. W. Purssell, Jr.: Behavior of Sodium Sulphite in High-pressure Steam Boilers, *Trans. ASME*, **60**: 469 (1938).
12. Powell, Sheppard T.: Chemical Deaeration of Boiler Feed Water, Part I, *Power*, **79**: 304 (1935); Part II, *Power*, **79**: 362 (1935).

* For more detailed instructions for feeding deaerating chemicals, see Chap. 11.

11 INTERNAL TREATMENT BY CHEMICALS: THEORY

The question of external feedwater purification vs. internal treatment has caused much controversy in the past. Competent specialists in both types of treatment today realize that one must complement the other. It is generally accepted that external purification should be as complete as can be justified by boiler operating conditions and economic factors. Regardless of the purity of the feedwater which the external purification gives, protective treatments are almost universally used as insurance against scale, corrosion, and other potential sources of trouble.

There are many cases where an investment in external softening equipment is not warranted because of (1) small size of the plant, (2) moderate pressure, (3) low make-up, (4) good water supply, or (5) a combination of these reasons. In such cases, internal treatment alone may produce entirely acceptable operating results.

Many organic and inorganic chemicals and compounds have merit for secondary treatment under certain conditions (Table 1). The type and quality of chemicals used for secondary treatment of feedwater and boiler water should be determined by careful study.

Inorganic chemical requirements are fairly clear-cut. For instance, any need for scale prevention, alkalinity adjustment, or chemical deaeration is revealed by a study of water analyses and of the steam and water cycle. Moreover, it will be confirmed by boiler inspections. The functions, dosage, known reactions, and results can be calculated from data in this chapter, using the equations and combining weights given in Tables 2 and 3 and elsewhere. Other items requiring consideration are the type of phosphate to use, control of boiler-water equilibria to produce the desired type of sludge, the optimum level of alkalinity, etc.

By contrast, organic water-treatment procedures must be judged on the basis of performance, often by trial and error. In general, it is best to consider any treatment material of indefinite composition and reaction to be of questionable value until proved by actual plant test. Comparison of boiler cleanliness, steam quality, and other conditions before and after thc use of a chemical is the best criterion. Since most organic water-treatment compounds are of indefinite composition and reaction and not subject to exact chemical calculations, the evidence in their favor consists of laboratory data and the experience of numerous users. Several of these experiences have been reported in technical journals.[1-3]

In addition to the use of organic and inorganic chemicals, various combinations of materials are used in the form of boiler compounds. Such products may be compounded as soluble briquettes or balls, concentrated liquids, or powders. Reputable manufacturers of these preparations inform the purchaser of the ingredients used so that their usefulness can be evaluated. The price of the compounded chemicals is usually higher when purchased as boiler compounds than when bought in the open market. However, the greater cost of the compounded materials can be justified frequently by simplicity and economy of treatment equipment and of control procedures and by technical guidance not otherwise available to operators of small plants.

There are four basic feedwater-treatment aims in addition to a number of more specialized ones developing from high-pressure boiler operations. These requirements are the prevention of (1) scale, (2) corrosion, (3) steam contamination, and (4) embrittlement of boiler steel.

Chemical treatment of feedwater and boiler water within the boiler cycle is designated as internal, or secondary, treatment to distinguish it from purification of boiler-feed make-up before it enters the cycle. It may vary from the use of simple additives in the form of boiler compounds for internal softening to the precise control of all constituents in the boiler water. The scope of internal treatment includes control of complex silicate and other types of scales, and corrosion, sticky boiler sludges, embrittlement, of steel, black iron oxide, adherent superheater and turbine deposits, and other operating difficulties. Some of these controls may conflict with others, and in such cases, a compromise must be accepted. Literature on theory and practice is too extensive for even a cursory review in one chapter, and many of the subjects related to these problems are discussed in detail elsewhere in this text.

HISTORICAL

Internal treatment of boiler water by means of chemical compounds was the earliest and most extensively used method of correcting feedwater troubles. A great variety of substances of widely different composition has been employed for this purpose. Many have little or no effect in retarding scale or corrosion and in preventing the priming and foaming of boiler water. Others have actually caused damage to boilers.

The use of reactive chemicals and inert substances as palliatives for scale formation and other boiler operating difficulties is an interesting subject with a long history. For many years before internal boiler-water treatment was recognized as a science, the technical press abounded in the publication of articles on the advantages and disadvantages of certain chemicals or prepared mixtures for proper conditioning of boiler water. It was difficult for those having no special training in this field to determine the real significance of many of these published statements. As a result, many attempted the treatment of water within the boiler on their own to correct bad feedwater and attain better efficiency. The materials used varied with the individual notions of the engineer. In some instances, the experience gained in actual practice resulted in fairly satisfactory treatment even in the absence of exact chemical knowledge of the substances used. Frequently, though, the treatment did more harm than good. The desire of the early engineers to correct bad feedwater led to the use of such diverse materials as dead animals, vegetables (particularly potatoes), potash, soda, etc. Advancement in this art has demonstrated the fallacy of the use of many of these products.

However, because of the demand for better boiler efficiencies, "boiler compounds" of little or no merit became a profitable business. These materials were readily sold to gullible operators who needed corrective treatments but who had no means of judging their effectiveness or value. An earlier text listed* 28 ingredients found frequently in boiler compounds, of which 12 were classed as frauds, some as objectionable, and 4 as dangerous. It was pointed out that sodium salts for internal softening were the effective ingredients in most of these. When evaluated on the content of the effective sodium salt (without water of hydration) most of these preparations were found to be highly overpriced.

The art of feedwater conditioning by the use of chemicals applied to the feedwater or to the concentrated boiler water is now well established on a sound scientific basis. However, some fraudulent and ineffective materials are still sold. In most cases where boiler compounds are used successfully, the treatment is applied by intelligent and informed operators and regulated by analyses of the raw and treated water.

CHEMICALS AND COMPOUNDS USED

Refined inorganic and organic chemicals are now used for an increasing number of specific functions formerly assigned to multiple-purpose boiler compounds. For orderly presentation, the more important corrective functions performed and the chemicals commonly used for each are shown in Table 1. Some of these chemicals serve purposes other than feed-

* In Sheppard T. Powell, "Boiler Feed-water Purification," Table 14, p. 212, McGraw-Hill Book Company, Inc., New York, 1927.

TABLE 1. CLASSIFICATION OF FUNCTIONS OF SECONDARY BOILER WATER AND FEEDWATER-TREATMENT CHEMICALS

Corrective Treatment Required	*Type of Chemical*
Maintenance of feedwater pH and boiler-water alkalinity for scale and corrosion control	Caustic soda Soda ash Sulfuric acid
Prevention of boiler scale by internal softening of the boiler water	Phosphates Soda ash Sodium aluminate Alginates Sodium silicate
Conditioning of boiler sludge to prevent adherence to internal boiler surfaces	Tannins Lignin derivatives Starch Glucose derivatives
Prevention of scale from hot water in pipelines, stage heaters, economizers, etc.	Polyphosphates Tannins Lignin derivatives Glucose derivatives
Prevention of oxygen corrosion by chemical deaeration of boiler feedwater	Sulfites Tannins Ferrous hydroxide Glucose derivatives Hydrazine
Prevention of corrosion by protective film formation	Tannins Lignin derivatives Glucose derivatives
Prevention of corrosion by condensate	Amine compounds Ammonia
Prevention of foam in boiler water	Polyamides Polyalkylene glycols
Inhibition of caustic embrittlement	Sodium sulfate Phosphates Tannins Nitrates

water and boiler-water treatment. To make the list historically accurate, it would be necessary to include many materials such as crude vegetable products. These are obsolete and have been omitted to avoid confusion.

TREATMENT WITH INORGANIC CHEMICALS

In Table 4 are listed the more important inorganic compounds generally used in internal treatment. The commercial purity and primary specifications are given. All are listed as sodium compounds for convenience, although other salts are employed. Hall[4] recommended the use of potassium compounds to correct certain operating troubles after examining solubility, concentration, and boiling-point data with reference to the film of water on evaporation surfaces.

FUNCTIONS ACCOMPLISHED BY INORGANIC CHEMICALS

Chemical treatment of feedwater and boiler water has a number of functions. Some of these, such as the control of corrosion, boiler-water foaming, and embrittlement, are discussed in detail elsewhere. However, they are summarized in this chapter in order to complete the subject of secondary chemical treatment.

CHEMICAL TREATMENT FOR CORRECTION OF pH AND ALKALINITY

The dissolving of iron in water to form ferrous ions, the initial step in corrosion, is markedly repressed as the pH of the water increases. Because of this, it is common practice to apply alkaline solution to correct the chemical deficiency of the water. The reaction rate of corrosion increases, and pH decreases as the temperature of the water rises. At the temperatures commonly employed for boiler feedwater, a pH of not less than 8.5 generally should be maintained for protection of boiler-feed pumps, piping, and stage heaters. Higher pH and alkalinity will effect better protection from corrosion but may not be permissible because of too high alkalinities in the boiler water or because of copper alloys used in station equipment.

There is a wide difference of opinion concerning the pH at which concentrated boiler water should be maintained. A great majority of plants operate satisfactorily in a range between 10.2 and 11.2, and at ordinary pressures and temperatures, a minimum of 10.6 may be recommended as the control point. However, some boilers are in successful operation where the pH of the boiler water is maintained at from 8.0 to 9.0. Such operation should not be attempted without expert advice and meticulous control. Where phosphates are used for internal softening, the pH must be high enough to assure the formation of insoluble tricalcium phosphate, $Ca_3(PO_4)_2$, or hydroxyapatite, $3\ Ca_3(PO_4)_2 \cdot Ca(OH)_2$.

At high pressures and temperatures, corrosion may occur if the pH is too high. In one case investigated by the author, the presence of hydrogen in the steam was found to rise sharply as soon as the pH exceeded 10.5. Such instances of corrosion have been explained[4] by progressive reactions:

$$Fe + H_2O \rightarrow FeO + H_2 \text{ (film formation)}$$
$$FeO + 2NaOH \rightarrow Fe(ONa)_2 + H_2O \text{ (film destruction)}$$

The maintenance of adequate sodium alkalinity in boiler water aids in keeping silica in solution and thus preventing calcium silicate scale. However, it is preferable to remove or prevent calcium scales completely by adding phosphates to the concentrated boiler water. In general, the alkalinity of the boiler water should be sufficient to keep it well buffered at the pH required for corrosion control and internal softening.

Alkalinity control must be based not only on pH but also on the titratable alkalinity which is reported in various forms.* Upper and lower limits for alkalinity vary with boiler pressure and other conditions. Foaming has occurred in some cases when the alkalinity of the boiler waters has reached 350 ppm. It is preferable to maintain alkalinity concentrations not higher than this value until it is demonstrated that greater amounts are permissible. Total alkalinity below 80 ppm may be too low to assure good buffering action. Added safety is obtained by holding the phenolphthalein alkalinity equal to 60 to 80 per cent of methyl orange reading. No generalization as to permissible maximum alkalinity should be made because the behavior of alkaline salts varies tremendously in the presence of other compounds and at different boiler pressures and operating pratices. It is imperative to maintain low concentrations of alkaline compounds if the boiler water contains soaps, saponifiable oils, or similar materials.

Boiler-feedwater pH and concentrated boiler-water alkalinity are usually increased by the continuous addition of caustic soda solution. This treatment should be added after deaeration to avoid fixing any carbon dioxide that normally would be removed by the deaerating heater. The dosage needed depends on the percentage of condensate returns and its carbon dioxide, CO_2, content. Soda ash should not be used to raise the pH of the feedwater or boiler water because it decomposes in the boiler to produce carbon dioxide, thus making the condensate more corrosive.

Where the make-up is softened by the hot lime-soda or phosphate processes, no caustic feed will be required. In fact, reduction of the pH may be necessary if the percentage of make-up is high enough to induce formation of carbonate sludge or phosphate scale in heaters or economizers. This is usually accomplished by feeding sulfuric, sulfurous, or phosphoric acids. Hydrochloric acid has been used where the sulfates are very high and sulfuric acid would produce calcium sulfate scale in economizers and elsewhere in the feedwater system.

INTERNAL SOFTENING BY CHEMICALS

The primary purpose of softening water within a boiler is to prevent the formation of scale by precipitating calcium and magnesium in the form of sludge. In the absence of internal softening treatment, the heating and concentration of boiler water causes the precipitation of a number of compounds, primarily those which decrease in solubility as the temperature is raised. Calcium sulfate, silicate, and carbonate and magnesium hydroxide and silicate all react in this way.

* For units of measurement, see Chap. 24, Analytical Procedures.

INTERNAL SOFTENING WITH SODA ASH AND CAUSTIC SODA

Calcium precipitation. The use of soda ash and caustic soda for boiler-scale control was primarily to prevent the formation of calcium scales by precipitating calcium carbonate sludge. Calcium bicarbonate in feedwater is converted by heat to calcium carbonate. Other soluble calcium compounds are precipitated by adding sodium carbonate. Table 2 shows the reactions and the combining weights of the chemical compounds involved. In this table the constituents representing hardness are listed as having the molecular weight of calcium carbonate (100), since hardness is reported as equivalent $CaCO_3$. The actual molecular weight of each compound so reported is given in parentheses.

TABLE 2. BOILER-WATER SOFTENING BY HEAT, CAUSTIC SODA, AND SODA ASH

Precipitation of calcium:					
$Ca(HCO_3)_2$		$\overset{\text{heat}}{\rightleftharpoons}$ $CaCO_3$	$+ CO_2$	$+ H_2O$	
100 (162)		100	44		
Calcium bicarbonate		Calcium carbonate	Carbon dioxide		
$Ca(HCO_3)_2$	$+ 2NaOH$	$\rightleftharpoons CaCO_3$	$+ Na_2CO_3$	$+ 2H_2O$	
100 (162)	80	100	106		
Calcium bicarbonate	Caustic soda	Calcium carbonate	Soda ash		
$CaSO_4$	$+ Na_2CO_3$	$\rightleftharpoons CaCO_3$	$+ Na_2SO_4$		
100 (136)	106	100	142		
Calcium sulfate	Soda ash	Calcium carbonate	Sodium sulfate		
Precipitation of magnesium:					
$Mg(HCO_3)_2$	$+ 4NaOH$	$\rightleftharpoons Mg(OH)_2$	$+ 2Na_2CO_3$	$+ 2H_2O$	
100 (146)	160	58	212		
Magnesium bicarbonate	Caustic soda	Magnesium hydroxide	Soda ash		
$MgCl_2$	$+ 2NaOH$	$\rightleftharpoons Mg(OH)_2$	$+ 2NaCl$		
100 (95)	80	58	117		
Magnesium chloride	Caustic soda	Magnesium hydroxide	Sodium chloride		
$MgCl_2$	$+ SiO_2$	$+ 2NaOH$	$\rightleftharpoons MgSiO_3$*	$+ 2NaCl$	$+ H_2O$
100 (95)	60	80	100	117	
Magnesium chloride	Silica	Caustic soda	Magnesium silicate	Sodium chloride	

* The actual substance formed corresponds to $3\ MgO \cdot 2SiO_2 \cdot 2H_2O$.

These reactions have been known for more than 100 years. However, exact knowledge of the solubility equilibria at elevated temperature has been established only within the last 25 or 30 years, following application of the solubility-product concept.* According to this principle, the solubilities of calcium sulphate and calcium carbonate are defined by the equations

$$(Ca^{++}) \times (SO_4^{=}) = K_{CaSO_4,t} \text{ and } (Ca^{++}) \times (CO_3^{=}) = K_{CaCO_3,t}$$

These equations establish the conditions necessary for the success of internal softening treatment. Fortunately if the concentration of the carbonate ion can be maintained sufficiently high, calcium carbonate sludge will be precipitated preferentially and calcium sulfate scale will be inhibited.

Following a long period of rule-of-thumb use of soda ash for internal softening, considerable research work was done to establish the carbonate to sulfate ratio necessary to prevent calcium sulfate scale. In 1927, Hall and his associates,[5,6] using the available values for the solubility of calcium carbonate, concluded that when CO_3 ppm > 0.0883 SO_4 ppm in boilers operating at 150 psig, adherent anhydrite (calcium sulfate) scale would be prevented.† Partridge, Schroeder, and Adams[7] subsequently made solubility studies which led them to conclude that the amount of carbonate necessary was much smaller, the value of the ratio probably being less than 0.01 instead of 0.0883. Straub[8] independently concluded that "the old calculated ratios of CO_3/SO_4 to prevent calcium sulphate scale are extremely high."

Magnesium Precipitation. In natural water, magnesium is generally present together with calcium, although in lesser amounts. It is essential, therefore, when considering scale-prevention treatments, that the magnesium content of the water be controlled to avoid formation of objectionable deposits of magnesium as well as those of calcium. Sodium hydroxide formed by hydrolysis of sodium carbonate in the boiler or added as a chemical treatment precipitates magnesium hydroxide. In the presence of silica, magnesium silicate is formed. Both are finely divided precipitates which usually appear as sludge rather than as scale. The reactions for magnesium precipitation are also shown in Table 2.

Use of Phosphates for Internal Softening. Soda ash is still used extensively as an ingredient of boiler compounds, but as a specific precipitant for calcium, it has been replaced to a large extent by phosphates. The equations for phosphate softening are given in Table 3, showing the

* The solubility-product principle may be stated thus: In saturated solutions of a sparingly soluble solid the product of the concentrations of its ions has a definite value, characteristic of the solute at a given temperature. The symbol Ca^{++} is read *concentration of calcium ion in mols per liter.*

† This recommendation was subsequently modified by Hall as a result of further study by himself and others.

use of trisodium phosphate as well as less alkaline phosphates, which need caustic to complete the reaction. The caustic is furnished by the caustic alkalinity of the boiler water. This, in turn, has its source in carbonate breakdown or in added caustic soda.

TABLE 3. BOILER-WATER SOFTENING BY PHOSPHATES

$2\,Na_3PO_4$		$+ 3\,CaCO_3 \rightleftharpoons$	$Ca_3(PO_4)_2$*	$+ 3\,Na_2CO_3$	
328 Trisodium phosphate		300	310	318	
$2\,Na_2HPO_4$	$+ 2\,NaOH$	$+ 3\,CaCO_3 \rightleftharpoons$	$Ca_3(PO_4)_2$	$+ 3\,Na_2CO_3$	$+ 2\,H_2O$
284 Disodium phosphate	80	300	310	318	
$2\,NaH_2PO_4$	$+ 4\,NaOH$	$+ 3\,CaCO_3 \rightleftharpoons$	$Ca_3(PO_4)_2$	$+ 3\,Na_2CO_3$	$+ 4\,H_2O$
240 Monosodium phosphate	160	300	310	318	
$2\,NaPO_3$	$+ 4\,NaOH$	$+ 3\,CaCO_3 \rightleftharpoons$	$Ca_3(PO_4)_2$	$+ 3\,Na_2CO_3$	$+ 2\,H_2O$
204 Sodium metaphosphate	160	300	310	318	
$2H_3PO_4$	$+ 6\,NaOH$	$+ 3\,CaCO_3 \rightleftharpoons$	$Ca_3(PO_4)_2$	$+ 3\,Na_2CO_3$	$+ 6\,H_2O$
196 Phosphoric acid	240	300	310	318	

* The actual precipitate is usually more nearly represented by $10\,CaO \cdot 3\,P_2O_5 \cdot H_2O$

The studies of Hall and Robb[9] on solubilities of calcium and barium sulfates and calcium phosphate led to a patent[10] covering the use of phosphates under certain conditions of equilibrium control.

The reaction

$$3\,Ca^{++} + 2\,PO_4 \rightleftharpoons Ca_3(PO_4)_2$$

is governed by the solubility-product equation

$$(Ca)^3 \times (PO_4)^2 = K_{Ca_3(PO_4)_2}$$

Figure 1, Chap. 12, is a graph of this equation showing how precipitation occurs if either calcium or phosphate is small, providing that the other is large enough so that their product exceeds K.

These investigators found that the constant $K_{Ca_3(PO_4)_2}$ is extremely small even at moderate boiler temperatures, indicating that soluble phosphate was preferable to soda ash for internal softening. From a practical viewpoint, the phosphate compound is superior, because it is stable, while carbonate decomposes in the boiler to produce high caustic alkalinity in the boiler water and introduce unwanted carbon dioxide into the steam.

The addition of phosphate to boiler water, when properly controlled, removes the calcium so completely that calcium sulfate and silicate scales may be entirely prevented.* The precipitate resulting from the addition of phosphate is tricalcium phosphate, or hydroxyapatite† if the hydrate content of the water in the reaction zone is sufficiently high. The latter compound is less sticky and, therefore, a preferable type of sludge. Magnesium phosphate, also precipitated by this treatment, adds to the stickiness of the boiler sludge. In some cases, it is necessary to modify the treatment to reduce the magnesium content of the feedwater, precipitate it as another compound, *viz.*, as a silicate, or add sludge-dispersing agents.

There are numerous commercially available chemicals which furnish the phosphate radical necessary for internal softening treatment. Compounds of the orthophosphate ion, PO_4^{3-}, are the most widely used, namely, mono-, di-, and trisodium phosphates and phosphoric acid. Other phosphate ions are meta, PO_3^-, and pyro, $P_2O_7^{4-}$. Some of these salts are commonly called "polyphosphates" because they form polymers. Among these are the glassy sodium phosphates (hexa- and heptametaphosphate) and crystalline tripolyphosphate.

Sodium metaphosphate, which was applied to water treatment in 1930 by Dr. R. E. Hall, is derived from monosodium orthophosphate by heating to remove one molecule of water:

$$NaH_2PO_4 \xrightarrow{\text{heat}} NaPO_3 + H_2O$$

In the boiler, this "molecularly dehydrated phosphate" recombines with water to form orthophosphate, which then precipitates calcium and magnesium. It will not precipitate a calcium salt and, hence, may often be added to boiler feedwater directly without forming deposits in feedwater pumps, heaters, and piping. On heating, it hydrolyzes to orthophosphate and then functions as an internal softening agent. The higher the temperature, the more rapidly metaphosphate changes to orthophosphate. Therefore, before adding it to boiler feedwater ahead of the boiler proper, the temperature conditions must be carefully studied.

An important criterion in the selection of the best phosphate chemical is the alkalinity added to the boiler water by the product. Depending on the need for decreasing or increasing the alkalinity, the selection of

* Phosphate treatment alone cannot always prevent certain silicate scales of the analcite group.

† A mixed phosphate and hydroxide.

TABLE 4. SODIUM COMPOUNDS USED IN WATER TREATMENT

Chemical name	Common name and trade name	Formula*	Commercial purity, %	Primary specification
Sodium hydroxide	Caustic soda	$NaOH$	97†–98.06	76% Na_2O
Sodium carbonate	Soda ash, "light"	Na_2CO_3	98.99	58% Na_2O
Sodium disilicate	60% silicate	$Na_2Si_2O_5$	99.5	18% Na_2O, 36% SiO_2
Sodium sulfate	Salt cake	Na_2SO_4	94.0	63.5% SO_4
	Glauber salt	$Na_2SO_4 \cdot 10H_2O$		
Sodium bisulfate	Niter cake	$NaHSO_4$	93.0	78.9% SO_4
Sodium sulfite	Santosite	Na_2SO_3	93.0	
Sodium bisulfite	....	$NaHSO_3$		
Hemisodium phosphate	....	$NaH_5(PO_4)_2$	99.7	64.0% P_2O_5
Monosodium phosphate	Primary sodium phosphate	NaH_2PO_4	98	58% P_2O_5
		$NaH_2PO_4 \cdot H_2O$	97	50% P_2O_5
Disodium phosphate	Secondary sodium phosphate	Na_2HPO_4	98	48.9% P_2O_5
		$Na_2HPO_4 \cdot 12H_2O$	98	19.4% P_2O_5
		$Na_2HPO_4 \cdot 2H_2O$	99.7	39% P_2O_5
Trisodium phosphate	Tertiary sodium phosphate	$Na_3PO_4 \cdot H_2O$	97.5	38% P_2O_5
		$Na_3PO_4 \cdot 12H_2O$	98	18.3% P_2O_5
		$Na_3PO_4 \cdot 6H_2O$	99.7	26.0% P_2O_5
Sodium metaphosphate	Glassy sodium phosphate, Calgon, Buromin, Hagan phosphate	$NaPO_3$	99.8	67% P_2O_5 (min) 23% Na_2O (max)
Sodium tripolyphosphate	....	$Na_5P_3O_{10}$	93	57.0% P_2O_5 (min)
Sodium nitrate	Chile niter	$NaNO_3$	99.3	
Sodium aluminate	....	$NaAlO_3$	75–92, dry 5–32, solution	

* Unless water of hydration is shown in the formula, the specifications should call for anhydrous materials.

† Electrolytic.

the type of phosphate to use may vary between phosphoric acid and trisodium phosphate. A second consideration in making a selection is solubility; and the third, ease of handling.

Monosodium phosphate is acid, and sodium hexametaphosphate, although neutral, will attack chemical tanks, pumps, and piping unless acid-resistant materials have been used. To prevent attack on equipment, these phosphates should always have enough caustic added to make the solution alkaline, as shown by a faint pink color when phenolphthalein is added. Disodium phosphate is mildly alkaline, trisodium strongly so, and both are readily dissolved. The chemicals can be purchased from a number of manufacturers. Many boiler compounds marketed as briquettes, balls, or concentrated solutions contain phosphates.

Phosphate treatment is based on the concentration of soluble phosphate, PO_4, ion in the boiler water. Where the make-up is softened or is low in amount, this may be maintained safely at 20 to 40 ppm. Higher limits may be desirable if appreciable calcium enters the system continuously; lower limits may be quite adequate for boilers in high-pressure central stations or other plants where there is very little contamination of the feedwater.

Control is accomplished by testing a sample of the boiler water as frequently as the concentration may be expected to drop materially—from once a day for low make-up plants to once every 2 hr for high make-up plants. After the test, a phosphate injection is made to restore the soluble PO_4 to the upper limit of the desired range. If continuous or automatic intermittent feeding is provided, the controls are set to increase or decrease the rate of feed as the test indicates.

Methods of feeding phosphate are discussed in a subsequent chapter.

Silicates and Aluminates. Silicate and aluminate compounds have been widely used for internal softening of boiler water. Sodium aluminate in liquid form* has been fed to boilers either alone or compounded in other materials at a rate based on the quality of hardness in the make-up supply. Under these conditions, the soda ash dosage is reduced by the equivalent sodium content of the aluminate. Sodium aluminate combines with magnesium to form a flocculent precipitate of magnesium aluminate. If silica is present, calcium or magnesium aluminum silicates are formed. These are chemically similar to natural zeolites and are useful for coagulating boiler sludges. Aluminates are extensively used as aids in lime-soda softening systems.

There are a number of sodium and potassium silicates available, varying in the ratio of sodium oxide, Na_2O, or potassium oxide, K_2O, to

* The compound is sold in concentrations ranging from 5 to 32 per cent of available sodium aluminate. For a discussion of the value of sodium aluminate for coagulation, refer to Chap. 3.

silica, SiO_2, and the total concentration and specific gravity of the material. When sodium silicate is applied alone or as an ingredient in boiler compounds, it is important to use the form which contains two molecules of silica to one molecule of sodium oxide. The most concentrated solution now available contains 18 per cent sodium oxide and 36 per cent silica, and the specific gravity is 60°Bé. The preferred potassium salt to use consists of 12.6 per cent of potassium oxide and 26.8 per cent silica and has a specific gravity of 40.75°Bé.*

When the correct type and amount of sodium silicate is added to the boiler water, magnesium is precipitated in a flocculent form which may be blown down. The silicates containing a large percentage of silica, such as $Na_2O:3SiO_2$, tend to produce calcium silicate scale, which, however, can be avoided by complete removal of the calcium with phosphates. On the other hand, the more alkaline silicates, such as the metasilicates, are objectionable because they tend to increase the caustic content of the boiler water unduly and are known to accelerate embrittlement of boiler steel under some conditions.[11] Sodium silicate treatment is used extensively for inhibiting corrosion of piping and equipment in cold- and hot-water systems by forming a film which protects the metal.

Silica and magnesium are mutual precipitants and, according to one viewpoint, should be so balanced in the boiler feedwater as to eliminate the objectionable effect of either one. Thus, if sticky magnesium phosphate sludge is objectionable, a reduction in the phosphate treatment and the addition of either sodium or potassium silicate may be considered as corrective measures. Sodium silicates decrease in solubility and may form deposits at high temperatures corresponding to 250 psi or above, whereas potassium silicates do not exhibit this retrograde solubility and have been used in boilers operated at relatively high pressure.

Some complex silicate scales are not preventable by internal softening. Such compounds are formed under conditions of high temperatures and rates of evaporation, particularly when aluminum and iron compounds are present.† It is obvious that, when conditions favor these scales, aluminates and silicates must be rigorously excluded from feedwater and boiler water. In many cases, feedwater purification includes the removal of silica.

Because of the complex reaction which may occur when silica is present in boiler water, the use or avoidance of siliceous compounds should be evaluated and guided by broad experience.

* Any form of silicate for conditioning boiler feedwater should be selected only on the advice of persons informed as to the merits and disadvantages of this type of treatment. When silicates are improperly used, serious operating difficulties may result.

† Analcite, sodium aluminum silicate, or acmite, sodium iron silicate.

CHEMICAL DEAERATION

Recognition of the role of dissolved oxygen in corrosion led first to the use of deaerating feedwater heaters as standard practice and later to the use of chemical reducing agents to absorb oxygen.[12] In low-pressure boilers, all deaeration may be accomplished chemically, but in high-pressure boilers, deoxygenating chemicals are employed only to remove small quantities of oxygen not completely removed by thermal deaeration.

Historically, metallic iron was one of the earliest materials used for this purpose. The procedure of passing water over scrap iron in a "deactivator" was in use prior to and during the development of thermal deaerators.[13] Because of the slowness of the reaction and the small capacity of the units, these devices were not suitable for modern steam plants.

Ferrous hydroxide for chemical deaeration of feedwater for 1,300-lb boilers was employed by the Milwaukee Electric Railway and Light Company in 1930.[14,15] In this process, ferrous sulfate and caustic soda were added to hot condensate in an air-tight tank and stirred with a motor-driven mixer. The gelatinous precipitate of ferrous hydroxide was allowed to settle for 8 hr, and the supernatant sodium sulfate solution was decanted. The precipitate was washed twice by adding fresh condensate, settling, and decanting. Finally, the tank was filled with condensate and kept stirred while it was injected into the boiler feedwater.*

Various sulfur compounds absorb oxygen. A procedure has been used in which sulfur dioxide, prepared by burning sulfur, was absorbed in softened, alkaline make-up water and then added to deaerated boiler feedwater.[16,17] The treatment accomplished not only chemical deaeration but also the adjustment of the sulfate to alkalinity ratio for control of caustic embrittlement. Today, sodium sulfite is recognized as an effective inorganic chemical for deaeration and is used extensively for this purpose. The sulfite is added continuously to the boiler feedwater, usually by means of a small pump. Control is effected by testing the boiler water by a simple chemical procedure and adjusting the rate of feed to maintain the desired sulfite concentration. Ten to twenty ppm in the boiler water will usually suffice to give protection. Recent experiences have shown that it is desirable to limit the sulfite in high-pressure boilers to as low a concentration as is consistent with complete removal of oxygen. This practice is required to avoid corrosion in condensate lines and other equipment resulting from formation of acidic products by breakdown of the sulfite in the boiler.[18]

The reaction between oxygen and sodium sulfite and the weights

* Ferrous hydroxide under boiler temperatures and pressures can be converted to black iron oxide, which is an objectionable material in high-pressure boilers.

involved are shown in Table 5. The dosage figures are corrected for 90 per cent purity of the compound and 75 per cent efficiency* of the process.

TABLE 5. REACTION BETWEEN OXYGEN AND SODIUM SULFITE

	Na_2SO_3	+ $\frac{1}{2}O_2$	$\rightleftharpoons Na_2SO_4$
Combining weights..........	126	16	142
Equivalent to..............	7.875 ppm	1 ppm	
Dosage.....................	11.67 ppm	1 ppm	
	0.0972 lb/1,000 gal	1 ppm	
	16.71 ppm	1 ml/liter	
	0.139 lb/1,000 gal	1 ml/liter	

In most modern plants, the consumption of sulfite by oxygen is negligible, and the dosage is practically equivalent to the loss in the continuous blowdown. Typical feeding rates (where blowdown is less than 10 per cent) vary from 0.5 to 2.0 ppm. However, sulfite may be destroyed by auto-oxidation, which, in simplest form, is the oxidation of some molecules to sulfate by others which are reduced to elemental sulfur in the process. This may explain the slow disappearance of sulfite from boiler water in central-station systems from which loss by blowdown and oxygen consumption should be practically zero.

The rate of reaction of sodium sulfite with oxygen and of the auto-oxidation reaction is markedly accelerated by certain catalysts. Among these are salts of cobalt, copper, and other heavy metals. The catalysts are chiefly useful when it is desired chemically to deaerate at low temperatures or low sulfite and oxygen concentrations or both. Pye[19] described the application of sulfite to a clarified river-water supply used once through for cooling-process heat-exchange equipment in a chemical plant. He reported that the addition of 1 ppm of copper caused 80 ppm of sulfite to react with 8 to 10 ppm of oxygen in about 5 min. However, the reaction time was reduced to 1 min by 0.001 ppm of cobalt, and to 15 to 20 sec by 0.01 ppm of cobalt. The catalyst was added in the form of cobalt chloride solution to a tank in which sulfur dioxide gas was mixed with caustic solution.

Catalyzed sodium sulfite, stabilized against auto-oxidation, is commercially available.

Splittgerber,[20] reporting on German feedwater-treatment practice during World War II, stated that chemical deaeration was accomplished

* Applicable only to low concentrations of reactants. At higher concentration, a 10 per cent excess over theoretical is used by some operators.

by the use of hydrazine, N_2H_4. Hydrazine reacts with oxygen to form water and gaseous nitrogen:

$$\underset{\substack{\text{Hydrazine} \\ 32}}{H_2N{\cdot}NH_2} + \underset{\substack{\text{Oxygen} \\ 32}}{O_2} \rightarrow \underset{\text{Water}}{2H_2O} + \underset{\text{Nitrogen}}{N_2}$$

Mol.wt 32 32

The nitrogen gas is released harmlessly in the steam. It will be seen that the reacting weights are equal, so that, if efficiency and purity were 100 per cent, 1 ppm of hydrazine would remove 1 ppm, or 0.698 ml per liter, of oxygen. Because the decomposition products are harmless and do not increase the total solids in the boiler water, hydrazine should have much promise as a deaerating agent. Hydrazine is manufactured as hydrazine base, N_2H_4, the hydrate, and several derivatives for use as intermediates in the manufacture of organic compounds. Because hydrazine is explosive, an American manufacturer* is now offering the hydrazine phosphate and other salts for use in boiler-water treatment.

There is now considerable interest among engineers in this country as to the value of this material, and experimental studies are in progress.

CHEMICALS USED FOR INHIBITING CAUSTIC EMBRITTLEMENT OF BOILER STEEL

For many years, chemicals have been added to concentrated boiler water to inhibit caustic embrittlement as a cause of cracking of boiler steel. Although such treatment falls within the specification of secondary treatment, reference to Chap. 19 should suffice to acquaint the reader with the type of treatment for such service.

TREATMENT WITH ORGANIC COMPOUNDS

Practically all the organic chemicals useful in water treatment are obtained from plants. Some of the more important are produced mainly for use as tanning agents in the leather industry. Others are produced as a by-product in the manufacture of wood pulp, in which the wood is digested, with or without the aid of chemical agents, to free the cellulose fiber from the lignin and other materials which surround it. In addition, there are a number of organic extracts prepared from such vegetable matter as seaweed, sugar beets, cactus, etc., that are mixtures of unknown composition. They may contain starches, gelatins, and other colloids without any such identifiable major components as the tannins or lignins in the first two cases mentioned.

An outstanding characteristic of most organic water-treatment compounds is the large size and molecular weight of the basic molecule.

* Mathieson Chemical Corp., Baltimore, Md.

Many of these compounds have a tendency to polymerize into chains even many times larger. Such aggregates become colloidal in dimension and exhibit the usual surface-active behavior of colloids. The activity is characterized by absorption on the surfaces of boiler metal, sludge or scale crystals, bubble films, etc., depending on the specific properties of the colloidal particles present.

An undesirable common characteristic of organic water-treatment materials is that they are seldom obtainable as pure chemical compounds but only as mixtures, and in some instances, the molecular structure of even the major constituents is unknown. Exceptions are certain specially synthesized complex organic compounds, such as amides and glycols developed as antifoam agents.

Another characteristic that is a disadvantage of these treatment chemicals is the lack of quantitative analytical methods for determining their concentration in water.

On the other hand, some of the extracts, such as those used in tanning leather, can be purchased under fairly narrow specifications so that the results produced in water treatment are predictable and a dosage schedule can often be worked out satisfactorily to meet the conditions of any particular problem. The tendency is in the direction of producing more highly refined materials that can be used with confidence to obtain specific results.

FUNCTIONS OF ORGANIC TREATMENT MATERIALS

Tannins. These compounds are extracted from certain woods, notably quebracho, chestnut, and oak. Dried quebracho extract contains over 60 per cent of its characteristic tannins. There are many tannins, varying in structure, complexity, and molecular weight. Their classification, type, and chemical structure are beyond the scope of this discussion. Properties of the group, especially as related to water treatment, are exemplified by gallotannin, obtained from oak galls.[21]

Tannins appear to modify or distort crystal growth by "plating" the nuclei, preventing the formation of a closely knit structure, and causing the crystals to be weak. Thus, certain boiler precipitates are prevented from forming a hard, adherent scale in boilers and in feedwater systems. Others which would produce a sticky sludge are dispersed so that they can be removed in the boiler blowdown. Sludge conditioning is one of the most important functions of tannins.

A desirable characteristic of tannins is that they combine with dissolved iron to form a corrosion-resisting film at moderate boiler temperature, *e.g.*, at 150 lb pressure. Tannins of one group hydrolyze to form derivatives of pyragallol; another group, derivatives of catechol. This is one basis for classification. Gallotannin, in the first group, when added to boiler water produces alkaline pyrogallol, which is an active absorber of oxygen. All tannins of this group are, therefore, valuable for chemical deaeration.[22]

Tannins produce a brown color in boiler water which interferes with certain analyses, such as those for methyl orange alkalinity and chlorides. Tannins are unstable at high temperatures, and their usefulness is limited to pressures below about 300 psia.* Nevertheless, they have been more widely used than any other organic treatment material.[23]

The control procedure for tannin treatment depends on matching the brown color of the boiler water with prepared standards furnished by the boiler-compound manufacturers or prepared from solutions containing a known weight of the specific tannin used. This lacks the precision of the titration for sulfite. A more accurate colorimetric method[24] has been developed in conjunction with the use of tannins for embrittlement control.

Lignins. The woods used as sources of paper or rayon pulp contain 41 to 53 per cent of cellulose fiber, sheathed in cellular coatings of which lignin is a major constituent. Lignin must be dissolved by drastic thermal and chemical treatment, often forming a sulfonated sodium salt which is water-soluble. Tannins, resins, terpenes, oils, and other constituents are also present in the digester liquor or are volatilized and sometimes recovered for their commercial value.

The unit natural lignin molecule is quite complex, while processed lignins are simpler, more stable, and more soluble.[25] The large molecules have the colloidal surface-active properties described for the tannins.

Processed lignins are prepared by digesting sodium lignin sulfonate with caustic at temperatures up to 625°F or higher, according to Bird.[26] They are stable at higher temperatures than tannins—up to 650°F, the saturation temperature of 2,200-psi steam. However, evidence of decomposition has been found in boilers operating at lower pressures, probably because of localized high-temperature areas.

Lignin derivatives are not oxygen-absorbing but form protective films. They are useful in the prevention of feedwater-system deposits by the crystal-formation interference described for tannins and serve as sludge conditioners at temperatures beyond the range covered by tannins. Some beneficial effect in preventing adherent deposits of magnetic iron oxide has been claimed. They are now used to prevent caustic embrittlement of steel.

Glucose Derivatives. Glucose is a six-carbon chain that is a basic building block for many complex natural organic compounds. Plants contain many varieties of natural sugars formed by combinations of other

* From private correspondence with J. A. Holmes, of National Aluminate Corp. See also Bibliography.[25]

organic groups with this basic structure. Further, inorganic radicals, such as sodium phosphate, sulfuric acid, chromate, etc., may be combined with glucose derivatives. The resulting compounds have some of the properties of analogous inorganic salts. In addition, the large organic molecule confers surface-active colloidal properties on the compounds. Thus, in the case of chrome glucosates, the protective film consists largely of organic material, and lower concentrations, in terms of chromate, are often effective for corrosion control in water systems. Analogous behavior is claimed for other compounds.[27]

Glucosates are available for dispersion of sludge, scale control, and several allied functions.

Seaweed Derivatives. Seaweed is botanically defined as red or brown marine algae, although the term is loosely used for many aquatic plants. Commercial products are irish moss, agar-agar, and algin, a mixture of complex organic compounds.[28] *Alginic acid* is a product of four *mannuronic acid* molecules, derived from mannose, a configuration isomer of glucose.[29] Sodium alginate and mannuronate used in boiler-water treatment are sometimes called "reactive colloids," because they precipitate calcium and magnesium from boiler water.[23] They also have some of the dispersive properties of the "protective colloids" described for tannins and for lignin derivatives. Alginates and mannuronates are usually marketed as ingredients of formulated boiler compounds.

Starches. Starches are complex carbohydrates which yield sugars on hydrolysis or fermentation and have properties in common with numerous other organic colloids used in water treatment. These materials have been used widely for many years. Navy Boiler Compound consists of 47 per cent anhydrous disodium phosphate, 44 per cent soda ash, and 9 per cent cornstarch. The functions of the inorganic chemicals have been explained above. In general, the colloidal starch is considered to coat and envelop suspended particles with a thin film, reducing cohesion and conferring fluid properties. Other results claimed are the even distribution of soluble and insoluble substances, dispersion of oil, etc. Control of sludge, scale, foaming and priming, embrittlement, corrosion, and other objectionable conditions are said to follow from these and similar properties of starch properly combined with the other chemicals in the compound. The inventor recommended evaluating the starch feed by examining the boiler sludge for dispersion and fluidity and drying a portion of it in a beaker. If the requisite amount of starch is present in the boiler, it will be found that the sludge cracks up into small pieces about ½ to 1 in. on the side.

Trials of this compound were made on 46 ships representative of the fleet, and questionnaires covering the benefits listed above were circulated. The summarized results[30,31] showed much cleaner boilers and drier steam, in support of the claims. Subsequently, the Navy Boiler Compound has been used with good results in stationary boilers where pressure and other conditions warrant considerable reliance on internal softening.

Antifoam Treatments. A number of substances will markedly reduce the foaming tendency of boiler water caused by high concentrations of caustic and other sodium salts, particularly in the presence of saponifiable oils or other organic contaminants. Extensive use of these compounds has been made in railroad boiler-water-treatment practice. Certain organic acids and castor oil have a marked inhibiting effect on foaming of water. Because of this property, castor oil has been used in antifoaming compounds. The amount required varies over wide limits depending upon the substances in the water and the rate at which the oil is removed by reaction with alkali. The foaming tendencies of some waters are difficult to control permanently owing to the fact that the soluble salt responsible for this phenomenon concentrates at a rate greater than can be corrected by the antifoaming treatment. In these instances, foaming may often be retarded sufficiently by blowing down to control soluble and suspended solids.

Barium salts were once used as antifoaming compounds. By the use of barium, the sulfates were precipitated without increasing the sodium concentration of the water, which would result if soda ash, phosphate, or caustic soda were used.

During the testing of many substances for antifoam properties, certain long-chain amides and glycols were found to be especially effective.

These large molecules are essentially insoluble in water. To explain their antifoam properties, Jacoby and Thompson[32] suggest that they are absorbed on both surfaces of a bubble film and cause drainage of water from the film until the bubble bursts. Bodach[33] described experience showing that the life of the effectiveness of an antifoam specific could be increased by absorption on other colloids, such as tannins, silica, etc.

Polyamides and polyalkylene glycol derivatives[34,35] are produced commercially for boiler-water treatment. These substances are added either as specific antifoam treatments or as ingredients in boiler compounds which may also have the purpose of dispersion of sludge and control of scale.

Miscellaneous Organic Materials. In addition to the organic treatment materials described, there are starches, gelatins, and various vegetable extracts,[36] some of which are marketed simply as "colloids." Many of these accomplish some of the benefits described above but are subject to less precise knowledge of composition, mechanism of functioning, and control.

Bibliography

1. Holmes, J. A., and C. Jacklin: Experimental Studies of Boiler Scale, (1) at 880 Psi, *Proc. 4th Ann. Water Conf., Engrs. Soc. West. Penna.*, p. 97, 1943; (2) at 1500 Psi, *6th Ann. Water Conf.*, p. 1, 1945; (3) at 2500 Psi, *11th Ann. Water Conf.*, 1950.
2. Hull, Glenn R.: Stanolind No. 3 Uses Hot Process on 100% Make-up for 1500 Psi Boilers, *Power*, **93**: 100 (May, 1949).
3. Liddell, Robert W., and Louis C. Bishop: The Use and Misuse of Organic Water Conditioning Materials, *Midwest Power Conference*, **13**: 263 (April, 1951).
4. Hall, R. E.: A New Approach to the Problem of Conditioning Water for Steam Generation, *Trans. ASME*, **66**: 457 (July, 1944).
5. Hall, R. E., G. W. Smith, H. A. Jackson, and J. A. Robb: "A Physicochemical Study of Scale Formation and Boiler Water Conditioning, Part I: Principles of Boiler Water Conditioning," Bull. 24, Carnegie Institute of Technology, Pittsburgh, Pa., 1927.
6. Hall, R. E.: "Treatment of Boiler Water," U.S. Patent 1,613,701, Jan. 11, 1927. Assigned to John M. Hopwood (Hagan Corporation).
7. Partridge, E. P., W. C. Schroeder, and R. C. Adams: The SO_4-CO_3 Ratio for the Prevention of Sulphate Boiler Scale, FSP, *Trans. ASME*, Vol. 54, 1932.
8. Straub, F. G.: Solubility of Calcium Salts in Boiler Water, FSP, *Trans. ASME*, **54**: 221 (1932).
9. Hall, R. E., and J. A. Robb: "The Solubility of Calcium Sulphate, Barium Sulphate and Calcium Phosphate at Boiler Water Temperatures, Part II," Bull. 24, Carnegie Institute of Technology, Pittsburgh, Pa., 1927.
10. Hall, R. E.: "Treating Steam Boiler Water," U.S. Patent 1,613,656, Jan. 11, 1927. Assigned to John M. Hopwood (Hagan Corporation).
11. Schroeder, W. C., and A. A. Berk: Intercrystalline Cracking of Boiler Steel and Its Prevention, *U.S. Bur. Mines, Bull.* 443, 1941.
12. Powell, S. T.: Chemical Deaeration of Boiler Feedwater, *Power*, **79**: 304, 362 (1935).
13. Speller, Frank N.: "Corrosion: Causes and Prevention," 2d ed., p. 390, McGraw-Hill Book Company, Inc., New York, 1935.
14. Milwaukee Electric Railway & Light Company: Use of $Fe(OH)_2$ for Deaeration, *Natl. Elec. Light Assoc., Rept. Prime Movers Comm.*, June, 1930.
15. Baker, L. R.: Ferrous Hydroxide as a Corrosion Preventive, Discussion, *Power*, **75**: 70 (1932).
16. Cain, G. A., and J. B. Chatelain: New Low-capacity Sulphur Burner, *Chem. & Met. Eng.*, **46**: 637 (1939).
17. Powell, S. T., G. A. Cain, and J. B. Chatelain: Sulphur Burner Applied to Feedwater Treatment, *Power Plant Eng.*, Vol. 54, May, 1940.
18. Rummel, J. K., and R. C. Alexander: The Quality of Steam Condensate as Related to Sodium Sulphite in the Boiler Water, *Trans. ASME*, **72**: 519 (1950).
19. Pye, D. J.: Catalytic Action of Cobalt and Copper in the Reaction of Sodium Sulfite with Oxygen, *J. AWWA*, **39**: 1121 (1947).

20. Splittgerber, Arthur T.: European Experiments with Feed-water Treatment, Summary of Lectures, *Combustion,* **21**: 48 (1949).
21. Nierenstein, M.: "The Natural Organic Tannins," The Sherwood Press, Cleveland, Ohio, 1935.
22. Fager, E. P., and A. H. Reynolds: Absorption of Oxygen by Alkaline Tannates, *Ind. Eng. Chem.,* **21**: 357 (1929).
23. Maguire, John J.: Organic Agents in Boiler Scale and Sludge Control, *Proc. Midwest Power Conf.,* **13**: 246 (April, 1951).
24. Berk, A. A., and W. C. Schroeder: Determination of Tannin Substances in Boiler Waters, *Ind. Eng. Chem., Anal. Ed.,* **14**: 456 (1942).
25. Holmes, J. A.: The Development of Organics for Water Treatment, *Proc. Midwest Power Conf.,* **13**: 238 (April, 1951).
26. Bird, P. G.: "Method of Inhibiting Scale," U.S. Patent 2,505,457, Apr. 25, 1950. Assigned to National Aluminate Corp.
27. Haering, David W.: "Organic Methods of Scale and Corrosion Control," 5th ed., D. W. Haering & Co., Inc., Chicago, 1943.
28. Tseng, C. D.: Colloids from Kelp, *Chem. & Met. Eng.,* **52**: 97 (1945).
29. McIlroy, Robert J.: "The Chemistry of the Polysaccharides," pp. 53 ff., Longmans, Green & Co., Inc., New York, 1948.
30. Solberg, Thorvald A., and R. D. Adams, Jr.: A New Boiler Water Treatment for the United States Navy, *Combustion,* **5**: 24 (1933).
31. Adams, Robert C., Jr.: Results from the Change of the Navy's Method of Boiler Water Treatment, *Combustion,* **6**: 37 (1934).
32. Jacoby, Arthur L., and W. H. Thompson: Improvement of Steam Quality, *Proc. 7th Ann. Water Conf., Engrs. Soc. West. Penna.,* January, 1947.
33. Bodach, C. M., and L. O. Gunderson: Organic Synergists in Foam Control, *Proc. Midwest Power Conf.,* **13**: 257 (April, 1951).
34. Bird, P. G.: "Prevention of Foaming in Steam Boilers," U.S. Patent 2,428,-775, Oct. 14, 1947. Assigned to National Aluminate Corporation.
35. Gunderson, Lewis O.: "Conditioning Water for Steam Generation," U.S. Patent 2,391,895, Mar. 18, 1942. Assigned to Dearborn Chemical Company. See also U.S. Patents 2,040,127 and 2,328,551.
36. Sutermeister, Edw.: "Chemistry of Pulp and Paper Making," 3d ed., John Wiley & Sons, Inc., New York, 1941. For structure of lignins, author refers to Freudenberg: *Papier-Fabr.* (*Cellulosechemie*), **36**: 34 (1938).

12 EQUIPMENT FOR SECONDARY CHEMICAL TREATMENT

TYPES OF TREATMENT SYSTEMS USED

Equipment for adding chemicals to feedwater and concentrated boiler water ranges from simple chemical pots and injectors to elaborate pumping assemblies. For orderly presentation, these devices may be classified as:

1. Apparatus for treatment of feedwater at low or moderate pressures.
2. Apparatus for treatment of concentrated boiler water at higher pressures.

Protection of the preboiler system includes elevation of the pH to retard corrosion of the feedwater pumps and piping, lowering of pH to prevent scale in stage heaters and economizers, and the use of polyphosphates or organic compounds to inhibit deposits. With these exceptions, most of the chemicals added to feedwater are used for their effect in the concentrated boiler water.

Adding chemicals to the low-pressure part of the cycle can generally be done by means simpler and less expensive than direct boiler-water treatment, but there are some cases where this is not permissible. The feeding of soluble orthophosphates, for instance, in the presence of even slight traces of calcium will result in the slow formation of a hard crystalline scale in stage heaters and economizers. By contrast to the flocculent, soft precipitate which forms in the boiler, where phosphate concentration, pH, and temperature are high, the precipitate that forms from a solution very slightly above saturation is a dense scale very difficult to remove. The conditions responsible for the formation of dense, hard phosphate deposits are high pH (about 9.0), excess phosphates, which may be as low as 3 to 5 ppm, some calcium, and moderate elevation of temperature. Where such conditions are encountered, as may occur in economizer tubes and with certain types of water treatment, hard scale will be deposited. The surest way to correct such deposits is to reduce the pH of the water to 8.0 or lower (Fig. 1).

If phosphate is added to the boiler feedwater *intermittently*, the flushing action between treatments may avoid scale. However, it is generally best to inject phosphate solution directly into the drums of the individual

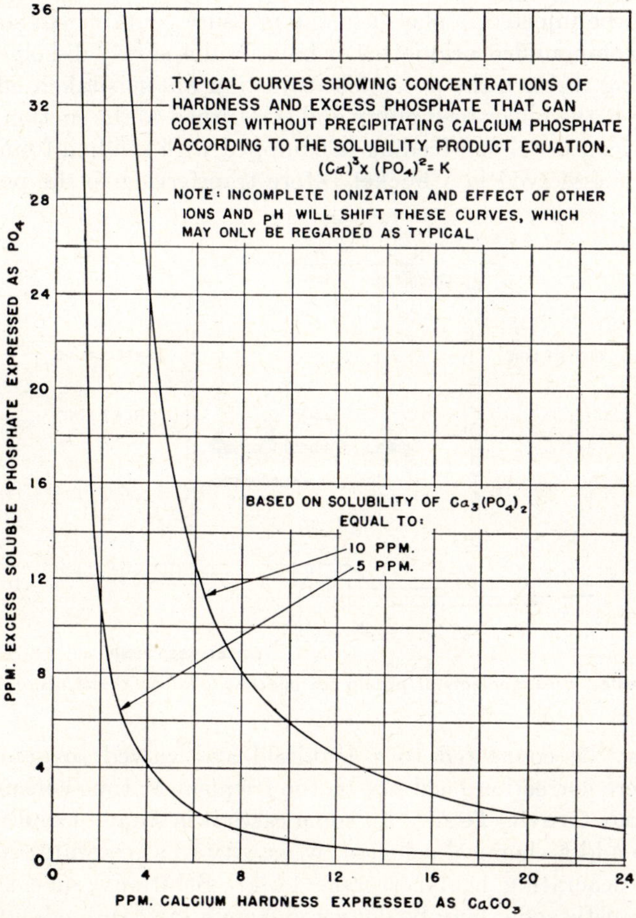

Fig. 1. Typical curves showing concentrations of hardness and excess phosphate that can coexist without precipitating calcium phosphate according to the solubility-product equation.

boilers. Another reason is the variation in phosphate consumption between boilers depending on load and blowdown and the need for correcting deficiencies in one unit without overdosing all the others.

Polyphosphates which hydrolyze slowly may be added constantly to the feedwater, provided that their retention in the preboiler system is not long enough for conversion and precipitation. This retention time depends on pH, temperature, and somewhat on the hardness in the feedwater.

EQUIPMENT FOR ADDING CHEMICALS TO THE FEEDWATER CYCLE

Pressure Pots. One of the simplest feeders is shown in Fig. 2. The chemical to be injected is placed in the pressure container in solid form, usually as compounded briquettes or balls, and is slowly dissolved by the water flowing through the pot. The injection water is taken off the discharge side of the boiler-feed pump and is returned to the suction side. A similar device for "shot" feeding is equipped with a filling funnel. The treatment is dissolved in a bucket before transferring to the pot. Such

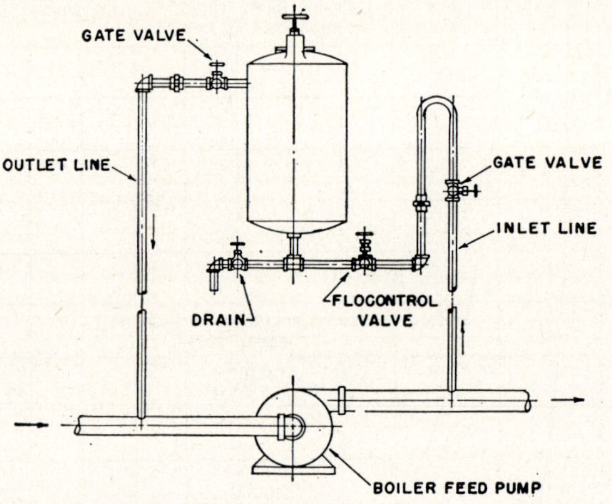

Fig. 2. Suggested arrangements of piping for injecting alkaline chemicals with pressure-pot feeders.

feeders may be connected to a manifold and valved to a number of boilers where correction need not be too frequent or time-consuming.

Continuous Gravity Feed. In some cases it is simplest and most economical to add a chemical solution by gravity to a feedwater surge tank or to the deaerating heater storage tank. Solutions, such as sodium sulfite or caustic soda, may be delivered from a tank through a float-controlled valve into a small constant-head box and thence through a calibrated needle valve to the point of application. Figure 3 shows an assembly consisting of porcelain flush tanks used for this purpose. The solution is delivered into a second tank which furnishes a larger flow of dilution water. This assures constant pressure conditions for the needle valve so that the feed may be adjusted precisely. All the vessels should have sufficiently tight-fitting lids so that diffusion of atmospheric oxygen over the liquid will be retarded.

Continuous Pumping Assemblies. Sodium sulfite used for chemical deaeration is best added continuously to avoid peaks and valleys of concentration. There is some evidence that shot feed of sulfite favors decomposition. Also, the addition of caustic to maintain a noncorrosive feedwater pH is necessarily continuous. The equipment in Fig. 4 has been designed for this purpose. The pumps are of the direct piston displacement, metering type, so that the delivery can be adjusted according to the requirements shown by boiler-water analyses and the feedwater

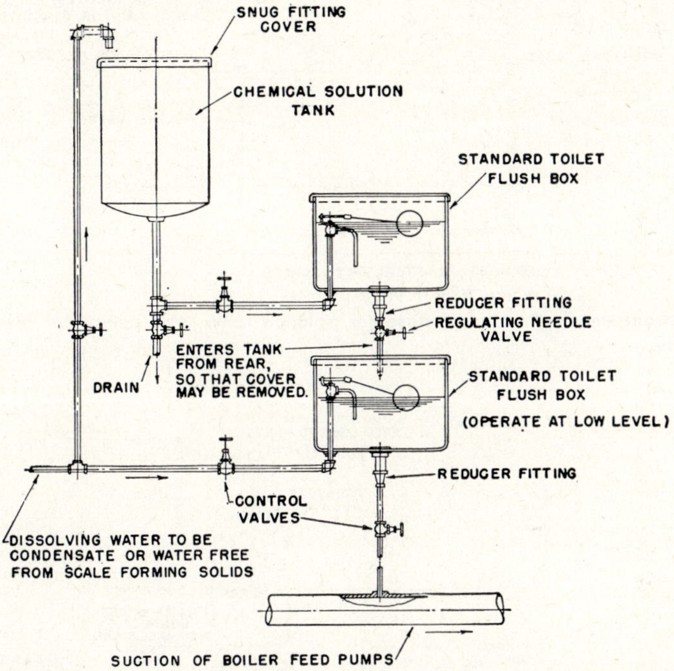

Fig. 3. Equipment for continuous treatment of boiler feedwater with chemical solution by gravity feed.

pH. The closed tanks avoid oxygen entrainment, and the displacement air for the sulfite solution is passed through a small bubbler seal on the side of the tank. The solution is drained and refilled each time the tank is charged.

This figure shows a dissolver which the author has found very satisfactory. The bottom opening is closed by a valve, and the desired weight of chemical is added. Hot water is then directed toward the apex of the pyramid by a high-velocity jet, which rapidly dissolves the chemical. The solution overflows through a strainer at the top of the hopper into the appropriate tank, and by the time it is about two-thirds full, all of the chemical will be dissolved, and the hopper should contain clear

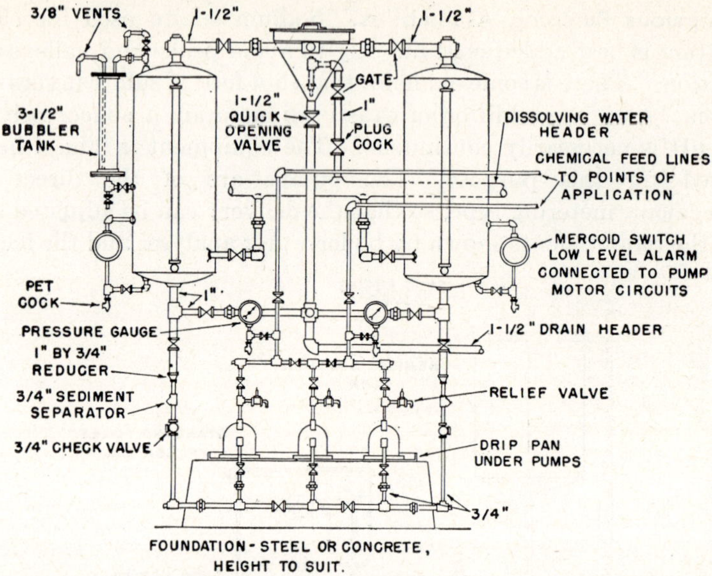

Fig. 4. Arrangement of chemical dissolver, solution tanks, and pumps for feeding sulfite and caustic.

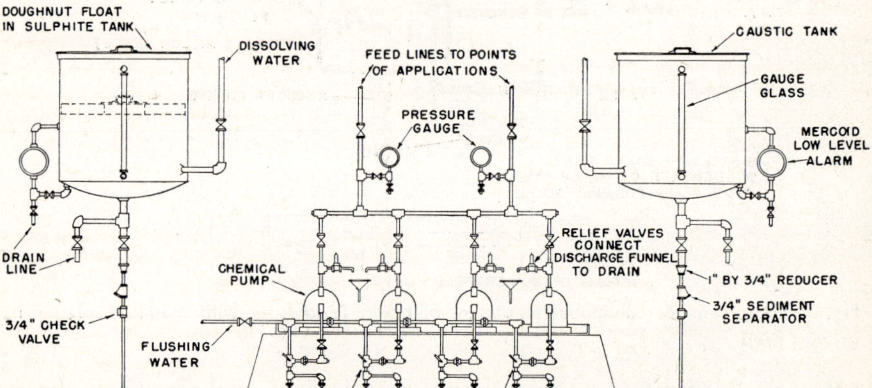

Fig. 5. Arrangement of chemical-solution tanks and pumps for feeding sulfite and caustic.

water. The water remaining in the hopper is then drained to the sewer, and the solution tank is filled through a tangential inlet line which dilutes the concentrated solution to the proper strength. Figure 5 is a comparable assembly using tanks with removable covers and a surface float to protect the sulfite.

Acid-feeding Equipment. If it is necessary to feed acids to lower the pH or alkalinity of boiler feedwater, extreme care must be used in designing equipment to safeguard the personnel from injury and the boiler-feedwater system from acid attack. Although it may be possible to proportion manually the acid feed to flow and keep it adjusted within a safe range, it is best to regulate the dosage automatically. A combination of volumetric primary control and supplementary pH regulation is often used if wide flow rates are experienced.

The equipment shown in Fig. 6 is a typical example of an automatic acid-control system. The feedwater line contains a flanged rubber-lined section with Monel metal baffles, which serves as an acid-mixing chamber.* A flowing sample is taken immediately downstream from the mixer

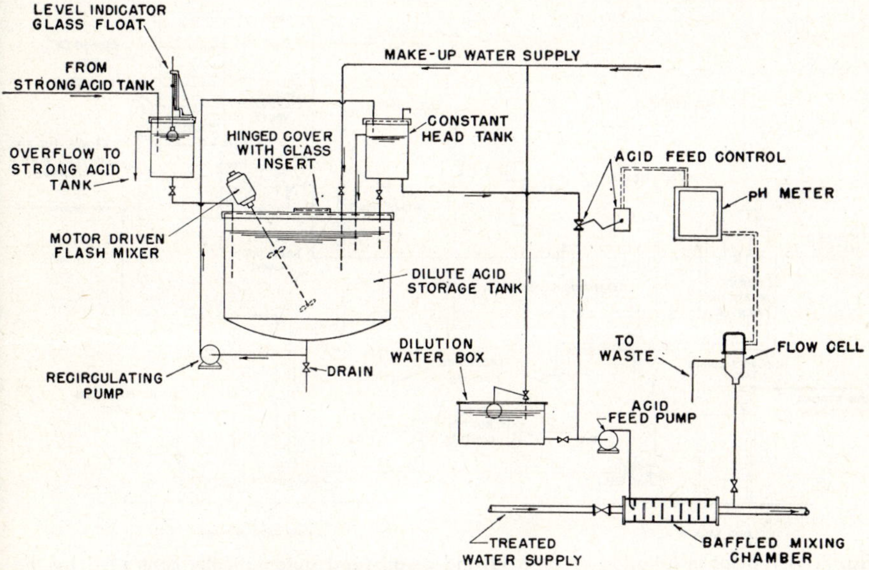

Fig. 6. Schematic arrangement of acid-feeding equipment.

to a vessel housing glass and calomel electrodes connected to a pH recorder-controller set for a control point of, say, 8.5 to 9.0. The controller operates a motor-driven mechanism connected to a Duriron plug cock. This, in turn, is supplied with dilute acid (5 per cent) from a constant-head box. Should the pH rise above the control range, the valve opens wider and delivers more acid, while if it falls below, the valve reduces the feed. The dilute acid recirculates continuously to a lead-lined constant-head tank and overflows to a larger lead-lined storage tank. The purposes of other parts of the equipment should be obvious from the drawing.

The success of a pH-controlled acid feeder in maintaining a steady non-hunting pH level depends on reducing to an absolute minimum the time

* Karbate mixing chambers are also satisfactory for such service.

interval between pH detection and the injection of the corrective acid feed. Otherwise, the control is constantly attempting to compensate for a condition which no longer exists. To reduce time lag, the sampling line to the electrode should be short and the velocity high, bypassing some of the sample, if necessary, to accomplish the desired control. The acid must be injected along with any needed dilution water at a velocity which will minimize the time interval between valve movement and change in water treatment. Also, the detention in the baffled mixing chamber must be the minimum required for thorough mixing. A total of 15 sec between sample take-off and acid injection will assure good regulation.

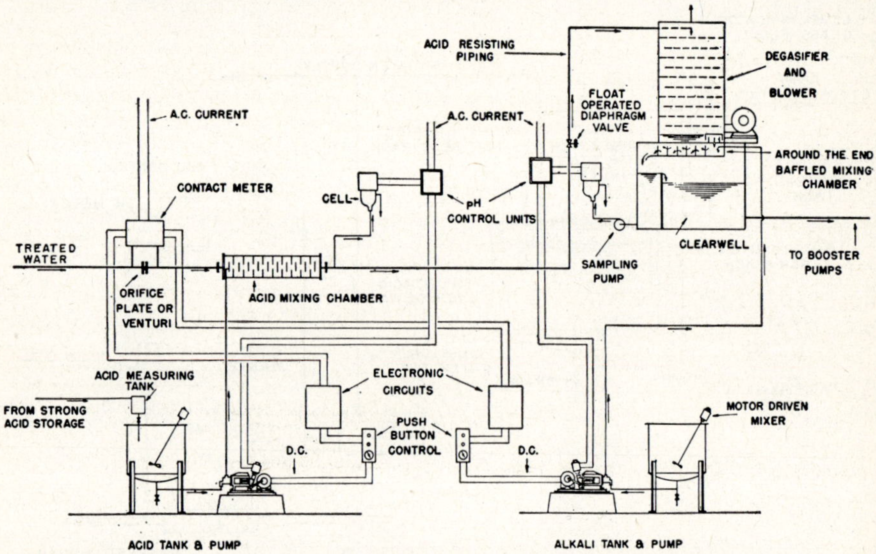

Fig. 7. A typical installation of an acid-and-alkali feed automatically controlled by the combined regulation of flow and pH adjustment.

Variable-speed Motor-driven Pumps. One automatic type of acid feeding is based on a variable-speed pump drive operated on direct current through an electronic rectifying circuit. The pumping assembly for automatically controlling the dosage of chemicals volumetrically and for secondary correction by pH regulation is shown diagrammatically in Fig. 7. The rate of chemical solution discharge is actuated by a direct-current variable-speed motor through an electronic circuit. The control is varied in proportion to the amount of water metered. For example, the assembly includes a variable-speed resistance or potentiometer to govern the speed of the pump so that the chemical feed is directly proportional to the flow. The feed of the chemical solution is adjusted by a secondary variable introduced by changing the stroke of the pump through a microdrive attachment operated by a small reversible motor.

A pH recorder-controller is used to provide this secondary regulation. When feeding acid where the change in the flow of the treated water is too rapid for response by a pH controller alone, the compounding of flow-proportioning control, in addition to the pH regulation, assures accurate and positive adjustment and feeding of the chemical solution used.

EQUIPMENT FOR ADDING CHEMICALS DIRECTLY TO BOILERS

Solution-displacement Apparatus. This type of equipment requires a pressure vessel heavy enough to withstand full boiler pressure. It is

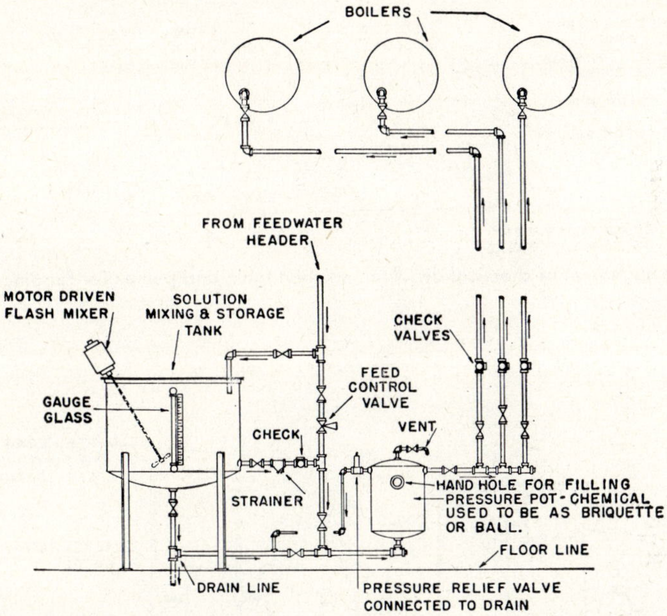

Fig. 8. Chemical-solution pressure pot for feeding directly to individual boilers.

charged with chemical solution while isolated from the boiler circuit. This vessel is designed to receive boiler feedwater at full pressure and to discharge it into the boiler drum or the high-pressure feedwater-inlet header. The equipment is cut into the cycle intermittently by manual or automatic control. When automatic control is used, the regulation is effected by motor-operated or solenoid valves. These may be actuated from the feedwater or steam flow meters to inject the chemical in proportion to the water entering the boiler (Fig. 8).

Chemical Pumping Assembly. Figure 9 shows a typical pump and tank assembly for feeding phosphate. It includes a closed tank and dissolver similar to those shown for sulfite or caustic and two high-pressure pumps, one of which is a spare. These feed directly to the boiler drums.

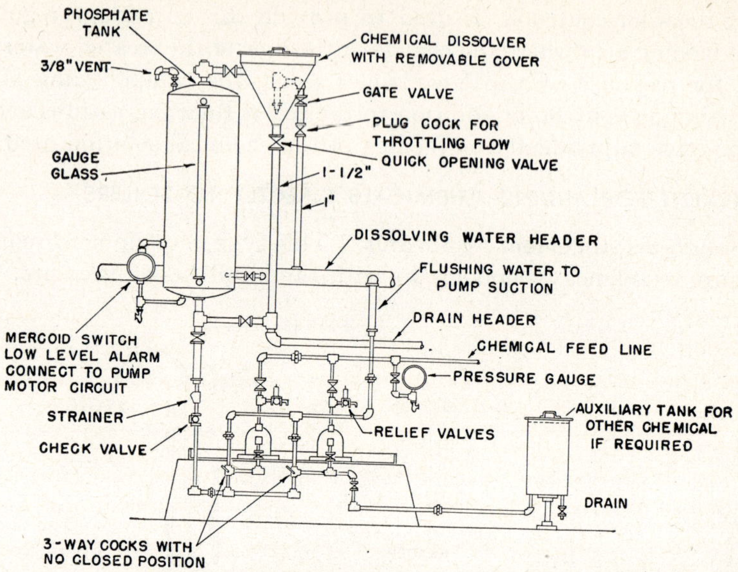

Fig. 9. Arrangement of chemical dissolver, solution tank, and pumps for feeding phosphate.

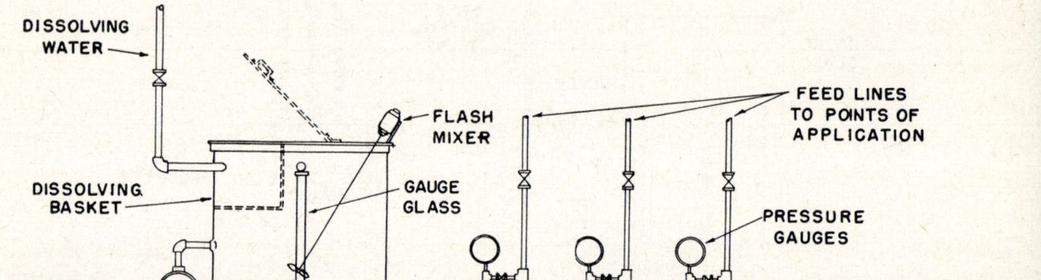

Fig. 10. Arrangement of chemical dissolver, solution tank, and pumps for feeding phosphate.

One set of pumps and discharge header will serve a number of boilers when the feeding is needed only once or twice a shift. This is usually the case if the make-up is not high and if it is completely softened.

An auxiliary chemical tank connected to the pump section is convenient for dosing idle boilers or injecting miscellaneous chemical treatments.

Figure 10 shows a similar design using a hinged-top tank with a motor-driven mixer. This is a more common type of installation where the phosphate requirements are fairly high.

Continuous Circulating System. One method of feeding phosphates and other chemicals is to provide a high-pressure chemical header adjacent to all the boilers with a feed to each boiler through a calibrated V port or needle valve. The pump header contains a release valve which discharges back to the solution tank if the throttling is excessive. After the boiler-water phosphate tests are made, the individual valves are reset in accordance with deficiencies below the desired phosphate range.

Automatic Feeding Equipment. Where the consumption of phosphate is continuous and fairly high, as in the case of a large percentage of lime-soda-softened make-up water, it is desirable to feed at fairly short intervals and, if possible, in proportion to hard water consumed. One phosphate pump is furnished for each boiler, with one extra serving as a spare. A motor- or solenoid-operated three-way valve at the suction of each phosphate pump alternately supplies phosphate solution or flushing water (condensate) to the pump suction. A timer system actuated by the flow meter or a program clock carries out the feeding cycle. One timer keeps the phosphate pump in operation for a predetermined limit of time, say 20 min. A second timer energizes the three-way valve at the beginning of the cycle and turns it to the phosphate-feeding position. A third timer, by which the dosage is adjusted, changes the valve back to flushing water after a preset period of time, for instance, 8 min. For the balance of the period, flushing water is pumped through the piping to push the phosphate solution into the boiler. The entire system then goes out of service until the next impulse.

Figure 11 is a photograph of a portable pumping apparatus that can be moved around from one boiler to another for emergency treatment.

HAZARDS INVOLVED IN USING ACID AND PRECAUTIONS REQUIRED

Several kinds of acid are fairly extensively used for the treatment of boiler feedwater and miscellaneous auxiliary water conditioning.

The use of acids should be undertaken only by those qualified to handle these corrosive chemicals, and the equipment for storing and applying the acid should be carefully designed to minimize the hazard of injury to personnel or equipment. Unless these precautions are taken, serious damage may occur in the equipment, or operators may be injured.

There are two cardinal principles to follow in connection with this type of water conditioning: (1) the need for such treatment should be unqualifiedly justified, and (2) the use of acid should be undertaken only in stations where the safety precautions necessary will be carried out.

All mineral acids are highly corrosive and operators should be furnished with equipment which will minimize injury. Furthermore,

Fig. 11. Portable chemical-feeding assembly.

operators should be instructed carefully as to the dangers involved in such treatments. Personnel should not be permitted to handle acid unless equipped with approved goggles, rubber gloves, and rubber aprons. An ample supply of fresh water and properly designed showers and flooding eye fountains should be installed in the vicinity where the acid is handled.

The preferred type of shower is one that is operated from a platform connected by a spring to a quick-opening valve. By stepping on the platform the operator can be completely flooded.

An approved type of eye fountain is shown in Fig. 12. This is standard equipment and is so designed that the injured person's eye may be

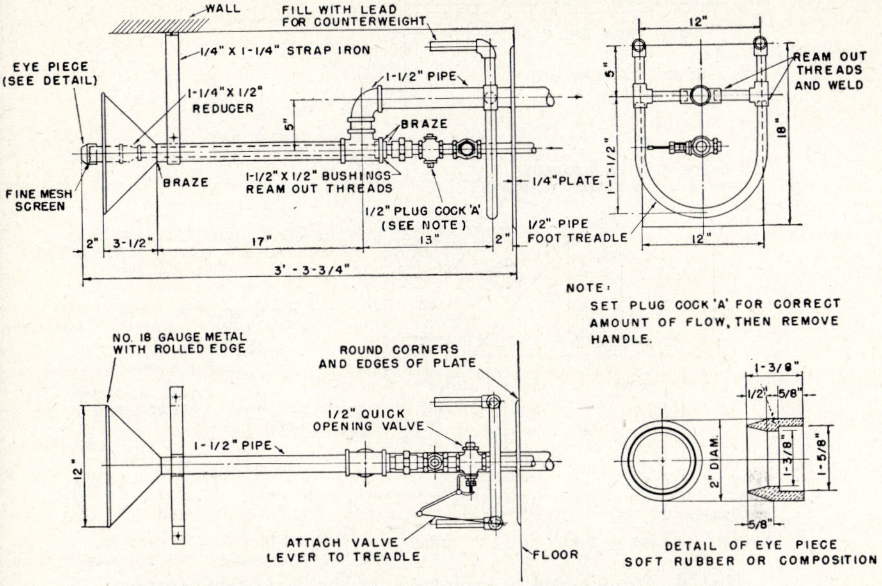

Fig. 12. Details and arrangement of emergency eye fountain.

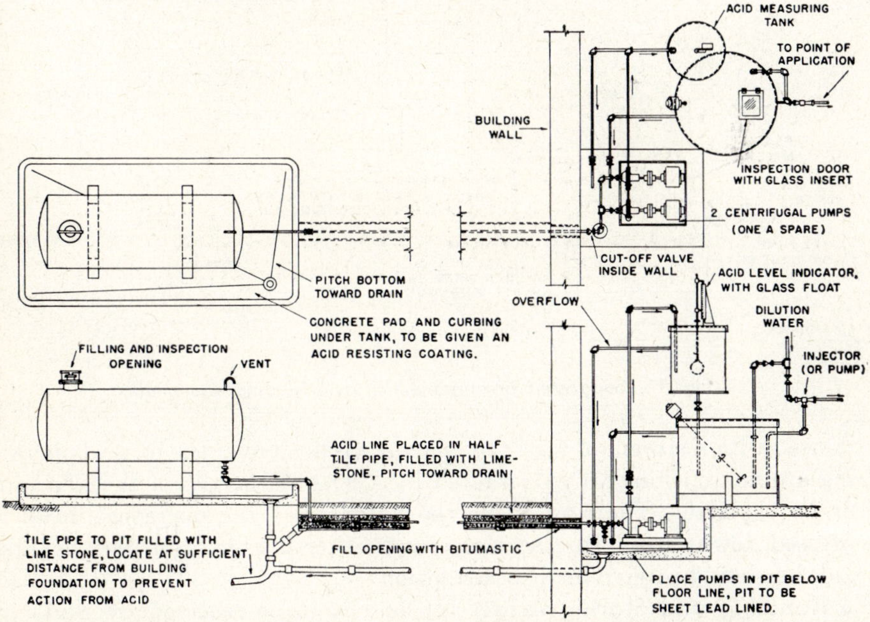

Fig. 13. Arrangement of acid-storage and -feeding equipment.

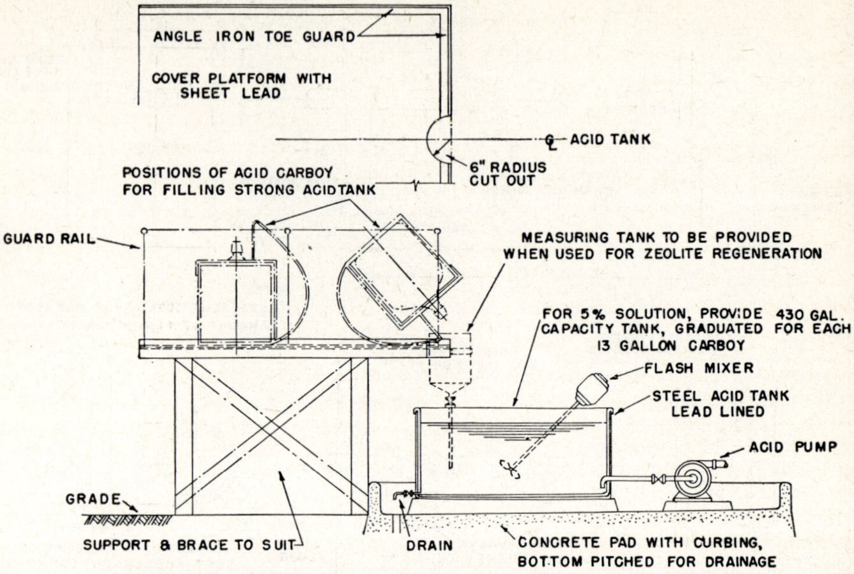

Fig. 14. Arrangement of equipment for handling acid in carboys.

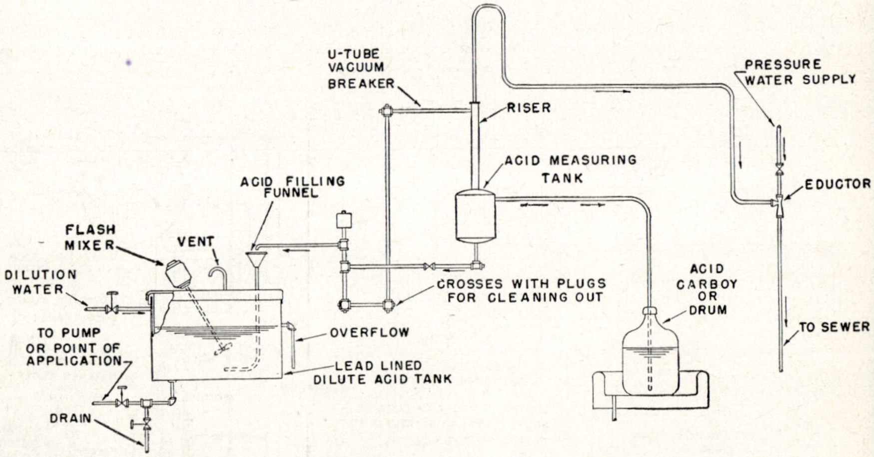

Fig. 15. Suggested arrangement of acid-feeding equipment.

doused with a stream of water by placing his head over the jet and operating a quick-opening valve actuated by a ring at the top of the fountain or by a pedal. The ring opens the valve when the operator's head is pressed downward. In case of accident, after the operator receives first aid, he should receive medical attention.

Handling and Storing Acids. If acid is to be received by tank-car shipment, it should be transferred to a storage tank adjacent to the location where the acid is to be used. Transfer of either hydrochloric or sulfuric acid from tank cars to storage should be done in accordance with safety rules recommended by the Manufacturing Chemists' Association* or other recognized authority. The design of the storage facilities should be such as to provide all the necessary safeguards to protect the work and equipment. A typical arrangement of satisfactory acid-storage facilities is shown in Fig. 13.

When acid is received in carboys, only approved means for handling and transferring of acid should be permitted. A simple design for transferring acid from a carboy into the dilution tank is shown in Fig. 14.

When the acid is to be delivered by pumping from a carboy to the dilution tank, arrangement of equipment for this operation may be provided safely in several ways. The design shown in Fig. 15 has been used successfully in a number of plants and provides the necessary precautions to protect operators.

* Manual of Standard and Recommended Practice," Manufacturing Chemists Association, Washington, D.C.

13 WATER CONDITIONING FOR RECIRCULATION SYSTEMS

The protection against corrosion and scaling of metals in recirculating fresh-water cooling systems is a problem of enormous economic importance. Some of the many corrective measures tried have been partially or wholly successful; others, complete failures. A critical investigation of final conditions often reveals the basic reasons for the results obtained by such treatments.

CORROSION PREVENTION

In the final analysis, protection of any metal from corrosion by water depends upon conditioning the supply to correct its corrosive characteristics or creating and maintaining a physical dam between the anodic or cathodic part of the metallic surface and the electrolyte (water). Protective coatings may contain inorganic or organic substances or both; they may be the product of anodic or cathodic corrosion reactions or may be precipitated selectively from the liquid phase on all areas. For complete protection, these deposits must be nonporous and continuous. When this favorable condition is attained, all corrosive attack will be prevented, since there will be no contact between the vulnerable metal areas and the corroding medium.

In the absence of coatings, metal surfaces can be protected against corrosion by miscellaneous forms of water treatment and by deaeration to remove gases which accelerate aggressive attack.

MECHANISM OF CORROSION AND THE ACTION OF INHIBITORS

A discussion of the causes of corrosion and of preventive measures must necessarily be based on the concepts and terminology of modern theories concerning the corrosion process. For simplification, consideration is limited here to corrosion of steel by dilute aqueous solutions typical of circulating cooling water or process water used in power stations and in miscellaneous industrial plants.

A complete study of a specific corrosion problem is often involved and complex. It must determine the cause and magnitude of the initial potential differences, the primary anodic and cathodic reactions, and those that may ensue at each electrode.* Polarity differences over the metal surface may be created by partial or complete screening by deposits, by differences in temperature, by stress, or by metal composition. Concentration cells involving liquids or gases and many other causes have a marked influence on corrosion rates.

Metallic surfaces are classified as anodic and cathodic. These classifications are of value in explaining the requirements for successful use of various inhibitors. An understanding of these classifications is essential in evaluating the type of corrective treatment required and its adjustment to effect satisfactory corrosion-control procedures.

CONTROLLED CALCIUM CARBONATE SCALE

Protective coating in the form of a calcium carbonate scale develops by regulating and adjusting the pH, alkalinity, and calcium ions to maintain a slightly supersaturated solution of calcium carbonate. By such control, a thin, impervious, and adherent layer can be deposited on the surfaces requiring protection. Calcium carbonate is a general inhibitor that covers both anodic and cathodic areas because its precipitation does not depend on the products of electrochemical reactions, although an alkaline film on the cathodic surfaces is an added factor favoring precipitation of the calcium carbonate.

The creation and rate of formation of this type of protective coating is influenced by a number of factors. Any chemical condition of the water which tends to increase the solubility of the calcium ions or prevent the deposition of a crystalline deposit will reduce or prevent the formation of the desired coating.

There are a few natural water supplies in which calcium carbonate exists in a state of equilibrium. For such waters, little or no chemical treatment is required, since these tend to precipitate a thin and uniform protective coating. The waters of many of the Great Lakes are of this type and are, therefore, relatively noncorrosive or can be made so with minor readjustment of their pH value. Unfortunately, most supplies are not so readily and easily controlled, and readjustment of the hardness or alkalinity of the water is required to develop and control the deposition of scale. The success or failure of such treatment depends to a great extent on the adequacy of the control and the maintenance of specific conditions favorable to the formation of a satisfactory coating over metal surfaces.

* Refer to Chaps. 16, 17 for an amplified discussion of corrosion.

Extension of the use of calcium carbonate scale for prevention of corrosion and the control of excessive scale deposition has been encouraged by the constructive and able research of Prof. W. F. Langelier.[1] He advanced the idea—now widely accepted—of a calculated saturation index as a means of predicting the corrosive or scaling behavior of a natural water. Previously, other investigators[2–8] had done work along somewhat similar lines but had employed different methods for calculating and expressing their data. Since Langelier's work was published in 1936, additional papers[9–11] have been presented which attempt to explain and correct certain discrepancies between predicted and actual scale deposition or corrosion. An understanding of this saturation index is necessary to condition water intelligently to produce a protective carbonate scale.

USE OF THE LANGELIER CALCIUM CARBONATE SATURATION INDEX

The dissolved mineral matter in most natural water supplies consists mainly of calcium and magnesium in the form of bicarbonate hardness and noncarbonate hardness, principally chlorides and sulfates. Sodium salts, dissolved carbon dioxide, oxygen, and other gases are usually present. The tendency of a water supply to deposit scale when heated or made more alkaline or to attack metals corrosively depends on the balance of these various constituents. By controlling them with corrective treatment, both scale and corrosion can be minimized.

Increased temperature accelerates the rates of both of these processes. For instance, the rate of corrosion in a water normally corrosive increases about the same number of times that the rate of scale deposition would increase if the water were scaling.

The scale that forms under moderate temperature conditions is due to the bicarbonate hardness being converted to normal calcium carbonate upon heating or is affected by an increase in alkalinity sufficient to cause supersaturation with respect to calcium carbonate. The solubility relations of calcium carbonate minimize corrosive attack because the aggressiveness of the dissolved carbon dioxide in the water is greatly reduced as the saturation equilibrium is approached. Exactly at equilibrium, the various forms of carbon dioxide (free carbon dioxide, bicarbonate, and carbonate) are so balanced that they cause neither scale nor corrosion. The adjustment of the composition of the water to control these conditions is, therefore, guided by the solubility-equilibrium relations of calcium carbonate.

A convenient method of interpreting a water analysis for this purpose utilizes logarithmic expressions of the calcium hardness, the methyl orange alkalinity, and a factor for temperature and total solids. In

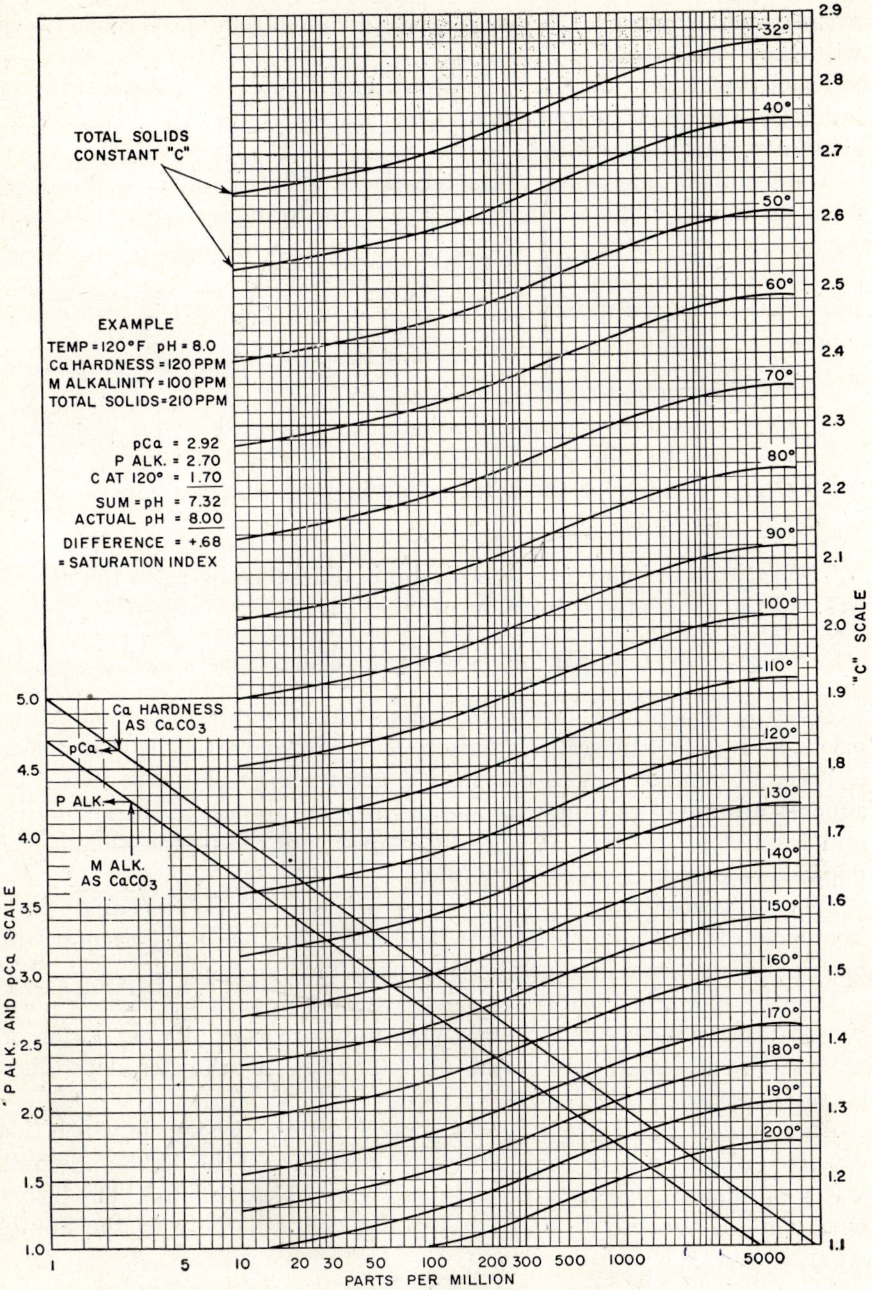

Fig. 1. Langelier saturation-index chart.

actual practice, however, it is advisable to supplement these calculations with the marble test.*

From the graph (Fig. 1), the saturation pH (pH_s) and the saturation index (I_s) can be calculated when values are known for calcium hardness, alkalinity, and total solids.

The scaling or nonscaling characteristics of water can be predetermined from the family of curves shown in Fig. 1, for example:

$$pH_s = pCa + pAlk + C$$

where pH_s is the saturation pH

pCa is log factor of the calcium hardness, expressed in ppm of equivalent $CaCO_3$
pAlk is log factor of the M alkalinity, expressed in ppm of equivalent $CaCO_3$
C is log factor of the total solids in ppm at the temperature of the water

The saturation index I_s is the algebraic difference between the actual pH and the calculated saturation pH, *i.e.*,

$$I_s = \text{pH (actual)} - pH_s \text{ (calculated)}$$

When I_s is zero, *i.e.*, when pH equals pH_s, a saturation equilibrium exists, and there is no scale formation, and corrosive attack is minimized.

When I_s is positive, *i.e.*, when pH is greater than pH_s, a condition of supersaturation of calcium carbonate exists with respect to the alkalinity and total solids content at existing temperature conditions, tending to deposit scale on heat-transfer surfaces.

When I_s is negative, *i.e.*, when the pH is less than pH_s, the equilibrium is unbalanced in the opposite direction, and corrosion of bare metal will occur, and any scale previously formed will be dissolved.

In order to inhibit corrosion, it is necessary to maintain a slightly positive saturation index I_s sufficiently high to insure a definite deposition of scale. The desired I_s may vary from +0.6 to +1.0 or higher, depending on the presence of inhibiting substances.

As a typical condition, consider a cooling tower system in which the dissolved solids are concentrated by evaporation and the make-up is well water containing high alkalinity due to sodium bicarbonate. The latter compound is present in many water supplies, especially in the South Atlantic and Gulf Coast areas.

The following conditions are assumed for the make-up and the circulating water; and from these data, a typical example has been calculated:

* Instructions for making this test are given in Chap. 24.

	Make-up water	Circulating water
Temperature, °F	80	140
Constituents, ppm:		
Calcium hardness, as $CaCO_3$	80	102
M alkalinity, as $CaCO_3$	196	255
Total solids	392	510
pH	7.5	8.0

Referring to the make-up water in the above table and the Langelier saturation-index chart (Fig. 1), the value for the saturation index (I_s) of -0.14 is calculated in the following manner. First locate the calcium hardness of 80 ppm on the parts per million scale, read vertically up to the diagonal line for Ca hardness as $CaCO_3$, and then horizontally to the left to the pAlk and pCa scale. The value for pCa is 3.10. In a similar manner, locate the *M* alkalinity of 196 on the parts per million scale, read vertically up to the diagonal line for *M* Alk as $CaCO_3$, and then horizontally to the left to the pAlk and pCa scale. The value for pAlk is 2.41. To find *C*, locate the total solids of 392 ppm on the parts per million scale and read vertically until the 80°F line is reached. From this point read horizontally to the right to the *C* scale. The value for *C* is 2.13. As illustrated in the table below, pH_s is the sum of the values for pCa, pAlk, and *C*. The saturation index is the value calculated by subtracting pH_s from pH.

	Make-up water	Circulating water
pCa	3.10	3.00
pAlk	2.41	2.30
C	2.13 (at 80°F)	1.57 (at 140°F)
pH_s	7.64	6.87
I_s	−0.14 (slightly corrosive)	+1.13 (scaling)

It would be expected that the make-up would be aggressive, and the circulating water scale-forming. The circulating water would deposit scale in heat-exchange equipment, but the resulting scale formations could be controlled by the addition of sufficient acid to reduce the alkalinity to approach the desired saturation index. Normally, a saturation index of +0.6 in the circulating water could be expected to be satisfactory, and this could be obtained by reducing its methyl orange alkalinity to 80 ppm, corresponding to a pAlk of 2.82.

It is often preferable to add the sulfuric acid to the make-up rather than to the circulating water. To maintain a saturation index of +0.6 in the circulating water in the foregoing example, the methyl orange alkalinity of the make-up supply should be reduced to 62 ppm, corresponding to a pAlk of 2.91.

It is obvious that such treatment requires careful analytical control and adjustment of acid or alkali required to maintain the desired chemical balance. Water supplies whose chemical composition varies widely are not well suited for such treatment. Likewise, water whose temperature varies over a wide range is difficult to control to insure proper scaling throughout the entire system. This type of treatment is best suited to water of relatively uniform chemical composition whose temperature throughout the circulating cycle is not over a wide range. However, by proper control and when adapted to the proper set of conditions, such treatment has proved satisfactory and more economical than many other forms of control.

MODIFIED MARBLE TEST FOR SATURATION-INDEX DETERMINATION

When inhibitors are present in the water, the calculated Langelier index indicates a higher scaling value than is actually the case. For cases such as this, it is advisable to use the marble test method to determine the saturation index. The following method can be used to determine the saturation index directly, the only test required being that of pH.

The theory of the method is that, if a water not in equilibrium is brought into contact with calcium carbonate, the water will attain equilibrium. If it is undersaturated with respect to calcium carbonate, which is the same as having a negative saturation index, the water will dissolve sufficient calcium to saturate the solution, making the saturation index zero. If the water is supersaturated, which is the same as positive saturation index, the excess calcium carbonate will precipitate, causing the saturation index to be zero. The testing procedure for control of scale by the marble test is given in Chap. 24.

CONDITIONS INHIBITING SCALING

Much work has been done at plants that have used a controlled carbonate scale, and many practical observations have been made by operators. It is worthwhile to discuss these in detail, and it is also important to consider one objection that has been advanced to the controlled-scale technique, *viz.*, the alleged loss of heat transfer because of such scale, particularly in the hotter passes of condensers and coolers. Observations by a number of operators lead to the conclusions that satisfactory scale formation and avoidance of heavy corrosion products are obtainable only by control of a number of conditions. The more important factors inhibiting good scaling are:

1. Organic inhibitors which disperse calcium, retard precipitation, and prevent crystalline scale formation

2. High sodium alkalinity or other solids which tend to increase the solubility of calcium ions
3. Oil on the metal surfaces
4. Algae and slimes
5. Excess phosphates

There are definite steps that can be taken to bring these inhibiting factors under control. For instance, with high sodium alkalinity, use lime instead of caustic soda or soda ash to adjust the alkaline content of the water in order to raise the saturation index. In some cases where part of the make-up is softened, raw water is bypassed for adjustment of alkaline content. The temperature coefficient of scale may be objectionably high where caustic is used to adjust the saturation index, and the deposits are likely to be porous and bulky. Moreover, manipulation of the "scaling force" or "descaling force" of the water can be accomplished with a smaller adjustment in composition when the calcium content is varied than when the sodium alkalinity is varied.

Much of the heat-transfer loss that has been attributed to scale has been due at times to organic slimes, which may have a heat-insulation effect greater than that of scale. Algae control by chlorination or by other means is imperative for a successful calcium carbonate controlled-scaling program in plants located in areas where algae growths in the circulating waters are prolific.

It is preferable to lay down scale rapidly before any corrosion occurs. After equipment is cleaned, especially if acid is used, the metal tends to corrode at a rapid rate. A high scaling index set up immediately will prevent the initial attack.

Where the operating temperatures or other conditions make it difficult to scale equipment, heating of the circulating water with steam or hot water is desirable until a good protective scale is laid down. In Fig. 2 are a family of curves for readily determining the saturation indices at temperatures other than test temperatures.

EFFECTIVE AND INEFFECTIVE SCALING-CONTROL OPERATIONS

In Fig. 3 is shown a test specimen indicating severe tuberculation with a covering of scale laid down over corrosion products. This specimen was in service for several weeks with no treatment of any kind. Serious corrosion took place prior to the treatment with lime and soda ash. The saturation index was thereafter maintained near +1. Scale was not formed, however, and corrosion continued at practically the same rate as that without treatment until high-rate chlorination was started. Improvement was noted immediately as a result of the formation of a protective coating after the inhibiting effect of the organic matter was eliminated by the use of chlorine. The scale deposition was rapid as long as sufficient chlorine was fed to maintain a positive residual. This experience demonstrates the need for chlorination to effect scale formation when circulating water contains algae or other organic materials.

Figure 4 illustrates the type of scale deposited when treatment is satisfactorily controlled. Figure 5 is an example of an undesirable bulky

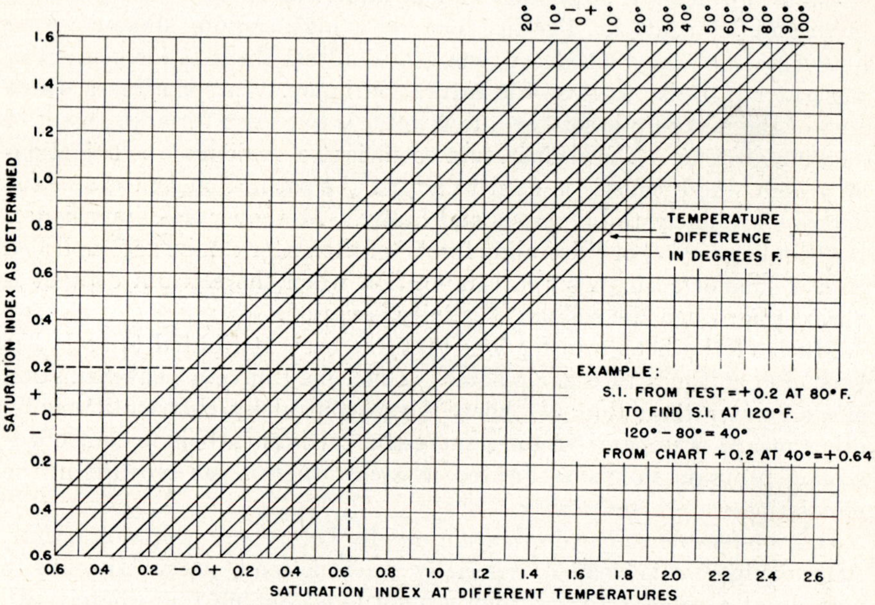

Fig. 2. Chart for determining saturation index at temperatures other than test temperatures.

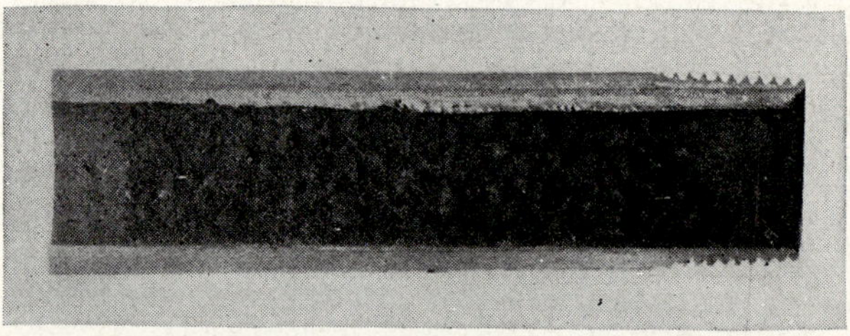

Fig. 3. Appearance of test specimen showing scale deposited on corrosion products.

deposit caused by the presence of algae entrained with the calcium carbonate precipitate. This deposit took place with a saturation index of +1.5 to +2.0 in a seven-day period without chlorination. Analysis showed it to consist of 29.7 per cent organic matter and 25.3 per cent calcium, as calcium carbonate.

Figures 6 and 7 show the effect of chlorine treatment in reducing the bulk of this type of deposit after 15 days of exposure to the same water

Fig. 4. Test specimen showing satisfactory protective scale laid down before corrosion has started.

Fig. 5. Bulky deposit caused by algae growths: (top) test specimen uncleaned; (bottom) test specimen cleaned.

Fig. 6. Deposits reduced by continuous chlorination of cooling water for two weeks.

Fig. 7. Deposits still further reduced after one month's chlorination of cooling water.

but with chlorine treatment sufficient to maintain a residual of from 0.3 to 0.5 ppm. The specimen had been in contact with treated water for 25 days without chlorination.

The specimen shown in Fig. 8 illustrates the detrimental effect of high sodium alkalinity. The make-up water supplied to the system in question was a blend of sodium zeolite softened water and hard water, and, consequently, the sodium alkalinity of the water in the system was high. The saturation index was +0.9 to +1.75, and the residual chlorine was 0.3 to 0.5 ppm. The specimen shown was in service for seven days. The scale was noticeably softer than when hard water only was used or when the alkalinity of the system was later reduced by an acid feed. The

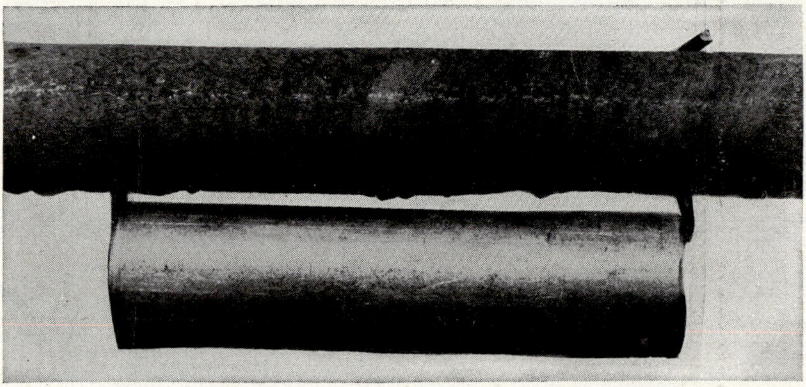

Fig. 8. Soft deposit caused by high sodium alkalinity: (top) test specimen uncleaned; (bottom) test specimen cleaned.

coating had become objectionably heavy two weeks later and, despite its thickness, did not give adequate protection against corrosion attack. The corrosion of the metal surfaces is noted where the overlying scale had been removed. These experiences demonstrate the ineffectiveness of deposits if they are not of the proper type or texture or if they are not continuous and well bonded to the surfaces.

The ideal scale coating for best protection with a minimum increase in the heat-transfer rate is a thin, closely adherent, crystalline deposit that can barely be seen. In fact, some of the best protective scales cannot be easily observed when wet and are only detected when a drop of hydrochloric acid is applied, revealing carbonate by an effervescence of carbon dioxide.

CONTROL OF TEMPERATURE EFFECTS

The most frequent criticism of the use of controlled scale has been that the deposit is too heavy on hot surfaces and not heavy enough on cold surfaces. However, when scaling control has been adequately regulated, there occurs an intermediate condition of acceptable uniformity of scale thickness over the entire temperature range in a system, provided the temperature range is not great.

METAPHOSPHATES

The "threshold" treatment of water with concentrations of metaphosphate far too low to combine with all the calcium present has been used successfully for several years to prevent scale deposition and, more recently, to hold iron in solution and to prevent corrosion. An impressive number of municipal water systems successfully apply the treatment to prevent "red water." A discernible film of iron metaphosphate is deposited on the surface of the metal—presumably at anodic areas where the ferrous ion is released—and this inhibits corrosion.

When properly applied, this type of treatment has considerable success in main circulating cooling-water systems. The treatment functions best at a pH below 7.0, a condition which some operators are reluctant to tolerate for fear of rapid corrosion. The failure to meet these low pH requirements has contributed to the failure of this form of treatment in some cases.

Metaphosphates are intermediate in cost between inorganic and organic chromates and the alkaline salts usually required for controlled carbonate scale treatment. They can be justified in many cases for use in closed systems which do not have excessive losses of water.

One reason for ineffective control by the metaphosphate has been improper maintenance of the meta salts, resulting from undertreatment or conversion of the metaphosphates to orthophosphates. The latter type of phosphate salt does not possess the corrosion-inhibiting properties of the metaphosphates. The conversion to orthophosphate results from too high temperature or slowly occurs if the solution stands for too long a time before it is introduced into the water under treatment.

CHROMATE COMPOUNDS

Chromates and dichromates have been used for years to inhibit corrosion, particularly as additives to refrigerating brines. The chromate or dichromate ions react with the ferrous ions which are released at anodic areas, forming a mixture of ferric and chromic hydroxides as a tough, adherent, self-healing film over the anodic areas.

The consumption of the chromate is directly proportional to the rate of release of ferrous ions, *i.e.*, to the rate of corrosion. Therefore, when the concentration is maintained high enough to give complete coverage of the anodic areas, the consumption of chromate by the inhibiting reaction may be zero and the only loss will be that which is due to windage or blowdown. Unless losses from the system are high, it is best to maintain chromate concentrations at a magnitude which will ensure complete protection, which may require 300 to 500 ppm. Fair protection is claimed at lower doses of chromate by the addition of small doses of metaphosphate in conjunction with the chromate.

As the corrosion rate increases because of higher temperatures and high electrolyte concentrations (notably chlorides), the concentration of chromate necessary to maintain a protective film also increases. Failures in preventing corrosion by the use of chromate have occurred because of concentrations insufficient to overcome conditions encountered in the cooling systems.

TYPES OF SALT USED AND pH RANGE OF SOLUTION

The selection of the type of chromate to use depends on the cost, since both mono- and dichromates serve to inhibit aggressive attack. Effective protection is secured over a wide pH range, varying between 7.5 and 9.5 when treating circulating water. It is preferable to maintain a zero or negative index to avoid scale formation. No scale is necessary, since the chromate film serves as a physical dam and prevents contact at the water and metal interface.

At times the alkalinity of the water may be very high, and treatment may be required to bring the pH of the solution within the desired range. Darrin[12] suggested:

> It is sometimes desirable to substitute chromic acid (CrO_3) for part of the bichromate. This is preferable to adjusting with hydrochloric acid because the chromic acid itself is an inhibitor. One pound of chromic acid will neutralize excess alkalinity equivalent to 0.8 lb. NaOH, while 1 lb. of sodium bichromate will neutralize only 0.27 lb. NaOH. When neutralized, it requires 0.671 lb. of chromic acid to be equivalent in corrosion-inhibiting value to 1 lb. of sodium bichromate. Similarly, the equivalent quantities of potassium salts are 0.987 lb. $K_2Cr_2O_7$ or 1.303 lb. K_2CrO_4, each equal in corrosion-inhibiting value to 1 lb. of sodium bichromate. All of these chromium chemicals are commonly supplied in an anhydrous form with the exception of sodium bichromate which is a dihydrate.

QUACHROM GLUCOSATE

A number of proprietary compounds containing chromates, organic compounds, alkalies, or all three have been used in water treatment. One of these is quachrom glucosate. This material is a complex organic compound incorporating the glucose structure combined with four chromate groups. Presumably, the chromium enters into the anodic inhibiting reaction as it does when used in the form of inorganic chromates. It is claimed that the attached organic molecule makes the film elastic and much greater in area per unit of chromate. The concentrations recommended for inhibition are about a tenth as great as those required when sodium salts of chromium are used. Control is based on titrations to determine the chromate content of the water.

Some operators have chosen this method of treatment because it appears to offer a ready-made solution to their problems and requires a minimum of technical help to control.

One of the main disadvantages of quachrom glucosate has been its relatively high cost. Only when a system can be closely bottled up and losses reduced to a very small amount, can this type of treatment be economically justified. Under such conditions, if concentrations are

Fig. 9. The absence of corrosion deposits on a condenser tube sheet resulting from chromate treatment of the circulating water.

sufficiently high to maintain a film under the temperature and the salt concentrations existing, the treatment has been very effective for inhibiting corrosion. Figure 9 is a photograph of a tube sheet showing the excellent protection against corrosion deposits that is afforded by the use of this type of treatment. Prior to the inauguration of the treatment, the surfaces were heavily encrusted with products of corrosion.

COMBINED INHIBITORS

For several years, commercial manufacturers of water-treating chemicals have used a combination of film-forming chemicals. Probably the most widely utilized dual treatment to date has been that with the metaphosphates and inorganic chromates. The formulation of the percentages of chemicals to be used was determined initially by pilot plant studies and subsequently by actual treatment of recirculated systems. It has been established that metaphosphates are most effective, in many cases, at a pH value of approximately 6.0.

Recently, two investigators conducted extensive studies on the use of sodium chromate and a molecularly dehydrated phosphate. Various combinations of the phosphates and chromates were employed within different pH ranges. According to the published data, this "dianodic"[13] treatment was most effective at a pH of 6.0 and with specific phosphate to chromate ratios. It was concluded that each inhibitor alone is effective but that the combined treatment provides a team of inhibitors effecting beneficial protection of iron and copper alloys at concentrations of the respective chemicals much lower than those of either chemical alone used as the inhibitor.

FACTORS AFFECTING PROTECTIVE FILMS PRODUCED BY CHROMATES, POLYPHOSPHATES, OR ORGANIC COMPOUNDS

In the discussion of controlled calcium carbonate scale, there was included a summary of requirements for successful application of this treatment based on experience for many years at various plants. Comparable experience with other types of protective film indicates several factors requiring attention, which are summarized below:

1. Units with high operating temperatures should be watched, especially if chromates (inorganic or organic) are used. The protection may fail in the hotter sections of a tubular exchanger. This has been pointed out by several operators who mentioned that, for this reason, inspections of tube sheets, exchanger heads, and tube ends which indicate good protection may not give a true picture of the condition of the tube surfaces subject to higher temperature.

2. The rate at which pit corrosion takes place at points where films are broken is extremely rapid. One reason for this is that oxygen corrosion is more aggressive at these points than at an unprotected surface because of the formation of concentration cells. Another is that the total corrosion rate has not been reduced and is concentrated on the small exposed area.

3. The formation of a protective film is greatly influenced by pH value. Generally, the phosphate films are more effective at low pH values (about 6.0), and organic chromates are more effective at a pH of 7.0 or higher. Under conditions of high pH (over 8.5), high temperature (approximately 120°F), and high saturation index (1.0 to 1.5), metaphosphate has been found to revert to orthophosphate, resulting in the formation of sludge, which deposits on tube surfaces and prevents renewal of the film.

4. Before starting the treatment, reasonable attention should be given to flushing the system and removing all the mud, wood chips, and paper that have collected during the construction period. The presence of this

foreign matter will result in incomplete film protection either by actually blanking off parts of the tube surfaces or by causing extremely high velocities in certain sections of the heat exchanger, thus preventing renewal of the film as rapidly as it is removed.

5. The effectiveness of film-type protective treatments depends on the continuous renewal of film on exposed surfaces which are being continuously swept by water, thus introducing the velocity factor for consideration. It has been proved conclusively that corrosion increases rapidly with water velocity when ferrous hydrate films are involved. Because these films must be renewed continuously, it is imperative that the concentrations of protective chemicals be maintained without interruption.

6. The surfaces on which protective films are to be used for coating should be clean and free from corrosion before the process is started, otherwise corrosion cells already established are more likely to be accelerated than retarded.

PHYSIOLOGICAL EFFECTS OF CHROMATES ON PERSONNEL

With reasonable care, chromates can be used without hazard to persons handling these chemicals. However, skin disease (dermatitis) and serious eye and nasal irritation can result from the use of the chemical dry or in solution. Persons handling these chemicals should be warned to avoid any contact of the material with the skin, and special precautions should be taken to prevent breathing the dry chemicals or getting them or solutions of them in the eyes. These precautions should be taken by those handling the chemicals; and precautions should be taken to prevent personnel from being sprayed by water containing chromates, such as that from vapor and sprays discharged from cooling towers in which the circulating water is treated with these chemicals.

The contact of chromates with skin which has cuts or abrasions causes chrome sores, a type of skin ulceration. Such a skin disease is difficult to cure, and the open ulcers are likely to be reinfected.

All persons handling these chemicals should be provided with rubber gloves, goggles, and inhalators. There should also be installed suitable eye showers adjacent to the point where the solutions are handled. Darrin[12] advises the following:

> Accidental entry of bichromates into the eye should be treated as quickly as possible by flushing the eye copiously with clean water for at least five minutes, after which a physician should be consulted without delay. Failure to act quickly in flushing with water may add substantially to the degree of eye irritation.
>
> Under no circumstances should bichromates be allowed to enter the nose. The possibility of nasal septum irritation is greatest where the bichromate is in air suspension and perforation of the nasal septum ultimately may occur if protective measures are not taken.

Chromates carried in water spray will stain painted structures and are particularly objectionable when they come into contact with automobiles, since they leave a yellowish-brown stain on the paint.

Reasonable precaution in the handling and preparation of chromate-treated waters will prevent the foregoing hazards, of which operators should be aware. These difficulties occur not only with individual chrome salts but also with chrome salts compounded with organic material.

CONTAMINATION OF WATER SUPPLIES BY CHROMATES

The manner of disposal of the blowdown water from cooling towers must be considered when corrosion is controlled by chromates. Such disposal poses a problem if these wastes are to be discharged into surface waters which are used for drinking purposes. Care also should be taken to avoid contamination of well water through seepage into the aquifer supplying wells.

PROTECTION OF EQUIPMENT BY OIL

It is well known that little corrosion takes place in a condensate system where the steam contains oil. In such a system, the oil film laid down on the metallic surfaces serves as a physical dam and prevents corrosive attack by water.

If the condensate is not to be reused or if it is to be treated for the removal of oil, excellent protection of condensate lines can be effected by spraying oil with the exhaust steam. One objection to this type of treatment is that, when the water is recirculated over towers, the wood is likely in time to become saturated with oil, and a fire hazard is created whenever the towers are taken out of service for any appreciable length of time. In systems where towers are not used, the addition of oil or other filming substances may be justified, provided that oil in the water is not objectionable in processing or wherever in the system it will come into direct contact with the product manufactured.

WETTING AGENTS

Wetting agents have been found effective when used along with corrosion inhibitors. Such agents make it possible to effect penetration of scale, corrosion products, or other materials which have formed on the metal surfaces and which normally prevent contact between the dissolved inhibitor and the metal. They have been employed most successfully where it is practically impossible to clean the system. In general, the use of wetting agents does not involve an appreciable expenditure, especially in recirculated systems where relatively small quantities of water are treated. The expenditure is limited to losses due to blowing down and windage. Only a few parts per million of the wetting agent are required to penetrate the scale, corrosion, and other deposits.

CORROSION PREVENTION BY VACUUM DEAERATION IN RECIRCULATING WATER SYSTEMS

Waters recirculated over cooling towers or cooled in spray ponds are generally corrosive to ferrous and nonferrous metals because of constant reaeration, which results in a high degree of saturation with dissolved oxygen. Where the chemical composition of the water is such as to lay down scale, the presence of oxygen causes no trouble, but where reaerated

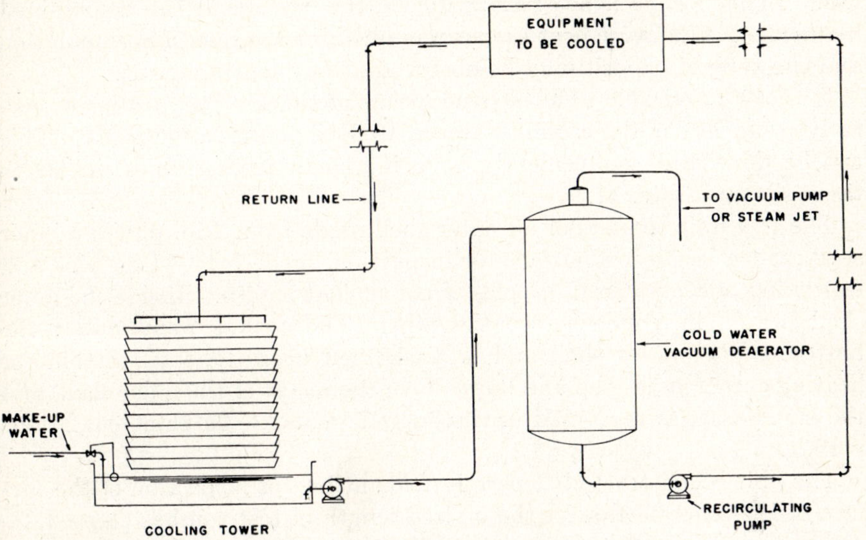

Fig. 10. Schematic arrangement of equipment for vacuum deaeration of circulating cooling water.

water is in contact with bare metal, rapid corrosion occurs. The rate at which the attack develops depends on the type of metal, temperature, pH value of the supply, and other factors. If conditions are such as to preclude conditioning of the water to lay down a controlled scale or to form film by chromates, phosphates, oil, or other filming products, deaeration of the recirculated water offers a means for inhibiting corrosive attack. Where such systems are used, the circulating water, after passing through the cooling towers or ponds, is deaerated under vacuum. Such systems are capable of greatly minimizing the corrosive attack of the water in contact with the metal surfaces throughout the cooling system.

A typical flow sheet of a recirculating system in which the cooling water is deaerated at each cycle is shown in Fig. 10. The aerated cooling water, after being cooled by sprays or in a tower, is pumped or sucked into a vacuum chamber, and by maintaining a vacuum of 25 to 28 in. Hg or higher, the noncondensible gases are largely removed.*

CONTROL PROCEDURES: COOLING SYSTEMS

Since the deposition of protective coatings depends on maintaining constituents in the water within specified limits, it is obvious that these protective measures can be adequately controlled only by routine testing of the water and adjustments of the treatment. This holds true regardless of the type of water conditioning practiced. However, it is necessary to have some indication of the effect of certain variables on the equipment used. The most satisfactory guidance of the treatments can be obtained by means of miniature heat exchangers in which the rate of accumulation and the type of deposit may be observed at frequent intervals.

It is often desirable to have some means of checking the scaling or corrosive tendency of the cooling water at various points in the plant. This can be done quite conveniently by using small testing units similar to those shown in Fig. 11.

The test unit consists of an outer shell made from 2 in. pipe, an inner shell, which is the test surface for inspection, made from ¾ in. pipe and consisting of a number of nipples of convenient length. Inside the inner shell is a ¼ in. condensate outlet pipe. The cooling water enters the bottom of the outer shell and is discharged near the top. Steam for heating enters at the top and flows down the inside of the inner shell, and the condensate is recovered or discharged to waste, as plant conditions permit.

The unit is constructed to permit dismantling for inspection of the surface of the inner shell after the desired length of test run has elapsed.

The test run consists of a period of operation during which the flow of cooling water into the unit and the flow of heating steam are so regulated that the outlet temperature of the cooling water corresponds to the temperature of the water discharged from the piece of equipment under study. These units can be installed any place in the plant where critical pieces of equipment are located and a close watch can be kept on the effect of water on the heat-transfer surfaces.

A simple procedure for determining the rate of scaling or corrosion can be carried out with testing spools. These consist of pieces of pipe inserted in the plant system which normally have a flow of water through them; they can be bypassed for removal without shutting down the plant equipment. A typical arrangement of the test spool and bypass around it is shown in Fig. 12. The advantage of this arrangement is that it affords an indication of the protection being effected, although this result may

* For more detailed information on vacuum deaeration, refer to Chap. 10.

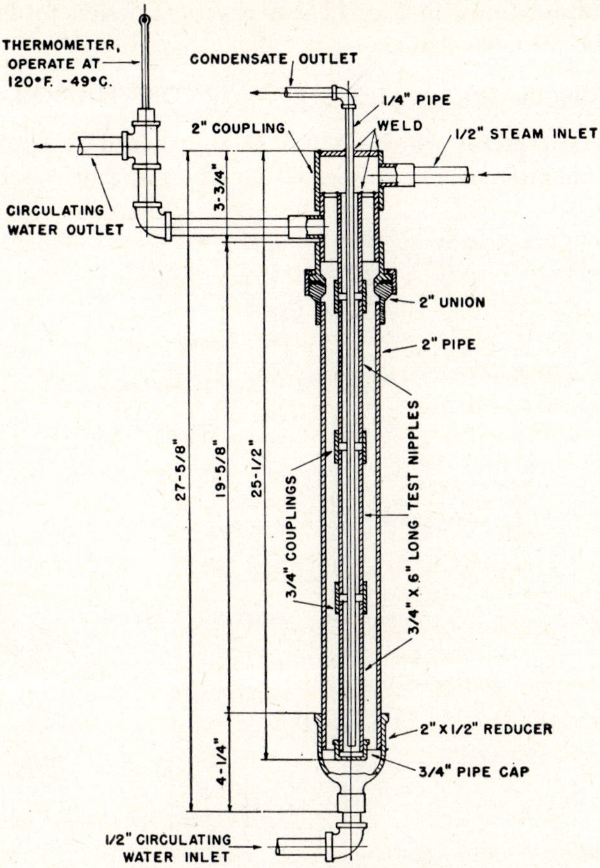

Fig. 11. Corrosion-testing unit. *(Used by permission of Dow Chemical Co., Freeport, Tex.)*

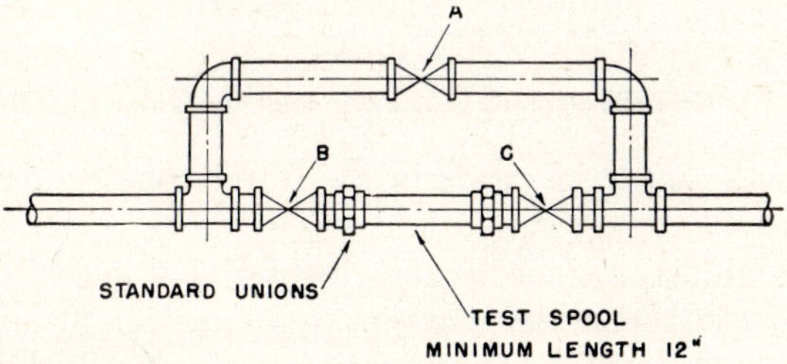

Fig. 12. Typical arrangement of valves and fittings for installation of corrosion test spools.

not be actually comparable with scale in the plant equipment. Such test equipment serves as a guide to aid control of scale but cannot be solely relied upon.

If it is advantageous to observe the rate of attack without dismantling the apparatus or if loss of weight of metal is to be determined, the testing assembly shown in Fig. 13 is a practical design to use for these requirements.

EFFECT OF CIRCULATING WATER ON TOWER-CONSTRUCTION MATERIAL

An important factor which should be taken into account to assure satisfactory maintenance of towers is the effect of any water treatment

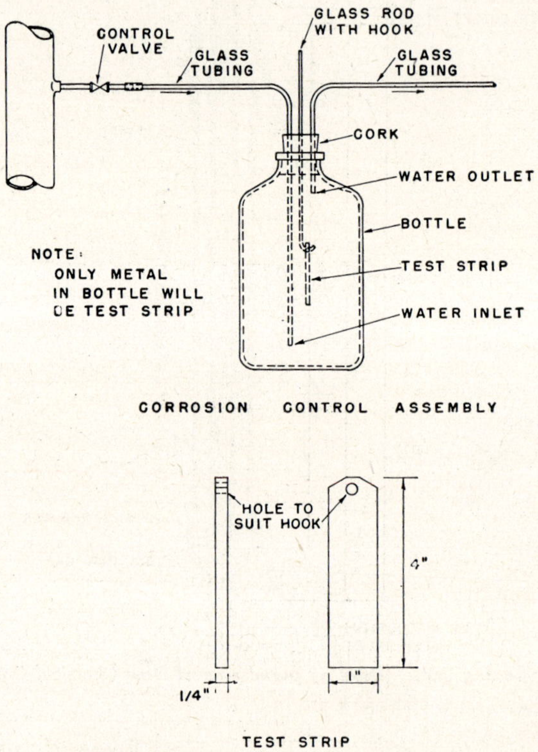

Fig. 13. Corrosion-control test equipment for water pipelines.

employed upon the materials used in the tower construction and in auxiliary equipment.

As stated previously, since waters recirculated over towers become reaerated, they are generally aggressive to metal, principally ferrous material, because of the high concentration of air absorbed during passage through the unit. This occurs in natural as well as forced-draft towers.

Because of this condition, it is preferable to employ nonferrous materials in the construction of units, but where this may not be done, metal parts should be protected by paints or other corrosion-resisting coatings.

Frequently, the wood used in the construction of towers is seriously damaged by the action of concentrated salts deposited on the wood by

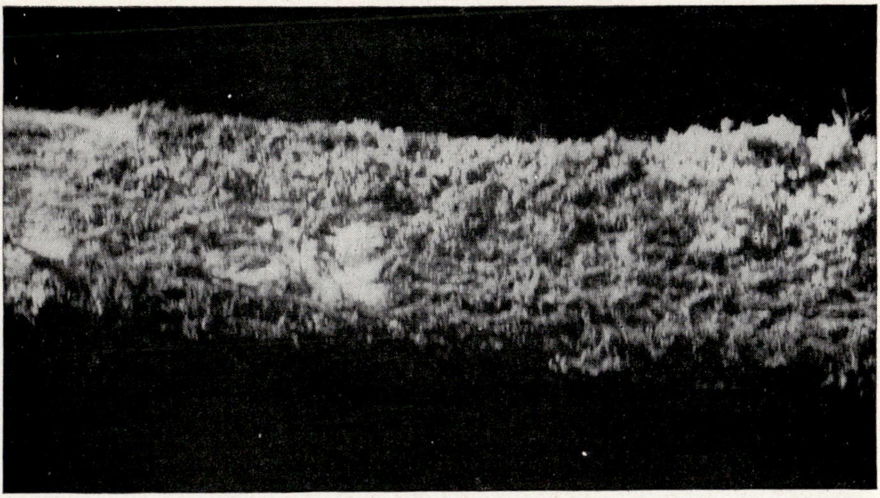

Fig. 14. The appearance of wooden slats from cooling towers heavily encrusted with sodium salts which cause disintegration of the wood fiber.

Fig. 15. Wooden slats showing delignified cooling-tower wood resulting from chemical attack.

evaporation and by fungus growth. The concentrated sodium salts are particularly destructive to many types of wood. When such attack occurs, the resins are leached from the wood, the material softens, and fibers are frayed. Figures 14 and 15 are photographs of wooden slats from towers which were badly damaged by permitting high concentrations of solids, mostly sodium compounds, to salt out on the wooden surfaces. When such action occurs, the strength of the structure may become so seriously weakened that it will require replacement. The action can be minimized if the towers are washed down at intervals whenever there are indications of salting out of alkalies. Caution should be used in order to avoid flushing fibers into the circulation system. In some cases observed by the author, serious operating difficulties occurred when large amounts of fibrous material were flushed into the pump and other equipment. Adequate conditioning of the circulating water and the proper selection of material used in the construction of towers will greatly reduce the rate of destructive chemical and biological attack on towers.

R. D. Baker,* in a comprehensive article, has discussed in detail the principles leading to an understanding of wood deterioration.[14] One point brought out in the investigation is the effect of fungi on various types of wood. The effect of high alkaline salt on the destruction of wood fiber has been known, but knowledge of the influence of fungi has not been widespread. Baker points out that the decay due to this type of attack "is not confined to any one make or type or location. Attacks have been found in atmospheric deck towers, forced draft and conventional induced draft towers and in double flow towers."

In describing wood constituents and the nature of the chemical and biological attack, Baker states as follows:

> Pulping, or delignification, is probably the most important commercial process dealing with the chemical decomposition of wood. This process removes lignin, pectins and resins from the unbroken or partially disrupted cellulose. It should be understood that no process will remove any one constituent without altering or attacking some of the others.
>
> Nitric acid delignifies wood by a combination of nitration, oxidation and hydrolysis. Alkaline salts produce delignification while neutral salts at high temperature and pressures act as acids to degrade cellulose. All oxidizing agents attack lignin. Drying of alkali-impregnated wood is serious because of the swelling action of alkali on cellulose, and because it promotes oxidation. Drying wood impregnated with mineral acids, or salts that hydrolyze to acids, makes wood brittle because of the formation of hydro-cellulose.
>
> Destruction of wood by biological attack may, like chemical deterioration, serve a useful purpose or it may be undesirable.
>
> Bacteria will attack sugars, starches and some forms of cellulose; but it is doubtful if any of them decompose wood. The simple fungi, or molds, decompose cellulose in soils and composts and attack various parts of living plants and decompose hemi-cellulose, especially in partially decayed woods. The fleshy fungi are the true wood destroyers, and most of them attack cellulose and produce "brown rot." Some species destroy lignin and produce the rarer "white rot."
>
> Fungi will not attack wood that is either too wet or too dry. Fungi require a moisture content in excess of fiber saturation, and 40 to 50 per cent moisture is

* R. D. Baker, research engineer, The Marley Company, Inc.

considered optimum. Decay does not occur in submerged wood, but it has been found in redwood eliminator slats removed from a cooling tower when the moisture content was 265 per cent of the dry weight. Optimum temperature for fungi is between 68 and 86.

Cellulose resists most chemicals but is readily decomposed by many of the micro-organisms. While fungi are selective in their attack on cellulose or lignin only a small minority prefer the lignin. Fungi will attack pure cellulose, but even those that destroy lignin in wood will not attack isolated lignin. This indicates that lignin serves to protect the cellulose from decay.

Fungi are efficient in utilization of nitrogen and other minerals so there is some question about the effect of the phosphate used in water treatment. Activity is affected by pH value. Alkaline salts seem to favor decomposition, possibly because organic acids formed by the fungi are neutralized. Some fungi are quite sensitive to tannin and other extractives present in wood. The more durable woods such as black locust, cedar, cypress and redwood are rich in volatile oils that are toxic to fungi or contain resins that afford protection.

Effects of fungus attack is acid hydrolysis followed by oxidation. Rate of degradation of cellulose or lignin depends on the fungus, and cellulose is the part usually attacked. Removal of lignin renders cellulose more digestible by microorganisms or higher animal life. All lumber is subject to attack when conditions of moisture and temperature are favorable, but the extractives impart natural durability to various species.

Baker lists a number of woods in the order of the relative durability of their heartwoods exposed to outdoor conditions, but notes that "there is no similar comparison available concerning the relative durability of wood used in cooling towers."

The author claims that redwood contains natural extractives that make it highly resistant to decay, and it is by far the most commonly used lumber for cooling towers in this country. Water cypress is also in high demand because of its resistance to decay; however, as pointed out by Baker, there are different types of this wood growing under different conditions and varying greatly.

CONTROL OF SOLIDS IN CIRCULATING SUPPLY

Regardless of the type of conditioning of the circulating water to correct its corrosive or scaling properties, it is imperative to maintain and control both soluble and insoluble salts at some predetermined maximum concentration. The permissible concentration of salts in the water depends on the type of service of the cooling water, the design of the heat-exchange equipment, and a number of other factors. The control of concentration of solids in the water will be governed by the type of tower, solids in the water after treatment, windage and blowdown losses, etc. The curves in Figs. 16 and 17 will be found convenient for determining the required blowdown with different tower operations.

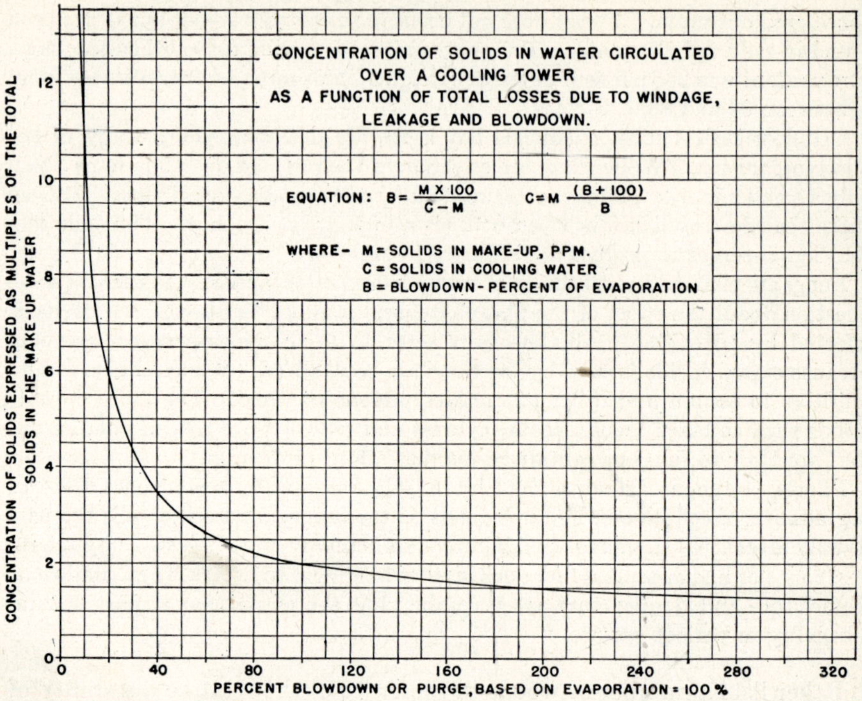

Fig. 16. Concentration of solids in water circulated over a cooling tower as a function of total losses due to windage, leakage, and blowdown.

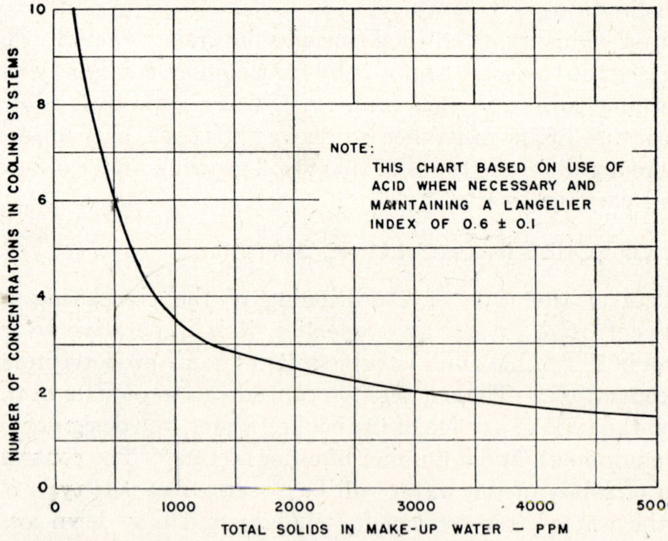

Fig. 17. Estimated allowable concentration of make-up water in cooling system for various amounts of total solids in the make-up water.

Bibliography

1. Langelier, W. F.: The Analytic Control of Anti-corrosion Water Treatment, *J. AWWA*, **28**: 1500 (1936).
2. Massink, A., and J. A. Heymann: Significance of Hydrogen-ion Concentration in Drinking Water and Particularly for the Business of Water Supply, *J. AWWA*, **8**: 239 (1921).
3. Tillmans, J.: "Die chemische Untersuchung von Wasser und Abwasser," 2d ed., W. Knapp, Halle, 1932.
4. Baylis, John R.: Treatment of Water to Prevent Corrosion, *J. AWWA*, **27**: 220 (1935).
5. Hale, Frank E.: Pipe Corrosion Experiments, Catskill Supply, New York City, *J. AWWA*, **26**: 1315 (1934).
6. Hale, Frank E.: Effect of Excess Lime Hydrate upon Corrosive Soft Water, *J. AWWA*, **27**: 1199 (1935).
7. Baylis, John R.: Prevention of Corrosion and Red Water, *J. AWWA*, **15**: 598 (1926).
8. Hoover, Charles P., and James M. Montgomery: Reduction of Carbonate Hardness by Lime Softening to the Theoretical Limit, *J. AWWA*, **21**: 1218 (1929).
9. Hoover, Charles P.: Practical Application of the Langelier Method, *J. AWWA*, **30**: 1802 (1938).
10. Powell, S. T., H. E. Bacon, and J. R. Lill: Corrosion Prevention by Controlled Calcium Carbonate Scale, *Ind. Eng. Chem.*, **37**: 842 (1945).
11. Powell, S. T., H. E. Bacon, and E. L. Knoedler: Corrosion Prevention by Controlled Calcium Carbonate Scale, II, *Ind. Eng. Chem.*, **40**: 453 (1948).
12. Darrin, Marc.: Corrosion Inhibition with Chromate, *Oil Gas J.*, Jan. 13, 1949, p. 83.
13. Kahler, H. Lewis, and Charles George: A New Method for the Protection of Metals against Pitting, Tuberculation and General Corrosion, *Corrosion*, **6**: 331 (1950).
14. Baker, D. R.: Fundamental Principles Leading to an Understanding of Wood Deterioration, *Industry and Power*, **59** (3): 94 (1950).

14 STEAM CONTAMINATION AND MEANS FOR CORRECTION

The production of clean dry steam has occupied the attention of engineers in the stationary- and railroad-boiler fields for many years, but it has become a much more urgent problem as the efficiency of steam-generating equipment has been increased by the use of high-pressure boilers. The improvements in boiler design which have met the demand for economies in steam production have, in many cases, worked against steam quality. As a result, there has been an intensified interest in the mechanism of steam contamination, its measurements, and its correction.

As in many other problems in boiler operation, there is much controversial evidence on the subject due, in part, to lack of scientific facts and to the formulation of conclusions based on insufficient and unreliable data.

When boiler water is entrained in steam, it may leave deposits in steam mains and valves, superheaters, and turbines. If such deposits occur to an appreciable extent, they are soon reflected in loss of efficiency or damage to the equipment.

DEFINITION OF TERMS

The phenomena causing boiler water to be carried over with steam have often been loosely defined, but as a result of the work of a number of investigators, there is now general agreement on definitions.

"Priming" is the term usually reserved for the propulsion of liquid into the steam drum by extremely rapid, almost explosive boiling of the water at the heating surfaces. It is a spasmodic action in contrast to the steady carry-over which may characterize normal operation. When priming occurs, the entire column of water in a tube may become so filled with steam bubbles that the density of the mixture is decreased and it expands rapidly into the drum, expelling water into the steam space and the steam outlet tubes.

The phenomenon of "foaming" occurs when bubbles of steam rise from the heating surfaces, forming a layer of foam on top of the water. The foam may fill the steam space to such an extent that some of it enters the steam tubes. Water can be carried into the steam space by means other than priming and foaming. The bursting of bubbles, even in the absence of much foam, will result in a relatively small and fairly steady entrainment of moisture into the steam.

A third condition responsible for deposits in the steam cycle is due to certain constituents, principally silica, dissolving in the steam and later redepositing on turbine blades when critical reduction of pressure and temperature occurs.

INFLUENCE OF BOILER CONSTRUCTION ON PRIMING AND FOAMING

The construction of the boiler has a marked effect on the priming and foaming tendencies of boiler water. Insufficient steam-disengaging space, for example, is one of the most important factors. Size of the steam header, velocity of the steam leaving the boiler, and other conditions accelerate both priming and foaming. The more obstructions such as stay bolts, braces, baffles, etc., within the boiler, the greater will be the tendency for these operating difficulties to occur.

It is obvious that the larger the amount of steam which must break through a given area of water surface on rising through the tubes, the more turbulent will be the water in the steam-disengaging drum and the greater will be the opportunity for priming and foaming. Similarly, the larger the steam space above the water, the more difficult it will be for moisture to reach the steam tubes. Moreover, when the steam-storage space is small, a sudden demand for steam will result in a sharp pressure drop, followed by sudden evaporation at the heating surfaces, that may cause priming. Production of dry steam will, therefore, be facilitated by boiler design providing the maximum water surface and steam space, *i.e.*, the largest drum diameter.

The steam-disengaging area is greatest when the water level is at the middle of the drum, and since lowering the water level increases the steam-storage space at the expense of steam-disengaging area, it is of interest to consider which of these factors is more important. H. Vorkauf[1] contends that the decisive factor is the steam-storage space, and that if the water level were dropped until it was in the tubes, reducing the disengaging surface to zero and giving maximum steam-storage space, the best steam quality would be obtained. He cites experiments at the Technical University of Darmstadt in which the allowable load increased by more than 150 per cent when the water level in a drum 3 ft in diameter was lowered 12 in. Vorkauf also emphasizes the importance of drum size, pointing out that, as the doubling of the diameter of the boiler drum doubles the steam-disengaging surface, it multiplies the steam storage space by four.

Although there appears to be justification for attaching major importance to steam-storage space, it is obvious that the total area available for passage of steam through the water is also important. While the author agrees in general with the principles laid down by Vorkauf, diametrically opposite results may be obtained under some conditions. An extensive investigation under the direction of the author to determine carry-over in one boiler plant showed that, by lowering the water level in the drum, carry-over was increased, whereas, when the water level was at the center of the drum or slightly higher, carry-over was reduced. The explanation is probably found in the position of the steaming tubes entering the drum. These tubes were so designed that, when the water level was lowered, circulating water was jetted into the drum violently because of the low head of water over the tube outlet. It is interesting to note in the case mentioned that concentrations of the boiler-water salines were of relatively minor importance and that the extent of moisture carried into the steam was not dependent upon concentration of solids in the particular boiler. However, the actual amount of solids entering the steam cycle will, of course, be greater under given conditions if the concentration of boiler-water salines increases.

The author observed two similar cases of severe steam contamination with respect to the control of water level in the two top drums. The steam which was disengaged in the front drum was carried by tubes to the rear drum, which was at the same elevation, and thence to the superheater header. However, the water-level regulation was applied to the rear drum, where the density was greatest, and the water in the front drum rose until at times it flooded the steam overcomers. These tubes accumulated deposits of solids from the boiler water at the tube end which discharged into the rear drum, as indicated in Fig. 1. The high velocity resulting from this obstruction caused a jetting against the water in the rear drum, and the spray thus created was carried into the superheater header. Many similar cases have been observed and recorded by other investigators.

Although certain aspects of modern boiler design do not favor production of steam of good quality, boiler manufacturers are turning their attention to features of design which will overcome this difficulty and are including baffles, steam purifiers, and other devices.

As is well known, contamination of steam increases sharply with rate of evaporation. Vorkauf emphasized this by making the "permissible load" the criterion of performance in his experimental work. The relation of quantity of steam to be produced to the physical facilities for releasing it can be upset by design deficiencies in these facilities and by evaporation at an excessively high rate. The latter condition may be due to deliberate overloading, forcing the production of more steam than the design contemplated, or to unavoidable incidents of operation. Sudden opening of a steam valve or cutting in a boiler quickly may draw excessively upon the stored steam, and the resultant pressure drop will cause violent evolution of bubbles. If a sudden rise in temperature is caused by removal of scale from a heating surface, the evolution of steam

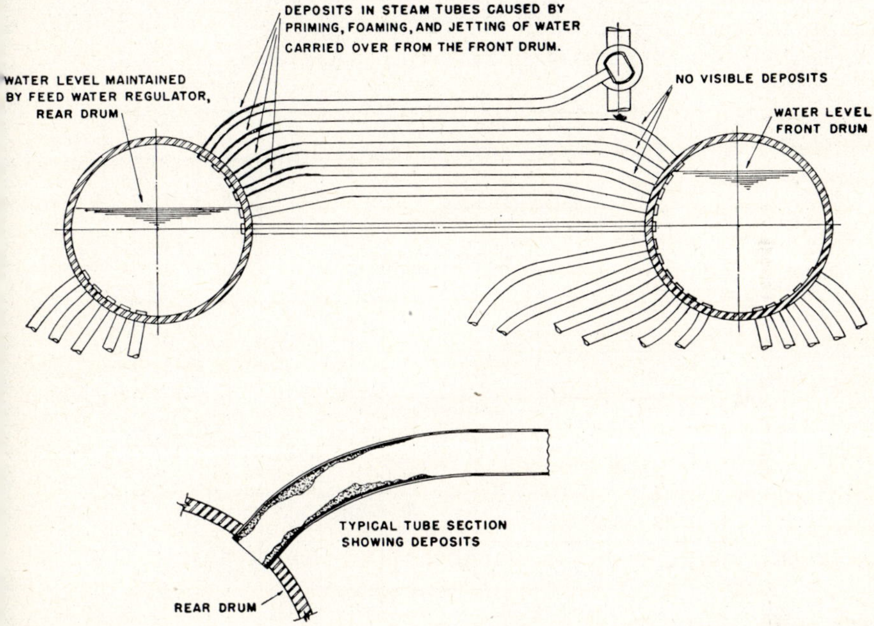

Fig. 1. Section of Erie boiler showing location of deposits in steam tubes connecting front and rear drums.

may be almost explosive in its violence. Obviously, the more fluctuating the load, the more frequent will be pressure drops that may cause priming.

In considering the phenomenon of priming, it is of interest that studies have been made of the velocity of rising gas bubbles in liquids. The relative velocity with which the bubbles pass upward through the water rising from an air-lift pump has been determined by Hoefer[2] and Pickert[3] from measured weights of air and water. The latter author states as follows: "Relative air velocities of 20 meters per second and more have thus been demonstrated." The velocity is expressed by Pickert as a function of the quantity of air released per second divided by the cross section of the tube and is not directly proportional to this fraction but to a higher power of it:

$$V = \frac{C}{D_w - D_m}\left(\frac{W}{A}\right)^{1.23}$$

where V = relative velocity of air bubbles, meters per sec
C = constant depending on temperature
D_w and D_m = densities of water and the mixture of water and air, respectively
W = weight of air blown in, grams per sec
A = area of tube, sq meters

Hypothetical values applied to this formula, using a 4-in. tube, give the curve shown in Fig. 2, where the relative velocity of the air bubbles is plotted against quantity of air released per second.

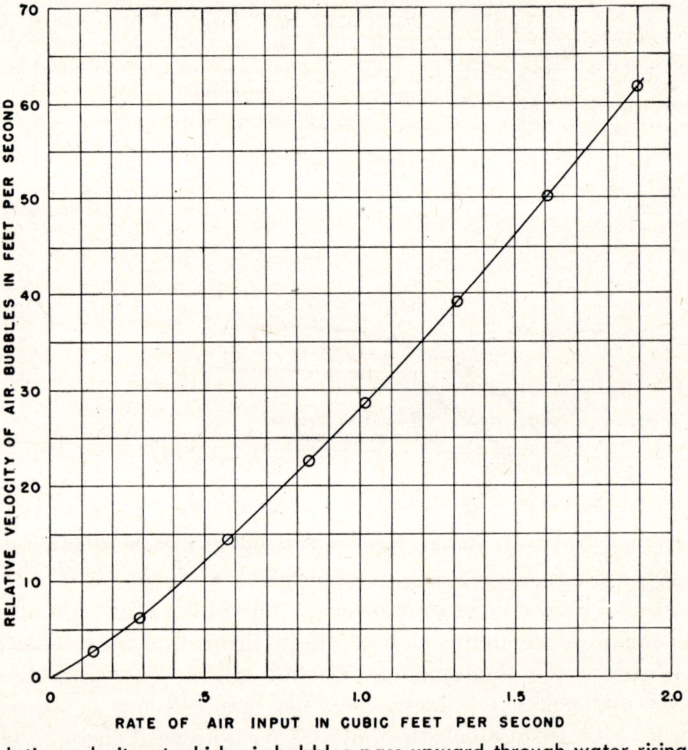

Fig. 2. Relative velocity at which air bubbles pass upward through water rising from an air-lift pump, based on hypothetical values using a 4-in. tube and the Hoefer and Pickert formula.

The relations derived from the performance of an air-lift pump are analogous to those that might be expected to hold for the rise of steam bubbles in nearly vertical boiler tubes, although changes in temperature, pressure, viscosity, surface tension, and other factors would have to be taken into account.

Priming and foaming are less likely to occur at high pressures than at low because the bubbles apparently are produced near the water surface and their size is much reduced. Working with an experimental boiler, Joseph and Hancock[4] found that, as the pressure was reduced from 155 to 70 lb, four to five times as much boiler saline was carried over with the steam. In an experimental boiler studied by J. A. Holmes,[5] it was possible to look through the drum by means of opposite sight glasses, an incandescent light being installed behind one of these. It was found that the steam was opaque at high pressure because of the bursting of bubbles, which seemed to be confined almost entirely to the surface, whereas the water was transparent. As the pressure was lowered, the water became opaque because of the bubbles formed, culminating in a thick foam on the surface, and the steam became transparent.

The influence of pressure on steam quality was demonstrated at a power station where 400-lb boilers and turbines had been added to the original 200-lb equipment. An attempt was made to operate the 400-lb boiler at 200 lb while part of the high-pressure station was down for inspection. The steam-generating space of the boiler had been designed for satisfactory operation at 400 lb, and at the lower pressure, severe carry-over and fouling of the superheater were encountered. It is a recognized fact that the steam drums on high-pressure boilers are made smaller than those for lower-pressure boilers because of the difference in density of the steam released under different operating pressures. Frequently, this is ignored by operating engineers, and serious troubles from carry-over have occurred which could have been avoided if the basic facts concerning the boiler design and operating pressures had been taken into consideration.

LEAKY BAFFLES

Frequently, minor leaks through baffles result in heavy carry-over into superheaters and turbines. Too little attention is given by operators to such defects. Very minor cracks often are sufficient to permit serious contamination of the steam. A typical failure of this kind is shown in Fig. 3.

CHEMICAL COMPOSITION OF BOILER WATER AND ITS EFFECT ON CARRY-OVER

Theoretical Concepts. Boiler operators are aware that priming and foaming are promoted by an excess of soluble salts and precipitated sludge in boiler water, but the manner in which they exert their influence is not entirely established and has been the subject of much study. Attention has been directed mainly to seeking an explanation of the mechanism by which dissolved and suspended solids assist the formation of foam or are carried by the steam in solution. The most extensive published writings on this subject are those of Prof. C. W. Foulk[6–12] of Ohio State University and his associates. The work of Niehaus[13] and Vorkauf[1] has added greatly to our knowledge of this problem, and a review of these investigators' findings is recommended to those encountering such problems. Extensive experimentation has been conducted at the Armour Research Foundation, in many universities and colleges, and by private water-conditioning companies.

It is apparent that the ability of bubbles to form on the surface of boiler water is related in some manner to substances in the water. If a bubble of steam rises in perfectly pure water, it will penetrate the surface and pass into the air without forming a film. When two bubbles collide

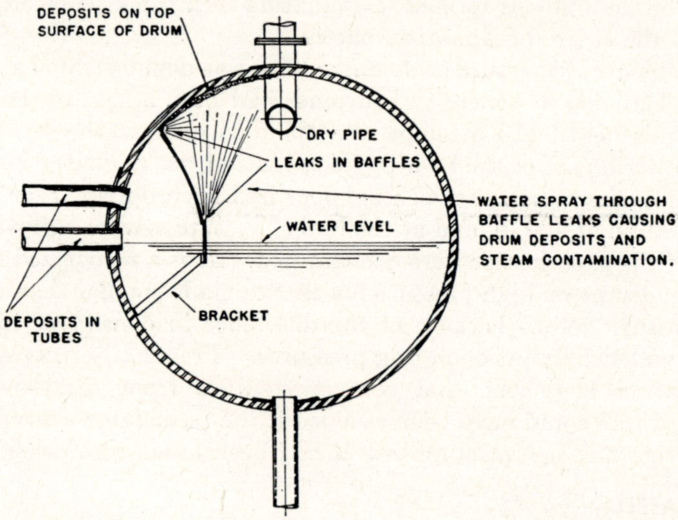

Fig. 3. Contamination of steam resulting from carry-over through leaks in baffles.

under the surface of pure water, they merge instantly to form a larger bubble, and this also passes to the space above the water without remaining on the surface. On the other hand, solutions such as boiler salines have the ability to form films that constitute bubble walls, and even though these have only a momentary existence, they permit the formation of bubbles on the surface of the water sometimes in such numbers as to produce foam. When steam bubbles are formed in boiler water at the heating surfaces in the tubes, they are small because a film separates them, and a film persists when they push up the surface of the water. There is thus built up on the water a layer of small bubbles characterizing a foam.

Investigators in the field of physical chemistry long ago established a theoretical basis for understanding foam formation and have demonstrated experimentally many facts about the physics and chemistry of surfaces and thin films. Bancroft,[14] in discussing the essential condition of foam formation, stated as follows:

Freundlich[15] considers that no pure liquid can foam and in this he is apparently right. He then deduces that a low vapor pressure, a low surface tension and a high surface viscosity are essential to foam. . . .

To get a foam the only essential is that there shall be a distinct surface film; in other words, that the concentration in the surface layer shall differ perceptibly from that in the mass of the liquid. All true solutions will therefore foam if there is a marked change of surface tension with concentration, regardless of whether the surface tension increases or decreases. All colloidal solutions will foam if the colloid concentrates in the interface or if it is driven away from the interface. To get a fairly permanent foam the surface film must either be sufficiently viscous in itself or must be stabilized in some way. . . .

Professor Foulk sought to apply these concepts to the foaming of boiler water. According to this theory, the stability of a bubble wall, however brief, depends on the fact that the wall consists of three layers, one at the outer surface, one at the inner surface, and one between the other two. The concentration of dissolved salts in the surface layers is different from that in the liquid between them, and as a result, both the inside and outside surfaces of the film are acted upon by forces which distinguish them from the thin middle layer. There is a difference in surface tension, resulting from the difference in concentration, which restrains the liquid of the bubble wall between the two surfaces and gives the film a brief stability.

Foulk demonstrated a correlation between foaming, the failure of bubbles to coalesce under water, and differences in surface tension such as those described above. Although these experiments were made at room temperature and fairly low pressure, the effect of dissolved solids on foaming in a boiler is also a matter of practical experience and has been studied by a number of other investigators. Holmes, working with the experimental boiler, established the relative concentrations of various salts that could be tolerated without foaming. Giving sodium carbonate, believed to be the worst foam producer, a value of 100, he obtained by his measurements the following relative permissible concentrations:

Sodium bicarbonate	110
Disodium phosphate	140
Disodium phosphate + sodium aluminate	160
Sodium bicarbonate + sodium aluminate	195
Caustic soda + sodium aluminate	205
Sodium aluminate	275

On the other hand, A. R. Mumford,[16] in an investigation of a large number of boilers, was convinced that hydroxide was the sole chemical influence in steam contamination. According to Hall,[17] a few parts per million of calcium will retard foaming. Niehaus believed that the sodium phosphate concentrations must be limited to produce dry steam, and

Vorkauf summarized his experiments by saying, "These results justify the assumption that the ability to cause foaming is about the same for all solutions investigated."

There has long been a general belief that sodium salts were related to carry-over and evidence justifies the conclusion that in excessive concentrations they promote foaming. There is little doubt that, where such difficulties have been experienced, high concentrations of these salts have occurred. There are ample reliable data, however, proving that it is possible to have excessively large amounts of sodium salts present in boiler water without these troubles. Taplock and Thompson[18] showed that boiler water containing sodium salts in concentrations of 50,000 ppm did not prime.

C. Herschel Koyl,[19] reporting on the foaming of water in locomotive boilers, stated that no foaming occurred at very high concentrations of sodium salts until suspended matter was present in the water. Recently, experience and special studies have confirmed the theory that priming and foaming are due not only to concentration of sodium or other soluble compounds but also to the presence of suspended matter in addition to these substances in relatively high concentrations. The author has observed boiler operation where no appreciable steam contamination occurred although the sodium chloride concentration in the boiler water was over 14,000 ppm, a similar condition reported by Elliott.[20] Many others have operated boilers containing sodium chloride concentrations at values between 15,000 and 25,000 ppm without any difficulties. However, it is not recommended that such high concentrations should be carried since the detrimental results from high concentrations are often reflected in damage to equipment not related to the quality of the steam.

EFFECT OF SOLUBLE AND INSOLUBLE SOLIDS ON STEAM QUALITY

It has been generally believed that boiler sludge has some relation to carry-over, and a series of studies of the effect of suspended solids on foaming of boiler water was made by Prof. Foulk and his associates several years ago. They found that various kinds of pulverized material caused foaming of saline solutions when boiled in glass receptacles in the laboratory. On the basis of these tests, minimum foaming concentrations of dissolved and suspended matter were established. It was observed that particles floated on the surface of the water and entered into the film, apparently giving it structural strength. However, the importance of this phenomenon in actual boiler operation was discovered to be doubtful when experiments were extended to a small laboratory boiler operating at 150 lb. The pulverized solids lost their ability to float and to stabilize the foam after boiling for a few minutes. As soon as the powder was completely wetted, it had no effect on foaming.

The experiments were then modified by precipitating the suspended solids in the boiler in a manner identical with the formation of sludge. The results confirmed the conclusion that boiler sludge, whatever its other influence on steam quality, does not stabilize foam to a great extent by floating on the surface and entering into the bubble wall. The foam-promoting behavior of the solids used in the first experiments was thought to be due to a coating of absorbed oil or other impurities which was removed by boiling at high temperatures. Holmes observed that sludge lost its foaming-producing properties by "aging" in the boiler. If it is possible for oily pulverized matter to stabilize foam, the question arises whether oil in feedwater might not give calcium and magnesium precipitates the same characteristics. There appears to be the possibility that this would occur if the entrance of oil and the precipitation of sludge took place more rapidly than the cleaning of the sludge by boiling.

As in the case of dissolved salts, various suspended solids differ in their effect on the carry-over from the boiler. Foulk has demonstrated experimentally that calcium carbonate increases carry-over and that magnesium carbonate decreases it in a laboratory boiler. He also noticed a great difference dependent on the conditions under which precipitates are formed. If crystalline magnesium hydroxide appears, carry-over is decreased, but if the normal gelatinous form is injected into the boiler, it promotes foaming.

It is recognized in water-purification work that the nature of floc particles may vary considerably, depending on the composition of the solution in which they are formed. Thus, we are accustomed to regard alum floc precipitated at one pH value as pure aluminum hydroxide, whereas at another pH value, adsorbed sulfate ions will give an electrical charge to the particles. Professor Foulk has suggested that a hypothesis might be developed that relates the charge on the particles of boiler sludge to their tendency to attach themselves to bubbles of foam.

EFFECT OF SEWAGE AND INDUSTRIAL WASTE CONTAMINANTS IN FEED-WATER ON STEAM QUALITY

When boiler feedwater is drawn from a supply polluted with industrial wastes, serious steam contamination may result. This is particularly the case where sewage or industrial waste containing vegetable oils constitutes a portion of the polluting substances. The presence of such products results in the formation of soaps and greatly increases the foaming characteristics of the boiler water. Similar effects may be encountered when the feedwater supply is contaminated with wool-scouring wastes, tannery effluents, meat-packing wastes and other types of industrial by-products. Frequently, these materials will volatilize at relatively low temperatures so that, even where evaporators are used, feed-waters will still possess foaming characteristics as a result of greases or oils passing over with the vapor from the evaporator or in the effluent from any other type of treatment. The author has observed that finely divided clays and other types of surface washings may, in the presence of concentrated alkaline salts, result in violent foaming.

SELECTIVE CARRY-OVER OF SOLIDS FROM BOILERS

There has been much discussion among power plant engineers and designers as to the phenomenon of selective carry-over of solids contained in concentrated boiler-water salines. In so far as the author is aware, there is little definite evidence to confirm or to disprove such occurrences. It is a fact, however, that selective deposition of solids does occur, regardless of whether or not such solids leave the boiler-water surfaces as pure salts. According to the theory of selective carry-over suggested by some observers, this phenomenon is the result of the concentration of solids in bubbles formed at the water surface within the boiler. It is contended that salts deposit out selectively in these surface films, in accordance with the solubility equilibria of the various salts present in such high concentrations as may occur when the film of solution is in contact with the steam above the boiler-water surface. When the films explode, these salts are propelled into the steam space and are carried along with the steam. However, this explanation is, so far, purely theoretical.

On the other hand, the selective deposition of certain salts on turbine blading and valve seats is an established fact. The author has analyzed solids collected from seats of turbine control valves which consisted of 95 per cent sodium sulfate and only traces of caustic soda, sodium chloride, silica, and other constituents present in the boiler salines in concentrations equal to, or only slightly lower than, the concentration of sodium sulfate. The apparent explanation is that the sodium sulfate, in passing through the superheater tubes, dried out more rapidly than the other salts and, as a dry powder, was thrown out more readily than the other contaminants in the steam either by changes in direction of the gas stream or by centrifugal force. This also accounts for deposition of silica on turbine blading. Similar conditions were experienced in another plant in which sodium chloride was deposited preferentially.

Straub[21] attributes the adherence of solids to turbine blades to sodium hydroxide in the boiler water carried over with the steam. The effect of sodium hydroxide is explained by the fact that this substance can remain in solution at concentrations of more than 90 per cent and does not reach saturation and precipitate as a dried dust, as is the case with other soluble salts. Thus, when a droplet containing sodium hydroxide is carried into the superheater, it retains a moist, pasty consistence even at very high temperatures and cements to the turbine blading other salts that would otherwise be carried through entirely dry. If, however, other salts are present in sufficient amounts, they will overcome the effect of the sodium hydroxide by their total bulk as a dry powder which absorbs the moisture.

Straub carried out laboratory experiments, impinging steam of varying composition against a specimen blade, and installing test nozzles and blades drawing steam from the main steam line of a central power station. These experiments led to the conclusion that a ratio of sodium sulfate to total alkalinity in the range between 3.5 and 4.6 markedly reduced turbine-blade deposits. This occurred even though the total solids in the boiler water were increased from 260 to about 1,200 ppm. Concerning the actual performance of the station turbine, he states as follows: "The deposit did not appear to be causing any more difficulty at the end of the ten weeks than at the end of the first four weeks." The first four weeks were prior to the increase in total solids and in sulfate-alkalinity ratios. A summary of conclusions drawn from the results of this work follows:

1. The basic material causing turbine-blade fouling is sodium hydroxide.
2. Inorganic salts, such as sulfate, chloride, and carbonates, if present in sufficient amounts, will prevent turbine-blade fouling.

Others who have tried to eliminate deposits by the maintenance of these ratios had less favorable results.

EXTENT OF DEPOSITS CARRIED OVER

Ordinarily, in well-operated boilers, a significant amount of carry-over, although troublesome because it deposits in superheaters and turbines, accumulates only over a fairly long period of time. In some cases, however, where priming occurs, the rate of accumulation of salts may be rapid and much damage can be done. A photograph (Fig. 4) shows the deposits of salines which occurred around the nonreturn valve on a 400-lb pressure boiler. This accumulation resulted from a short period of priming of the boiler. The deposit removed from the valve was about 6 in. long and about 3 in. wide. The accumulation was so great that the valve had to be completely dismantled and cleaned. Similar occurrences of this kind, although rare, have taken place in a number of boiler plants.

STEAM-PURIFICATION EQUIPMENT

Mechanical Steam Purifiers. When boilers are properly designed for the generation of steam at specified pressure and rating and when extreme changes in rating can be avoided and for composition of the boiler water controlled to prevent foaming, steam contamination may be reduced to a low value. However, fluctuating loads and other conditions frequently make it impossible to meet these requirements, and even where the best control procedure is practiced, there is always some contamination of the steam leaving the boiler. Because of these limitations and the marked trend to high pressures and temperatures, supplementary steam purification is now recognized as an essential feature of the design and operation of boiler plants. The need for steam purification is apparent when it is realized that a part per million or less of contaminating solids in the steam may deposit many pounds of solids in superheaters and turbines because of the high capacity of the steam-generating units and the necessity for long, continued service without outage of boilers and turbine equipment.

Fig. 4. Deposit accumulating in a nonreturn valve because of priming of the boiler.

Many devices for the purification of steam have been developed. Steam purifiers fall into three general groups. The first group effects the removal of entrained solids by changes in direction of the steam flow whereby the droplets of water are separated from the steam and are removed by gravity. The second group provides washing of the steam by intimate contact with boiler feedwater or condensed steam. The third type separates steam and boiler water by tangential whirling of the water in a "cyclone separator." By such action, the steam is released and the water is returned to the boiler-water cycle.

Dry Type. Figure 5 shows the construction of a familiar form of steam separator which may be installed in the steam drum of a boiler by bolting the flanged connection to the steam outlet nozzle. In the unit which is illustrated, the entering steam must pass through several rows of vertical, gutter-shaped baffles placed in a staggered arrangement so that the steam describes a tortuous path. Thc steam is split into small ribbons, thus exposing a large surface to the scrubbing action of the baffles. Because of the curvature of the path, a centrifugal separation of particles is effected. Expansion occurs as the steam emerges from each row of baffles, and the change in velocity brings about further removal of suspended particles. The steam passes into a central compartment and discharges through the steam nozzle; the particles which have impinged

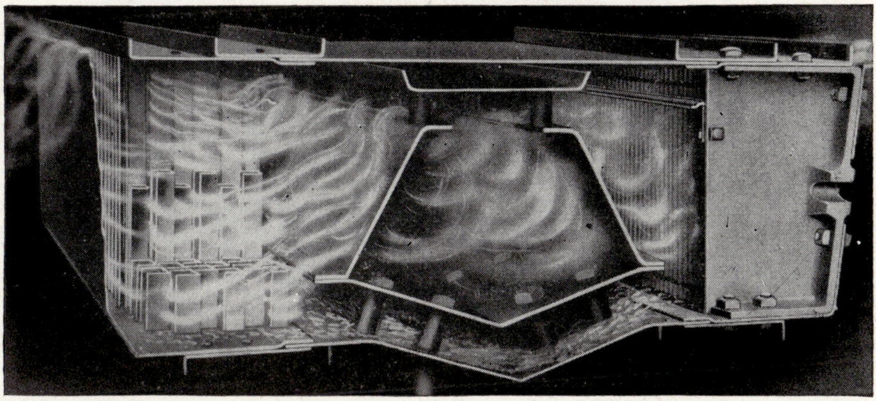

Fig. 5. Steam purifier employing baffles, centrifugal separation, and change in velocity. (*Courtesy of Blaw-Knox Company.*)

on the gutters are drained into a collecting trough at the bottom whence they may be discharged to the feedwater supply or to waste through a suitable trap.

Another type of steam separator designed to be placed inside the steam drum at the outlet nozzle is shown in Fig. 6. The steam enters the external annular section tangentially through radial blades which impart to it a vigorous whirling motion. Centrifugal force drives the moisture and solids toward the periphery to a separating zone of maximum concentration, and they fall to the bottom of the purifier. An upper collecting race is provided for particles which rise higher. The water may be returned to the boiler drum or discharged outside through a trap.

A receiver type of steam purifier, discharging to the superheater inlet, may be installed above the steam drum of the boiler. This is illustrated in Fig. 7. The steam expands into the separator drum and experiences a reduction in velocity which causes part of the suspended matter to drop out. The velocity is increased as the steam passes through the central nozzle, which directs it against a baffle. Moisture is collected by impinging on the baffle and by the change in the direction of the flow, and the

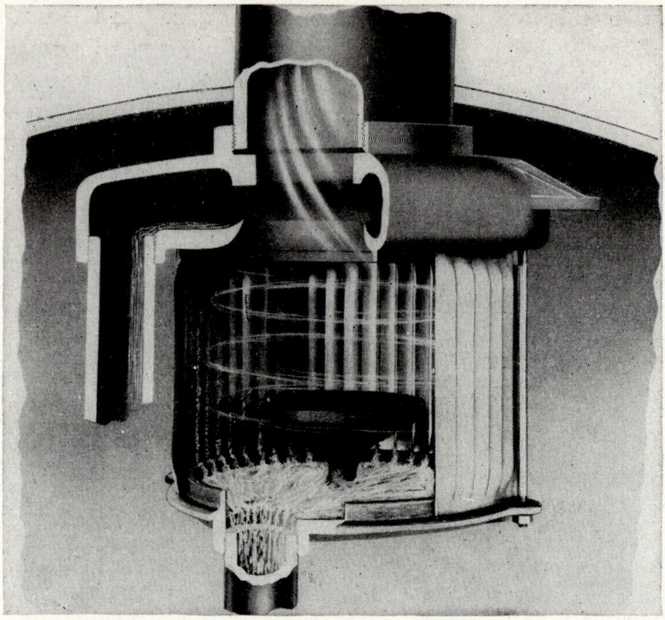

Fig. 6. Internal upflow purifier. (*Courtesy of Centrifix Corporation.*)

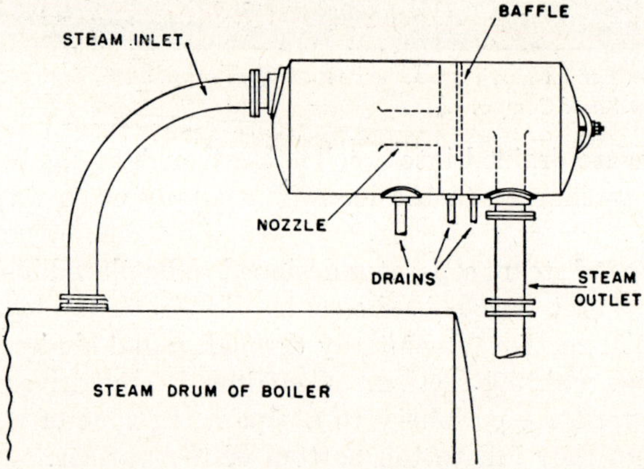

Fig. 7. Receiver-type steam purifier. (*Courtesy of Cochrane Corporation.*)

dry steam passes through an outlet located beyond the baffle. The water may be drained back to the boiler drum if the steam contamination is not caused specifically by excessively high concentrations of impurities, in which event the water is best discharged to waste through a trap. The discharge of highly concentrated water in this manner automatically provides a supplementary blowoff under conditions of high ratings and increased carry-over. Because of its large storage space, this separator is especially applicable to stopping slugs of water and, like those described previously, may be obtained in several forms for installation at any point in the steam and water cycle.

Figure 8 illustrates a centrifugal separator. The wet steam enters through directing vanes which are placed around the periphery of the main separating chamber. Large droplets are broken up in passing through the vanes, and the steam is directed against the wall of the chamber and is thus given a rotating motion. Centrifugal force removes a large part of the moisture, which drains down the wall into the bottom of the compartment. After impinging on the bottom of the bowl, the

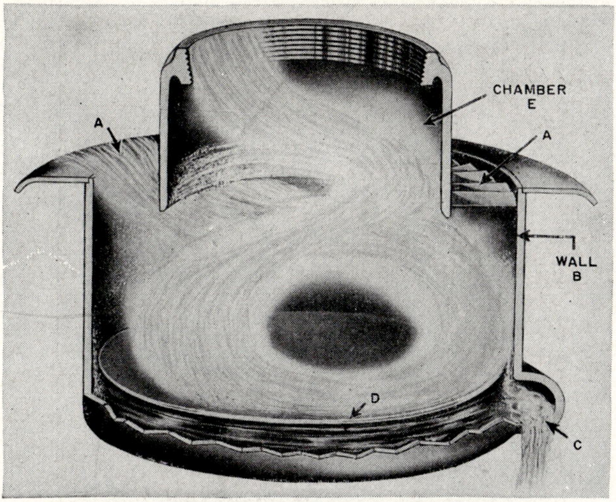

Fig. 8. Hagan steam separator, internal type.

steam is directed upward through the chamber and thence out of the nozzle through the steam outlet. It is claimed that centrifugal separation of moisture occurs at all points in the path traced by the steam. Moisture depositing on the wall of the upper chamber forms a thin film and is unloaded by an ejective process at the lower edge of the wall. The moisture collected in the annular ring is discharged through a tangential take-off to the boiler water or to a steam trap outside the boiler. In Fig. 9 is shown a separator of similar design for installation in a pipeline.

Steam Purifiers of the Washer Type. There are a number of steam-purifying devices which provide for intimate contact between the steam and the incoming feedwater by bubbling, spraying, or jet action through "contact tubes." It is the purpose of this step in the process to absorb into the feedwater droplets of boiler water entrained in the steam. The moisture in the steam may, of course, be relatively high after this process, but instead of the highly concentrated boiler water, it will be much more dilute feedwater. The wet steam is then dried, as much as possible, by screens of various separating devices, operating on the principles of those described above. The efficiency of this type of separator depends on the moisture content of the steam and the concentration of solids in the feedwater used for washing. It is apparent, therefore, that good removal of water from the steam is necessary.

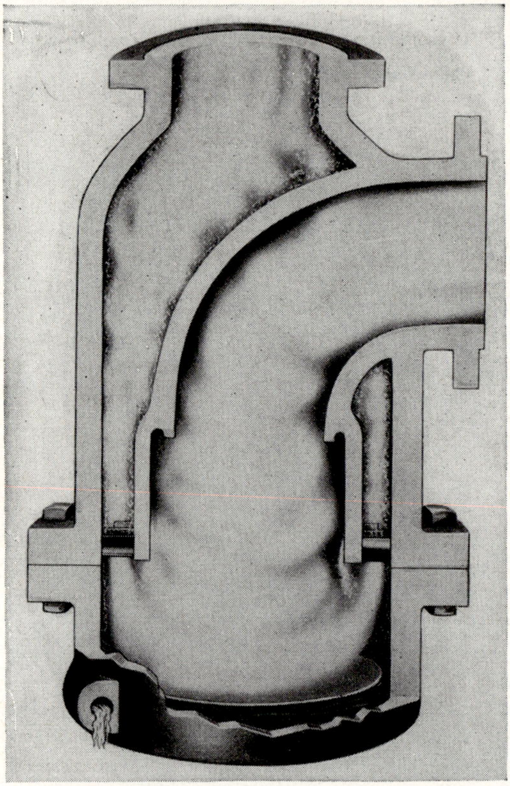

Fig. 9. Line steam separator. (*Courtesy of Hagan Corporation.*)

A combination centrifugal separator and bubble-type steam washer installed in the boiler-steam drum is shown in Fig. 10. The steam and water entering the drum from the steam-generating tubes pass through multiple, primary centrifugal separators and then through the upper centrifuge compartments, supplied individually with feedwater for washing the steam. After preliminary separation, the partially clean steam bubbles through wash water as it passes through holes in the inverted U-shaped, perforated plates immersed in the water. The steam then follows a tortuous path between corrugated plates on top of the scrubber unit. After leaving the scrubber and traversing the space above the drum water level, the steam flows finally through a series of drying screens in the top of the drum and into the steam outlet nozzles.

In this process, the bulk of the boiler water is separated from the steam by centrifugal action in the lower part of the washer. Following

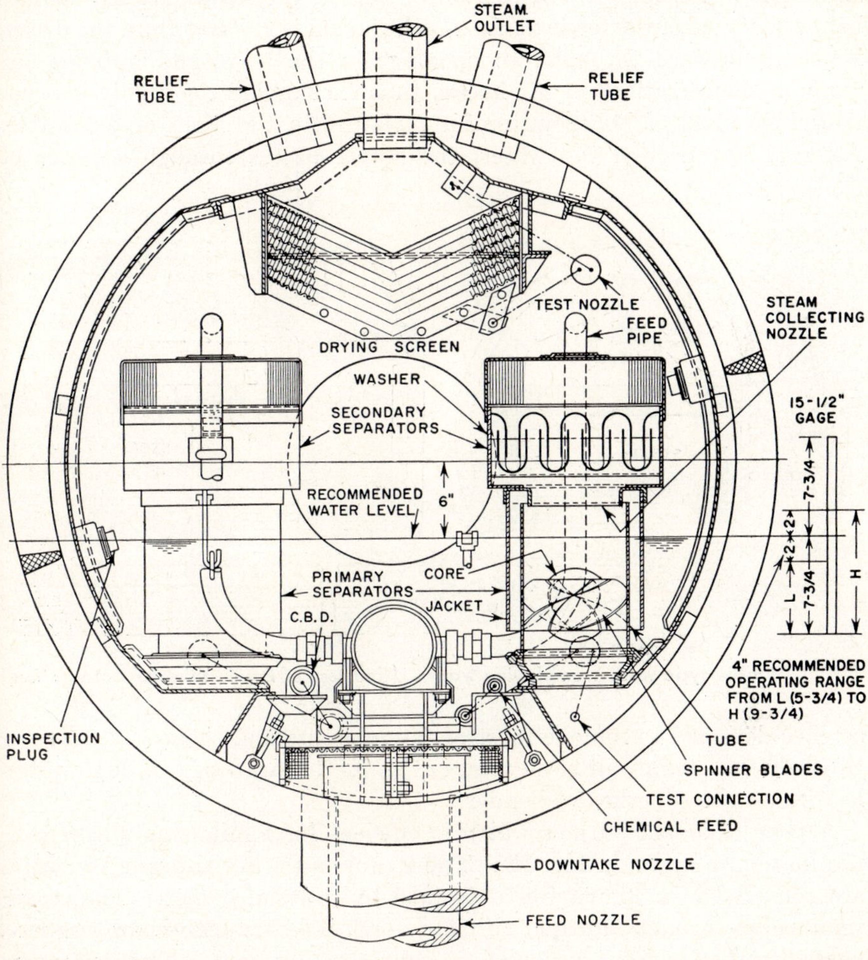

Fig. 10. Bubble-type washer. *(Courtesy of Combustion Engineering–Superheater, Inc.)*

this, most of the remaining boiler water is dropped out and absorbed in the wash water. Finally, any other entrained water in the steam is removed in the passages of the corrugated elements over the washers and in the screens at the top of the drum.

Another bubble-type washer is shown in Fig. 11. The washing elements are a series of cast-iron trays with vertical sides to which feedwater is supplied by risers from the main feedwater line. In the center of each tray and between adjacent trays, a long rectangular aperture with vertical sides to retain the water provides a means of entry for the steam. Each of these rectangular openings is covered with a metal hood with sawtooth edges, and the angular openings in the hoods are submerged below the level of the feedwater. As the steam is released from the drum, it passes upward through the openings in the trays and bubbles out through the serrations in the hoods, thus exchanging the highly concentrated droplets of boiler water for moisture in the form of feedwater. After rising through the water, the steam passes through a series of

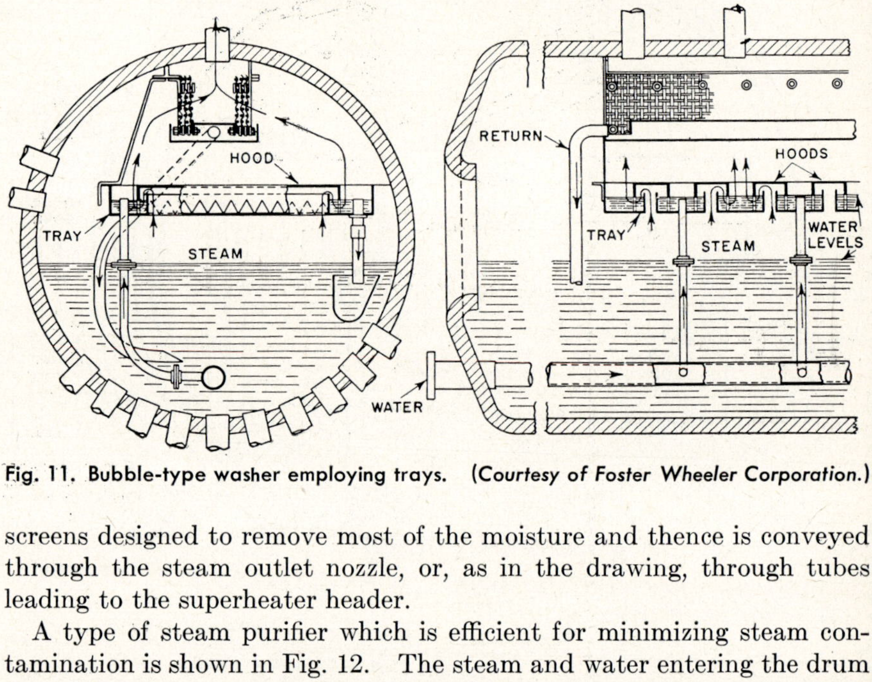

Fig. 11. Bubble-type washer employing trays. *(Courtesy of Foster Wheeler Corporation.)*

screens designed to remove most of the moisture and thence is conveyed through the steam outlet nozzle, or, as in the drawing, through tubes leading to the superheater header.

A type of steam purifier which is efficient for minimizing steam contamination is shown in Fig. 12. The steam and water entering the drum are collected in a compartment formed by internal baffles. The steam then passes through multiple, tubular, vertical centrifugal separators connected to the lower inner side of the inlet compartment. The separators impart a circular motion to the steam-water mixture, causing the water to travel to the outside wall, where it is taken off to the main body of water in the drum through a skim-off lip and outer annular ring. Simultaneously, the steam travels upward through the center of the separator to the top, where it takes a tortuous path through two banks of corrugated metal sheets. The tortuous path and resulting contact of water with the wetted metal have the effect of removing the water, which runs down the corrugated surfaces to the main body of water below. Finally, the steam flows upward through the drum to the drying screens at the top where any remaining moisture is removed before the purified steam passes on through the outlet nozzles.

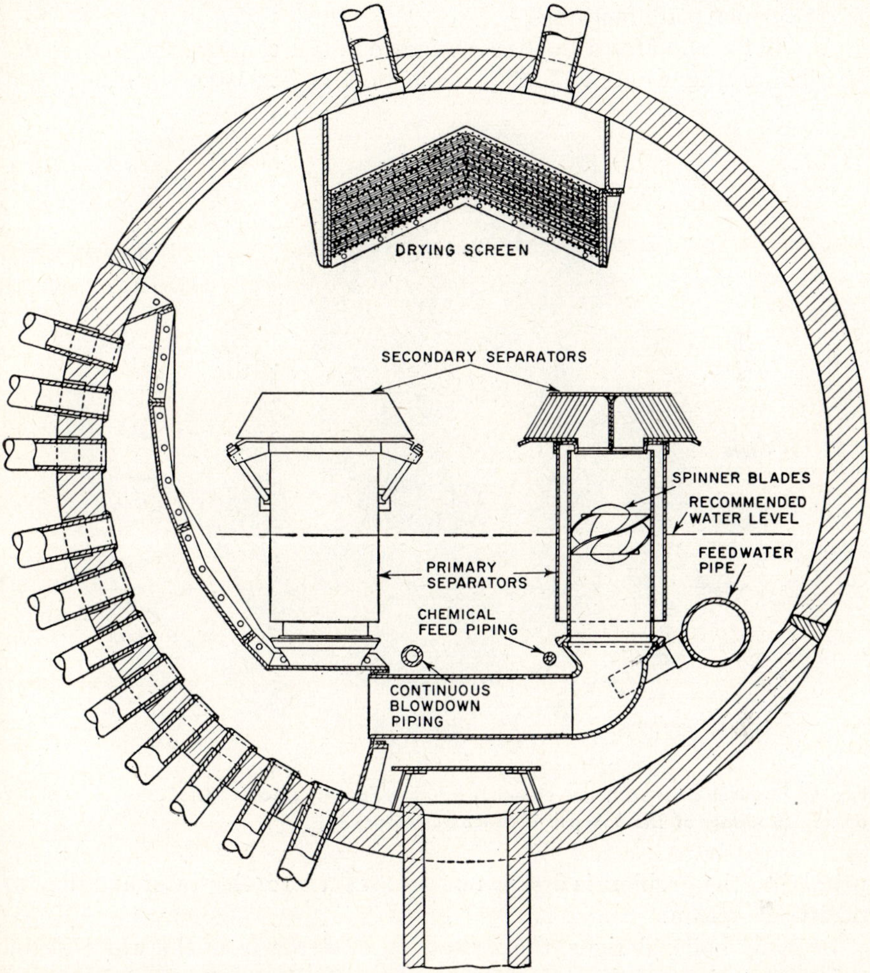

Fig. 12. Arrangement of steam and water separator and steam purifier. *(Courtesy of Combustion Engineering–Superheater, Inc.)*

The design provides that feedwater enters the steam and water space below the primary separators through nozzles extending from the internal feedwater pipe. In this case, the primary separators can serve as a direct-contact feedwater heater. In the arrangement shown, the steam and water enter the drum on one side only, but with boilers of different design, double steam and water inlet compartments, one on each side, may be provided. This general design of steam purifier, which provides separation of steam from the water followed by removal of water from the steam by multiple reversal of flow through a special arrangement of baffles and screens, or baffles only, is common to a number of arrangements developed by the boiler manufacturers.

As will be noted from the designs shown in this chapter, there are many possible arrangements of baffles, washers, and final steam purifiers, but

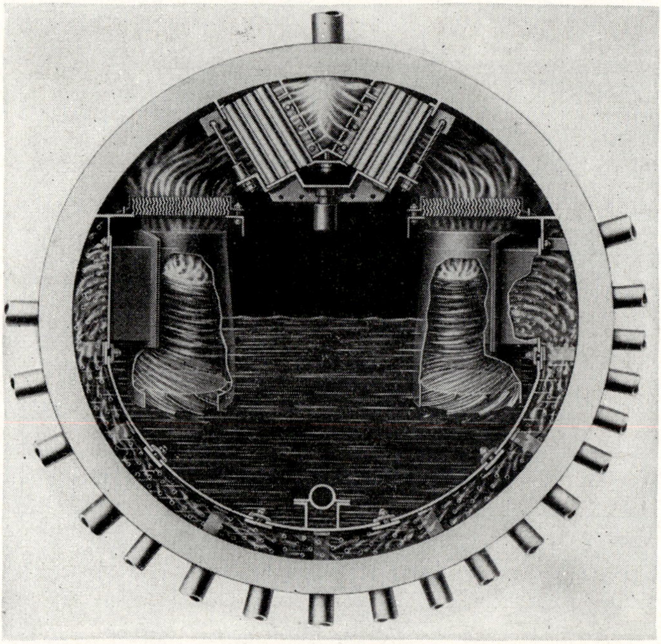

Fig. 13. Typical cross-sectional elevation of a cyclone-separator installation in a boiler drum. (*Courtesy of The Babcock & Wilcox Company.*)

primarily, the problem is first to remove steam from the water and then to purify the steam.

In many boiler designs, the separation of steam from the water in the primary sections of the apparatus is important because when the steam is removed, the water has a higher density. Therefore, true water levels are more easily seen, and circulation in the boiler is improved. The purification of the steam is almost invariably accomplished by baffles or the equivalent in the form of screens.

Cyclone Separators. The principle of separating steam and water and thereby reducing solids carried in the steam has been accomplished very effectively by the type of equipment shown in Fig. 13. By this arrangement, the steam and water mixture enters a compartment in a manner which causes the water to whirl through the separator units until the steam and water are separated, the steam flowing through the hollow space in the center of the apparatus and passing out of the separator through corrugated plates at the top of the cylinder. Any moisture carried from the centrifugal separator is removed and returned to the

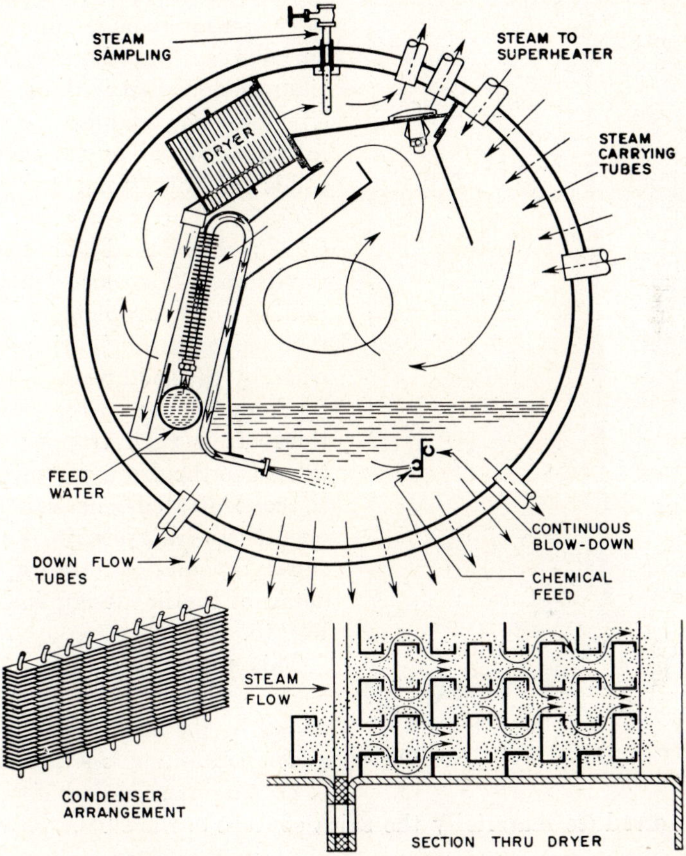

Fig. 14. Steam purifier, condenser type. (*Courtesy of Riley Stoker Corporation.*)

cylinder. The steam, before entering the steam outlet tubes, is passed through perforated distribution plates where any residual water is finally separated. The water so removed is then collected in a central compartment and drained back into the water in the drum. A large number of this type of separator have been installed in boilers operated under varying pressures, and the equipment has proved sufficiently effective to reduce the residual solids in the steam to a very low level. In many cases, these separators, tested by the author, have produced steam having a residual total solids of 0.5 ppm or lower.

Condenser-type Steam Purifier. The design of this steam-purification apparatus is ingenious in that a portion of the steam released in the steam drum is washed by condensate produced by fin coils through which the feedwater flows. The feedwater usually has a temperature lower than that of the saturated steam by approximately 75 to 100°F. This results in condensation on the fins and tube surfaces, so that the steam is continuously washed with condensate. The washing reduces the solids, including silica, by transfer of the solids in the steam to the condensate. The wash water with its entrained solids drains back into the drum. The washed steam containing some droplets of water is then passed through drier elements, whereby the solid water is removed by frequent changes in flow direction. It is then trapped and returned to the drum and mixed with the concentrated boiler water. The operation of a typical installation is shown schematically in Fig. 14. A detail of the fin design is shown in Fig. 15.

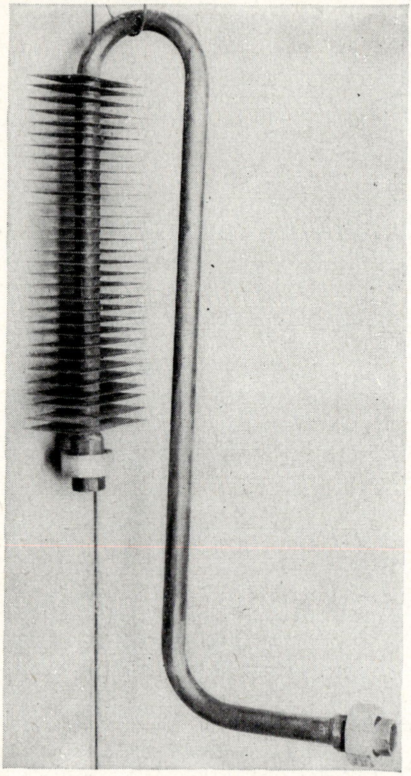

Fig. 15. Fin element used for condensing vapors.

This steam purifier is quite efficient and, in some installations tested by the author, has produced steam containing less than 0.5 ppm of contaminants. It has been also possible to reduce materially the silica content of the steam delivered to turbines.

EFFECTIVENESS OF THE APPLICATION OF STEAM-PURIFICATION EQUIPMENT

Equipment for producing clean steam has now been developed to a point where assurance can be given that the total solids carried by the steam after passing through mechanical equipment can be held, generally, at not more than 1 ppm. The guaranteed purity of steam is influenced by many conditions, but if the concentration of solids in the boiler waters is maintained within specified limits at different boiler operating pressures and if the foaming characteristics of the water are controlled, most of the boiler manufacturers will make a firm guarantee for steam quality. Such guarantees are qualified, however, by the maintenance of a maximum concentration of solids in the boiler waters. The limiting concentrations of solids at various operating pressures are given in the following table:

Pressure at Outlet of Steam-generating Unit		*Total Solids*
0–300	psig	3,500 ppm
301–450	psig	3,000 ppm
451–600	psig	2,500 ppm
601–750	psig	2,000 ppm
751–900	psig	1,500 ppm
901–1,000	psig	1,250 ppm
1,001 and higher	psig	1,000 ppm

Further, to meet these guarantees, the amounts of oil or grease or other organic materials are limited by the manufacturers to a maximum of 7 ppm. These standard guarantees of performance have been specified by the American Boiler Manufacturers' Association and Affiliated Industries, 1954 Manual.

Bibliography

1. Vorkauf, H.: *Forsch. Gebiete Ingenieurw., Forschungsheft*, 1931, p. 341.
2. Hoefer: *Forsch. Gebiete Ingenieurw., Forschungsheft*, 1913, p. 138.
3. Pickert: *Forsch. Gebiete Ingenieurw.*, **3** (6): 310 (1932).
4. Joseph, A. F., and J. S. Hancock: *J. Soc. Chem. Ind.* (*London*), **46**: 315 T (1927).
5. Holmes, J. A.: *Trans. ASME, Research Papers* 54, 55, p. 110, 1932 (discussion of ref. 9).
6. Foulk, C. W.: Foaming of Boiler Water, *Ind. Eng. Chem.*, **16**: 1121 (1924).
7. Foulk, C. W.: A Theory of Liquid Film Formation, *Ind. Eng. Chem.*, **21**: 815 (1929).
8. Foulk, C. W., and V. L. Hansley: Solid Matter in Boiler-water Foaming. I. Experiments at Atmospheric Pressure, *Ind. Eng. Chem.*, **24**: 277 (1932).
9. For a review of studies prior to this, see Foulk, C. W.: Foaming and Priming of Boiler Water, *Trans. ASME, Research Papers* 54, 55, p. 105, 1932.
10. Foulk, C. W., and S. F. Whirl: Solid Matter in Boiler-water Foaming. II. Loss of Foam-stabilizing Properties at Higher Pressures, *Ind. Eng. Chem.*, **26**: 263 (1934).
11. Foulk, C. W., and H. C. Brill: Solid Matter in Boiler-water Foaming. III. Effect of Calcium Carbonate and Magnesium Hydroxide Precipitated inside the Boiler, *Ind. & Eng. Chem.*, **27**: 1430 (1935).
12. Ulmer, Richard: "Abstracts of Doctors' Dissertations 21," The Ohio State University Press, Columbus, Ohio, 1937.
13. Niehaus, Fritz: Physical-Chemical Considerations on the Question of the Foaming of Boiler Waters, *Vom Wasser*, **4**: 140–153 (1930).
14. Bancroft, W. D.: "Applied Colloid Chemistry," 1st ed., McGraw-Hill Book Company, Inc., New York, 1921.

15. Freundlich, H.: "Kapillarchemie," p, 302, 1909.
16. Mumford, A.A.: *Trans. ASME, Research Papers* 54, 55, p. 111, 1932 (discussion of ref. 9).
17. Hall, R. E.: Present Tendency of Boiler-Water Conditioning, FSP 50–58, *Trans. ASME*, Vol. 65, 1927–1928.
18. Taplock, and Thompson: *J. Soc. Chem. Ind.* (*London*), **52**: 1 (1890).
19. Koyl, C. H.: *Railroad Gaz.*, Oct. 12, 1900.
20. Elliott, Orin M.: Brackish Feed Water Treatment, presented at the 8th Annual Water Conference of the Engineers Society of Western Pennsylvania, in Pittsburgh, Nov. 12–14, 1947.
21. Straub, F. G.: Cause and Prevention of Turbine-Blade Deposits, *Trans. ASME*, **57**: 447–454 (1935).

15 DETERMINATION OF STEAM PURITY

In order to regulate boiler operation to produce steam of good quality and to evaluate the efficiency of steam-purifying equipment, it is desirable to measure continuously the solids carried over with the steam. The most rapid and effective way of securing such information is to condense the steam and determine the conductivity of the condensate. However, silica and some other compounds which do not change the conductivity of the solution will not be revealed and gravimetric check tests may be required.

DRAWING A REPRESENTATIVE SAMPLE OF STEAM

Impurities present in steam are contained in fine droplets of boiler water, or of feedwater if the steam is washed, and are present as volatilized gases. The solids present in the vapor may lose moisture until the residue is a highly concentrated paste or a dry powder. Whatever the state of the contaminating particles, their density is greater than that of the steam, and because of the influences of gravity, centrifugal force, impact, or changes of velocity, they may not remain homogeneously distributed throughout the steam as it passes from the boiler to various parts of the steam cycle. It is difficult, therefore, to secure a sample of steam containing a quantitative measure of the contamination which will be accurately representative of the total quantity flow from the boiler.

The ASME Power Test Code, Part II, "Determination of Quality of Steam,"[1] describes the standard steam-sampling nozzle which has been adopted as most suitable for securing a representative sample. The apparatus consists of a pipe of ¼ or ⅜ in. inside diameter with a closed end and with a series of ⅛-in. holes drilled in a straight line; these holes face the steam flow when the nozzle is in place. Cognizance is taken of the danger of uneven distribution of the contaminating particles, the code specifying that, when possible, the nozzle should be inserted in a pipe in which the steam flow is vertically downward and as far removed from disturbing restrictions as possible. A sample from a vertical line with upward steam flow is believed to give high results for moisture, and, according to the Code, the "pipe bend and horizontal pipe should be avoided, if an accurate determination of the quality of the steam is to be obtained."* An improved ASME sampling nozzle preferred for high-pressure installations is shown in Fig. 1.

It is often difficult to carry out these recommendations completely, for example, when sampling steam delivered to the saturated header of a

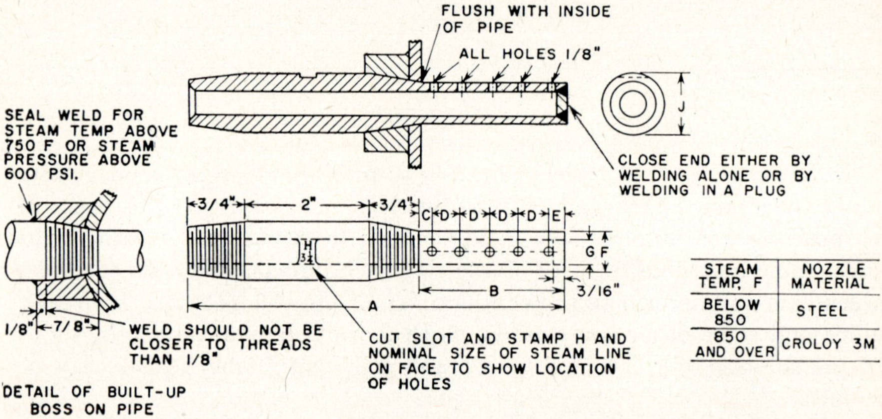

STEAM TEMP, F	NOZZLE MATERIAL
BELOW 850	STEEL
850 AND OVER	CROLOY 3M

Fig. 1. Dimensions of steam-sampling nozzles. *(Courtesy of The Babcock & Wilcox Company.)*

superheater through a number of tubes from the steam drum. For such samples, small nozzles may be inserted in representative tubes and connected to a common tubular header, a nozzle may be attached to each side of a tee located at a handhole plate on the header, or a pipe may be mounted on one end of the header to extend a convenient distance across it. Lateral variations in the circulation of the boiler or in the operation of steam-purifying equipment may result in corresponding variations in the quality of the steam drawn from various tubes or sections of a superheater header, and it is necessary to establish experimentally the representative nature of the sample. In some instances, boiler manufacturers install several nozzles for sampling saturated steam in the drum; the nozzles are connected to a common steam-sampling header outside the drum.

The rate of sampling should be great enough to permit the steam to enter the holes in the nozzle with the velocity at which it flows past the nozzle although some available evidence obtained by the author and

* More recently, the American Society for Testing Materials method of sampling steam (D-1066-49T) has been tentatively accepted and possibly will be adopted as standard.

others indicates that this may not be necessary, especially when good steam-purification equipment is in use. It has been found, however, that, when the apparatus for measurement of condensed steam is limited in size and can utilize only a portion of the sample drawn (as may be the case in conductivity determinations), it is preferable to condense the entire sample and to draw off the desired portion as condensate instead of as steam. Figure 2 shows data collected recently by the author's

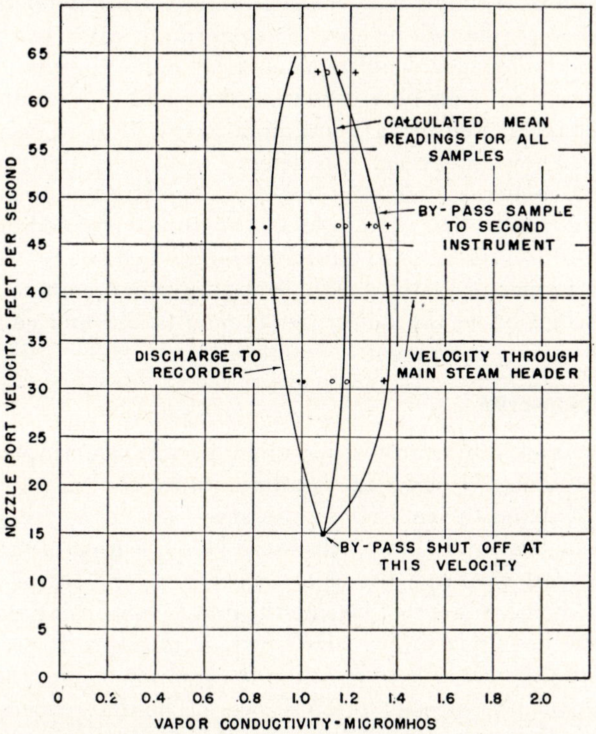

Fig. 2. Effect on conductivity measurement of separating a steam sample into two portions before condensation.

staff. It is evident that the quality of the steam withdrawn from each end of a tee, one supplying a recorder and one a second instrument, varied with total and relative sample flow rates, but the average was nearly constant. Therefore, when the steam sample is to be condensed to a low temperature, the condenser should be large enough to condense the entire quantity of steam delivered by the sampling nozzle.

PRECAUTIONS REQUIRED TO AVOID CONTAMINATION WHILE SAMPLING

Extreme care should be taken when drawing a representative steam sample to avoid contamination after the sample leaves the nozzle. For calorimeter measurements, the material selected for the nozzle should be chosen mainly for the desired structural strength, but when the steam is condensed for conductivity measurements, certain metals and other construction materials may influence the accuracy of the results. Copper tubing is frequently used to carry the steam sample to the condenser, but this material should not be used when the steam temperature exceeds 406°F. Stainless-steel tubing is preferable for use at or above this temperature. Copper cooling coils are satisfactory if the temperature is not too high, since the metal reaches a stable state after several days of operation, and under such conditions, no measurable contamination of the condensate occurs. Iron pipe, soldering flux, and pipe compounds constitute a persistent source of contamination, and their use should be eliminated whenever possible. In general, a rapidly flowing stream of condensate at moderate temperatures will not attack ordinary resistant materials at a rate sufficient to influence conductivity readings.

Provisions for representative sampling should be made for each installation and the proper procedures established by experiment so that homogeneous distribution of the impurities is maintained and contamination from the measuring equipment is avoided.

STEAM CALORIMETRY

The most familiar steam-testing apparatus is the throttling calorimeter. This device is designed to admit steam through an orifice and then to let it expand to atmospheric pressure. Since, at pressures at and below 400 psia, the total heat of saturated steam is greater than the total heat of saturated steam at atmospheric pressure, the steam on the low-pressure side of the orifice will be superheated by absorbing the difference between these heat quantities. However, if moisture is present in the steam, part of the heat will be required to vaporize it, and the superheat temperature will be correspondingly lower. The difference between the theoretical and observed superheat is, therefore, proportional to the moisture in the steam. At pressures above 400 psia, the steam desuperheats as it leaves the calorimeter, and the evaporation of entrained water further lowers the theoretical temperature. This gives a calibration curve of somewhat different shape than for pressures below 400 psia. Radiation losses and thermometer inaccuracies must be determined by operating the boiler so as to assure 100 per cent dry steam and establishing the correction, or "normal," of the apparatus. An accuracy of ± 0.2 per cent of moisture may be obtained with proper care.

The simplified description of a throttling calorimeter given above does not adequately cover the numerous types of steam calorimeters. For more complete details, the operator should refer to the ASME Power Test Code[1] which contains instructions for operating recognized types of apparatus.

It should be noted that the possible error which the code ascribes to the steam calorimeter, when applied to a carryover of boiler water containing 2,500 ppm of total solids, would correspond to 5 ppm of solids in the steam. In many cases, especially when testing the performance of highly efficient steam purifiers, such measurements are inadequate. Thus it is not a satisfactory procedure for detecting very small amounts of water in steam, nor does it assist in revealing the nature of solids or gases contained in steam. However, steam calorimetry is widely used and serves a useful purpose for determining the moisture content of steam when the moisture exceeds the probable error of the measurement.

LABORATORY ANALYSIS OF CONDENSED STEAM

When gravimetric check results are required, samples of condensed steam are evaporated to dryness and the solid residue weighed on an

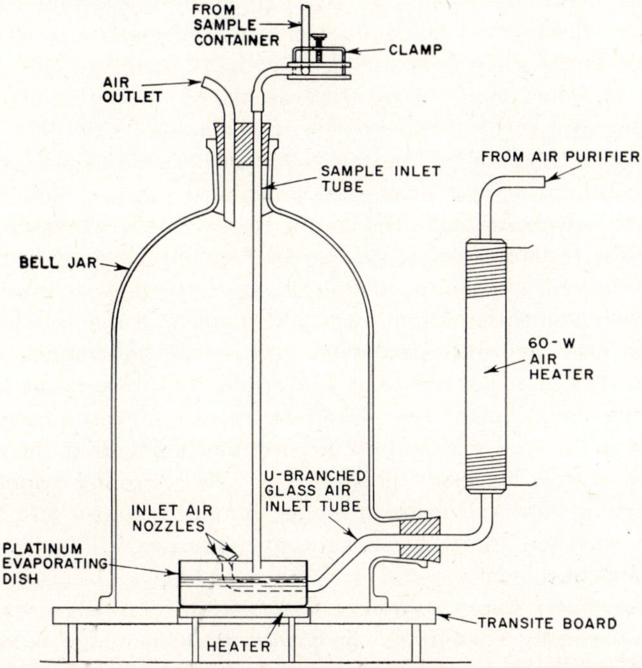

Fig. 3. Apparatus for determination of total solids by evaporation.

accurate balance. However, if a liter of condensate containing 1 ppm of total solids is evaporated in a dish weighing 15 grams, the increase in weight that must be measured is only one part in 15,000, and extreme care in evaporating and weighing the evaporating receptacle is required to obtain satisfactory accuracy on so small a sample. If possible, a platinum evaporating dish should be used, since it is difficult to heat and dry porcelain or glass dishes to constant weight. To obtain accurate results, 5 to 20 liters of the sample should be evaporated to secure a residue large enough to minimize errors in handling and weighing. Extreme precautions should be taken to avoid contamination from the atmosphere or sampling containers. Apparatus of the type shown in Fig. 3 has been developed for this purpose; its operation is apparent from the drawing.

DETERMINATION OF STEAM QUALITY BY ELECTRICAL CONDUCTIVITY METHODS

The soluble salts in the boiler water which are carried over in the steam increase the electrical conductivity of the condensate, and this property furnishes a means of estimating the percentage moisture as boiler water and the quantity of these contaminating materials. Such procedures for steam testing have been in use for many years. They have been employed for detecting surface-condenser leakage and in other applications for a much longer time. In 1931, Hecht and McKinney[2] published an extensive description of conductivity measurements, and a bibliography of earlier work was included. Probably owing to the concurrent use of steam calorimeters, emphasis was placed on expressing results in terms of per cent moisture carry-over in the steam. In 1931, Rummel[3] published a method for estimating solids rather than moisture in steam from the conductivity readings. In addition, he described a method of degassing the steam to make the results for soluble solids more accurate.

The ohmic resistance of a sample of condensate may be measured simply by dipping two metal rods into the sample and connecting them in series with a sensitive ammeter or voltmeter and a source of current. The meter may be calibrated with samples of known concentration. Such assemblies are described by Tray[4] and by Ulmer[5] and have been used in some plants. However, when a current sufficient for these meters flows through an aqueous solution, a potential is set up at the electrodes and results in polarization. This opposes the externally applied potential and gives a false indication of higher ohmic resistance, *i.e.*, lower salt concentration, than the solution actually possesses. These errors may be of serious consequence.

The most effective way to reduce polarization is to measure resistance with an alternating current of the lowest possible magnitude and the highest convenient frequency passing through the cell. For normal test purposes, this may be done with a properly designed Wheatstone bridge circuit for 60-cycle alternating current in which the resistances of the various parts are so balanced that a very small current passes through the liquid.* Usually, cells with constants of 0.05 to 0.1 cm^{-1} are the

* For further specifications dealing with suitable instruments for specific purposes, reference is made to the tentative ASTM test method (D-1125) for electrical conductivity of industrial water.

most satisfactory for measuring water solutions of low conductivity, since 1×10^{-6} mho corresponds to about 0.5 ppm. Conductivity meters of the preferred type with compensating dials for temperature and cell constant are commercially available from a number of instrument manufacturers. For continuous readings, fully automatic recording conductivity meters are available. For special work requiring a high degree of accuracy, such instruments may be interconnected with a temperature compensator.

The electrodes which are used for determining the conductivity of condensed steam should be inserted into a continuously flowing sample and should be constructed of a highly resistant material which will not be dissolved sufficiently to influence the results. Platinum and gold are satisfactory for this purpose, although for the less accurate ammeter methods, monel or nickel rods have been used. Specific conductivity refers to conductivity of the solution between two parallel electrodes, each 1 cm square and 1 cm apart, but in practice these dimensions may vary. To test solutions of low conductivity, the electrodes are spaced closer together and the conductivity reading multiplied by the cell constant, which is the distance in centimeters between plates, divided by the cross-sectional area in square centimeters of the solution between the plates. The cell constant is determined by use of a standard potassium chloride solution of known specific conductivity. For example, a 0.001-*N* solution has a specific conductivity of 146.93×10^{-6} mho at 25°C (77°F).

Polarization may be reduced by using platinum electrodes covered with an electrically deposited layer of platinum, which has a velvety black appearance when properly applied. When the deposit becomes contaminated or marred, it is necessary to remove it, replace the electrodes,* and redetermine the cell constant. For steam-condensate samples of low conductivity, plain platinum electrodes will not become polarized, and in some cases, they are used in place of those coated with platinum black. Figure 4 shows a typical glass flow cell in which the electrodes are inserted. The temperature of the sample should be brought to the specified temperature for which the instrument is calibrated, or the temperature of the condensate should be recorded each time a measurement is made and the results multiplied by the factors

* Dip the electrodes in aqua regia (one volume concentrated nitric acid with three volumes concentrated hydrochloric acid), which will quickly remove the platinum black, and rinse them in distilled water; then immerse in a 3 per cent solution of platinic chloride containing 0.025 per cent neutral lead acetate and connect the electrodes to one dry battery. Continue plating 3 min, reversing the polarity several times a minute. Rinse well in distilled water and wait 30 min for stabilization before determining the cell constant. Electrodes should be kept immersed in water at all times.

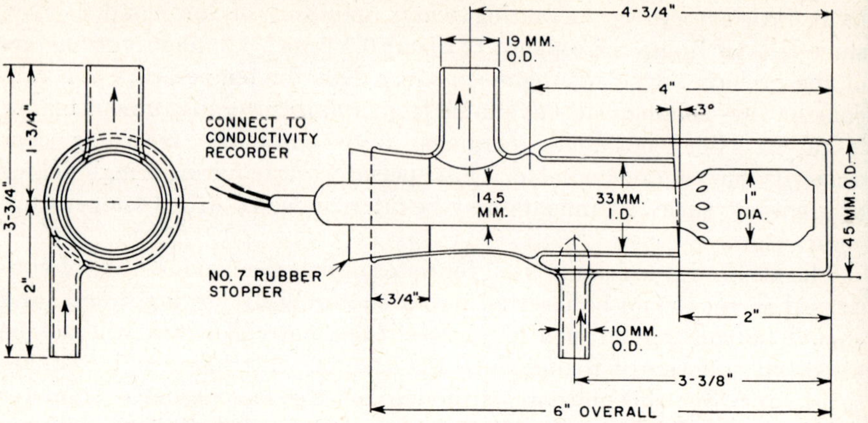

Fig. 4. Detail of continuous-flow cell for measuring conductivity.

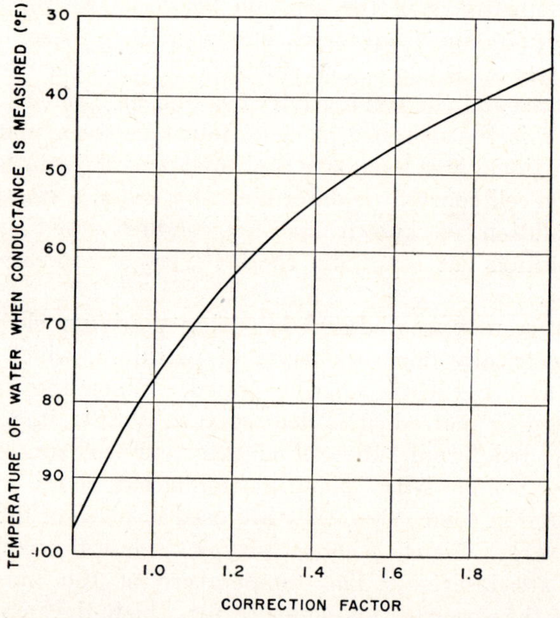

Fig. 5. Temperature-correction factors for specific-conductivity reading.

shown in Fig. 5 to correct them to standard (77°F). Resistance thermometers may be used with manual or recording temperature bridges for reading at remote points.

CORRECTION OF ERRORS IN CONDUCTIVITY MEASUREMENTS DUE TO DISSOLVED GASES

Steam contains appreciable concentrations of dissolved carbon dioxide derived from the decomposition of carbonates in the boiler feedwater.

When the steam is condensed, this gas goes into solution as carbonic acid and may be firmly held by converting hydroxides to carbonates or carbonates to bicarbonates. This condition can cause the relative composition of the solids in the dissolved steam to deviate from those in the boiler water.

Boiler feedwater often contains ammonia either from chloramine, amine, or other compounds for the treatment of the water supply or from the natural decomposition of organic nitrogenous matter present in water from deep wells or from polluted sources. Although feedwater is highly alkaline and is usually deaerated at 212°F or higher, the ammonia may not be completely eliminated. Later, some of it is released in the boiler and appears in the steam. The gas passes into solution again in the condensate, possibly in combination with carbon dioxide as ammonium carbonate. It reenters the feedwater system and may be recycled indefinitely. Even where a very small percentage of make-up is used, the author has encountered 0.07 ppm of ammonia nitrogen in samples of condensed steam, and as much as 0.40 ppm or more has been observed in plants using large percentages of make-up or make-up water supplies high in ammonia.

Since neither carbon dioxide nor ammonia represents entrained solids in the steam but is a highly conducting electrolyte when dissolved in water, the presence of one or both of them introduces an error into the measurement of solids carried by the steam when determined by conductivity methods. Since the necessary corrections may comprise nearly all the measured conductivity, it is often necessary to eliminate these gases as effectively as possible and to make corrections only for the gases remaining after degasification. These residual contaminants can be evaluated by refined chemical testing procedures.

APPARATUS FOR DEGASIFYING STEAM SAMPLES

The first apparatus for degasifying a sample of condensed steam before measuring its electrical conductivity was designed by J. K. Rummel[3] and was based on flashing out dissolved gases at an elevated temperature after the sample was collected. The patent[6] covering this apparatus also includes the method for estimating dissolved solids from the measurements. A modified form of this apparatus, by Gurney,[7] is shown in Fig. 6. The steam entering this apparatus is divided at a tee, part going through a "reboiling" coil in the bottom of the degasifier chamber and part passing through a condenser. The hot condensate is discharged into the degasifier chamber just above the steam coil, which raises the temperature to the boiling point so that dissolved gases are flashed out of solution. These are vented out around the top of the chamber. Above the sample inlet is located a coil identical with the lower one but carrying cooling water so that this section serves as a vent condenser. Only a little vapor should issue from the apparatus when it is properly adjusted. Part of the sample is discharged from the bottom of the degasifier through a heat exchanger which cools it and then passes through a flow vessel con-

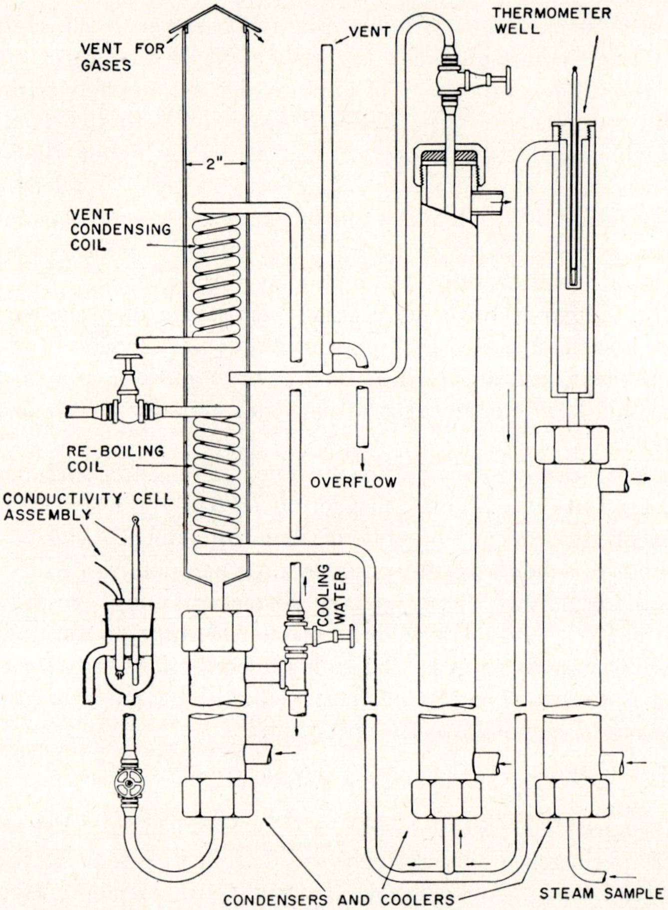

Fig. 6. Arrangement of apparatus for removing carbon dioxide and ammonia from condensed steam.

taining a conductivity cell and thermometer. An overflow on the sample inlet to the degasifier maintains the maximum water level just below the vent condenser coil, and by closing a valve in the overflow, the apparatus can be flooded to wash the surfaces free of deposits.

Another apparatus, in which the vent condenser was omitted, introduced positive errors because sufficient steam escaped to concentrate the sample. The Rummel degasifier is capable of reducing the carbon dioxide content of steam to negligible amounts, but present experience indicates that it will not eliminate ammonia, although some reduction has been observed. The apparatus requires careful adjustment of the several rates of flow of steam and cooling water, and because of the difficulties

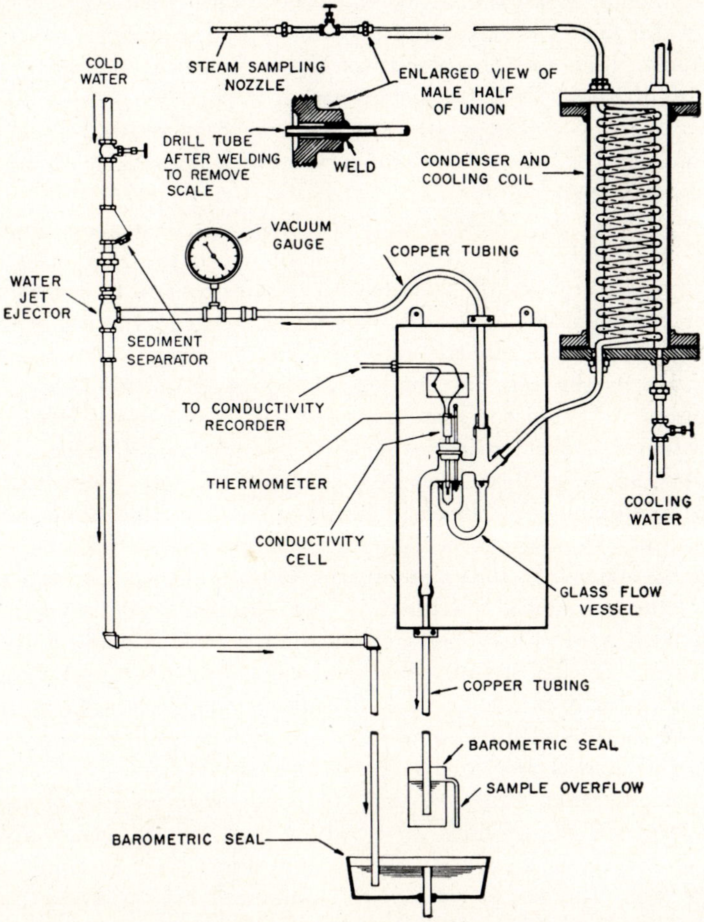

Fig. 7. Apparatus for sampling steam and measuring conductivity under vacuum.

inherent in maintaining a small, constant flow of high-pressure steam, it may not retain its adjustment over long periods without attention from the operators.

Several of these difficulties led to the development at plants where steam testing was inaugurated under the author's supervision of procedures for drawing off soluble gases by means of a vacuum pump, or equivalent, and measuring the conductivity of the sample while under vacuum.[8,9] The essential features of the design are shown in Fig. 7.

A suitable length and diameter of stainless-steel tubing is used between the sampling nozzle and the condenser so as to deliver a representative flow of sample without throttling with a valve. The condenser is a coil of block-tin tubing of ¾ in. inside diameter and should have sufficient surface to condense all the steam and cool it to a constant temperature. With a water jacket consisting of an 8-in. pipe containing a central 4-in. pipe baffle to increase the velocity of the cooling water, a heat-transfer coefficient of about 300 Btu per hr per sq ft per °F has been observed. However, using a water jacket in the form of an outer coil of copper tubing of larger diameter, the higher velocities obtainable should yield a higher coefficient.

The condensed and cooled sample is delivered to a glass flow vessel made of chemically resistant glass through which the condenser is connected to a vacuum pump. This reduces the pressure to several inches of mercury at the inlet end of the condenser, while at the lower end, the vacuum gauge should indicate only the vapor pressure of water at the temperature in the cell, noncondensible gases in measurable amounts being absent. In effect, the steam is condensed and cooled in an atmosphere which is almost entirely free from ammonia and carbon dioxide, and, except as these gases may combine to form ammonium carbonate in the sampling line or at the early stages of condensation, they should be entirely absent from the cooled liquid. Carbon dioxide may be reduced to negligible concentrations, but during investigations observed by the author, some ammonia remained in the sample. The ammonia can combine with acid constituents other than carbon dioxide and aid in neutralization of the steam. For example, Alexander and Rummel[10] have shown that it can combine with sulfurous acid released by decomposition of sodium sulfite, but an acid condition is obtained only when the sulfurous acid is present in sufficient excess.

The flow assembly which receives the condensate may be of any convenient form which provides an overflow and the necessary pressure-equalizing tubes, as shown in the design, Fig. 7. The liquid which overflows from the vessel should be discharged into a water leg constructed of copper tubing and leading to a seal at barometric height, *i.e.*, about 35 ft below the apparatus. Connections between the glass apparatus and metal parts are made with rubber tubing, and all joints are made vacuum tight with a sealing wax prepared by melting one part of black pure tar with four parts of flake shellac.

F. G. Straub, in an effort to remove dissolved gases more effectively, developed the apparatus shown in Fig. 8, known as the "Straub degassing condenser." It functions as follows:

Steam from a sampling tube containing an orifice for regulation of flow enters the reboiling coil at the bottom, and the condensed, partially cooled sample flows through a loop seal to an orifice in the side of the reboiling and condensing chamber above the stripping column. Any excess sample is discarded through the overflow opening in the tube above the level of the orifice. The sample entering the stripping column meets rising steam from the reboiling section at the bottom of the cylin-

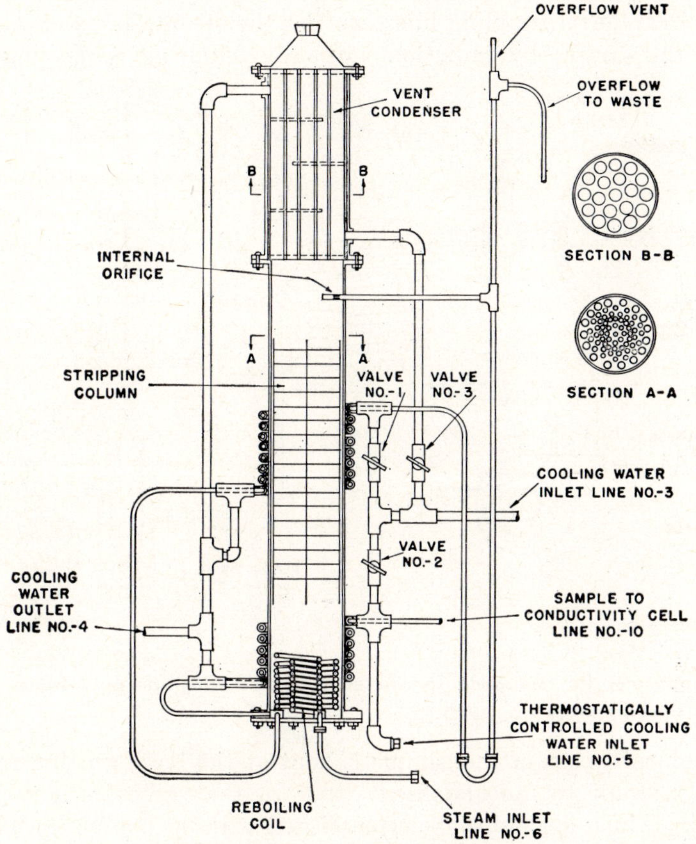

Fig. 8. Straub degassing condenser.

der, and the gases are scrubbed out as the sample flows downward. A vent condenser at the top of the cylinder condenses most of the rising steam from the stripping column, but a certain small amount goes to waste at the top along with the gases released from the sample. Finally, the sample is taken from the cylinder at the bottom of the reboil section and cooled for delivery to the conductivity flow cell. The height of this outlet automatically maintains the desired level of water in the reboil section.

As a laboratory apparatus or under the supervision of a good instrument man, this equipment is effective in producing a gas-free steam sample. However, it requires appreciable attention and, therefore, is not entirely satisfactory in all installations.

Powell and McChesney developed an apparatus, shown diagrammatically in Fig. 9, which is used in measuring the electrical conductivity of condensed steam from boilers or of vapor from evaporators.[11] A stainless-steel capillary sampling line throttles the steam flow and condenses part of the sample by radiation, normally requiring no cooling water.

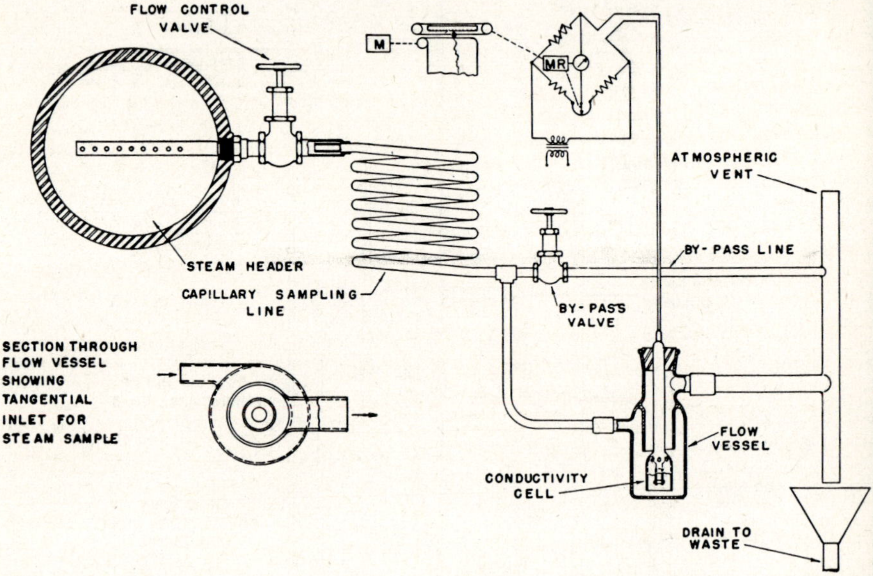

Fig. 9. Apparatus for measuring specific conductivity of high-pressure steam samples, Powell-McChesney design.

The condensed part of the sample is kept at the boiling point and continuously swept by uncondensed vapor so that gases cannot go into solution. Thus, degassing is accomplished without complete condensation and reboiling.

The sample entering a standard sampling nozzle in the steam main passes through a high-pressure stainless-steel gate valve and through a sampling line of stainless-steel capillary tubing of suitable length. This tubing may be 50 to 150 ft long, since it serves as an orifice to limit the total sample flow and as an air-cooled radiating condenser which will deliver a mixture of steam and condensate to the apparatus. In practice, any excess of sampling line between the nozzle and conductivity unit is formed into a helix or flat serpentine on hangers or other suitable mounting. A typical arrangement is shown in Fig. 10.

The apparatus is designed primarily for use when sampling steam in a range from somewhat above atmospheric pressure to extremely high pressures. It discharges at constant atmospheric pressure. Because of the constant discharge pressure, the temperature is nearly constant, and automatic temperature compensation is not required. The flow of sample to the cell is adjusted by using the proper length of sampling tube or by sampling at a rate above the requirements for the condensate cell and, if necessary, passing the excess through a bypass line to a vent.

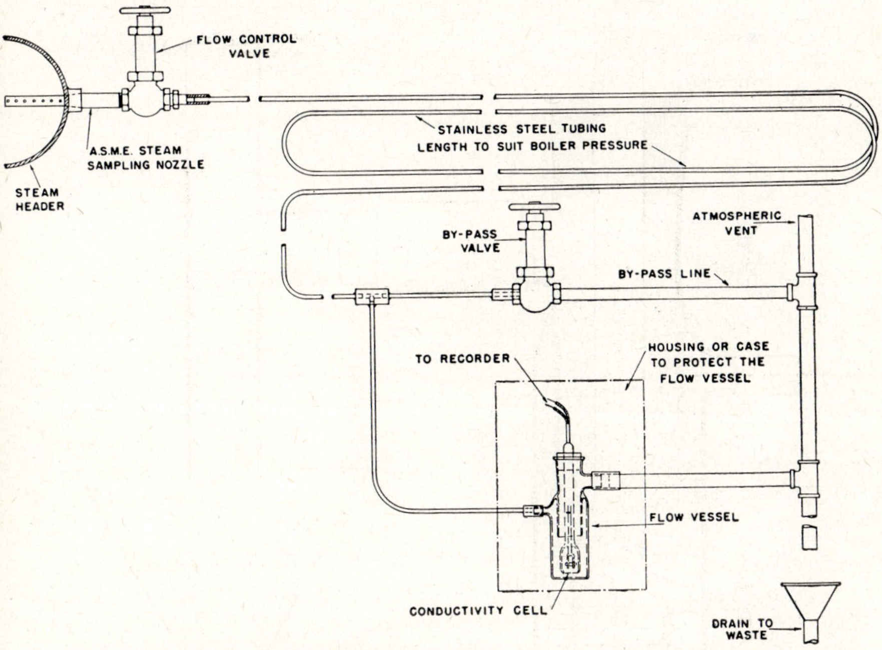

Fig. 10. Installation of Powell-McChesney steam-conductivity testing apparatus.

The partially condensed sample of steam and vapor enters the condensate cell where the electrical conductivity is measured and is finally discharged to waste. The design of the condensate cell is shown in Fig. 11. The sensitivity of the instrument is very high.

Steam, condensate, and gases enter this cell tangentially, and while the condensate particles are collected and flow down the walls, their exposed surfaces are being continuously scrubbed by steam. Level regulation in the cell is automatic because of water being thrown from the condensate surface out through the discharge opening at the top. As the level of the condensate approaches the bottom of the vertical outlet passage, an orifice is formed which causes an appreciable pressure drop at the surface of the condensate. High turbulence and surface boiling is the result.

The foregoing action not only carries away excess condensate but also prevents absorption of gases in the condensate that is temporarily collected within the cell. The turbulence caused is useful in thoroughly mixing condensate held in the cell. The length of travel through the discharge and pipe is sufficient to prevent the diffusion of gases from the atmosphere back to the cell. Thus, the electrodes inserted in the cell are

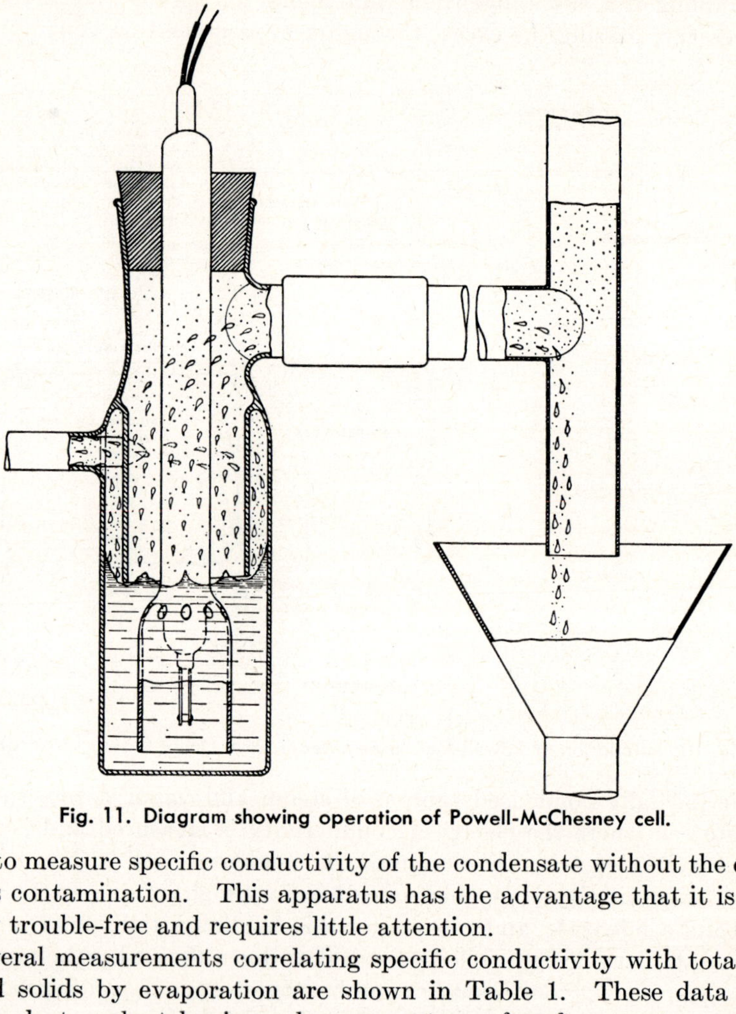

Fig. 11. Diagram showing operation of Powell-McChesney cell.

used to measure specific conductivity of the condensate without the effect of gas contamination. This apparatus has the advantage that it is relatively trouble-free and requires little attention.

Several measurements correlating specific conductivity with total dissolved solids by evaporation are shown in Table 1. These data were obtained at a plant having a low percentage of makeup water and are applicable to steam samples from boilers. Table 2 shows comparable data for evaporator-vapor samples.

Another novel and accurate apparatus used for effectively degassing and determining the concentration of dissolved solids in condensed steam samples has been developed by Hall Laboratories, Inc. This apparatus

TABLE 1. COMPARISON OF TOTAL SOLIDS IN STEAM, DETERMINED BY EVAPORATION OF CONDENSED STEAM SAMPLES AND BY POWELL-MCCHESNEY CONDUCTIVITY CELL (TESTS MADE AT RUSSELL STATION, ROCHESTER GAS AND ELECTRIC CORPORATION)

Date, 1948	Specific conductivity, mhos × 10^{-6}	Total solids by evaporation, ppm	Total solids by conductivity, ppm	Difference, ppm
10/1 10/5	1.65	0.50	0.57	+0.07
10/6 10/11	1.80	0.60	0.63	+0.03
10/12 10/16	1.65	0.70	0.57	−0.13
10/25 10/28	1.43	0.50	0.50	0.00
10/29 11/3	1.70	0.50	0.59	+0.09
11/4 11/8	1.52	0.60	0.53	−0.07
11/8 11/12	1.43	0.50	0.50	0.00

TABLE 2. COMPARISON OF TOTAL SOLIDS IN EVAPORATOR VAPOR, DETERMINED BY EVAPORATION OF CONDENSED STEAM SAMPLES AND BY POWELL-MCCHESNEY CONDUCTIVITY CELL (TESTS MADE AT RUSSELL STATION, ROCHESTER GAS AND ELECTRIC CORPORATION)

Date, 1949	Specific conductivity, mhos × 10^{-6}	Total solids by evaporation, ppm	Total solids by conductivity, ppm	Difference, ppm
7/27 8/5	0.70	0.38	0.29	−0.08
8/8 8/15	0.60	0.20	0.25	+0.05
8/15 8/22	0.63	0.19	0.26	+0.07
8/26 9/2	0.60	0.31	0.25	−0.06
9/7 9/15	0.58	0.27	0.24	−0.03
9/16 9/23	0.73	0.24	0.31	+0.07
9/23 9/30	0.67	0.32	0.28	−0.04
10/3 10/13	0.78	0.30	0.33	+0.03

has now been operated in a number of stations where exact quantitative measurements of total solids in a continuously flowing sample of steam are desired. Dissolved gases in the sample are first removed from solution by a process of condensing and reboiling the sample. To accomplish degasification, the apparatus automatically divides the sample into two approximately equal portions: one portion contains all the total solids in

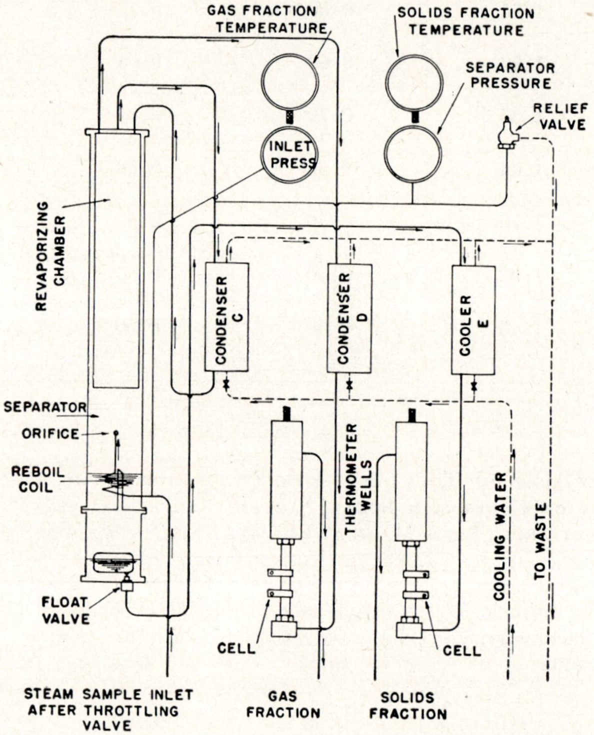

Fig. 12. Diagrammatic arrangement of equipment for measuring conductivity of degassed samples of steam or condensate. (*Courtesy of Hall Laboratories, Inc.*)

the steam; the second portion contains all the gases. Heat-exchange elements in the apparatus are so designed that the weight ratio of both streams of condensed steam stays practically constant.

A schematic arrangement of the apparatus is shown in Fig. 12. The functioning of the apparatus is described by Rivers[12] as follows:

1. Steam, throttled to between 15 and 100 psig, enters the double reboil coil A immersed in a condensate pool in the separator. Heat exchange takes place between the fluids inside and outside of the coil. Steam in the coils is partially condensed before discharge through orifices, and condensate in the pool around the reboil coils is continuously boiled to eliminate traces of gases.

2. Uncondensed steam passing upward in the separator contacts the outside of the revaporizing chamber B. Partial condensation occurs, with water returning to the pool around the reboil coil A. The residual steam, containing gases, goes out the top of the separator.

3. The steam containing the gases ("gas fraction"), after leaving the separator, is completely condensed in condenser C.

4. The condensate leaving C is passed into revaporizing chamber B, where it absorbs heat from the steam on the outside and is completely vaporized, effecting the partial condensation in the separator, as mentioned in 2.

5. The gas fraction leaves the revaporizing chamber in the separator, is then condensed and cooled in condenser D, passes through conductivity flow cell G, and overflows to waste.

6. The water in the pool surrounding the reboil coil A contains the solids originally present in the steam but essentially no gases. This "solids fraction" overflows into a float chamber from which it issued continuously through float valve F.

7. The solids fraction is then cooled in cooler E, passes through conductivity cell S, and overflows to waste.

Experience in a fairly large number of installations has demonstrated that the equipment insures good removal of gas, and reliable recording of solids in the degassed sample can be obtained.

This equipment is somewhat more complicated than some other equipment available for measuring steam quality, but where conditions can justify the need for automatically recording the conductivity of degassed samples and those containing gases in the steam, this equipment is satisfactory.

TESTING FOR DISSOLVED GASES AND CONDUCTIVITY

The procedures for determination of dissolved gases, such as carbon dioxide and ammonia, which affect the conductivity of condensed steam are given in the analytical methods.

TESTING OF MISCELLANEOUS CONSTITUENTS IN STEAM

Determinations are sometimes made for alkalinity, chloride, carbon dioxide, ammonia, silica, iron, copper, and pH of the steam. Usually, such tests are made only for investigation of special problems and are not always justified for routine control; however, tests for pH and dissolved gases have such justification where close control is required.

Bibliography

1. "Instruments and Apparatus, Part II: Determination of Quality of Steam," Series 1929, ASME Power Test Code.
2. Hecht, Max, and D. S. McKinney: Fuels and Steam Power, *Trans. ASME*, **53**: 139 (1936).

3. Rummel, J. K.: Estimation of Solids in Steam by Conductivity, *Ind. Eng. Chem., Anal. Ed.,* **3**: 318 (1931).
4. Tray, S. E.: Treat Feedwater to Prevent Carry-over, *Power,* **80**: 365 (1936).
5. Ulmer, R.: Foaming and Priming of Boiler Water and Condenser Leakage, *Ind. Eng. Chem., Anal. Ed.,* **9**: 172 (1937).
6. Rummel, J. K.: "Fluid Treating Device," U.S. Patent 2,046,583, July 7, 1936.
7. Gurney, W. B., M. C. Schwartz, and T. E. Crossan: Determination of Purity of Steam by the Electrolytic-conductivity Method, *Trans. ASME,* **62**: 728 (1940).
8. Powell, S. T., H. E. Bacon, I. G. McChesney, and F. Henry: Design and Development of Apparatus for Measurement of Steam Quality by Electrical Conductivity Methods, *Trans. Am. Inst. Chem. Engrs.,* **33**: 116 (1937).
9. Powell, S. T.: Steam Contamination—III: Determination of Steam Quality, *Combustion,* **9**: 25 (1937).
10. Alexander, R. C., and J. K. Rummel: The Quality of Steam Condensate as Related to Sodium Sulphite in the Boiler Water, *Trans. ASTM,* **72**: 519 (1950).
11. Powell, S. T., and I. G. McChesney: Simplified Process for Determining Steam Purity, *Trans. ASME,* **73**: 849 (1951).
12. Rivers, H. M., W. H. Trautman, and G. W. Gibble: An Automatic Degasser for Steam Sampling in Power Plants, *Trans. ASME,* **72**: 511 (1950).

16 CORROSION OF METALS: CAUSE AND CONTROL

Corrosion of ferrous and nonferrous metals in all industries is enormous despite the earnest endeavor of a great number of investigators to eliminate such losses. The factors that influence corrosion are numerous and for this reason difficult to control. However, recent investigation of this subject has thrown much scientific light upon the fundamental principles upon which corrosion reactions proceed. As a result, many industries now can cope with corrosion problems more effectively than formerly. While we may not be able to eliminate corrosion completely, the problem can be minimized in most cases by following these five steps: (1) careful processing in the manufacture of the metal, (2) cathodic protection, (3) preservation of the surfaces by use of protective coatings, (4) control of the substances coming into contact with the metallic surfaces, and (5) selection of suitable materials for specific conditions. In this chapter, the last three phases of the problem are discussed, since these items are intimately related to feedwater treatment.

The factors influencing corrosion in steam-station practice are practically the same as those occurring elsewhere. In the case of boiler and auxiliary equipment, high temperatures and pressures, which are not present to so great an extent under atmospheric conditions, influence the rate of acceleration of corrosion.

Noncondensible gases released from water in boilers are probably the greatest single cause for the corrosion of boiler tubes, drums, economizers, and superheaters. The corrosion of nonferrous metals is due to the same causes that bring about these conditions in ferrous materials. In steam stations, the most important nonferrous corrosion problems are those occurring in condensers. How well condenser-tube metals resist corrosion is related closely to the composition of materials employed. In fact, the material is the most important single factor requiring consideration.

THEORIES OF CORROSION OF IRON

From the earliest times, man has been cognizant of the destruction of metals by various forms of corrosive attack. In the effort to develop means for preserving ferrous and nonferrous metals, inquisitive investigators have sought for a logical explanation of how corrosion proceeds. Some of the theories advanced have proved entirely inadequate and have been discarded. Time has shown other concepts to be reasonable, although not a complete answer for all types of corrosion. In order for the reader to be informed of the historical background on the subject, there is included in this text a brief account of the various theories which have been propounded.

Acid Theory. The acid theory was put forward more than 50 years ago and for a time had many supporters. It was contended by those who favored this theory that three factors influence the phenomenon, namely, an acid, water, and oxygen. It was believed that water containing an acid, even one so weak as carbonic, reduced iron to the ferrous condition and produced hydrogen in accordance with the following reaction:

$$\underset{\text{Iron}}{2Fe} + \underset{\substack{\text{Carbonic}\\\text{acid}}}{2H_2CO_3} \rightarrow \underset{\substack{\text{Ferrous}\\\text{carbonate}}}{2FeCO_3} + \underset{\text{Hydrogen}}{2H_2}$$

The ferrous carbonate was then thought to react with the oxygen and water to form ferric hydroxide and then to release carbonic acid again as follows:

$$\underset{\substack{\text{Ferrous}\\\text{carbonate}}}{2FeCO_3} + \underset{\text{Water}}{5H_2O} + \underset{\text{Oxygen}}{\tfrac{1}{2}O_2} \rightarrow \underset{\substack{\text{Ferric}\\\text{hydroxide}\\\text{(rust)}}}{2Fe(OH)_3} + \underset{\substack{\text{Carbonic}\\\text{acid}}}{2H_2CO_3}$$

Since carbonic acid was again released, it attacked more iron and the cycle continued indefinitely. Thus, it was contended that a cyclic reaction in which a relatively small amount of any acid, no matter how weak, was present might result in considerable corrosion provided that there was also present sufficient oxygen to combine with the iron.

Experimental studies and practical experience have shown that the acid theory is an inadequate explanation, because corrosion can proceed rapidly in the absence of carbonic acid or other acid, provided that the metal is in contact with water containing oxygen.

Peroxide Theory. The peroxide theory of corrosion of iron[1] was suggested more than 40 years ago. According to this theory, rusting of the metal takes place, first, by water coming into contact with metal and uniting with it as follows:

$$\underset{\text{Iron}}{Fe} + \underset{\text{Water}}{2H_2O} \rightarrow \underset{\text{Ferrous hydroxide}}{Fe(OH)_2} + \underset{\text{Hydrogen}}{H_2}$$

The hydrogen released in this way then combines with the dissolved oxygen in the water to form hydrogen peroxide:

$$\underset{\text{Hydrogen}}{H_2} + \underset{\text{Oxygen}}{O_2} \rightarrow \underset{\text{Hydrogen peroxide}}{H_2O_2}$$

Some of the peroxide thus formed was believed to unite with ferrous hydroxide to form ferric hydroxide, or rust:

$$\underset{\text{Ferrous hydroxide}}{2Fe(OH)_2} + \underset{\text{Hydrogen Peroxide}}{H_2O_2} \rightarrow \underset{\text{Ferric hydroxide (rust)}}{2Fe(OH)_3}$$

This conception of corrosion has now been discarded and proved to be erroneous.

Colloidal Theory. The colloidal theory of corrosion, advanced by J. N. Friend,[2] had some ardent supporters. He suggested that the action of colloidal hydroxide in promoting further corrosion might be due to this compound's acting as an oxygen carrier or catalyst passing alternately between the ferrous and ferric conditions. However, Friend has stated, "The colloidal theory was not intended to supplant the electrolytic. It is an auxiliary theory." This theory does not offer a complete or satisfactory explanation of the phenomenon of iron corrosion. Speller[3] has probably given the true significance of the effect of colloidal ferrous hydroxide when he states as follows:

> Of course, it is conceivable that under some conditions, ferric hydroxide might act as a depolarizer or oxygen carrier for the hydrogen film and thus indirectly accelerate corrosion. In any case, however, this is a secondary factor which might possibly accelerate corrosion in certain media, but it is not essential or fundamental in character.

In addition to the foregoing, the biological, the film, and the direct oxidation theories have been conceived by different investigators. These may account for the corrosion phenomena which take place in specific cases. They are not sufficiently broad to cover the basic or fundamental conditions under which the corrosion of metal proceeds.

Electrochemical Theory. The electrochemical theory was suggested in 1903 by Whitney[4] and has since been confirmed by a number of other investigators.[5] It is now accepted generally as a satisfactory explanation of the basic reactions responsible for the corrosion of ferrous as well as nonferrous metals. The theory is somewhat complicated and involved. It may be briefly explained as follows:

Water, H_2O, when ionized, forms hydrogen ion, H^+, and hydroxyl ion, OH^-. Iron in contact with the water goes into solution as ferrous ions, and an equivalent number of hydrogen ions (the hydrogen behaving as a metallic element) are discharged and plated out upon the surface of the iron.

The reaction for the dissociation of water is

$$H_2O \rightleftharpoons H^+ + OH^- \qquad (1)$$

All metals in contact with water have a specific solution tension, that is, they exert the power to go into solution. When this reaction takes place, the metal that has dissolved in the liquid is in the form of ions.*

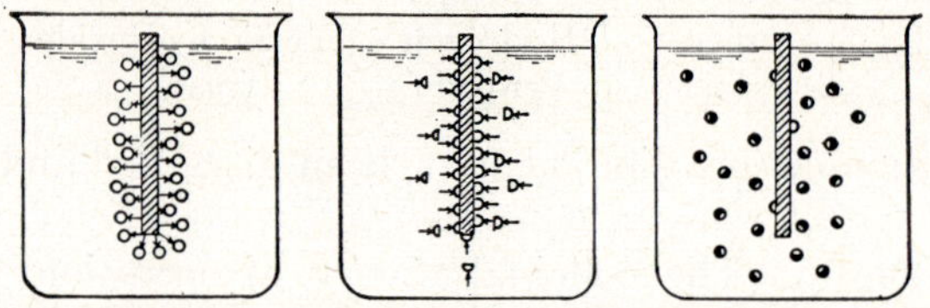

Fig. 1. Graphic illustration of the electrolytic theory of corrosion.

When the metal and water are chemically pure, the rate at which the metal dissolves is very slow. The solution rate is influenced greatly by impurities in the metal and the conductivity of the water due to dissolved electrolytes. The solution of pure iron in water, according to this theory, takes place in the following way (Fig. 1):

Each atom of metal, M, in dissolving forms a positively charged metal ion, M^+, and gives up one or more negatively charged electrons, e. This reaction can be written

$$M \rightleftharpoons M^+ + e \qquad (2)$$

The positively charged metal ion is attracted to the negatively charged hydroxyl ion and forms the metal hydroxide as follows:

$$OH^- + M^+ \rightleftharpoons MOH \qquad (3)$$

This metal hydroxide, which is relatively insoluble, may deposit on the metal surface as a permeable or impermeable film.

In the presence of oxygen, the iron that has gone into solution to form ferrous ions in the water in contact with the metal is readily oxidized, forming the insoluble compound ferric hydroxide. When such deposits form on iron, the corrosive action is somewhat retarded, since they may act as a protective coating on the metal. However, corrosion of iron is

* Atoms or atom groups carrying an electric charge. If the ions have a positive charge, they are termed "cations," and if they carry a negative charge, "anions."

not entirely prevented by these deposits and may, and usually does, proceed beneath them.

When electrons are released in reaction (2), an equivalent number of hydrogen ions, H^+, are immediately neutralized (reduced) and form an invisible layer of neutral hydrogen atoms on the cathode as follows:

$$e + H^+ \rightarrow H \tag{4}$$

Atomic hydrogen plating out on the cathode builds up to a point where the counterelectromotive force (emf) of the hydrogen opposes the flow of current in the cell and thus stifles the reaction. This phenomenon is called polarization. When dissolved oxygen is present in the electrolyte, oxygen diffuses to the cathodic area and reacts with the hydrogen to form water. This reactions tends to clear the cathodic area of hydrogen, prevent polarization, and permit the corroding action to continue. The equation for this reaction is

$$2H + \tfrac{1}{2} O_2 \rightarrow H_2O \tag{5}$$

Reactions (2) and (3) occur at the anode, and reactions (4) and (5) take place at the cathode. For every hydrogen ion neutralized and converted to water, a corresponding hydroxyl ion combines with the metal to form insoluble metallic hydroxide ions, leaving a condition of over-all neutrality.

From reaction (5) it will be noted that dissolved oxygen is an important factor in the mechanism of corrosion, and in the absence of this gas, inappreciable quantities of metal are able to pass into solution from anodic areas. However, when the pH* value of the solution drops, the hydrogen-ion pressure is sufficient to allow hydrogen to be liberated at the cathode; when this condition exists, corrosion may continue in the absence of oxygen.[7]

Practically all natural waters contain dissolved oxygen and as such are corrosive to a greater or lesser degree. As has been stated previously, the rate of solution is affected by the impurities in the metal and in the electrolyte. Because of this condition, the solution tension of the metal is variable at different areas of the metal. The solution tension is increased in some areas and decreased at others so that the surface is an intricate checkerboard of areas anodic and cathodic to every other area in differing degree. Owing to these conditions, electric currents may flow from one area to another, the electrolyte serving as the conducting medium. This is equivalent to the action of an ordinary wet-cell battery.

The electrolytic theory of corrosion to explain the mechanism of attack on metals has been confirmed by reasearch conducted by a number of investigators in recent times. Thornhill's[8] studies proved by delicate

* For explanation of the term "pH," refer to Chap. 24.

galvanometer measurements that there was a flow of current from the anodic to the cathodic areas and that corrosion occurred only at the anodic pole of such areas.

U. R. Evans[9] has reported on unpublished work of J. N. Agar further substantiating extensive experimentations on the flow of current and has prepared charts indicating the magnitude of the flow of current. A very comprehensive explanation of the mechanism by which the attack occurs is given by Evans in a recent discussion.[10]

DISSOLVED OXYGEN AND CORROSION

From the foregoing discussion, it is evident that oxygen dissolved in water has considerable effect in accelerating the rate of corrosion (Fig. 2). The effect of dissolved gases on corrosion rates and the problem of oxygen

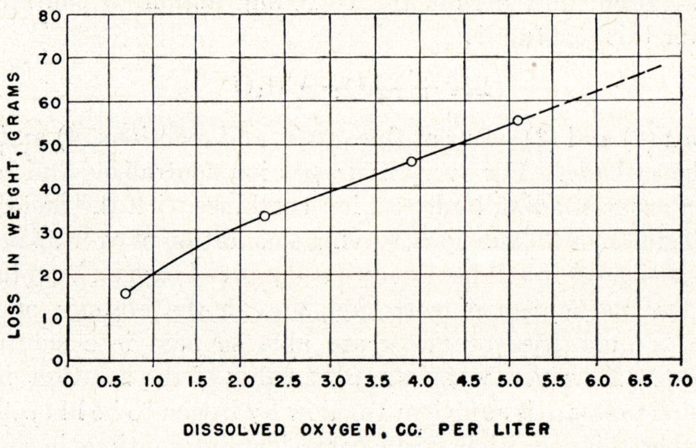

Fig. 2. Corrosion at various oxygen concentrations. (*After Speller.*)

removal are discussed in Chap. 10. The rate at which corrosion proceeds in waters containing a specific amount of oxygen will not be the same in all cases. This condition is influenced by the hydrogen-ion concentration, pH, of the water (Fig. 3), the incrusting solids in the water supply, temperature, service conditions, and many other factors, all of which affect to a marked degree the rate at which the attack proceeds.

CORROSION DUE TO MINERAL ACID

Some boiler-feedwater supplies are contaminated by mineral acids. Hydrochloric, nitric, sulfuric, and acetic acids are occasionally present in surface water in appreciable quantities as the result of industrial waste discharged into surface or underground supplies. Organic acids may also be present. Mine drainage frequently is responsible for the acids in feedwater, since many mine waters have a high acid content. Unless these acids are neutralized, they will cause rapid destruction of boilers and auxiliary equipment, the corrosion appearing in the form of pitting, which, if not checked quickly, destroys the metal.[10]

Hall and Teague[11] state that the presence of ferric sulfate in acid mine wastes is probably the main factor in the corrosion rate of waters so contaminated, since it acts as a depolarizer. These investigators showed that, when the concentration of oxygen, the velocity, the temperature, and other factors are constant, the rate of corrosion is proportional to the amount of ferric sulfate present in the solution.

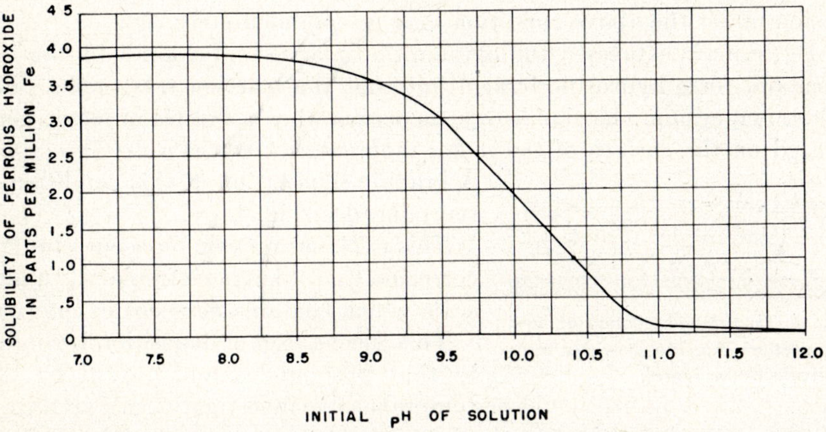

Fig. 3. Relation between pH and solubility of iron in deaerated water at room temperature.

Speller[12] has shown that, when the iron salt is present in the form of ferrous sulfate in an acid solution, the opposite effect is noted and the corrosion is inhibited. This is due to removal of dissolved oxygen and to increase in the metal-ion concentration.

An important salt affecting the corrosiveness of water is magnesium chloride. Investigators of this problem differ as to the exact reaction which takes place when magnesium chloride decomposes. The generally accepted reaction, however, is as follows:

$$\underset{\text{Magnesium chloride}}{MgCl_2} + \underset{\text{Water}}{2H_2O} \rightleftharpoons \underset{\text{Magnesium hydroxide}}{Mg(OH)_2} + \underset{\text{Hydrochloric acid}}{2HCl}$$

Since the magnesium hydroxide is removed as an insoluble precipitate, the reaction continues to go forward.

The acid generated then attacks iron as follows:

$$\underset{\text{Iron}}{Fe} + \underset{\text{Hydrochloric acid}}{2HCl} \rightarrow \underset{\text{Iron chloride}}{FeCl_2} + \underset{\text{Hydrogen}}{H_2}$$

A further reaction between the iron chloride and the magnesium hydroxide, forming iron hydroxide and magnesium chloride, follows:

$FeCl_2$	+ $Mg(OH)_2$ →	$Fe(OH)_2$	+ $MgCl_2$
Iron chloride	Magnesium hydroxide	Iron hydroxide (rust)	Magnesium chloride

As a result of this reaction, a relatively small amount of magnesium chloride in water may become a source of active corrosion, since the corrosion under the above condition may go on indefinitely.

Dewrance[13] expressed the opinion that magnesium chloride in solution does not form hydrochloric acid but that the reaction takes place only when magnesium chloride is dry, as occurs when a bubble of steam generated on the surface of the metal momentarily leaves a dry, hot spot. A graphic illustration of this condition is reproduced in Fig. 4.

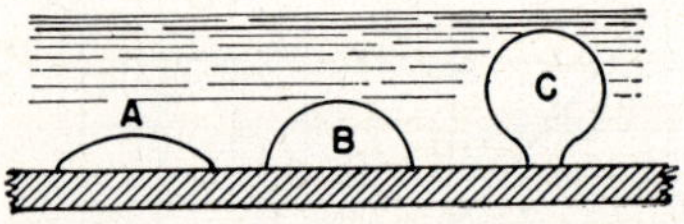

Fig. 4. Graphic illustration of how magnesium chloride in water may corrode boiler metal.

This explanation would account for the corrosion that takes place frequently under scale which contains magnesium chloride.

This mechanism of deposition is similar to that propounded and proved by the comprehensive investigation of Partridge[14] to explain the formation of anhydrite (calcium sulfate) salt. The findings of this investigator are important and are generally accepted. Partridge's explanation is given herewith:

Bubbles of steam evolved from an experimental evaporative surface in contact with a solution saturated with calcium sulfate were studied. Beneath each bubble, during its period of formation and attachment to the heated surface, a small ring of crystals was formed, evidently deposited in direct contact with the hot surface at the triple interface at which the solid surface, the solution, and the vapor of the bubble were in contact. The formation of crystalline deposits is logically explained. Within each interface, the hot surface was in contact with a gas film rather than with a liquid film. Because of the great increase in resistance to heat flow of the gas film over the resistance of the liquid film, the metal surface tended to overheat locally. This caused increased evaporation into the bubble from the very thin edge of solution surrounding the interface. As the body of the solution was saturated with calcium sulfate, this thin edge of solution was already supersaturated as a result of the increased temperature of the liquid film at the heated surface. This caused rapid deposition of the excess dissolved material on the hot surface.

Sodium chloride is only mildly corrosive but in the presence of magnesium sulfate reacts with the latter to form magnesium chloride as follows:

$$\underset{\text{Sodium chloride (common salt)}}{2NaCl} + \underset{\text{Magnesium sulfate}}{MgSO_4} \rightarrow \underset{\text{Magnesium chloride}}{MgCl_2} + \underset{\text{Sodium sulfate}}{Na_2SO_4}$$

The magnesium chloride thus formed becomes actively corrosive unless neutralized by an alkaline salt. If the alkalinity is due to caustic soda, the magnesium chloride will react with it to form sodium chloride and magnesium hydroxide, thus inhibiting the corrosive properties of the magnesium chloride according to the following:

$$\underset{\text{Magnesium chloride}}{MgCl_2} + \underset{\text{Sodium hydroxide}}{2NaOH} \rightarrow \underset{\text{Sodium chloride}}{2NaCl} + \underset{\text{Magnesium hydroxide}}{Mg(OH)_2}$$

Magnesium carbonate and magnesium nitrate in the presence of sodium chloride may generate acids, the former giving hydrochloric acid and the latter nitric acid. Silica and sodium chloride in the presence of steam will produce hydrochloric acid and sodium silicate. Although this reaction is possible, it is doubtful that the condition occurs in boiler waters except under unusual conditions.

Ammonium chloride may be responsible for the corrosive properties of some feedwater. Although this condition is found only infrequently, water high in ammonium chloride may cause corrosion of boilers and appurtenances. A case where boilers were seriously corroded from this source was called to the author's attention a number of years ago. Many feedwater problems relating to corrosion may be traced to the production of acids due to the reaction of the various salts concentrated in the boiler water.

CORROSION OCCURRING UNDER SCALE

As a rule, scale formation will inhibit corrosion, since the deposited salts mechanically protect the metal and prevent actual contact between the corrosive agent and the metal. Under some conditions, however, as described earlier, corrosion can take place beneath the scale or even under a protective coating applied to protect the metal surfaces. When this action occurs, the corrosion is generally in the form of pitting and often progresses without detection, since the corroded areas are not readily visible.

At one time, the author investigated a case in which the corrosion progressed beneath a protective paint which had been used to cover the drum of a large boiler. The particular coating had been used for a number of years and had given satisfaction. When the boiler was opened for inspection, it was found that serious corrosion had taken place just below the water line in one drum and that the corroded area extended from the center of the drum forward about one-half the distance from the center to the front of the drum. The investigation showed the following facts: the successive coats of paint had been applied each year without complete removal of the preceding layer, so that the thickness varied from 1/16 to 3/16 in. in spots. Under the microscope, the material showed distinct laminations with numerous hair cracks in the surface. A section of the material was boiled in caustic soda for 15 min, then washed with clean water, dried, and finally heated in an electric furnace. This treatment ruptured the section, splitting the lamination apart and opening the hair cracks on the surface so that they were plainly visible. It was indicated from these tests that the corrosion resulted when the protective coating became cracked, permitting the water, which had a fairly high magnesium chloride content, to concentrate between the laminations; the salt was dissociated at the metal surface. The magnesium chloride, removed from the neutralizing effect of the salts in the water of the drum, set up active corrosion of the metal. It is probable that reactions such as these occur beneath scale, although the active corroding agent may not always be magnesium chloride.

CARBONIC ACID

Carbonic acid, H_2CO_3, is the most important dissolved gas after oxygen to accelerate corrosion. When present in small amounts in feedwater, it has no appreciable corroding effect because there is usually sufficient alkalinity present in the concentrated boiler water to neutralize its acidity. A relatively high carbonic acid content in feedwater or bicarbonates which decompose and release carbonic acid under the influence of heat may cause severe pitting of the tubes or drum of the boiler. Even though no corrosion may take place in the boiler, the carbon dioxide discharged from the boiler with steam may cause active corrosion of turbines, bearings, and auxiliaries. Corrosion occurring in steam and condensate pipes may be traced usually to this source. Figure 5 is a photograph of corrosion occurring in a steam turbine due to a high concentration of carbon dioxide in the steam. This condition was due to condensation in the turbine caused by frequent outage of the unit, which permitted re-solution of carbon dioxide.

Some feedwater supplies contain organic acids or organic compounds which under high steam temperatures and pressures are converted into acids. These substances are usually the result of industrial pollution of the feedwater source, although they may be due to drainage from swamps or marshes. Since many of these compounds are volatile, they are particularly objectionable because they pass from the boiler with the steam, causing active corrosion of steam piping and other parts of the system. J. H. Paul,[15] in discussing the action of carbonic acid in boiler water, has attempted to prove that this acid under certain conditions is converted first into formic acid or acetic acid and then into aldehyde. Corrosion is due to these compounds. The author investigated a case of this kind where a feedwater supply was polluted by organic wastes discharged from a paper mill. Paul states that formyl cellulose in water yields a volatile acid when boiled and that this distillate, though very dilute, shows the characteristics of formic acid.

Fig. 5. Corrosion in a high-pressure turbine due to carbon dioxide in steam.

CORROSION OF STEAM-HEATING SYSTEMS AND CONDENSATE-RETURN LINES

Central steam-generating plants are employed to provide heating for large buildings or for apartment communities in urban districts. Condensate is collected and returned to the boiler house for reuse as feedwater. These returns contain oxygen and carbon dioxide in solution; thus they are very corrosive. Oxygen may be removed from make-up water by deaerating heaters.

The addition of alkaline materials may be employed to neutralize acidity. Ammonia is an effective neutralizer of acid but may corrode brass and copper equipment. Certain amines* have been used for preventing corrosion in condensate but, like ammonia, may cause corrosion of copper and brass alloys.

A. A. Berk[16] has investigated the use of cyclohexylamine in steam-heating systems. His summary and conclusions are quoted below:

> Corrosion of steam heating systems is due principally to carbon dioxide, which acidifies the condensate. One method of preventing corrosion is to introduce

* Any of a class of compounds derived from ammonia by replacement of hydrogen by one or more univalent hydrocarbon radicals.

sufficient amine into the steam to neutralize the acidity. The use of cyclohexylamine for this purpose has been studied both in the laboratory and in an operating system.

The amine* is very volatile from boiler water, so much so that it appears to leave the boiler with the steam practically as soon as it enters with the feedwater. To maintain an alkaline condition in the condensate at traps and heaters, frequent and preferably continuous additions of amine must be made to the system.

Losses of the volatile amine probably occur at every vent in the system. Large quantities of the chemical were found in the vented vapors from the deaerating heater. Loss was negligible in the blowdown water but appreciable in the condensate that was not returned to the system.

In a system that is amine-tested all of the carbon dioxide in the steam tends to be fixed in the condensate and returned to the boiler for recirculation. Condensate which is not returned to the boiler is important, not because of its amine content but because it must be replaced by raw water, the source of additional carbon dioxide. For this reason, loss of water as blowdown is almost as expensive in terms of make-up amine as loss of condensate although the former contains no cyclohexylamine.

A deaerating heater, operated at temperatures below 212°F, volatilized a large percentage of the amine recirculated with the condensate. The quantity of carbon dioxide vented simultaneously was relatively small. Above 212°F., the increased loss of amine was small, but much more carbon dioxide was removed. Requirements of make-up amine for the cyclohexylamine treatment would appear to be least if the heater were operated at as high a temperature as possible. This may mean no recirculation of amine to the boiler but at the same time there would be a minimum of carbon dioxide in the steam.

Cyclohexylamine treatment has helped reduce return-condensate corrosion in steam-heating systems. This result cannot be attributed entirely to its alkalinity, for in many instances the quantity of amine in the steam was small as compared with the carbon dioxide content. It is probable that the amine can inhibit corrosion in the manner peculiar to the nitrogenous bases to which it is related.

All of the defects in the use of cyclohexylamine in steam systems are due to its excessive volatility. This disadvantage is not unsurmountable, since it can be overcome through determining a less-volatile amine of the same desirable properties. Amine treatment has been shown to be useful in steam systems where return-line corrosion is a serious problem.

OIL TREATMENT FOR PREVENTION OF CORROSION OF CONDENSATE LINES

Oil has been used in some locations to provide a protective coating on the internal surfaces of steam lines and condensate-return lines. The

* A very comprehensive study of amines and their effect on inhibiting corrosion is given in a paper entitled Contamination of Condensate by Heat Exchanger Tube Alloys, by J. D. Ristroph and E. B. Powell, which was presented at the annual meeting of the American Society of Mechanical Engineers, December, 1952.

author has successfully utilized soluble oil for this purpose. If the application can be uniformly effected and sufficient oil added continuously to mend any breaks in the film, the treatment is quite satisfactory. In most installations where the oil has been used, it has been atomized directly into the steam ahead of locations where the dew point is reached.

If the condensate is discharged to waste, the oil is not a problem. However, if the condensate is recovered, a removal plant must be installed to provide oil-free condensate. This is expensive and generally precludes the use of oil for a protective film when condensate recovery is practiced.

Recently filming-type amines have been developed which can be applied in much the same manner as the oil. These are waxy materials which provide a protective film on the metal surfaces. In a number of low-pressure installations, with which the author is associated, the filming compounds give favorable results. Recent experience at one plant has shown that some of these waxy materials, when present in condensate added to the boiler feedwater, have baked onto the tubes, resulting in overheating of the metal. Careful evaluation of the use of these compounds is required where there is a possibility of contaminating process steam or steam which comes directly into contact with food products. Here again the filming compounds must be added continuously at a rate sufficient to mend any damage in the protective film. Otherwise accelerated corrosion will occur on unprotected area.

Deaerating heaters and cold-water vacuum deaerators can be used satisfactorily to remove corrosive gases from condensate which is collected at a central point and pumped through a long steel pipeline back to the boiler plant. This effectively protects the piping.

Alloy pipes and tanks lined with noncorrosive materials can be used advantageously in condensate systems. Further, the piping and related equipment can be so designed as to provide for flooding with condensate, release of gases, and minimizing of reaeration.

In general, it is sound practice to eliminate the bicarbonate and carbonate from the feed make-up by proper feedwater conditioning, thus removing the possibility of corrosion in the condensate lines due to carbon dioxide. The treatment to effect proper carbon dioxide removal from the treated feedwater is discussed elsewhere in this text.

European water-treatment practice has for many years provided for the application of ammonia to the steam-water cycle. This gas dissolves in the condensate and neutralizes carbon dioxide, thus effectively raising the pH value. Ammonia has all the objectionable features discussed in connection with the use of a volatile amine. Further, relatively low concentrations of ammonia cause damage to copper and brass fittings.

Compounds of hydrazine, which are efficient scavengers of dissolved oxygen, were used first in Europe and later in America to prevent corrosion in boilers and feedwater systems, when in service or laid up with water inside. Research and practical test work is being carried out with hydrazine hydrate and hydrazine salts to determine fully the most efficient and satisfactory methods of using these chemicals. As with sodium sulfite, the feeding rate should not be excessive, since, depending on service conditions, hydrazine decomposes to a greater or less extent in the boiler to form ammonia, which usually should be held to certain maximum concentrations when the condensate and dissolved ammonia come in contact with copper alloys in the system.

ALKALINITY OF WATER AND ITS EFFECT ON CORROSION OF IRON

The relation between the concentration of alkaline salts in water and its corrosive properties has been subject to much controversy. F. Lyon[17] stated that dilute alkaline solutions showed an initial increase in the amount of corrosion and a greater effect on piping when quiescent than that of distilled water exposed to the atmosphere. He stated further:

> As the concentrations increased, the rate of loss increased to a maximum, then fell rapidly to zero and remained there, the areas of local corrosion becoming smaller and more pronounced until they disappeared altogether at the concentration in which the loss was zero.

The effect of alkalies on the corrosion of iron is clearly stated by Speller:[18]

> Alkalies, as is well known, cause a retardation in the corrosion rate of iron, which diminishes to a certain minimum as the concentration of the base increases. This general fact has been well established by experience and tests. In the works of different investigators, there is some disagreement, however, about the extent of the inhibiting effect. The lack of agreement is not surprising, as the rate of corrosion is influenced by a number of factors other than alkalinity, such as the film-forming properties of the waters, the presence of other substances in solution, aeration, rate of motion and temperature. Hydroxide alkalinity is much more effective than carbonate or bicarbonate alkalinity, because a normal solution of a hydroxide gives a greater hydroxide-ion concentration than does a normal solution of carbonate or bicarbonate.

Figure 6 shows graphically the relative rates of corrosion occurring in caustic soda solution protected and unprotected from carbon dioxide absorption.

The foregoing explanation of the action of alkalies in retarding corrosion is not applicable under conditions where high concentrations of alkalies to the point of supersaturation are encountered, such as can occur due to film boiling. The term "film boiling" is described by Hall and Partridge[19] as a condition in which the alkaline salts greatly accelerate rather than retard corrosion of steel. The mechanism of attack follows a characteristic pattern and is brought about in the following manner: in tubes only slightly inclined from the horizontal, as in dry-bottom furnaces and in the floors of slag-bottom furnaces, the upper tube surfaces may be blanketed with steam causing high metal temperatures. Frequent wetting of these partially insulated surfaces by intermittent splashing of water may lead to concentration of sodium hydroxide in these areas. High caustic alkalinity tends to destroy the protective oxide film. The combination

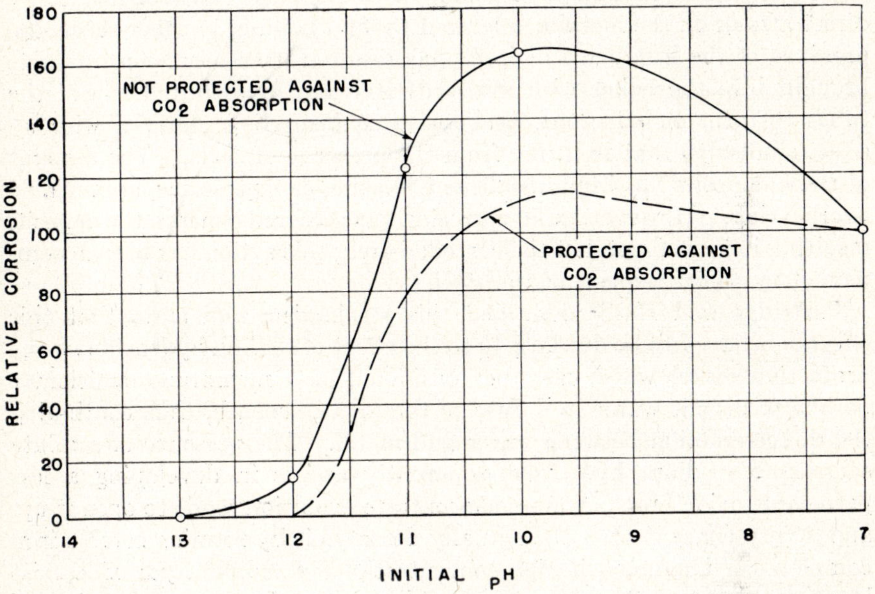

Fig. 6. Effect of exposure to the air on the corrosion in a sodium hydroxide solution. (*After Speller.*)

of higher temperatures and impairment of this film causes grooving of tubes and ultimate failure.

As remedial measures, Hall and Partridge suggest increasing the inclination of the tubes, controlling heat input, improving circulation, and reducing the caustic alkalinity of the boiler water.

PROTECTIVE FILM OF IRON OXIDE

Experience has demonstrated that, during early periods of operation of new boilers, the internal surfaces become coated with a protective film of dense, adherent, black iron oxide which forms under normal or desirable operating conditions. According to Corey,[20] the film usually has a thickness of 0.003 in., and it remains intact without appreciable change in thickness unless damaged by physical change or chemical or electrochemical action or a combination of these factors. When the film is broken or penetrated, the underlying surfaces may be corroded by the boiler water or steam. If the protective film is not healed, the corrosion continues at an accelerated rate.[21]

The mechanism involved in the damage of the protective film and in the subsequent corrosion is complex and not fully understood under all conditions. In many instances, severe corrosion has been associated with the excessive formation of black iron oxide, with or without the presence of metallic copper. In some cases, the oxide has been formed by direct attack on the metal accelerated by film boiling; in other cases, the presence of the iron oxide deposits has resulted from corrosion products brought into the boiler with the feedwater and precipitated within the tubes to form an adherent, heat-resisting, insulating material with an effect similar to that resulting from the presence of scale. The amount of these deposits has been insufficient in some cases to cause appreciable overheating. However, rapid corrosion has occurred beneath the deposit, resulting in loss of metal at the corroded area; this action has been accompanied by embrittlement of the steel.

Partridge and Hall[19] described this mechanism and state that this effect is believed to be due largely to the effect of sodium hydroxide on the protective oxide, which may be considered as a membrane sufficiently porous to permit water to diffuse to the metal beneath, thus continuing the direct reaction between water and metal. The presence of a dilute solution of sodium hydroxide apparently assists in developing a less permeable oxide film. A more concentrated solution tends to open it up, and a very concentrated one actually destroys it by forming soluble iron complexes. Therefore, if high concentrations of sodium hydroxide exist locally, the protective film is impaired, increasing the generation of hydrogen and the progress of hydrogen embrittlement of the metal.

INFLUENCE OF THE HYDROGEN-ION CONCENTRATION OF A SOLUTION ON CORROSION

The hydrogen-ion concentration* of a water is a measure of the ionized hydrogen, or the active acidity or alkalinity, of the water. High concentrations of hydrogen ions tend to increase corrosion of metal because, stated briefly, an increase of hydrogen ions means a higher acidity of the water. This, in turn, accelerates the deposition of atomic hydrogen as the iron ions go into solution in the water. For each ion of metal which passes into solution, there must be deposited, or plated out, one electrical equivalent of hydrogen.

Hydrogen-ion concentration affects the conductivity of the electrolyte and the solubility of the products of corrosion. The rate of corrosion is controlled by the rate at which hydrogen deposited on the cathode area

* For a more complete discussion of this subject, refer to Chap. 24.

can be removed by oxidation and by the rate of release of bubbles of the gas. In neutral or alkaline waters, removal of hydrogen by oxidation predominates. In acid waters, depolarization of the cathode area by the release of hydrogen bubbles is dominant. Both mechanisms can proceed simultaneously. This phenomenon is influenced by many factors which

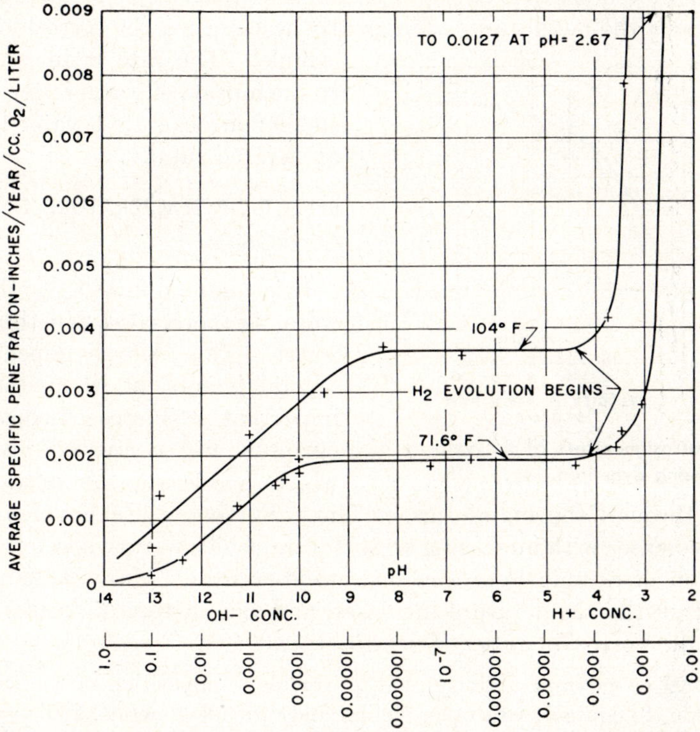

Fig. 7. Effect of hydrogen-ion concentration (pH) on the corrosion of steel.

cannot be discussed here, and the reader should refer to complete text references.[22,23]

It is interesting to note that the ionization of pure water increases with rising temperatures. Figure 7 shows the influence of the hydrogen-ion concentration on the corrosive properties of water at 71 and 104°F, reproduced from the studies of Whitman, Russel, and Altieri.[24] These tests show the rapid corrosion rate effected by increasing the hydrogen-ion concentration of water in contact with iron to the point of gas evolution of hydrogen.

VELOCITY OF WATER AND ITS EFFECT ON CORROSION

It has been accepted generally that an increase in velocity of water passing over metal greatly accelerated the rate of corrosion. There are a number of factors that may account for this condition. However, the increase in corrosion is due generally to the fact that the rapid movement of water over the surface brings much greater amounts of oxygen into contact with the metal than those resulting only from diffusion. In Fig. 8 is shown the rate of corrosion of boiler metal in tap water saturated with oxygen and having a hydrogen-ion concentration of 6.4 (slightly acid). It will be noted that relatively low changes in the velocity of the water passing over the metal resulted in greatly accelerating the corrosion of the sample. The rate of increase due to the motion of water is very much smaller in deaerated water than in an aerated supply.

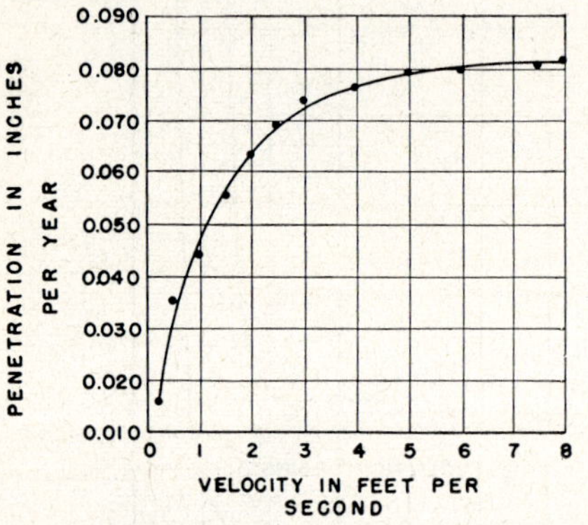

Fig. 8. Corrosion effect of the velocity of water passing over iron.

EFFECT OF TEMPERATURE ON CORROSION

It is common knowledge that hot water is more corrosive than cold water. This effect depends on several factors, of which the most important are increased convection currents and decreased viscosity, which accelerate the diffusion of oxygen through the metal surface film. Further, the pH value of the water decreases with increasing temperature, which in turn accelerates the deposition of atomic hydrogen on cathodic areas. Many salts are dissociated as the temperature increases, and this in time increases or decreases the corrosion rate. Friend[25] states:

> Electrical or galvanic activity, due to irregular composition of the metal, to segregation, to contact with other similar and dissimilar metals, and so on, and which may be capable of slowly exerting an influence upon the corrosion of the iron at ordinary temperatures, becomes greatly enhanced as the temperature rises, so much so as to yield quite astonishing results.

GASES OTHER THAN CARBONIC ACID AND OXYGEN DISSOLVED IN WATER

Hydrogen sulfide, ammonia, sulfur dioxide, and many other gases may be present in surface water employed for feedwater purposes. Of such gases, hydrogen sulfide and chlorine are the most important.

Although sulfides are occasionally found in surface water as a result of the deterioration of organic compounds or from miscellaneous pollution, these compounds occur most commonly in well waters. Because of their offensive odor, corrosive nature, and other objectionable characteristics, it is highly desirable to remove them from water which is used for sanitary purposes, processing, and cooling.

Greensand zeolites are damaged by the continuous application to the mineral of a water containing sulfide; as little as 2 ppm of sulfide, expressed as hydrogen sulfide, has been known to destroy such zeolitic material in a few months. Furthermore, many metals are rapidly attacked by relatively high concentrations of sulfides. Resistant alloys, such as stainless steel and brass, which are effective in an oxidizing medium are subject to aggressive attack by water containing sulfide, and the removal of this constituent is imperative if rapid deterioration of equipment is to be avoided. Sulfide is present in water generally as the result of industrial waste pollution or decomposition of sulfates by biological action.

Hydrogen sulfide, even in small amounts, reacts as an acid and may be actively corrosive to iron even in the absence of oxygen. This gas in the presence of oxygen may be oxidized to sulfuric acid under some conditions.

In Germany,[26] it was found that sodium sulfite in the presence of alkali is converted at around 525°F and above to sodium sulfide and sodium tetrathionate. The sodium sulfide is hydrolized into hydrogen sulfide, which, like the soda-separated carbonic acid, leaves the boiler in the steam. Even at pressures as low as 225 psi, hydrogen sulfide has been found in the steam as a result of superheated steam being cooled by injection of feedwater dosed with sodium sulfite. In turbines having nickel alloy parts, damage has been found due to action on nickel by hydrogen sulfide.

Alexander and Rummel[27] have correlated the effect of varying concentrations of sodium sulfite in boiler water with the quality of the steam produced. The data were taken from boilers operating at 900 psi and over. Increasing concentrations of sodium sulfite in boiler water lowered the pH and increased the conductivity of the condensed steam. It is probable that when the boiler water contains more than 5 to 8 ppm of sodium sulfite, the quality of the steam will be affected. The alkalinity and pH of the boiler water affects the steam quality to a minor degree. The reducing material in the steam condensate resembles sulfurous acid.

Removal of hydrogen sulfide from water[28] has been satisfactorily accomplished by reduction of the pH value of the water followed by aeration in a forced-draft tower. Aeration removes the sulfides by the scrubbing action of the air and by oxidation. Lowering the pH value of the water greatly reduces the size of the aerating equipment required. If it is desired to reduce the sulfides to zero, secondary chlorination is necessary.

METALLIC COPPER AND ITS EFFECT ON RATE OF CORROSION

Feedwater contains at times an unusually heavy concentration of ammonia, and this condition in the presence of oxygen in the feedwater results in pickup of fairly large amounts of copper from the heat-exchange equipment. The copper may be deposited in boilers in appreciable quantities as metallic copper. In one case investigated, copper was identified both in the boiler sludges and as an adherent layer plated out on the drums and tubes. In some parts of boilers, copper is deposited in a spongy mass; in other instances, it is laid down in sheets. Corroded areas are found adjacent to most of these deposits of spongy copper. Damaged tube sections showing the characteristic corrosion-embrittlement attack are found under layers of dense magnetic iron oxide, and considerable metallic copper may be mixed with the adherent deposits. It is the author's opinion that the copper present may not be the primary cause of the attack on the tubes but accelerates the action once the corrosion has started.

EFFECT OF FREE CHLORINE ON METAL

The author has noted rapid corrosion in the form of pitting of valves and pipes which were in close proximity to the point of application of chlorine solution. If a water supply is treated with relatively high doses of chlorine, there may be an appreciable residual chlorine content after 15 min of absorption. This is an unusual condition, however, and few chlorinated supplies are so heavily treated, except where the supplies are grossly contaminated or contain high concentrations of reducing agents. Where such conditions exist and where the water contains an appreciable amount of dissolved oxygen, corrosion may occur. The difficulty can be relieved or corrected by permitting sufficient time for complete reaction of the chlorine compound and by dechlorination.

Bibliography

1. Dunstan, W. R., H. A. D. Jouett, and E. Goulding: The Rusting of Iron, *J. Chem. Soc.*, **87**: 1548–1574 (1905).
2. Friend, J. N.: Recent Progress in the Study of Corrosion, *J. West Scot. Iron Steel Inst.*, **31**: 74–83 (1923–1924).
3. Speller, Frank N.: "Corrosion: Causes and Prevention," p. 46, McGraw-Hill Book Company, Inc., New York, 1926.
4. Whitney, W. R.: The Corrosion of Iron, *J. ACS*, **25**: 394–406 (1903).
5. Walker, W. H.: *J. ACS*, **29**: 1251–1264 (1907); **30**: 473–474 (1908). Cushman, and Gardner: "Corrosion and Preservation of Iron and Steel," pp. 44–45, McGraw-Hill Book Company, Inc., New York, 1910. Wilson, R. E.: The Mechanism of the Corrosion of Iron and Steel in Natural Waters and the Calculation of Specific Rates of Corrosion, *Ind. Eng. Chem.*, Vol. 15, 1923.
6. Baylis, John R.: Prevention of Corrosion and "Red Water," *J. AWWA*, **15**: 598 (1926).
7. Sudrabin, L. P.: Cathodic Protection of Steel Surfaces in Contact with Water, *Water Works & Sewerage*, **92**: 11, 51 (1945).

8. Thornhill, R. S., and U. R. Evans: *J. Chem. Soc.*, p. 2109 (1938).
9. Evans, U. R.: *J. Iron Steel Inst.* (*London*), **141**: 221 (1940).
10. Evans, U. R., in Herbert H. Uhlig (ed.), "Corrosion Handbook," p. 5, John Wiley & Sons, Inc., New York, 1948.
11. Hall, R. E., and W. W. Teague: "The Effect of Acidity and Oxidation Capacity on Corrosion of Metals and Alloys in Acid Mine Water," Bull. 15, Carnegie Institute of Technology Corporation, Coal Mining Investigation, Pittsburgh, Pa., 1924.
12. Speller, Frank N.: "Corrosion: Causes and Prevention," McGraw-Hill Book Company, Inc., New York, 1935.
13. Dewrance, John: Corrosion of Boilers, *Power*, Mar. 29, 1921.
14. Partridge, Everett P.: Formation and Properties of Boiler Scale, *Univ. Mich. Eng. Research Bull. No.* 15, 1930.
15. Paul, J. H.: "Boiler Chemistry and Feedwater Supplies," p. 18, Longmans, Roberts and Green, London, 1919.
16. Berk, A. A.: *U.S. Bur. Mines Rept. Invest.* 3754, June, 1944.
17. Lyon, F.: The Corrosion of Boiler and Piping on Shipboard, *J. Am. Soc. Naval Engrs.*, Vol. 24, 1919.
18. Speller, Frank N.: "Corrosion: Causes and Prevention," McGraw-Hill Book Company, Inc., New York, 1935, 1951.
19. Hall, R. E., and E. P. Partridge: Attack on Steel in High-capacity Boilers As a Result of Overheating Due to Steam Blanketing, *Trans. ASME*, **61**: 597 (1939).
20. Corey, R. C.: Corrosion by Liquids, *Combustion*, **15**: 35 (August, 1943).
21. Corey, R. C.: Corrosion by Hot Gases, *Combustion*, **15**: 34 (November, 1943).
22. Speller, Frank N.: "Corrosion: Causes and Prevention," McGraw-Hill Book Company, Inc., New York, 1935, 1951.
23. Evans, U. R.: "The Corrosion of Metals," Edward Arnold & Co., London, 1926.
24. Whitman, G. W., R. P. Russel, and V. Altieri: Effect of Hydrogen-ion Concentration on the Submerged Corrosion of Steel, *Ind. Eng. Chem.*, **16**: 665 (1924).
25. Friend, J. N.: "The Corrosion of Iron and Steel," p. 96, Longmans, Green & Co., Inc., New York, 1911.
26. Splittgerber, Arthur G.: European Experiences with Water Treatment, *Combustion*, **21**: 48 (1949).
27. Alexander, R. C., and J. K. Rummel: The Quality of Steam Condensate as Related to Sodium Sulphite in the Boiler Water, *Trans. ASME*, **72**: 519 (1950).
28. Powell, S. T., and L. G. von Lossberg: Removal of Hydrogen Sulfide from Well Water, *J. AWWA*, **40**: 1277 (1948).

17 CORROSION PROBLEMS IN HIGH-PRESSURE BOILERS

INTRODUCTION

Many types of corrosion occur in boilers and other power-station equipment because of the widely differing conditions encountered in the miscellaneous operations and the various kinds of metals in the system. To determine the cause of corrosive attack and to prescribe corrective treatment requires a comprehensive knowledge of operating conditions and an evaluation of all contributing factors involved in the mechanism of the corrosion. It is not possible to catalogue accurately all forms of corrosion experienced in boilers and in other power-station equipment because the corrosion may involve several contributing factors that make classification difficult.

Fig. 1. Section of boiler tube showing corrosion due to inadequate water treatment and defective circulation.

The installation of many high-pressure boilers in recent years has introduced operating problems which were not encountered in the generation of steam at low pressures. These developments have made it imperative to provide feedwater of the highest quality and to ensure meticulous care in the design and subsequent control of steam-generating equipment. Numerous high-pressure boilers have been operated with few or no difficulties from scale or corrosion. In some instances, however, severe corrosion has occurred either because of lack of proper feedwater conditioning or because of certain deficiencies in the design or operation of the boilers. In a few cases of serious boiler-tube trouble, it has been very difficult to develop a clear-cut analysis of the attack on the metal and to devise dependable corrective treatment.

Fig. 2. Surface of metal from ring showing large intercrystalline crack starting from inside surface, numerous smaller intercrystalline cracks, and complete decarburization.

Slater and Parr,[1] in a very complete and comprehensive discussion of various types of corrosion in marine boilers, have listed and classified a number of types of attack. Although their references were related to marine practice, the data submitted are applicable to similar attacks in stationary boilers. A tabulated summary of the classification of the various forms of corrosion observed by the authors is given in Table 1 (used by per mission of the authors).

TABLE 1. FORMS OF BOILER DETERIORATION*

Classification			Appearance of corrosion	Usual location	Cause
Internal deterioration	Pitting	Air-bubble pitting	Isolated hemispherical pits, often covered with a cap of loose, soft corrosion products, which, when dry, is of black iron oxide covered with a membrane of red dehydrated ferric hydroxide.	Mainly on the roof of steam and water drums, but likely to be found on any surfaces in contact with water, where air bubbles may collect.	Liberated air bubbles attaching themselves to metal surfaces, thereby initiating anodic attack by the generation of localized electrolytic action.
		Scab pitting	Deep localized pitting covered by a scab of hard, scaly, black magnetic iron oxide which cannot be removed by wire brushes.	In hotter areas of heating surfaces where corrosion products are formed and rapidly consolidated into a hard scab by the heat. This is possibly the main form of deterioration encountered.	Use of untreated impure boiler water containing dissolved salts and gases in comparatively small concentrations so that pH of water lies between 4.8 and 9.4.
		Soft scab pitting	Broad, shallow pitting covered with caps of fairly hard corrosion products, usually black magnetic iron oxide covered with a layer of red dehydrated ferric hydroxide.	In superheater tubes and headers and in main steam lines.	Deposits of salts by priming in superheaters, main steam lines, etc., followed by condensation when the boiler is shut down, resulting in the formation of droplets of electrolyte and the initiation of a pit by differential aeration.
	General wastage	Tube-wall thinning	A general reduction in tube-wall thickness over comparatively large areas, often without the presence of blankets of corrosion products.	Fairly evenly distributed over water surfaces with a tendency to be more pronounced in hotter areas.	Use of acid feedwater.
		Necking and grooving	A circumferential band of corrosion on the water side of fire tubes, stay tubes, and stay bolts in Scotch boilers.	Where stay bolts, stay tubes, and fire tubes enter end plates or combustion chambers.	The influence of adjacent end plates on expansion and contraction, resulting in stresses which cause repeated localized flaking of protective films.
		Galvanic attack	Local reduction in metal thickness.	Adjacent to nonferrous boiler mountings, in areas of nonhomogeneity of steel composition or at breaks in surface films.	Use of dissimilar metals in contact (or at areas of nonuniformity on the metal surfaces) in the presence of an electrolyte, when a difference in potential will result in sacrificial electrolytic attack at the anode.
		In tubes	Network or series of cracks over the water side of tube surfaces.	In boiler and superheater tubes on the hottest side of the bores. It is most frequently found in inherently curved tubes or tubes which have been buckled in service.	Irregular circulation in tubes causing frequencies of alternate overheating and cooling which induce pulsating thermal fatigue stresses in the walls. Cracking is accelerated by residual stresses and a corrosive environment.

	Corrosion fatigue	In steam and water drums	Fissures, usually difficult to detect owing to covering layers of corrosion products; usually associated with deep pitting.	Commencing at the arboring in tube plates and extending axially, eventually joining two or more holes.	Bending stresses due to lack of common neutral axis between tube and wrapper plate, and excessive cold working due to initial rough machining operation. Accelerated by thermal stress and corrosive environment.
		In main steampipes.	Lines of cracks on bores of pipes.	On the bores of main steampipes, especially in the base of corrugations.	Bending stresses due to movement during operation in the presence of a corrosive environment.
		Strain–age embrittlement	Spontaneous cracking in boiler components during detubing or cleaning operations.	Between tube holes in tube plate, under rivet heads, around top end of tubes adjacent to the rolling surfaces.	Reduction of impact properties by the combination of cold work and aging. Spontaneous failure occurs only at, and below, room temperature.
		Caustic embrittlement	Fissures in boiler plates, etc.	In boiler plates, radiating from, and often connecting, adjacent rivet holes, joints, arborings, etc.	Joint effect of a combination of nonuniform stressing, and high concentration of caustic soda in the presence of a catalyst at high temperatures.
	Deterioration at elevated temperatures	High-temperature oxidation	Continuous layers of hard black magnetic iron oxide scale.	Internal—in fire row and superheater tubes; occurring in drums and headers. External—on surfaces in contact with furnace gases.	Internal—poor circulation resulting in overheating in a steam atmosphere. External—local flame impingement or poor circulation or presence of heat-insulating scale on water side.
		Bursts	Longitudinal gaping split, which may or may not be associated with ballooning.	Hottest side of generator or superheater tubes.	Water starvation, resulting in overheating with corresponding reduction in mechanical properties of the tube until bursting occurs under working pressure.
		High-temperature creep cracking	Cracking and splitting associated with layers of high-temperature oxide scale on tube surfaces.	In superheater tubes.	A combination of stress due to restricted expansion and over-heating at temperatures around 550°C.
External deterioration		General wastage by damp soot	Localized wasting of boiler metal which can only be seen after removal of soot deposits.	Wherever soot collects, especially where tubes enter lower drums and headers.	Presence of sulfuric acid formed by action of moisture (when shut down) on oxides of sulfur present in the oil-fuel soot.

* *Inst. Mech. Engrs. (London)*, *Proc.*, Vol. 160, following p. 348, 1949.

Figure 1 is a section of a corroded tube where the active attack occurred beneath the formation of adherent, black iron oxide. The photomicrograph of the metal of the damaged tube is shown in Fig. 2; note decarburization and intercrystalline attack. This corrosion was caused principally by high heat input, but the presence of oxygen and copper also contributed to the failure of the metal.

EFFECT OF HIGH METAL TEMPERATURE ON RATES OF CORROSION

It is a recognized fact that water is dissociated in electric boilers, with the generation of hydrogen and oxygen by electrolysis. In electric boilers, the electrodes are submerged in the water and the water is the conducting medium for the current, thereby creating a condition favorable for gas production. A number of investigators have recently directed attention to the possible production of hydrogen and oxygen in fuel-fired boilers.

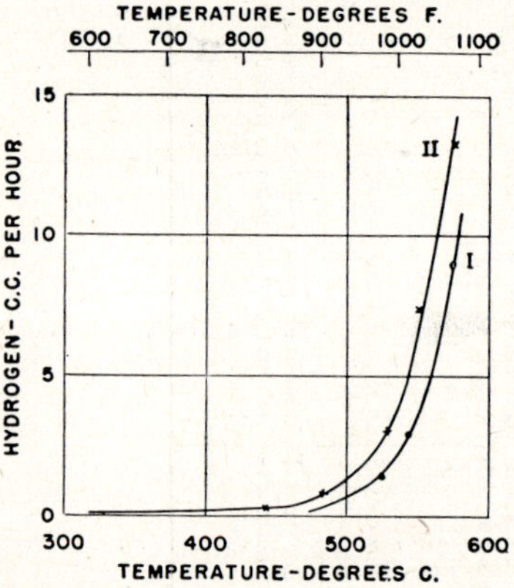

Fig. 3. Production of hydrogen from water at temperatures above 932°F (500°C).

Below 750°F, dissociation of water is probably negligible. Above this temperature, the degree of dissociation tends to be more significant.

Production of hydrogen by permitting steam to act upon hot iron is an old, well-established method of hydrogen-gas generation, which is used commercially. The reaction which takes place is as follows:

$$\underset{\text{Iron}}{3\,Fe} + \underset{\substack{\text{Water}\\\text{(steam)}}}{4H_2O} \rightarrow \underset{\substack{\text{Magnetic}\\\text{iron oxide}}}{Fe_3O_4} + \underset{\text{Hydrogen}}{4H_2}$$

This reaction is reversible, that is, if the hydrogen comes into contact with the magnetic oxide, it reduces the oxide and water is generated. When hydrogen is produced by this method on a commercial scale, the temperature is maintained at 1200°F or above, since below this point the production of the gas is relatively slow.

In an interesting paper by J. Porter, it was shown that stagnant steam at atmospheric pressure in contact with heated boiler tubes produced little hydrogen below 500°C (932°F, Fig. 3) but that above this temperature the rate of production was appreciable. The method of testing and the results obtained are quoted verbatim from an abstract of Porter's paper:

The following experiments were performed to show the action of stagnant steam, at atmospheric pressure, on a boiler tube heated to various temperatures. The apparatus employed is illustrated in the accompanying sketch [Fig. 4]. An electric furnace F was available, containing two porcelain tubes ($1\frac{5}{16}$ in. diameter) wound with nickel-chrome wire, and surrounded by heat-insulating material. The experimental boiler tube A ($1\frac{1}{8}$ in. diameter) was inserted in one of the porcelain tubes, and a duplicate tube B with a thermojunction T in the other. Preliminary experiments were made to test the equality of temperatures at corresponding positions in the two tubes, and to find the temperature gradients from the center toward the ends of the furnace. As the result of these experiments, it was estimated that the thermojunction would give the temperature of the central 1-in. length of tube A with an accuracy of about 20°C at temperatures

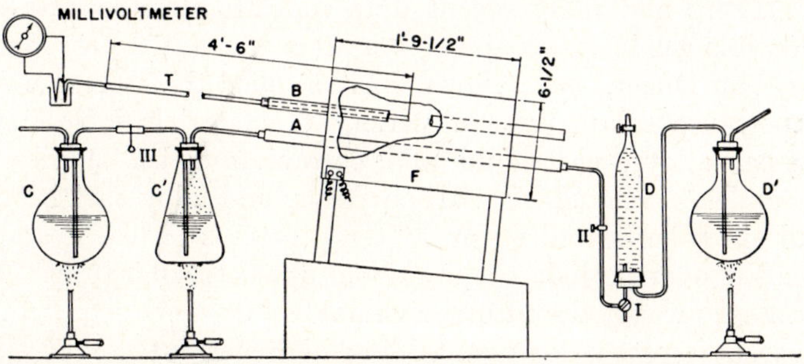

Fig. 4. Apparatus employed in experimental studies to determine the production of hydrogen from water at high temperatures.

such as 500°C. It should be remembered, however, that tube A was cooled by the passage of fresh steam into it about every quarter of an hour, so there is little doubt that the temperatures plotted in the accompanying diagram [Fig. 3] are rather higher than they ought to be.

The supply of air-free steam was obtained from the two flasks C and C′. C′ was filled with thoroughly boiled water and steam; C with thoroughly boiled water, steam and a little air. The receiver D was filled with thoroughly boiled water from the flask D′.

While the tube was at a temperature not very far above 100°C, the two-way tap I was opened to the air, tap II was opened and the clip III was closed. Steam from the flask C′ was then used to sweep all air (as far as possible) out of the tube A. The temperature of A was then gradually raised to the required value, and, usually at about quarter-hour intervals, the steam from A was passed into the receiver D at the top of which gases collected. These gases were transferred to a graduated tube, and the volume of hydrogen present was found, sometimes by adding oxygen and passing electric sparks, sometimes by absorbing the hydrogen by a palladium tube. Of course, the apparatus was never completely freed from air, so that other gases besides hydrogen collected in D. The water in D was being boiled by the steam passing into it, so little or no gas could be lost by absorption here.

The results of the experiments are illustrated by curves I and II of the accompanying diagram [Fig. 3]. No hydrogen was detected for temperature 400°C (750°F) even after, say about 1,000 cc. of steam had been passed through the tube, the experiment continuing for about five hours. Above 500°C (900°F) the rate of evolution of hydrogen increased very rapidly. Curve II was obtained from a second tube which had been heated in a current of carbon dioxide gas to over 600°C for a considerable time, for the purpose of removing any hydrogen which it was thought might possibly have been occluded in the steel.

The amount of iron required to produce a large volume of hydrogen is, of course, not large, about 0.1 g. being sufficient to produce 50 cc. of hydrogen measured at normal temperature and pressure.

Considering the present tendency toward higher pressures and temperatures in steam-boiler practice, these experimental tests are of much interest. These and more recent data indicate that routine tests for noncondensible gas in water and steam are a warranted procedure.

Rohrig, Van Duzer, and Fellows[2] have studied the corrosive effects of high-temperature steam upon a variety of materials including nickel, nickel-copper alloys, chromium and chromium-nickel stainless steels, medium- and low-alloy steels, carbon steels, and alloy cast iron. Tests were made in superheated steam at temperatures of 925°F and 1100°F and extended over periods ranging from 4,000 to 16,000 hr. Rates of corrosion were measured and other valuable data were acquired.

In general, ferrous alloys containing high percentages of chromium alone or of chromium and nickel are corrosion-resistant in steam at 1100°F. Nonferrous alloys containing high percentages of nickel and copper that were tested in either 925°F or 1100°F steam can be expected to corrode in steam at those temperatures. Although significant differences exist between the corrosion rates of carbon steel and medium- or low-alloy steels at 1100°F, at 925°F only a slight difference exists. The formation and maintenance of an adherent scale increases the corrosion resistance of materials exposed to high-temperature steam. The test results reported in this paper indicate that the weight loss of plain-carbon steel exposed to steam at 1100°F continues at a high rate, whereas the rate of loss of the alloyed materials decreases with time.

Solberg, Potter, Hawkins, and Agnew[3] have found that stress does not influence the corrosion rates of steel exposed to high-temperature steam.

TUBE FAILURES DUE TO CREEP AT HIGH TEMPERATURES

Failure resulting from high-temperature creep cracks is well known and has occurred on numerous occasions but is not common in most boiler installations. Some failures of this type were experienced in an installation recently examined by the author. The boilers had been in service for many years and no difficulty was experienced under normal operating conditions. Under increased demand due to wartime pressure, these boilers were greatly overloaded and shortly thereafter tube failures occurred.

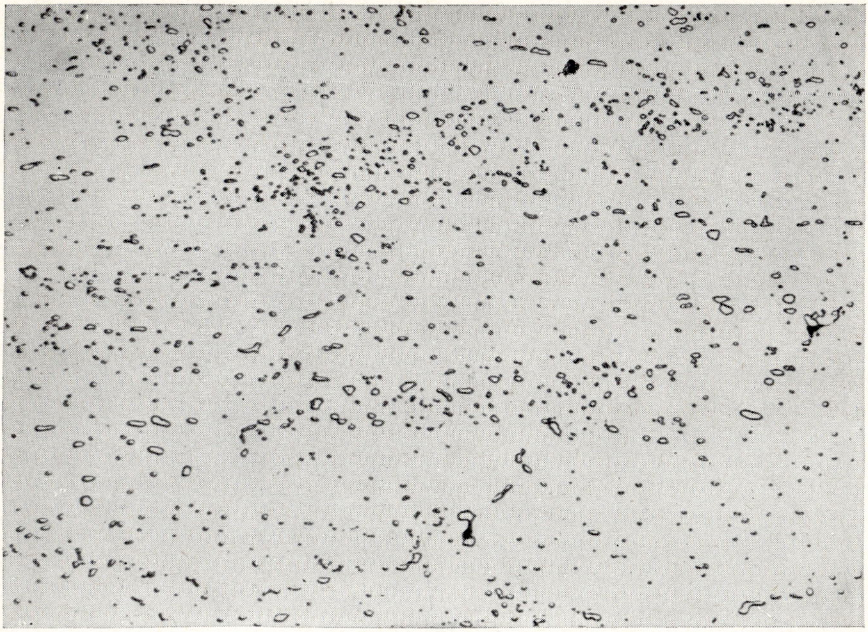

Fig. 5

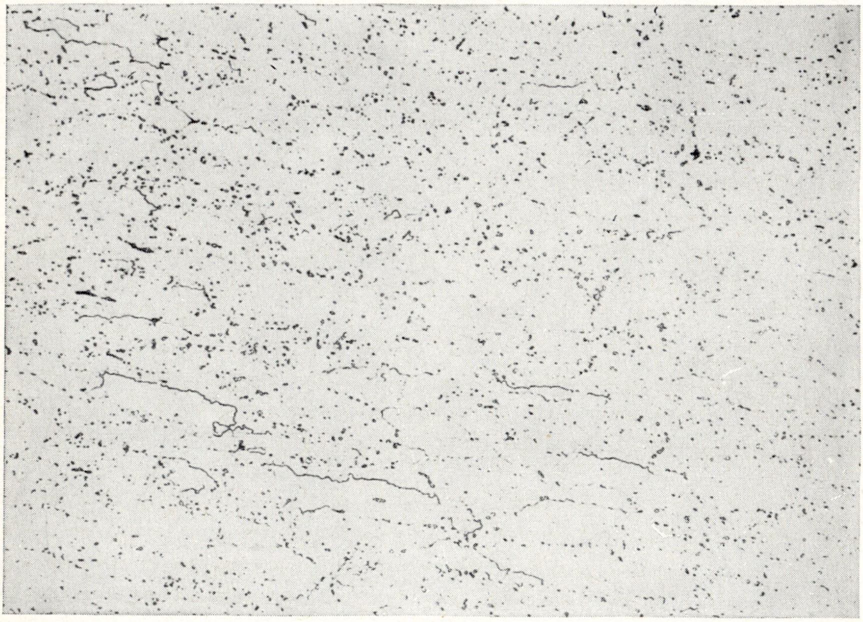

Fig. 6

Figs. 5 and 6. Structure at center, or point of greatest heating. Complete spheroidization is indicated in upper photo. Elongated (strained) ferrite grains are seen in lower photo.

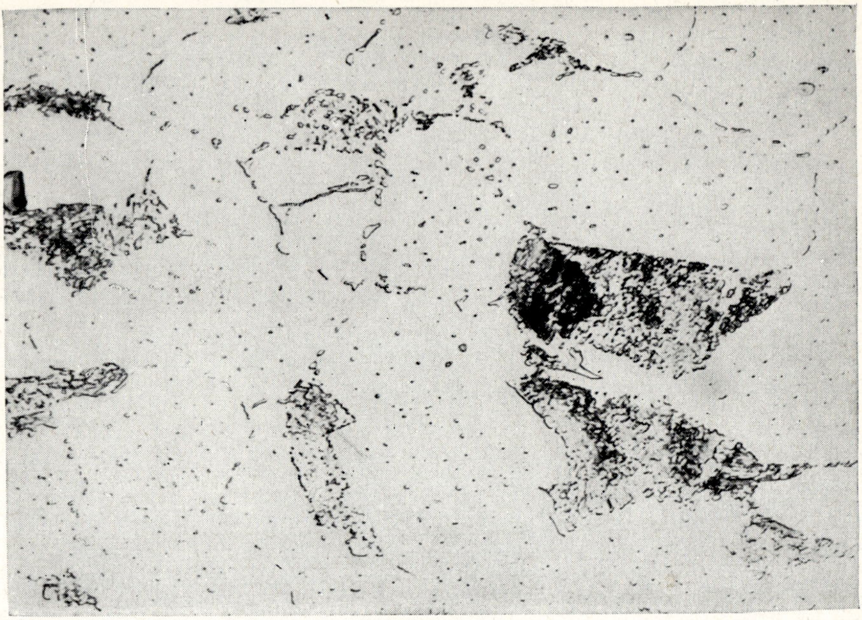

Fig. 7

Fig. 8

Figs. 7 and 8. Micrographs showing the structure three-fourths in. around the tube circumferences from the center of the bulge. Carbide spheroidization has just started.

It was established that at the higher ratings defective circulation developed in certain tubes, resulting in overheating of the metal. Photomicrographs (Figs. 5 and 6) show the structure of the metal at points of greatest heating in the hot zones. Figure 5 reveals the carbides to be completely spheroidized, and masses of spheroids which represent the original pearlite grains tend to be elongated in the direction of the greatest stress (hoop stress). A more pronounced condition is shown in Fig. 6. This grain elongation indicates plastic flow at some temperature below the recrystallization temperature of the steel but, considering the condition of the carbides, at a temperature high enough to accomplish spheroidization.

Figures 7 and 8 were taken about three-quarters of an inch from the point of greatest heating. Here the carbides are only partly spheroidized, and the tendency to grain distortion is slight. It is evident that the prevailing temperature-time relationship imposed a much less rigorous service. As indicated previously, the evidence collected in connection with the operation of the boiler showed that the failures were due to excessive creep at elevated temperature. The nature of the damage to the metal in the failed tubes also indicated steam blanketing at a point of critical circulation or high heat input.

A thorough study of the boiler operation indicated that the failures were due to overheating intermittently for periods of short duration. The narrow bands affected attest to the intermittency and the short duration of the high-temperature intervals.

Grain elongation is perhaps the best indicator of the temperature-time relationship. It has been well established that low-carbon steels or completely spheroidized steel will recrystallize, in the presence of a critical strain, at temperatures below the lower critical point. In view of the fact that no recrystallization took place, it is evident the tube temperatures during the intermittent heat periods were somewhat below 1200°F and possibly in the range of 950 to 1100°F.

Metallurgical studies were substantiated by thermocouple measurements and other data pertaining to the plant operation. The problem with regard to creep failures in these boilers was corrected by redesigning to effect positive circulation under all operating conditions.

INTERGRANULAR DISINTEGRATION

In several boilers operating at water temperatures of 540 to 685°F, intergranular disintegration on the water side of steam-generating tubes has been reported by Brister and Romer.[4] So far, corrosion of this character, in their experience, occurs only in high-pressure units. In the affected areas, metallic copper, oxides of copper, and black oxide of iron have been found. Under these deposits, large corrosion pits formed, and beneath these pits serious deterioration of the metal in the form of intergranular disintegration occurred.

Although metallic copper has been observed in many units free from this type of corrosion, Brister and Romer have usually found copper or copper oxide present at locations where such failures occur. The presence of copper is due to corrosion of the copper-alloy condenser and other heat-exchanger tubes. This action is accelerated by dissolved gases, such as ammonia and carbon dioxide, in the condensate.

The action of high-temperature steam on iron forms magnetic iron oxide, Fe_3O_4, and ionic hydrogen. The film of black iron oxide formed protects the metal, but this barrier may be destroyed by hydrogen or caustic soda.

Dissolved oxygen creates pits filled with black oxide, but it does not produce hydrogen. Copper may create an anodic condition beneath the black oxide resulting in corrosion of the metal and production of atomic hydrogen, which penetrates the steel, causing decarburization and intergranular disintegration. Brister and Romer[4] further claim that copper and oxygen are the principal agents causing this type of metal damage but point out that more research is necessary in order to correlate the potential effects with the many variables and factors present.

CORROSION EMBRITTLEMENT

Boiler-tube failures of the type in which corrosion of the metal is experienced beneath an adherent formation of black magnetic iron oxide accompanied by embrittlement of the adjacent steel have been designated by the author as due to "corrosion embrittlement."[5] Although this term may not be entirely accurate from a metallurgical point of view, it is quite descriptive of the conditions under which this type of attack has occurred.

The mechanism by which corrosion embrittlement develops is not clearly defined or understood and may depend on a number of factors. It can be explained by excessive heat or sluggish circulation resulting in metal temperature insufficient to overheat the metal but high enough to cause film boiling. When this condition is experienced, water vapor dissociates, and the metal combines with the oxygen; magnetic oxide is formed on the surface of the metal with a release of nascent hydrogen. As the attack proceeds, the wall of the tube is wasted away, resulting in thinning of the cross section. The hydrogen formed then penetrates the steel along crystallographic planes or grain boundaries and reacts chemically with oxides, nitrides, and carbides to form gaseous products within the metal. As a result of such action, tremendous pressures are created, which are sufficient to rupture the metal at the grain boundaries. If the action proceeds sufficiently, the weakened and embrittled sections completely fail; further, because of the high operating pressures of the boilers, these sections may be blown out of the tube walls by the internal steam and water pressure. Figure 9 is a typical example of the embrittled area of tubes so attacked. In such failures, free oxygen may be present. However, the same type of attack has been noted in the absence of dissolved oxygen.

In every instance where this type of corrosion embrittlement has been experienced, it has been associated with heavy deposits of dense magnetic

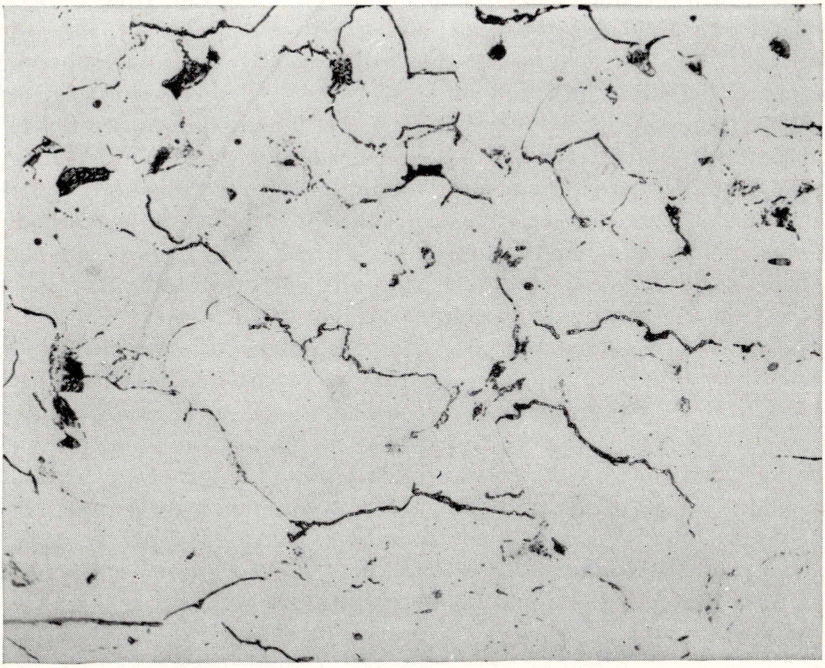

Fig. 9. Embrittled specimen from water-wall tube.

iron oxide layers on the inside of the tube at the point of attack. In some of the cases, copper has been present also, but failures of this type have likewise occurred in tubes where copper was absent or present only in minor amounts. It is indicated that the mechanism can and does proceed in the absence of copper, but it is conceivable that copper, when well bonded to the metal, has accelerated the rate of corrosion and hydrogen embrittlement.

It is particularly interesting to note that in the corrosion-embrittlement failures investigated, overheating of the metal has not occurred or at least has not been sufficiently severe to produce failure from this cause alone.

In some boiler failures examined, there has been evidence to indicate that much of the black oxide in the boilers was not caused by film boiling or oxygen corrosion that took place within the boiler itself. However, the presence of the oxide is often traced to products of corrosion carried into the boiler by feedwater.

Magnetic oxides, when allowed to accumulate, compact, and adhere to the protective oxide films, can damage the latter. A logical explanation of this type of failure conforms to the diffusion theory outlined by Richard C. Corey.[6,7]

Once a heavy oxide formation has been established, the protective film next to the metal is penetrated, and the resultant deposit, not being homogenous, no longer protects the steel. As a result, contact between metal and the steam or water can occur beneath the black iron oxide formation. The rate of attack changes from a parabolic function (characteristic of protective film) to a linear function when plotted against time. In other words, the protective film is destroyed substantially, and the corrosion proceeds rapidly. This theory seems to be a reasonable one and readily explains the tube losses caused, when overheating is not evidenced, by thinning of the tubes due to corrosion and simultaneous embrittlement from the hydrogen released.

Corrective measures have been effected by completely removing dissolved oxygen and limiting the quantity of copper and iron oxides permitted to form and accumulate in the boilers. In a few cases, redesign of boilers has been necessary to improve circulation and eliminate overheating of the metal. Water conditioning, both internal and external, can be controlled to eliminate dissolved oxygen, minimize deposits, control the volume of iron and copper oxides, control the pH value, and provide a fluid sludge. These conditions are essential as safeguards against corrosion embrittlement in high-pressure boilers.

FAILURES CAUSED BY CORROSION-FATIGUE

Corrosion-fatigue failures have been investigated many times; so the mechanism by which such failures occur is fairly well known. The phenomenon has been described in detail by A. E. White and others.[8] This type of failure is caused by alternating stresses set up by vibration or other periodic mechanical or thermal changes. As is well known, cracks which occur in metal because of fatigue are predominantly transcrystalline. Failures of this type are infrequent in stationary boilers but have often occurred in locomotive boilers.

When corrosion accompanied by fatigue of the metal takes place in boilers, the attack develops along a stress concentration at the surface of the metal and ultimately penetrates into the metal. A typical type of corrosion-fatigue failure, which occurred in the tubes of a high-pressure boiler, is shown in Fig. 10. This specimen is from a rolled-in section of a feedwater tube delivering the boiler feedwater from the rear economizer header into the rear drum of a three-drum boiler. The design of this boiler and the point of attack are indicated in Fig. 11. This unit was operated at a pressure of 700 psig, and the temperature of the saturated steam was 503°F. The temperature of the water leaving the economizer was approximately 300°F.

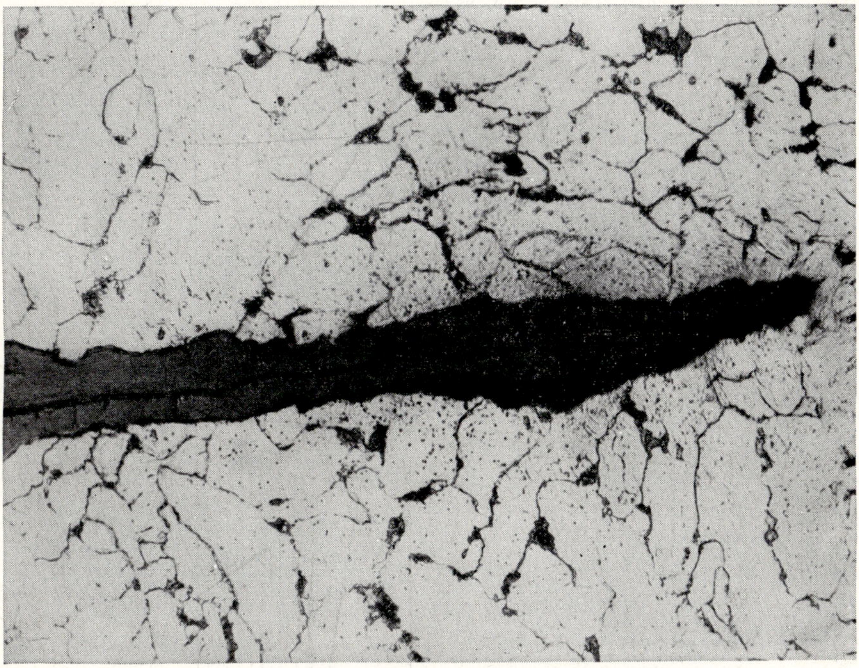

Fig. 10. Specimen cracked by corrosion fatigue.

The boiler design originally provided for a trough into which the feedwater discharged from the tubes and spilled into the drum by flooding the trough. The feedwater trough was later eliminated, incidental to changes in the steam-purifying accessories. The feedwater governed by the feedwater regulator, flowed intermittently into the drum, in accordance with the steam demand. This arrangement resulted in alternate discharge of the feedwater, at a temperature of 300°F, over the surface of metal which was in contact with the steam at 503°F. These widely fluctuating metal-temperature conditions occurred many times each day. Straight feedwater tubes were subject to high stresses exerted at the rolled-in section of the elements. Further, this was the point of most pronounced thermal change during the operation of the boiler. The combination of these conditions resulted in many failures of feedwater tubes due to corrosion-fatigue. Quality of the feedwater was not responsible for the primary or secondary attack.

The difficulty was corrected by the following:

1. The installation of a feedwater trough which kept the tubes so flooded with feedwater at relatively uniform temperatures at all times

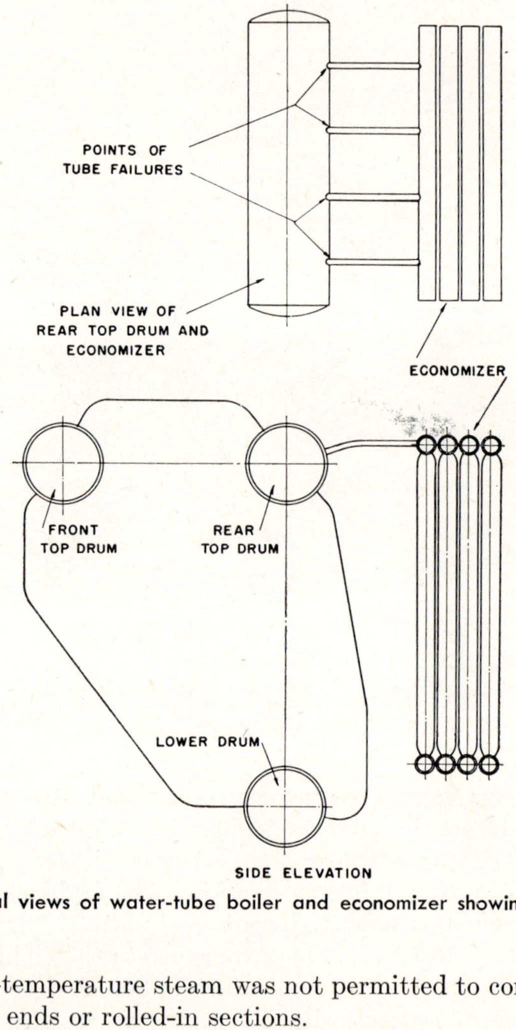

Fig. 11. Sectional views of water-tube boiler and economizer showing location of tube failures.

that the high-temperature steam was not permitted to come into contact with the tube ends or rolled-in sections.

2. The installation of S-shaped tubes to permit expansion, thereby relieving the high stresses at the ends of the tube where they were rolled into the drum.

A schematic drawing showing the rearrangement of the tubes and trough subsequent to effecting corrective measures is shown in Fig. 12.

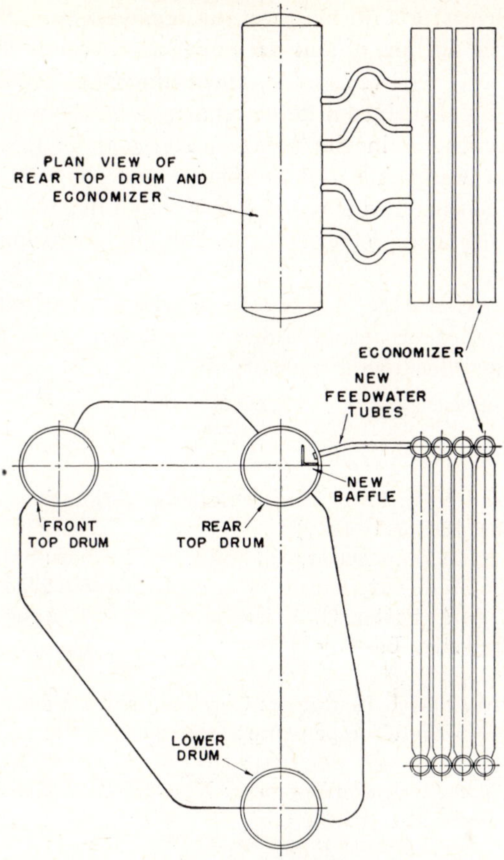

Fig. 12. Sectional views of water-tube boiler and economizer showing changes which eliminated tube failures.

DETECTION OF CORROSION BEFORE EXTENSIVE DAMAGE IS EXPERIENCED

A hydrogen recorder* for determining the concentration of hydrogen in the steam is an effective means for detecting the rate of attack which may be proceeding within the boiler.

The Cambridge Hydrogen Recorder, which is designed for measuring hydrogen in parts per billion or parts per million, has been employed by the author and others in special studies for indicating the rate of corrosion occurring in steam-generating units. The recorder is valuable because it gives a continuous record of the gas in the steam and because the results can be correlated with operating conditions in the plant and with

* For detailed description of this equipment and instructions for operation, the reader should refer to Chap. 24.

the hydrogen release for a normal, well-designed boiler. In one investigation, made by the author, of starved boiler tubes where heavy iron oxide accumulations occurred frequently, the Cambridge Hydrogen Recorder was found to be helpful since it detected attack on the wall tubes by showing excessive hydrogen in the steam. As soon as this condition was noted, the boiler was taken out of service and the tubes were cleaned. The hydrogen recorder has also proved of value in revealing overheating of tubes resulting from steam blanketing in corrosion-embrittlement types of attack.

For adequate evaluation of the data provided by a hydrogen recorder, it is necessary to measure, as a blank, the hydrogen in the feedwater delivered to the boiler, because a measurable amount of dissolved hydrogen may be present.

Bibliography

1. Slater, I. G., and N. L. Parr: Marine Boiler Deterioration, *Proc. Inst. Mech. Engrs.* (*London*), **160**: 341 (1949).
2. Rohrig, I. A., R. M. Van Duzer, Jr., and C. H. Fellows: "High-temperature-steam Corrosion Studies at Detroit," *Trans. ASME*, **66**: 277 (1944).
3. Solberg, H. L., A. A. Potter, G. A. Hawkins, and J. T. Agnew: The Corrosion of Stressed Alloy-steel Bars by High-temperature Steam, *Trans. ASME*, **65**: 47 (1943).
4. Brister, Paul M., and J. B. Romer: Corrosion and Oxidation Experiences in High Pressure and High Temperature Steam Service, *Trans. Electrochem. Soc.*, **91**: 655 (1947).
5. Powell, S. T., and L. G. von Lossberg: Unusual Corrosion in High Pressure Boilers, *Corrosion*, **5**: 71 (1949).
6. Corey, R. C.: Corrosion by Liquids, *Combustion*, **15**: 35 (1943).
7. Corey, R. C.: Corrosion by Hot Gases, *Combustion*, **15**: 34 (1943).
8. White, A. E.: Changes in a High-pressure Drum to Eliminate Recurrence of Cracks Due to Corrosion Fatigue, *Trans. ASME*, **61**: 507 (1939).

18 ELECTROLYTIC PROCESSES FOR CORROSION CONTROL

For many years, numerous electrolytic processes have been applied to protect metallic surfaces against corrosion. All these processes have been based on the principle of a counterelectromotive force created to maintain the surfaces to be protected in a cathodic state. This method of control is known as cathodic protection. When such a condition can be brought about and maintained, the corrosion activity is almost completely stifled, and the solution of the metal by the electrolyte is almost entirely prevented. One of the fundamental principles of cathodic protection is based on the difference in electrode potentials of metals. Because of this difference, any given metal is more or less anodic to other metals,* and it is therefore possible to arrange all metals in order of their electrode potentials in what is known as the electromotive series, Table 1.

As the table shows, zinc is anodic with respect to iron. When these two metals are (1) submerged in an electrolyte and (2) connected electrically, current flows from the zinc to the iron. The zinc will corrode, and the iron will be protected. In a similar manner, iron is anodic with respect to copper and will corrode while copper will be protected. It is apparent that any metal in the electromotive series may be protected by the presence of another metal higher in the series.

This principle has been put to work by installing so-called sacrificial methods for the purpose of protecting the major metals of which structures are composed. Sir Humphry Davy,[2] in a paper delivered in 1824, described a system of cathodic protection of ships by the use of zinc plates to prevent corrosion of copper sheathing covering the hulls. This method of protection has disadvantages. Chief of these is that the protective influence of the zinc or the metal serving as the anode extends to only a relatively small area of the metal to be protected.

* The fundamental theory of corrosion, based on the work of W. R. Whitney,[1] is described in Chap. 16.

All such electrolytic methods of treatment for the prevention of scale or corrosion may be classified into two groups: (1) those that create a cathodic surface by imposing an electric current on the metal, the current being supplied from an external source, and (2) those that generated a current by galvanic action, such as the Davy installation referred to previously.

The general principle upon which both processes function is the same. They impose a direct current of low voltage and low amperage on the

TABLE 1. ELECTROMOTIVE SERIES

Metal	Ion	Molal electrode potential at 77°F (25°C), volts
Magnesium	Mg^{++}	−2.34
Aluminum	Al^{+++}	−1.67
Zinc	Zn^{++}	−0.76
Chromium	Cr^{++}	−0.557
Iron	Fe^{++}	−0.44
Cadmium	Cd^{++}	−0.40
Nickel	Ni^{++}	−0.23
Tin	Sn^{++}	−0.14
Lead	Pb^{++}	−0.13
Hydrogen	H^{+}	Arbitrary zero point
Copper	Cu^{++}	+0.34
Silver	Ag^{+}	+0.80
Mercury	Hg^{++}	+0.80
Platinum	Pt^{++}	+1.2
Gold	Au^{+++}	+1.36

metal surfaces so that the metal requiring protection is made the negative pole of a battery and the anode of sacrificial metal is the positive pole. The cathode is protected at the expense of the anode, which in time is consumed.

This principle has been used with varying degrees of success in various systems in an attempt to prevent corrosion in low-pressure boilers, heaters, heat exchangers, condensers, and other power plant equipment. In so far as is known, such treatment has not been used in any high-pressure boilers.

CATHODIC PROTECTION BY GALVANIC ACTION

For many years it has been common practice to attempt to protect boilers, surface condensers, and other equipment by bonding plates of zinc or other metal to the equipment, much as in the system described by Sir Humphry Davy.[2] The effectiveness of this method diminishes as the distance from the anode increases. The areas at a distance from the sacrificial metal receive less protection than the ones close at hand. A further disadvantage is that in time the action is slowed down by the formation on the surface of the zinc of an insulating scale consisting of insoluble zinc salts.

Although fundamentally sound in principle, this method of producing a current flow from a more anodic metal to one that is less anodic has some disadvantages for furnishing protection. The sacrificial metal requires replacement, and the equipment must be carefully designed to obtain uniform protection.

Despite these difficulties, recent extensive experimental work and wide application in practice have demonstrated the practicability of this form of protection. It is being extensively adopted as a protective measure against corrosion for many services, such as pipelines, especially those buried in corrosive soils. This process has also been used with marked success in many installations for the protection of fresh- and salt-water storage tanks.

A fairly large number of cathodic-protection processes have been developed recently, both abroad and in the United States, but the results obtained have been by no means consistent. In some instances, the process used has given satisfactory results; others have been complete failures. Many factors influence the results, and it is the control of these conditions which is responsible frequently for the success or failure of the installations. The entire cathode surface must be reached by the current flow, and a uniform distribution must be effected. These conditions are frequently difficult to maintain.

CATHODIC PROTECTION BY AN EXTERNAL SOURCE OF ELECTRIC CURRENT

Instead of using a sacrificial metal to impress a counterelectromotive force that opposes the corrosion cells of the metal to be protected, it is possible and practicable to provide an external potential. This potential is applied to suitable electrodes to a degree that will reverse the normal currents prevailing between the anodic and the cathodic areas of the metal to be protected. The electrode operating at the controlling potential will then be anodic. All surfaces influenced by the current leaving the electrode will be cathodic and, consequently, protected if the current density, due to the external potential, is sufficient to stifle local action on the surface of the protected metal. With this arrangement, the positive side of a source of direct current is connected to the electrode or electrodes and the negative side of the system is attached to the metal which it is desired to protect. A simple wiring diagram for this type of protection is shown in Fig. 1.

Power Supply. The external potential source, operating at a pressure of 6 to 10 volts must be direct current and can be supplied by rectifiers, motor generator sets, storage batteries, or any source of direct current having adequate capacity for supplying sufficient current to protect the cathodic surfaces. Dry-type rectifiers are generally used for purposes of cathodic protection and may be of the copper oxide, copper sulfide, or selenium type. Where the quality of the water supply varies widely in total solids content, it is necessary to vary the applied voltage (and, consequently, the current) by controls that are based upon the electrical resistance of the water.

Electrodes. The materials of which electrodes are constructed should resist rapid deterioration, introduce no objectionable products into the

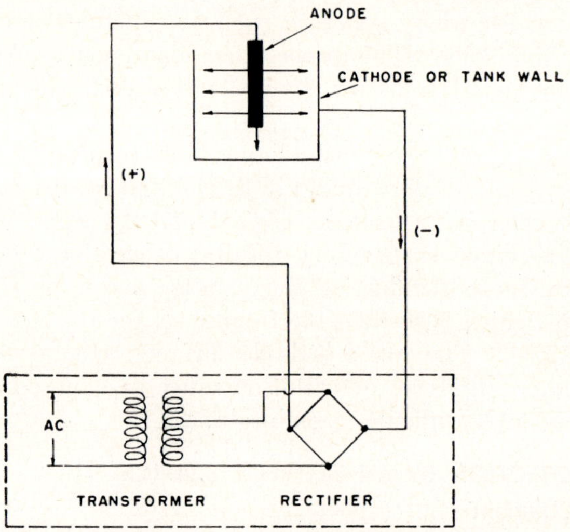

Fig. 1. Typical electrical circuit provided for cathodic protection of steel tank.

electrolyte, be mechanically strong, and have low electrical resistance. Many different materials are used for electrodes, including carbon, graphite, mild steel, Duriron, aluminum, magnesium, and stainless steel. Since the electrodes are anodic, all wastage of metal or material is supplied by these units if the cathodic area is properly protected. Consequently, the electrodes must be of ample size in order to avoid frequent replacement. Also, it is important to provide sufficient surface in order to decrease the resistance between electrode and electrolyte.

The approximate wastage of metal from electrodes can be estimated from Table 2. A metal having a low electrochemical equivalent obviously has the advantage of a low rate of dissolution per unit of current and time.

It should be noted that the proper size of electrode is based upon the amount of current delivered to the electrolyte. The metal consumptions listed below are generally exceeded in practice owing to local corrosion currents that diminish the efficiency of the electrode.

Arrangement of Electrodes. The number and arrangement of electrodes is based upon the necessity for providing ample current density to all areas of metal needing protection. Current density diminishes as the distance from the electrode to the area to be protected increases and is influenced by the specific resistance of the water in which the active surfaces are submerged. Higher voltages are required in cases where the resistance of the water to passage of current is high.

TABLE 2. ELECTROCHEMICAL EQUIVALENTS OF ANODES

Anode	Valence	Sp. gr.	Lb per amp per year	Cu in. per amp per year
Aluminum	3	2.70	6.5	66.8
Carbon (graphite)	4	2.26	2.1	26.0
Iron	2	7.86	20.1	71.0
Nickel	2	8.90	21.2	66.2
Zinc	2	7.14	23.5	91.4
Magnesium	2	1.74	8.8	140.0

Many water tanks have received satisfactory protection from the installation of a single electrode centrally located in the tank. In some cases, it has been necessary to have a number of electrodes in order to provide protection. In elevated water tanks, particular attention must be paid to protection of the bumped head of the tank, and in many instances, it is necessary to extend an electrode through the entire length of the riser connecting the tank with the water main.

Cathodic protection of steel vessels having simple geometric shapes, like pipelines or water tanks, can be carried out with a considerable degree of reliability. Protecting complicated apparatus, such as distillation columns, surface condensers, etc., by this method is much more difficult, and results are less easily predicted.

Protection of surface condensers is usually carried out by installing electrodes in the water boxes near the tube sheets. In all cases where an external source of power is used, the electrodes should be electrically insulated from the surface to be protected.

Current Density. The flow of current from electrodes connected to external sources of power will vary with the following:

1. The applied electromotive force
2. The area of the electrode
3. The active area of the cathode
4. The mineral content of the water
5. The length of path through which the current must flow

One of the principal difficulties in providing satisfactory cathodic protection is that of determining the proper current density needed to protect structural surfaces continuously. Current densities vary from 0.3 to 60 ma per sq ft when iron is the metal to be protected. Usually the density that maintains satisfactory conditions is of the order of 2 ma per sq ft of cathodic surface.

According to Sudrabin,[3] higher current densities are required for water having high electrical resistance, that is, for water low in solids content, and are influenced by the concentration of dissolved oxygen and by the velocity of flow past the protected surfaces. The higher the dissolved oxygen content and the velocity of the water, the higher are the requirements of current density.

When iron is to be protected, no harmful effects are likely to occur from excessive current densities, but in the case of amphoteric metals such as aluminum and zinc, excessive current may damage the metal because of the increase in alkalinity of the film on the metal surfaces. The pH of the film is increased by removal of hydrogen ions, which creates a preponderance of hydroxyl ions, and by migration of cations such as Ca^{++}, Mg^{++}, and Na^{+} to the protected cathode surfaces.[4]

The rate of plating out of atomic hydrogen on cathodic surfaces must be at least equal to the rate of reaction between hydrogen on these surfaces and dissolved oxygen diffusing from the main body of the electrolyte to the protected surfaces. Scale formed on metal surfaces before application of cathodic protection is frequently loosened and at times removed by building up the hydrogen film on cathodic surfaces.[5] The pressure of the hydrogen in this film exerts sufficient force to break the bond between scale and metal.

When calcium and magnesium salts are present in the water, the tendency is for calcium carbonate and magnesium hydroxide to deposit on protected surfaces, because the pH of the film next to the metal is high and bicarbonates diffusing to this zone are converted to insoluble carbonates. The deposit is usually chalky in appearance and may be rubbed off. The presence of such deposits is beneficial, particularly on surfaces that are alternately submerged and exposed to the atmosphere, as, for example, those of a tank in which the water level varies through a considerable range. In these cases, it is generally recommended that surfaces which are frequently exposed to the atmosphere be painted in order to provide adequate protection during the uncovered periods, when cathodic protection is interrupted. In the absence of paint, calcium carbonate deposits are helpful in protecting the uncovered metal but should not be depended upon exclusively for this purpose.

The required amperage per square foot of surface area will vary widely with water of different chemical composition. If the current used is too high, corrosion of the anodes may be excessive, and as a result, the increase in oxygen generated may accelerate corrosion of the metal outside the protected zone.

The author has investigated a number of cathodic-protection installations. Some of these appeared to be fairly efficient in reducing scale formation and preventing corrosion. In other cases, no apparent improvement could be noted. In one instance, corrosion of the tubes and drum of the boiler was undoubtedly accelerated.

In some large-diameter standpipes containing untreated water, it has been found desirable to cover the bottom of the tank with a thin layer of limestone, thus tending to reduce the over-all current and power requirements. This simplifies the electrode design since only the side walls require the higher current densities.

Precleaning Surfaces. Surfaces to be protected do not require special cleaning, such as sandblasting or wire brushing, to remove old nonmetallic scale. The presence of such scale may actually be beneficial because of its tendency to reduce the current density required.

The removal of mill scale, however, before the application of cathodic protection is sometimes advisable. This is not essential for the proper working of the process, and in time, the scale will be forced off by hydrogen plating out on cathodic surfaces. In the first few months, the scale will deposit in the bottom of the tank and may be carried into the distribution system.

Effectiveness of Treatment. Periodical inspection of cathode-protected surfaces is desirable in order to determine the effectiveness of the protection provided. Areas at some distance from the electrodes should receive careful attention to ascertain whether current densities at remote areas are sufficient to stifle corrosion by local currents.

The formation of a white deposit consisting chiefly of calcium carbonate on the protected metal surface is an indication that corrosion is being prevented.* The loosening of previously formed scale is another indication that ample current density is being provided at the areas under observation. In fact, such loosening of scale may indicate a current density in excess of that actually required and, therefore, uneconomical in both use of power and electrode consumption.

Cathodic Protection for Tubing. A method for prevention of corrosion has been developed by inserting inner tubes or liners of an aluminum-zinc alloy into standard aluminum tubes.[6] The inserted tube, which may be metallurgically bonded to the outer tube, is anodic to the outer tube. Corrosive water or chemicals pass through the tube, and in due time, the inner tube is perforated to a degree that permits the solution (electrolyte)

* This condition will not occur in tanks containing condensate or completely softened water because of the absence of such minerals in the water.

to make contact with the aluminum tube. When this occurs, the anodic inner tube provides protection for the outer tube. This protection may persist until as much as 50 per cent of the inner tube has dissolved into the electrolyte. It is claimed that the life of tubes so constructed is greatly lengthened. Up to the time of penetration of the anodic tube, protection consists in preventing contact of the solution with the outer tube. At the present time, this arrangement is adapted for use in condensers where the cooling water flows through the tubes.

Zinc has been used successfully for protecting aluminum equipment from corrosion by cooling water. Zinc is usually anodic to aluminum in neutral or slightly acid solutions and sometimes in weakly alkaline solutions. However, when solutions are strongly alkaline, aluminum becomes anodic to zinc, and the latter material cannot be used for protection of the aluminum.[7]

Mears and Fahrney[7] state:

> On several occasions the authors have been questioned regarding the feasibility of protecting aluminum by zinc. The questioners pointed out that aluminum was higher than zinc in the electromotive series and, therefore, it might be predicted that zinc would be cathodic, not anodic, to aluminum. Of course, it should be pointed out that, while many calculations of fundamental importance are based on the electromotive series, this series has no great value in predicting the corrosion behavior of couples of dissimilar metals under service conditions. In service, the solutions employed are different from the special solutions used in preparing the electromotive series. Since the solution potential of each metal depends on the composition of the liquid to which it is exposed (as well as on other factors such as temperature, degree of agitation, gas pressure, etc.), it is not surprising that the solution potentials of the metals are quite different under service conditions from what they are under the carefully controlled arbitrary conditions employed while making measurements for the electromotive series.

The reason that aluminum does not develop a high solution potential in many solutions, as would be expected from the electromotive series, is probably that it rapidly coats itself with a protective oxide (or other) film. In alkaline solutions, this protective film is so unstable that the aluminum remains active.[8]

No reliable data are as yet available for determining the size and distribution of zinc strips installed for the purpose of providing cathodic protection for aluminum. The protective influence of the zinc usually extends about 18 in. from the electrode, but in some cases, protection is afforded at distances several times greater.[7]

Since aluminum is an amphoteric* metal, it has been found safer to employ zinc to protect aluminum equipment than to use inert anodes and potentials from outside sources. Excessive currents due to high poten-

* Can be electropositive or electronegative or function as an acid or a base.

tials will cause high alkalinity in the liquid film adjacent to the metal surface and thus endanger the metal.[7]

Cathodic Protection for Pipelines. Extensive inhibition of pipeline corrosion in underground pipelines, such as the "Big Inch," transporting gas and oil from sources to distant parts of the country, is afforded by cathodic protection. Such protection* is applicable to all types of underground piping systems and particularly to those in damp soils, as in marshes or at stream crossings.

When galvanic currents are employed, zinc, aluminum, or magnesium anodes are used for this purpose because these metals are strongly anodic

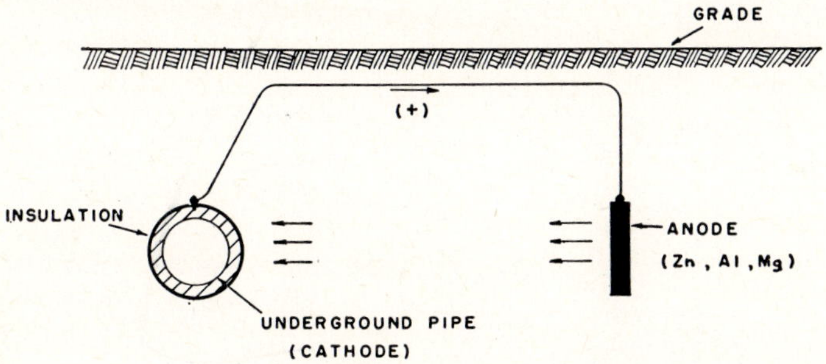

Fig. 2. Typical electrical circuit provided for cathodic protection of an underground pipe (galvanic action).

to the pipelines for which protection is required. Conductivity of the soil, location of anodes, and current densities must be considered in order for such installations to function successfully. Figure 2 shows schematically cathodic protection applied to a buried pipe using current generated by galvanic action.

Where aluminum electrodes have been used for the protection of underground pipelines, it is usual to bury these electrodes in a mixture of clay and common salt. The addition of chloride ions has been found to reduce the electrical resistance not only of the backfill but also of the film immediately adjacent to the metal surface. Lime is sometimes used to increase the alkalinity of the backfill, thus increasing the anodic potential of the metal and the solubility of the products of corrosion. Aluminum alloyed with 5 per cent of zinc has been found suitable for use in electrodes employed for this purpose.[9]

The conditions essential for successful design of cathodic protection of underground pipelines when applying direct current from an external source are the following:

* G. I. Rhodes, "Method of and Means for Protecting Pipe Lines and Other Buried Metallic Structures from Corrosion," U.S. Patent 1,962,696, June 12, 1934; "Composite Electrode for Protecting Buried Metallic Structures from Corrosion," U.S. Patent 2,204,823, June 18, 1940.

1. A continuous electrical circuit must be maintained throughout the pipeline to be protected by bonding with copper wire around all flanges, valves, and fittings. When welded construction is used, bonding is not necessary.

2. Pipelines to be protected must be insulated by coatings having a high electrical resistance to ground. The insulating coating reduces the power requirements since the necessary negative potential can be maintained on a well-insulated line with a minimum of current flow. This

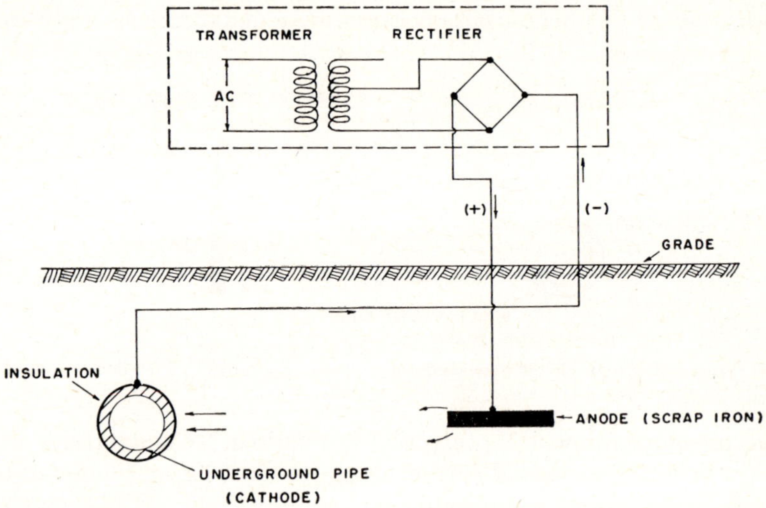

Fig. 3. Typical electrical circuit provided for cathodic protection of an underground pipe (external source of current).

coating is sometimes composed of asbestos felt dipped in hot asphalt and carefully wrapped around the pipe. It is equally important to insulate fitting and valves in order to avoid a heavy inflow of current at non-insulated points.

3. All auxiliary pipelines for which protection is not required must be joined to the protected line by means of insulated connections.

The bonded and insulated pipeline is connected to the negative side of a rectifier or other source of direct current. The positive side of the rectifier is connected to scrap pipe or metal buried in the soil. The amount of metal in the anode must be sufficient to provide ample area for electrical contact with the soil and to provide sufficient weight for wastage of metal equivalent to the flow of current. Sufficient metal to provide for an operating period of 2 years is usually installed. Sometimes the soil adjacent to the scrap metal used at the anode is impregnated with salts in order to increase the electrical conductivity of the film between the metal and the soil. Figure 3 shows schematically cathodic protection of a buried pipe using an external source of current.

Sections of pipe line from 5 to 10 miles long may be protected by current provided by a single rectifier station. Distances between stations, which serve as a source of potential, depend upon the electrical resistance of the pipeline to ground and upon the amount of current which must be provided in order to protect adequately the most remote parts of the pipeline.

It has been found that pipelines are adequately protected when the potential of the pipe is maintained at -0.85 volts when measured to a copper sulfate electrode placed in the earth. The potential difference can be measured with a McCollum earth-current voltmeter.

When parallel lines in the same trench are to be protected, electrical bonds should be installed frequently between the two systems in order to prevent any difference in potential between the two pipelines.

In any electrically protected system, the current flowing into the circuit will be proportional to the number of breaks in the coating insulating the pipe from the soil. Therefore, for reasons of economy, it is necessary to apply the coating carefully and to avoid punctures or breaks when installing the pipeline.

APPLICATION OF CATHODIC PROTECTION

At present, specialized experience is necessary in order to design a cathodic-protection installation properly. There are still unexplained features of this type of protection which sometimes preclude definite predictions as to the completeness of treatment. This uncertainty is due in part to a lack of sufficient data upon which to base reliable assumptions concerning current densities necessary for complete protection and the position and number of electrodes that must be installed. Great progress has been made in the art. Cathodic protection has now been established as a practical means for protection of many types of equipment subject to corrosion by water.

For details of design and construction, the reader should refer to Cathodic Protection Symposium, prepared by the Electrochemical Society and the National Association of Corrosion Engineers in 1949.

Many processes have been patented covering protection of miscellaneous types of equipment and operated under varying conditions. It is not possible to mention all these methods here but several which have been more extensively developed and applied are described.

THE CUMBERLAND PROCESS

This process was developed in England and, for a time, was extensively studied by both the British and French admiralties. It was also applied for the protection of condensers and boiler and miscellaneous power-plant equipment. The process requires a dynamo generating continuous current at a pressure of 6 to 10 volts to supply current to iron electrodes suspended in the water contained in the vessel to be protected and suitably insulated from that vessel. In the application of the process to condenser-tube protection, iron anodes* supported by steel studs and insulated from the condenser shell are connected through an adjustable resistance and a two-way switch to the positive terminal of the generator. The condenser itself is connected to the negative terminal of the generator; thus current can flow continuously from the electrodes through the water in the condenser to the metal surfaces of the condenser and back to the generator. This superior electromotive force neutralizes or overcomes the minor destructive currents produced by galvanic action between the condenser tubes and any dissimilar metals as well as those currents generated by corrosion cells occurring on tube surfaces.

The amount of current required to protect condenser surfaces varies slightly, but in extreme cases, 2 ma per square foot is found to afford protection. In the case of boilers where removal of scale must be taken into account, the amount of current found necessary is somewhat larger. It is estimated that 3.3 ma per square foot is sufficient for this service.

THE KIRKALDY SYSTEM

The Kirkaldy electrolytic system operates by supplying low-voltage current from an external source to anodes immersed in water. The anodes are installed in the boiler or condenser but are properly insulated. The water is used as the electrolyte. The boiler or other metal surface thereby becomes the cathode. As the current flows, the scale-forming solids pass to the anode. The current density is small, but proper control is effective in preventing scale and inhibiting corrosion. A diagram of a characteristic installation is shown in Fig. 4.

The working principle of electrolytic protection may be stated briefly as follows:

A low direct current of small amperage supplied from a secondary battery or a current generator is led to metallic anodes submerged in the boiler circulating water, the anodes forming the positive pole of the circuit. The anodes and the electrical connections thereto must at all

* Bruce B. Greaves, of The Cumberland Engineering Co., Ltd., of Liverpool, England, in a private communication to the author, stated that the use of iron electrodes is not true in this system, "as we only use iron in certain special cases. The majority of our installations are equipped with "Cumbolite" in one or other of the 3 grades at present available. "Cumbolite" anodes give from three to nine times the life obtained from iron—the actual life being influenced by the grade of anode selected and its environment."

times be insulated from the boiler shell. The current is transmitted through the water (the electrolyte) to the boiler surfaces, which thereby become negatively charged, completion of the circuit being made by earthing the boiler shell and the negative pole of the current-supply source. The amount of current supplied to any particular boiler can be varied and controlled by means of adjustable resistance.

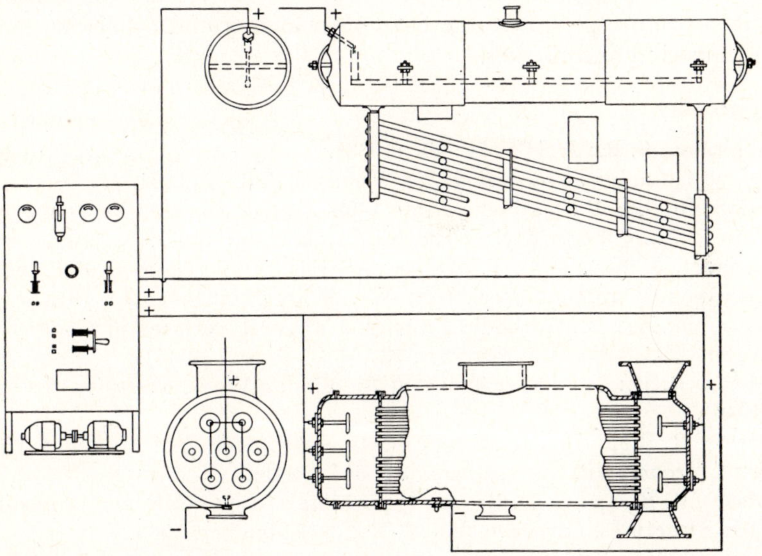

Fig. 4. Diagrammatic sketch and wiring diagram of a Kirkaldy system installed in a boiler for corrosion and scale control.

GUNDERSON PROCESS

The Gunderson electrolytic system for the prevention of corrosion consists in passing an electric current of low voltage to boiler water to which an arsenic salt has previously been added. This system differs from the Cumberland process primarily in the addition of the metallic salt by means of which the inventor claims the process overcomes much of the difficulty experienced by other electrolytic polarization systems. Briefly, the Gunderson process consists in the installation of two insulated electrodes in the boiler shell, one in the upper part and one in the lower part of the boiler. The conducting members of these electrodes are rigidly connected to a Duriron anode which extends the entire length of the boiler and is submerged in the boiler water. The anode is securely attached by insulated clamps to the shell, and the electrodes are connected to insulated wires connected to the turbogenerator with suitable fuses and resistances to provide 2 volts and 2 to 3 amp to each electrode. The current flows from the anode to the boiler water and back to the boiler metal and to the negative pole of the generator. Sodium arsenate or arsenic peroxide is added once or twice a week at the rate of 3 lb per 1,000 gal of water. Since arsenic is poisonous, extreme care must be used in handling it and in use of steam or disposal of blowdown water.

The inventor claims that the soluble arsenic plates out on the boiler sheets and tubes, polarizing the metal and thus inhibiting the solution of iron and preventing corrosion. It is also claimed that the coating of arsenic prevents local galvanic currents which would otherwise tend to cause corrosion and pitting.

Bibliography

1. Whitney, W. R.: *J. ACS,* **25**: 394 (1903).
2. Davy, Humphry: *Trans. Roy. Soc.* (*London*), Jan. 22, 1824.
3. Sudrabin, Lee P.: Cathodic Protection of Steel Surfaces in Contact with Water, *Water Works & Sewerage,* **92**: 11 (January, 1945).
4. Clement, J. K., and L. V. Walker: An Electrolytic Method of Preventing Corrosion of Iron and Steel, *U.S. Bur. Mines Tech. Paper* 15, 1913.
5. Sudrabin, Lee P.: Cathodic Protection of Steel Surfaces in Contact with Water, *Water Works & Sewerage,* **92**: 51 (February, 1945).
6. Mears, R. B.: Condenser Tubes of Aluminum Alloys, *Corrosion,* **2** (3): 165 (1946).
7. Mears, R. B., and H. J. Fahrney: Cathodic Protection of Aluminum Equipment, *Trans. Inst. Chem. Engrs.,* **37**: 911 (1941).
8. Mears, R. B., and C. D. Brown: Light Metals for the Cathodic Protection of Steel Structures, *Corrosion,* Vol. 1, No. 3, September, 1945.
9. Hoxeng, R. B., E. D. Verink, and R. H. Brown: Galvanic Aluminum Anodes for Cathodic Protection, *Corrosion,* **3** (6): 263 (June, 1947).

19 CAUSTIC EMBRITTLEMENT OF BOILER STEEL

In the past, many boilers have been damaged by intercrystalline cracking of the boiler steel. Such failures are commonly known as "caustic embrittlement cracking." The term is used to distinguish the phenomenon from cracking due to corrosion fatigue and other types of failure. Damage to boiler metal resulting from caustic attack is indicated by cracks in the metal, which always occur below the water line (Fig. 1).

Substitution of welded drums for riveted design, improved methods for tube-hole construction, and improved tube-rolling techniques have greatly minimized this type of failure. Further, the chemical mechanism of the attack on steel is now well established as a result of extensive research conducted during the past 35 years. This knowledge has led to positive means of inhibiting attack by more effective chemical conditioning of feedwater and concentrated boiler waters.

Regardless of these developments, caustic embrittlement of boiler steel still occurs and is always potentially possible where water of an embrittling type is used and where crevices exist adjacent to highly stressed areas in which caustic soda can concentrate.

The general characteristics of metal which has become embrittled may be summarized briefly as follows:

1. The cracks occur below the water line and appear to start on the dry face of the joints.

2. The failures occur in rolled plates, cast steel, or steel boiler plate of good quality, depending on the material used in the boiler.

3. The cracking is irregular, and changes in direction of the cracks are marked and generally do not join one another.

4. The cracks do not appear to occur at areas which might indicate the greatest stress as a result of boiler pressure, although the cracks occur usually in the region of high stress.

5. The cracks occur at riveted or at lap joints, affecting both plates. They occur also at butt joints in both straps and plates.

6. There is marked brittleness of the metal under shock; static tests, however, do not indicate this phenomenon.

7. The brittleness of the metal appears generally to be localized at the areas adjacent to the joints.

8. Black iron oxide or sulfide is present at embrittled areas.

Fig. 1. Boiler cracked by caustic attack.

Figure 2 is reproduced from a photomicrograph of a boiler plate taken from an embrittled boiler. It will be noted from this photograph that the cracks do not occur in any one direction and change sharply from one direction to another.

THEORETICAL CONCEPTS

As early as 1866, Graham[1] discovered that red-hot iron was penetrated by hydrogen and that a portion of the hydrogen was retained by the metal after it had cooled down.

From that time on, many investigators[2–8] studied this type of embrittlement of steel and deduced several theories concerning the mechanism by which highly stressed steel failed when in contact with high concentrations of caustic soda.

Parr's experiments were directed toward determining the effect of caustic soda upon steel or the indirect effect of hydrogen resulting from such action. He reported as follows:

The evidence seems to show that the embrittling effect of sodium hydroxide on steel is due to the evolution of hydrogen and the absorption by the steel of the

Fig. 2. Section of shell plate cracked by caustic attack.

hydrogen in the nascent state. However, there are two other possibilities which may be considered:

First, there is the possibility that the alkali actually works into the metal eating out the "amorphous intercrystalline cement," which, being thermodynamically less stable than the crystals, would be first attacked. This would cause a brittleness akin to that caused by burning or overheating and fracture would be intercrystalline.

Second, Andrews claims that sodium hydroxide causes a "crystallization," by which he means a coalescence of the smaller crystals into larger ones. This effect would also be the same as that caused by overheating.

Fuller,[9] Rosenhain and Houser,[10] Merica,[11] Kriegsheim,[12] and Bauer[13] either rejected the theory of caustic attack as the primary cause of embrittlement of the steel or minimized its importance.

Professor A. Thiel's[13] research studies have shown that 77.5 per cent of caustic soda may concentrate in the clearances between the boiler plates, resulting in reducing the tensile strength of the boiler. This action, according to Thiel, is due to the absorption of hydrogen which is released from the caustic soda when the sodium unites chemically with the iron.

The extensive investigation of Williams and Homerberg[14] on the action of caustic soda on boiler steel sheds much light on the problem.

The results obtained by these investigators led to the following conclusions:

It seems evident that hydrogen acts in three ways to produce embrittlement: First, the temporary brittleness caused by absorbed hydrogen (as in acid pickling); second, it acts to reduce oxides; third, its effects due to the change in volume at the grain boundaries, resulting because of the production of water. This latter volume increase would create a stress which, added to those originally present, may cause cracking, especially at those points where these stresses are at a maximum.

The extensive early work of Parr and, subsequently, that of Parr and Straub have produced a mass of scientific data assembled over a period of more than 30 years. The conclusions reached by these investigators concerning the fundamental cause of caustic embrittlement of boiler steel have not been disproved. Their general conclusions, expressed in 1927,[15] are as follows:

The results indicate that two conditions must be present simultaneously to produce the embrittlement effect: first, stressing of the metal above the yield point and, second, concentration of sodium hydroxide to a point in excess of 350 grams per liter. Parallel tests omitting the sodium hydroxide show no effect. Tests upon various types of metal from the purest obtainable to those with a high percentage of impurities show that impurities do not modify the conditions.

Inhibition of embrittlement might be accomplished, so far as the experiments have gone, by the elimination of localized stresses, which is assumed to be impossible. The modification or control of the chemical condition of the water has been found to be effective.

The extensive work of the Joint Research Committee on Boiler Feedwater Studies has added greatly to our store of knowledge and has filled in many of the gaps in the explanation of conditions not recognized by earlier investigators.

THE EFFECT OF MINUTE AMOUNTS OF SILICA ON THE MECHANISM OF INTERCRYSTALLINE CRACKING

Schroeder, Berk, and O'Brien[16] have pointed out that the effect of alkaline solution causing intercrystalline cracking of steel is enormously accelerated by the presence of silica and some oxidizing agents, even when these compounds are present in very small amounts. They have pointed out further that the action of these compounds is related to concentration and temperature. These investigators have shown that when the concentration of nitrate is sufficiently high, cracking is prevented. At high temperatures, the nitrate is shown to be very effective as an inhibitor in preventing embrittlement of steel and cracking.

HYDROGEN EMBRITTLEMENT VERSUS CAUSTIC EMBRITTLEMENT

Some investigators in recent times have minimized the importance of caustic embrittlement as a primary factor in the cracking of boiler steel and related this phenomenon to entrained hydrogen. Zapffe[17] has presented many data in support of his contention that hydrogen in steel would account for the failures generally attributed to caustic embrittlement. From these studies, this investigator has concluded that embrittlement of boiler steel normally reported as caustic embrittlement is actually hydrogen embrittlement. The explanation for this theory is that the steam-iron reaction, accelerated by the presence of caustic solution in high concentration, results in the continuous penetration of hydrogen through the steel. The hydrogen reacts with nonmetallic materials to form gaseous products whose pressure within the grain boundaries exceeds the cohesive strength of the grains and causes rupture.

Zapffe has further concluded that investigation of types of boiler-plate metal immune to hydrogen embrittlement from caustic attack and chemical and metallurgical study of the role of inhibitors and accelerators in controlling hydrogen absorption would prove fruitful as a means for controlling such failures.

There are many investigators who are not in accord with this theory. Boetcher,[18] in commenting on Zapffe's conclusions, believes that three types of embrittlement exist, which should be differentiated, namely, (1) hydrogen embrittlement of equipment in various chemical apparatus, (2) pickling at low temperatures, and (3) embrittlement occurring at high temperatures.

Regardless of the theories of the fundamental factors involved in causing failures of this type, there is a preponderance of evidence that high concentrations of caustic soda in boiler waters in the presence of silica can be held responsible for intercrystalline cracking of boiler steel in highly stressed areas.

CHEMICAL TREATMENT OF WATER TO INHIBIT EMBRITTLEMENT

Parr and Straub[19] and others have demonstrated that a number of chemicals added to boiler water will inhibit, if not completely stop, this type of failure. The general conclusion reached by Straub, in 1930, was that "the presence of sodium sulphate in the boiler water tends to retard the embrittling effect of the sodium hydroxide and, if in proper proportions, will stop it entirely." This has not been entirely confirmed by subsequent experience. However, it has not been proved that sodium sulfate has no inhibiting effect on caustic-embrittlement attack. Straub's[19] further conclusion, namely, "The presence of phosphates, tannates, chromates, nitrates, acetates, etc., will also inhibit the embrittling action of caustic soda if these salts are present in the boiler water in proper amounts," has been confirmed recently.

TABLE 1. OVER-ALL SUMMARY OF STATIONARY-PLANT TESTS WITH THE EMBRITTLEMENT DETECTOR

Principal testing chemical	Specimens not cracked				Specimens cracked, inhibitor ratio* high			
	30 days	60 days	90 days or more	Total	30 days	60 days	90 days or more	Total
Sodium nitrate†	190	140	404	734	1	1	2	4‡
Quebracho extract	78	46	49	173	4	4	6	14
Zero caustic	7	21	38	66	...	...	...	0
None of above§	194	100	104	398	305	152	129	586‖
				1371				604

* Ratio of sodium nitrate or quebracho tannin to alkalinity (carbonate plus hydroxide), as NaOH = 0.3.

† Includes tests in which the boiler water had a high nitrate concentration.

‡ There were good reasons for believing that the nitrate concentrations had not been maintained at the indicated level, especially during the early days of the test; no cracks resulted when the tests were repeated under more careful supervision, sometimes with somewhat higher nitrate to alkalinity ratios.

§ Includes approximately 50 test specimens not cracked when the boiler water was known to contain a high concentration of tannin-type organic matter from natural sources, silica equivalent to or greater than, otherwise "free sodium hydroxide" or high concentrations of sodium chromate; also includes many tests for which the analytical data were incomplete.

‖ In a total of 984, 586 cracked (60%).

The most extensive work that has been carried on in recent years is that conducted at the U.S. Bureau of Mines under the sponsorship of, and in cooperation with, the Boiler Feedwater Studies Committee.* In the final report of the work completed in 1942 and reported by J. H. Walker,[20] the following statement was made:

* Sponsored by the following groups: American Boiler Manufacturers' Association and Affiliated Industries, American Railway Engineering Association, American Society for Testing Materials, American Society of Mechanical Engineers, American Water Works Association, and Edison Electric Institute.

Lignin-containing materials, such as waste sulphite liquor, and tannin materials, particularly quebracho, have been found to be inhibitors of embrittlement. Waste sulphite liquor is a by-product of the wood-cellulose industry. This material is not as effective an inhibitor as quebracho extract, a phlobatannin obtained from the heartwood of the quebracho tree of South America and imported in large quantities for the tanning and dyeing industries. Quebracho is stable and has been effective (as tested by the detector) in boilers operating at pressures up to 700 psi. A ratio of quebracho to total alkalinity in the boiler water of not less than 0.4 to 1 is required for satisfactory results.

TABLE 2. LOCOMOTIVE BOILER CRACKING ON THE C. & O.

Year	Total engine* miles, millions	Boilers found cracked	Detector test specimens cracked†
1930–1937	21.9 (avg.)	161	
1938	19.2	31	
1939‡	20.6	33	
1940	22.6	19	34
1941§	24.6	13	8
1942	27.8	10	8
1943	30.0	6	5
1944	31.2	8	2
1945	30.7	5	0
1946	28.6	5	2
1947	33.2	6	1
1948	31.9	4	0
1949	23.4	1	2

* Steam locomotives only.
† Includes slightly cracked test specimens.
‡ Organic treatment begun.
§ Nitrate treatment substituted for organic treatment.

Sodium nitrate has been found to be stable up to 400 psi. It has proved to be an effective inhibitor, when tested by the detector, if the ratio of sodium nitrate to total alkalinity is not less than 0.4 to 1. It may be possible to use sodium nitrate at higher pressures but we do not have sufficient data to be sure of this at the present time.

Tannic acid and sodium nitrate had previously been found by Straub to have inhibiting properties.

Subsequent experience has demonstrated the soundness of the committee's findings.

The high effectiveness of nitrate treatment in conditioning water to prevent boiler-seam cracking in railroad boilers was reported by Berk[21] in 1948. This investigator reported additional test data in 1951,[22] clearly substantiating the value of nitrate treatment. The statistical data reported by Berk on the value of various types of inhibitors are given in Tables 1 and 2. Berk's general summary is quoted verbatim:

Data have been presented to show that nitrate, quebracho tannin, and zero-caustic treatments have been applied successfully in stationary plants to prevent cracking in embrittlement-detector specimens.

On one large railroad, prevention of cracking in detector specimens has paralleled a spectacular decrease in boiler-seam cracking in locomotive engines.

The principal conclusion from a survey of cracking in stationary boilers in the last 10 years is that embrittlement is not a frequent occurrence.

Bardwell and Laudemann[23] reported that tests with waste sulfite liquor indicated that this material was satisfactory for the treatment of many waters to inhibit caustic-embrittlement attack but that its protective action failed in some cases. However, it is known definitely that this type of inhibitor has brought about a notable reduction in locomotive-engine cracking for the period of 2 years in which it was used.

MAINTENANCE OF ZERO ALKALINITY TO INHIBIT EMBRITTLEMENT OF BOILER STEEL

Purcell and Whirl[24] have recommended conditioning of the boiler water to eliminate practically all alkalinities yielding sodium hydroxide. This was effected by the use of phosphate alkalinity only. The basis of this treatment is that, by the use of the phosphate salts, the pH value of the water can be increased without developing caustic alkalinity. This relationship is indicated by the Purcell and Whirl curves reported in their paper on this subject.[24] These authors caution, however, that this type of treatment may not be feasible in all cases. A statement in regard to this matter follows:

Due to variable feedwater quality, in some stations, the use of tri-sodium phosphate alone may not be feasible. An acid feedwater would require excessive amounts of this chemical for neutralization of the acid, resulting in high boiler-water salines; an alkaline feedwater would tend to defeat the purpose of the treatment by developing caustic alkalinity.

It is possible, however, to adhere to the principle of the method of treatment by controlling the pH of the boiler water with the requisite amounts of either alkaline or acid chemicals, depending upon conditions. The relationship of pH and tri-sodium-phosphate concentration in pure water is shown in Fig. 3. As long as the pH and phosphate concentration conform with this curve, tri-sodium-phosphate salt will be obtained on evaporation; no caustic soda will be present. Should the pH be above that corresponding to the phosphate concentration, a mixture of tri-sodium phosphate and caustic soda would result. Therefore, to assure complete protection against embrittlement in accordance with the theory, the measured pH should never be above the curve. Should the pH of the boiler water fall below the curve value, acid-phosphate salts, which are not conducive to embrittlement, would be obtained. The optimum condition, therefore, with respect to embrittlement, corrosion protection, and boiler-water salines is the maintenance of the pH value slightly below that on the curve corresponding to the desired phosphate concentration.

The ASME Suggested Rules for the Care of Power Boilers"[25] has specified for many years the maintenance of definite ratios of sulfate to alkalinity in boiler waters as a means for inhibiting caustic embrittlement of boiler steel. In the 1954 edition of the code the recommendations have been modified somewhat, leaving the selection of treatment within some boiler-pressure ranges and some types of construction to the judgment of the operators.

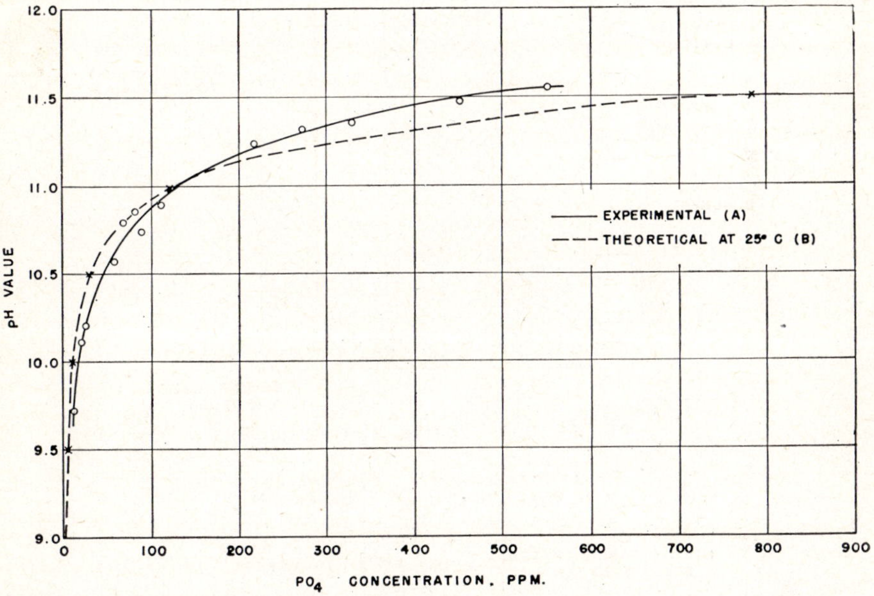

Fig. 3. Relation of pH to trisodium phosphate concentration, expressed as PO_4 ion.

EMBRITTLEMENT DETECTOR

A number of investigators have designed apparatus to detect the embrittling effect of water on boiler steel in service. Many of these appliances have proved of great value in research laboratories. The most practical apparatus which has been developed for actual use in boiler plants is known as the "embrittlement detector."[26]

The details of the design of the detector are shown in Fig. 4. The purpose of the equipment is to predict potential danger of cracking of the steel in the actual operation of boilers. The equipment is attached to a boiler, and the steel, under stress condition, is exposed to the concentrated boiler water. Actually, an artificial leak is created in the specimen, thereby permitting the accumulation of high concentrations of boiler-water salines in the crevice. The steel specimen is highly stressed by means of an adjusting screw.

If the specimen under test cracks after a given period of time, it is an indication that the water is embrittling in character and that there is danger of cracking of the boiler steel if a similar crack or crevice exists

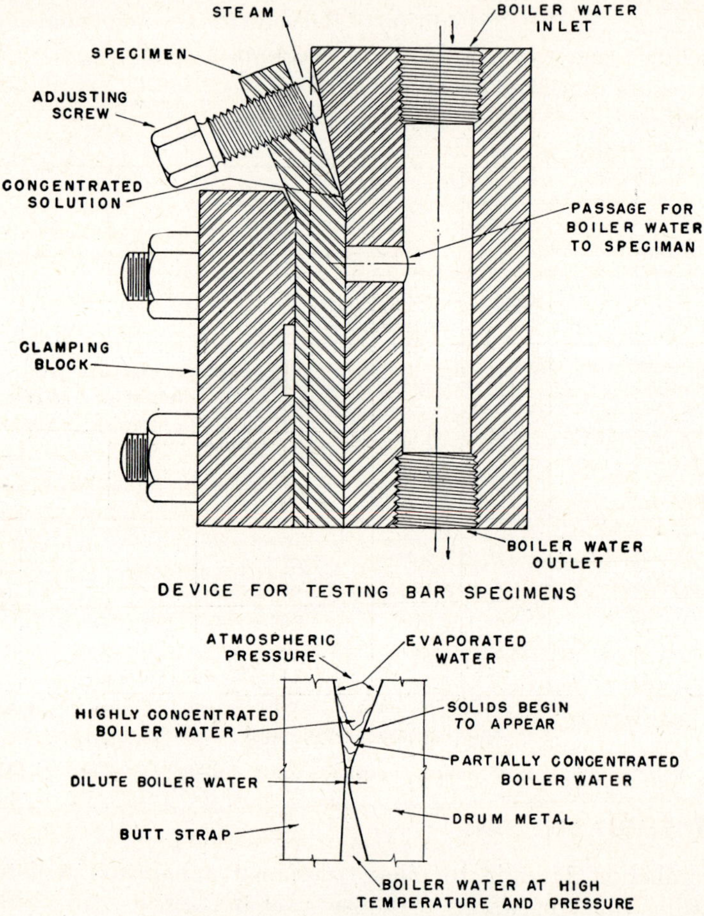

Fig. 4. Device for testing bar specimens.

in the boiler, where high concentrations of caustic soda can occur in the absence of an inhibitor. This type of detector[27] has been installed on hundreds of boilers, and at present, a mass of data has been assembled from the operation of such test units.

Data obtained by such tests must be carefully and intelligently interpreted by experienced persons. It frequently happens that the specimens will crack in service due to excessive strains and not because the water is of a highly embrittling nature. On the other hand, some of the specimens may fail to crack even though the water is of an embrittling type. Such conditions can occur if the adjustment screw is not properly operated and evaporation fails to take place in the crevice, thereby limiting the concentrations of salts in the seam. In general, however, the embrittlement detector is a very practical apparatus and has been of much value for classifying embrittling and nonembrittling types of water.

PROCEDURES FOR DETECTING CAUSTIC EMBRITTLEMENT

The destruction of steel by embrittlement is insidious and when discovered, has frequently advanced so far as to require extensive replacements of tubes and drums. If the caustic action is limited to riveted holes and is discovered in time, the holes may be reamed and oversize rivets used. However, the steel has been so seriously damaged in some cases as to require complete replacement of the boiler drums or the entire unit. The characteristic appearance of steel subject to caustic embrittlement shows fine hair cracks radiating into the metal from the periphery of rivet holes. In the advanced stages, the cracks may extend from one rivet hole to another. In one installation investigated by the author, a section of plate was cracked for the entire distance between 48 rivet holes.

It is a costly undertaking to dismantle even partially a boiler suspected of damage by caustic embrittlement. It is customary, where embrittlement is indicated, to remove one or more rivets and to examine carefully the suspected damaged area of the plate.

A number of procedures have been developed over the years to facilitate the inspection of a metal to determine the presence and extent of the cracking. Where rivet heads crack off in service or break under hammer testing, embrittlement of the steel is indicated. Although this may not always be due to caustic embrittlement, such failures indicate a suspicious condition requiring careful investigation.

A further indication of difficulty from caustic attack is the presence of repeated leaks which cannot be stopped by caulking. A number of years ago, before the development of other methods for detecting cracks, it was common practice to remove rivets and carefully examine the surfaces by polishing with emery and etching with acid. This is a painstaking and not altogether satisfactory operation. To expedite the inspection of this type of failure, a number of procedures have been developed which more clearly reveal small cracks. The use of a magnifying periscope,* of electrical apparatus, and of refined photograph techniques has aided in revealing small cracks which are difficult to detect by mere visual inspection.

* Designed by J. P. Morrison, former Assistant Chief Engineer, Hartford Steam Boiler Inspection and Insurance Company, Hartford, Conn. The need for this type of inspection is less necessary at present, and other methods of inspection are used.

DETECTION METHODS

For detection of cracks at or below the surface, there are available a number of methods of the nondestructive type, usually employing electrically operated equipment. These include X-ray or film radiography, sonic-vibration techniques, magnetic-particle tests, and the use of fluorescent-dye indicators. All require some experience and frequently an expert analysis is required to obtain an accurate interpretation.

Some use has been made of sonic or ultrasonic methods[28,29] for detecting cracks, but such procedures have not been fully evaluated for detection of embrittlement cracks such as occur in boilers.

In the magnetic-particle method[30,31,33] developed by Magnaflux Corporation for detection of cracks in boilers or other power-plant equipment, a specially prepared ferromagnetic powder applied to the surface of the metal aligns itself with cracks inside the magnetic field. R. F. Cavanaugh[32] described this process and its value for detecting embrittlement cracks in steel.

Bibliography

1. Graham: *Proc. Roy. Soc. (London)*, **17**: 219 (1869).
2. Caillet: *Compt. rend.*, **80**: 319 (1875).
3. Johnson: *Proc. Roy. Soc. (London)*, **23**: 186 (1875).
4. Heyn, E.: Eisen und Wasserstoff (Iron and Hydrogen), *Stahl u. Eisen*, **20**: 837–844 (1900).
5. Thompson: *J. Soc. Chem. Ind. (London)*, 1894.
6. Richards, T. W., and G. E. Behr: *Z. physik. Chem.*, **58**: 301 (1907).
7. Parr, S. W.: The Embrittling Action of Sodium Hydroxide on Soft Steel, *Univ. Illinois Eng. Expt. Sta. Bull.*, 94, p. 55, 1917.
8. Andrews, D. W.: *Trans. Faraday Soc.*, March, 1914, p. 316.
9. Fuller, T. S.: The Penetration of Iron by Hydrogen, *Trans. Am. Electrochem. Soc.*, **36**: 113 (1919).
10. Rosenhain, and Houser: *J. Iron Steel Inst. (London)*, 1920.
11. Mercia, P. D.: The Embrittling Action of Sodium Hydroxide on Mild Steel and Its Possible Relation to Steam Failures of Boiler Plate, *Chem. & Met. Eng.*, **16**: 496–505 (1917).
12. Kreigsheim, H.: Bulletins, The Permutit Co., 1921, 1924.
13. Papers read at the meeting of the Association of Owners of Large Boilers at Darmstadt, Germany, December, 1925.
14. Williams, R. S., and V. O. Homerberg: Why Caustic Solutions Make Steel Brittle, *Chem. & Met. Eng.*, **30**: 589–591 (1924).
15. Parr, S. W. and F. G. Straub: The Cause and Prevention of Embrittlement of Boiler Plate, *Univ. Illinois Eng. Expt. Sta. Bull.* 155, 1946; American Society for Testing Materials, 1926–1927.
16. Schroeder, W. C., A. A. Berk, and R. A. O'Brien: Intercrystalline Cracking of Steel in Aqueous Solution, *Metals & Alloys*, **8**: 320–330 (1937).

17. Zapffe, Carl A.: Boiler Embrittlement, presented at the Annual Meeting of the American Society of Mechanical Engineers, December, 1942.
18. Boetcher, H. N.: Comments on Boiler Embrittlement, by Carl A. Zapffe, paper presented at the Annual Meeting of the American Society of Mechanical Engineers, December, 1942.
19. Parr, S. W., and F. G. Straub: *Univ. Illinois Eng. Expt. Sta. Bull.* 94, 1917; 155, 1926; 177, 1928; and 216, 1930.
20. Walker, J. H.: Caustic Embrittlement Research Brings Results, *Mech. Eng.*, December, 1942, p. 891.
21. Berk, A. A.: Railroad Boiler-seam Cracking Controlled by Nitrate Treatment, *Ind. Eng. Chem.*, **40**: 1371 (1948).
22. Berk, A. A.: The Prevention of Embrittlement Cracking, presented at the Annual Meeting of the American Society of Mechanical Engineers, December, 1950.
23. Bardwell, R. C., and H. M. Laudemann: Experience with Intercrystalline Cracking on Railroads, *Trans. ASME*, **64**: 403–407 (1942).
24. Purcell, T. E., and S. F. Whirl: Embrittlement of Boiler Steel—Experiences with the Schroeder Detector, *Trans. ASME*, **64**: 397 (1942).
25. "Suggested Rules for the Care of Power Boilers," The American Society of Mechanical Engineers, New York, 1949.
26. U.S. Patents 2,283,954 and 2,283,955. The use of this apparatus under the patents can be secured by application to the Bureau of Mines, Washington. Walker, J. H.: "Caustic Embrittlement Research Brings Results," work conducted at the Bureau of Mines, Washington, 1942.
27. Berk, A. A., and W. C. Schroeder: A Practical Way to Prevent Embrittlement Cracking, *Trans. ASME*, **65**: 701–711 (1943).
28. Moriarty, C. D.: *Combustion*, **22**: 49 (October, 1950).
29. Tash, J. A.: *Combustion*, **23**: 37 (July, 1951).
30. Marks, Lionel S.: "Mechanical Engineers' Handbook," 5th ed., p. 1705, McGraw-Hill Book Company, Inc., New York, 1951.
31. Rathbone, T. C.: Detection of Fatigue Cracks by Magnaflux Method, *Mech. Eng.*, **59**: 147 (March, 1937).
32. Cavanaugh, R. F.: Magnaflux Inspection of Boiler Drums and Unfired Pressure Vessels, *Mech. Eng.*, **59**: 153 (March, 1937).
33. McMaster, R. C.: Nondestructive Testing, *Proc. ASTM*, **52**: 617 (1952).

20 BOILER SCALES AND THEIR IDENTIFICATION

The deposition of solids within boilers under steaming conditions results in an accumulation of sludge or in the production of scale. The deposits in the form of scale are highly objectionable, since they are poor conductors of heat, cause reduced efficiency, and are often responsible for burned tubes or plates (Figs. 1, 2, and 3). The mechanism of scale formation is

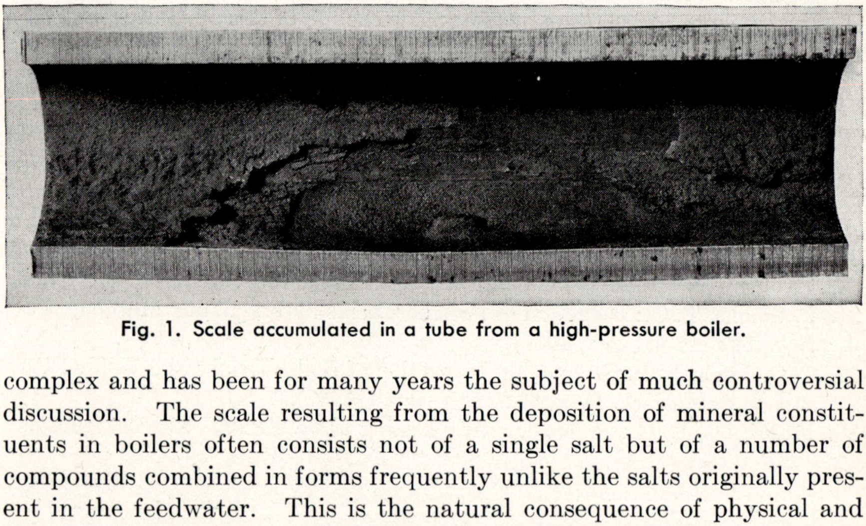

Fig. 1. Scale accumulated in a tube from a high-pressure boiler.

complex and has been for many years the subject of much controversial discussion. The scale resulting from the deposition of mineral constituents in boilers often consists not of a single salt but of a number of compounds combined in forms frequently unlike the salts originally present in the feedwater. This is the natural consequence of physical and chemical phenomena resulting from the reactions at high temperatures and pressures existing in steam boilers.

Much material has been published relating to fuel losses from scale. Many of the early investigators' results are misleading, however, since their conclusions were based upon insufficient and unreliable data. The heat-insulating properties of scale may not be measured in terms of thickness only. This fact was stated tersely by Prof. S. W. Parr* many years

* *J. ACS*, **28:** 640 (1906).

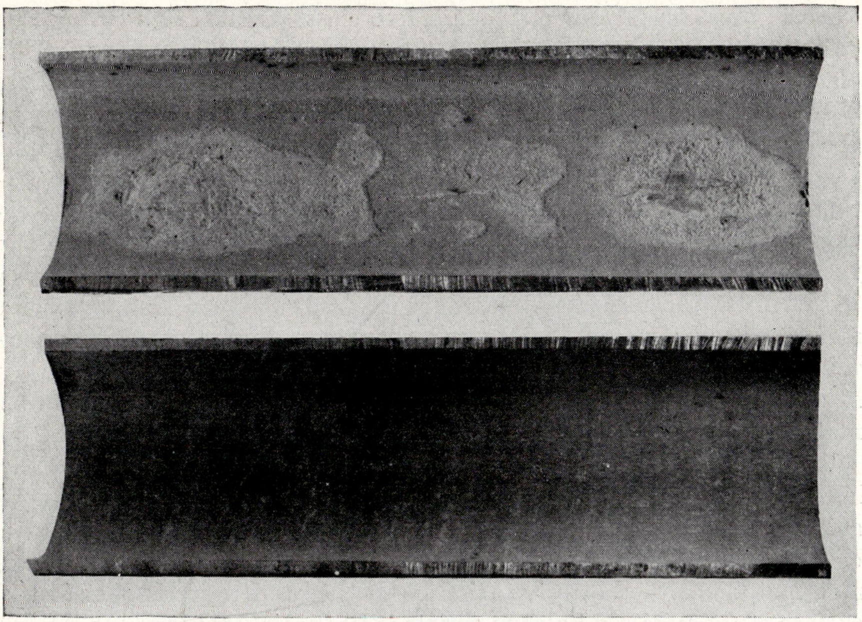

Fig. 2. Spotty scale accumulated in a low-pressure boiler.

ago in summarizing the results of the studies made by Prof. Schmidt at the University of Illinois.

Nordell,[1] in discussing the conductivities of scales, concludes from a study of a number of earlier investigators:

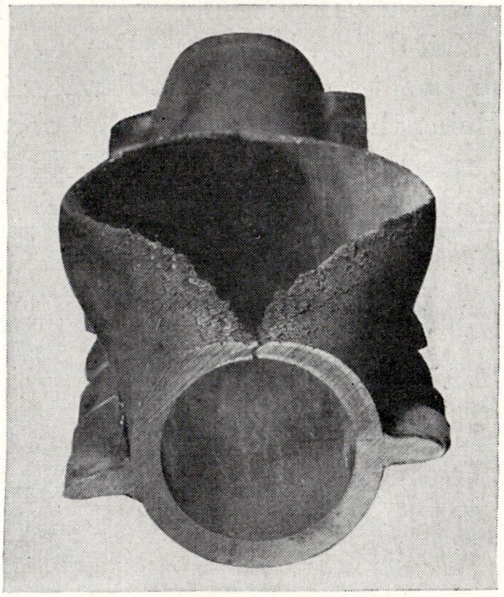

Fig. 3. Tube rupture resulting from scale and corrosion in a high-pressure boiler.

The thermal conductivities (K) of adherent scales expressed in B.T.U. per square foot of area, per foot of thickness per hour, per 1°F temperature difference, have been found to be between 0.66 and 2.06, with a mean value of approximately 1.5. The degree of porosity of the scale affects the thermal conductivity, as apparently these pores may be filled with steam instead of water under boiler operating conditions, thus reducing the thermal conductivity in much the same manner as air cells in the commonly used heat-insulating materials. There is some evidence that such porous scales may have thermal conductivities of only 0.2 B.T.U. per square foot per hour per 1°F.

Some investigators have attempted to classify scale as either hard or soft and thus to predict the loss of energy resulting therefrom. This classification is a difficult one because hardness is a relative term and is, therefore, indefinite. The hardness of a scale depends primarily upon the condition responsible for the deposition of the hardening salt, and this, in turn, involves the chemical phenomena of solubility at given temperatures and pressures. Since scale is generally the result of the deposition of a number of salts, the density of the structure is influenced by the reaction of one salt with another. As a rule, dense adherent scale results from the use of untreated boiler feedwater containing calcium sulfate, certain forms of silicates, or carbonates and silicates. Water containing bicarbonate hardness results in the formation of a relatively softer scale. These conditions, however, as has been indicated, will be materially influenced by many other factors.

The mechanism of scale formation has been demonstrated by R. E. Hall,[2] as a result of carefully controlled experimental studies and of actual operation of boilers with different water and under varying load conditions. Hall's studies have directed attention to the well-known fact that, since the hottest portion of the body of water is next to the surface of the tube, the calcium sulfate is least soluble there and that the precipitation of this salt first takes place there. From this, it is concluded that dense adherent sulfate scales are built up by crystallization on the metal surfaces and not by precipitation due to supersaturation of these salts in the main body of water in the boiler.

Hall deduced a system based on this premise, of internal boiler-water treatment that conditioned water by adding sodium salts and by maintaining the carbonate to sulfate or carbonate to phosphate ratio such as to ensure the precipitation of the salt as nonadherent calcium compounds, which could be blown from a boiler as a sludge.

The ratio of the amount of carbonate radical to that of sulfate radical, which must be maintained to inhibit the formation of adherent (calcium sulphate) scale, varies with the pressure at which the boilers are operated. Curves were developed by Hall showing these ratios for various pressures and temperatures at which boilers are operated.*

A report of an interesting study entitled "The Influence of Scale on the Efficiency and Safety of Heat Appliances" was made at the Bavarian Research Society by E. Reutlinger more than 40 years ago. Three series of tests were conducted with scales of different thickness and composition:

1. A hard, dense scale 1.4 mm thick obtained by evaporating a calcium sulfate water

* This type of treatment has now been largely replaced by more efficient phosphate treatment.

2. A scale 5.5 mm in thickness of much the same composition
3. Tar paint 0.3 mm in thickness

The insulating properties of these three deposits were determined. The results are reported graphically in Fig. 4. These graphs are of considerable interest, since they demonstrate not only the relatively high

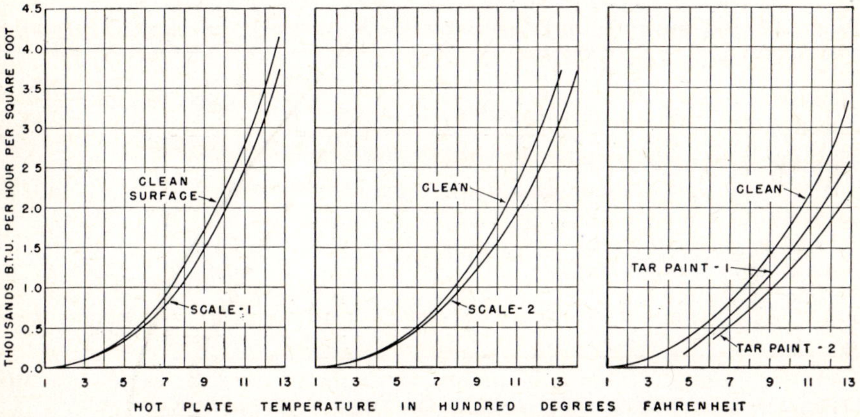

Fig. 4. Curves showing the effect of deposits on heat transmission through boiler tubes.

insulating properties of scale but also those of tarry substances, which at times may contaminate feedwater.

The composition of scale varies greatly with different water supplies, and it is difficult to predict with any degree of accuracy the specific characteristics of the deposits from similar or even identical water sources under different steaming conditions. The density, or texture, of the

TABLE 1. ANALYSES OF SAMPLES OF SCALE DEPOSITED FROM THE SAME WATER SUPPLY

Constituent	1	2
Silica (SiO_2)	2.12	3.24
Iron and aluminum oxide ($Fe_2O_3 \cdot Al_2O_3$)	0.11	1.21
Calcium carbonate ($CaCO_3$)	85.31	68.00
Calcium hydroxide $Ca(OH)_2$	1.26	0.98
Calcium sulfate ($CaSO_4$)	9.86	25.20
Magnesium hydroxide (MgOH)	0.15	0.90
Undetermined	1.19	0.47
Total	100.00	100.00

deposits is influenced greatly by boiler-operation conditions. This is illustrated by analyses of scale taken from boilers using the same raw-water supply (Table 1). Both plants function practically the same with respect to load and percentage of make-up water. Scale 1 was taken from a boiler blown down at frequent stated periods; scale 2, from a plant where control in this respect was haphazard and lax. Both samples were from the first row of tubes.

An even more pronounced change in the composition of boiler scale is shown by Paul, who reported analyses of scale deposited at the front and at the back of a boiler (Table 2).

TABLE 2. ANALYSES OF TWO SAMPLES OF SCALE DEPOSITED IN THE SAME BOILER

Constituent	Scale from front of boiler	Scale from back of boiler
Silica (SiO_2)	2.28	7.08
Iron and aluminum oxide ($Fe_2O_3 \cdot Al_2O_3$)	2.00	5.52
Calcium sulfate ($CaSO_4$)	75.38	30.63
Calcium carbonate ($CaCO_3$)	13.28	41.55
Magnesium hydroxide (MgOH)	6.09	10.64
Organic matter, etc	0.97	4.58
Total	100.00	100.00

TYPES OF SCALE AND MECHANISM OF THEIR FORMATION

If heat is applied to water containing calcium bicarbonate, the carbon dioxide is expelled from the compound, and, since the relatively high solubility of calcium bicarbonate is due to this gas, calcium carbonate is precipitated rapidly as soon as the gas is removed. Should the carbon dioxide be removed by chemical means, the same results are obtained, and the precipitation of the monocarbonate of calcium follows. As a result of this phenomenon, calcium carbonate is precipitated within boilers, forming scale.

Calcium sulfate, owing to its insolubility at high temperature, is precipitated rapidly, and unless converted into some other form by chemical treatment or removed from the water as it is formed, it causes a hard, dense scale, which adheres tenaciously to the internal metal surfaces of the boiler.

The problem of deposition of boiler scales is primarily one of the solubility of salts at elevated temperatures. Although there are many equilibria data available on the solubility of salts in aqueous solutions at relatively moderate temperatures, there is a dearth of such data at elevated temperatures. Even when the data are available, it is seldom possible to relate their values to those of concentrated boiler-water salines. The necessity for the collection of such information has been responsible for the inauguration of the several research programs.

The control of the deposition of incrustations on the heating surfaces of high-pressure boilers requires an intimate knowledge of thermodynamics and crystallography, with special reference to the influence of temperature, pressure, and solubility on the kinetics of crystal formation in complex solutions. The most plausible explanation of the deposition of solids in boilers is that advanced many years ago by Dewrance[3] and later by Partridge.[4] The latter investigator found the following:

Scale deposition and growth on the heating surfaces of boilers is due chiefly to the fact that certain slightly soluble substances become less soluble with increase in temperature. Some initial deposition of scale crystals by sedimentation, or by trapping in surface irregularities may take place in boilers of the horizontal return tubular or marine types, but it is probable that scale formation in the tubes of water-tube boilers is due to an entirely different action.

The initial deposition of scale crystals is generally the result of crystallization at the boiler surface from a solution which is supersaturated with respect to the scale-forming substance. The whole body of a boiler solution constantly tends toward a state of supersaturation as steam is removed; in addition, since scale-forming substances decrease in solubility with increase in temperature, the film of boiler water adjacent to a heating surface tends to be supersaturated even when the body of the solution is only at the saturation point. Supersaturation in the liquid film at the heating surface may be relieved by spontaneous crystallization, possibly promoted by irregularities in the surface, even when no vaporization is taking place. When bubbles are being evolved at the heating surface, however, the excess solute is deposited as minute crystals in rings directly at the solid-liquid-vapor interfaces formed by the metal surface, the boiler water, and the bubbles forming in contact with the surface. This latter action undoubtedly accounts for the deposition of practically all of the initial scale nuclei on boiler surfaces at which actual vaporization of water is taking place.

The growth of scale following initial deposition of crystal nuclei may proceed in either of two ways. The existing crystals may grow by contact with the supersaturated liquid film with which they are continuously in contact. If these crystals are removed from contact with the solution, as by the adsorption of a film of a colloidal substance present in the boiler water, further deposition of new crystals superimposed on the previous deposit may take place as the result of bubble evolution. In the first case, the resulting scale will consist chiefly of relatively large continuous crystals oriented more or less perpendicularly to the heating surface; in the second case, the scale will consist of relatively small crystals with a random orientation.

The growth of such deposits is undoubtedly accelerated by the effect of skin friction at the internal surface of the tubes, since the velocities along the tube wall are much smaller than those near the center of the tubes. This latter phase in the mechanism of scale growth has not been studied, in so far as the author is aware, but his experiences would indicate that it has a marked accelerating effect. This accounts for the deposition of many forms of silicate scale which might not otherwise develop in the presence of very small amounts of siliceous materials in feedwater.

Since scale laid down in boilers is the result of the deposition of minerals at high temperatures and pressures, the phenomena are comparable to geologic formation, confirmation of which is clearly apparent through petrographic identification. The application of petrographic procedure to critical studies of industrial water has revealed a close similarity between many sedimentary and igneous deposits occurring in nature and synthetic deposits in modern boilers.

TYPES OF SCALE AND PROCEDURES FOR IDENTIFICATION

The decided trend in recent years toward higher boiler pressures and temperatures has introduced many new factors and makes the control of scale an engineering problem of major importance from the standpoint of reliability and continuity of boiler service.

To those whose experience has been limited to the operation of low-pressure stations, the significance of the situation may not be readily apparent; consequently, opinions within the engineering profession as to the importance of the problem differ materially. This has been aptly pointed out by one observer,[5] who states, "This cleavage of opinion is, broadly speaking, a matter of past and present."

Recent experiences of many operators tend to confirm this opinion and have directed attention to the desirability of future research on the subject. In the past, the effort toward scale prevention was largely in the hands of groups interested commercially in the sale of equipment or scale-prevention processes, and relatively little time or effort has been devoted to fundamental research concerning the mechanism of scale formation. Further advance toward a solution of this troublesome problem must develop through such research with due consideration of the inadequacy of current procedure to cope with the difficulties. This in turn requires complete revision of stereotyped analytical procedures, essential improvement in the accuracy of existing sampling techniques, and critical study and interpretation of heat-transfer rates through metals and other influencing factors. A few investigators have undertaken comprehensive studies along these lines, and although the results obtained are by no means complete, they have served a most useful purpose in directing attention to the necessity for future study in this field.

There are many data in the literature concerning the solubility of calcium and magnesium and other compounds normally present in water, but the majority of these data relate to solubilities in pure solutions and at relatively moderate temperatures. It is only within recent years that such investigations have been extended to include solubility data of mixed salts in aqueous solutions and at temperatures and pressures comparable with contemporary practice in stations generating high-pressure steam.

Analytical Procedure for Scale Identification. Determining the composition of boiler scales by quantitative chemical analysis has always been, and still is, widely practiced. Such analytical procedure serves a useful purpose in determining the type and quantity of various constituents present in the deposits under examination, but, beyond this, little accurate information is to be gained by such analysis. It is frequently impossible from these data only to determine the density or porosity of the deposits, the exact mineralogical classification of the material, or any other essential information relating to its formation, structure, or physical properties. Quantitative analyses merely provide arbitrary mathematical combinations of the specific elements recorded. They may or may not indicate the specific material deposited, depending upon chance and the intelligence and experience of the observer. Such analytical reporting, unless confirmed by physical, microscopic, or crystallographic examination, will give little specific information concerning crystal formation of the specific material or variation of crystal growth of one type of deposit upon another.

Three samples of the same scale deposit were submitted by the author to three chemists for analysis. There was complete disagreement as to the character of the scale and none was able to identify the material accurately. The reason for this wide diversity of opinion was that the scale submitted was in reality a double crystal formation consisting of calcite crystal and crystals of calcium-magnesium phosphate and carbonate, presumably species resembling collophane or dahllite. According to Rogers:[6]

> Collophane is amorphous calcium carbonphosphate. This mineral bears the same relation to crystalline dahllite ($3Ca_3P_2O_8 \cdot CaCO_3$) that opal does to chalcedony or quartz. Collophane, however, contains a small amount of calcium fluoride. Actually collophane may be regarded as a solid solution of calcium carbonate, calcium fluoride, calcium sulphate and water in tricalcium phosphate. Dahllite as occurring in nature is closely associated with collophane.
>
> Concerning the conditions which bring about the formation of these deposits in engineering practice, future data are needed. These should include accurate analysis of the feedwater used and crystallographic studies of the compounds formed. With such data available the control of scale formation will be indicated.

The failure to identify this and many other scales by chemical analysis is in no way a reflection on the ability or integrity of the analysts in question but demonstrates the inherent inadequacy of chemical analytical procedures as a reliable differential method for identification of complex crystalline deposits occurring in high-pressure boilers.

Silicate scales are particularly troublesome, and their control and elimination are frequently difficult. Probably there are no constituents in water so little understood as silicates. Few accurate scientific data have been acquired concerning the behavior of siliceous compounds in water at high temperatures and pressures, although the need for such data is becoming increasingly important.

It is commonly supposed that, when silica deposits as boiler scale, it is generally in the form of calcium silicate or magnesium silicate; these silicates undoubtedly do occur. There is no certainty, however, that silicate deposits are always of this type. Other silicates have been identified, indicating that many forms of silicates may occur under critical boiler-operating conditions.

A comprehensive study of silicate scales often reveals the deposit to be analcite. There are several types of analcite which are not readily identified and may not be revealed except by refined analytical techniques. These scales are highly important because they have a relatively low coefficient of heat conductance and may form when the concentrated boiler waters contain only a few parts per million of silica. The author has often found such compounds in boilers, but their occurrence has always been associated with high heat input on tubes or other high-temperature surfaces.

Petrographic Study of Boiler Scales. Petrography is the systematic and descriptive study of rocks. The identification of boiler scales by means of the petrographic microscope may be accomplished by study and observation of the minerals in thin sections or by the immersion method. Larsen[7] advocates identification of minerals by immersion in preference to investigation of thin sections.

The immersion method consists in immersing the powdered mineral in a liquid medium of known index of refraction and determining the optical constant. Larsen, in 1921, prepared a set of tables for systematic determination of minerals from their optical constants and set forth the optical constants for more than 500 species for which data were not previously available. In discussing this microscopic procedure, Larsen stated:

> The science of mineralogy needs also good connected and consistent data on minerals. Chemical analyses, crystallographic studies and determination of physical properties, optical constants, and paragenesis should be made on identical materials. Greater care should be taken in examining minerals for lack of homogeneity whether it is due to zonal growths or to admixture or included foreign material. The optical method might be used with advantage in work done in a number of branches of science other than mineralogy and in some industries. It should be much more generally used in chemistry, especially in analyses of artificial crystalline products to determine rapidly the exact nature of the material and its homogeneity, for artificial products have a definite optical constant as minerals.

This suggestion has particular pertinence in studies of boiler scales owing to the marked similarity of these synthetic forms to minerals occurring in nature. The identification of the structure and growth development of boiler scale by petrographic methods has special merit, since it not only makes possible the accurate determination of substances deposited but reveals the presence of concurrent deposits and contaminating material occurring or crystallized simultaneously with the major crystal formation. By correlation of such crystal growth with boiler-water-conditioning and operation data, procedures for inhibiting scale deposits may be developed.

Except concerning the sulfates, carbonates, and hydroxides of calcium and magnesium and, to a lesser degree, sodium salts, there are few definite crystallographic data on various complex forms of boiler scales. Calcium carbonate is almost always present as calcite; calcium sulfate, as anhydrite; and magnesium hydroxides, as burcite. These forms may readily be identified by petrographic analyses and X-ray diffraction procedures. Many of the more complex deposits, especially silicates, require broad experience and adequate microscopic equipment for accurate identification.

Typical Scales Identified by Petrographic Analyses. A number of scales and miscellaneous deposits occurring in boiler pipelines and feed pumps have been observed in studies of difficult scale problems. The photomicrographic records have revealed many crystal structures of unusual interest. Some of these are reproduced here to illustrate the potential value of such studies as a control procedure to indicate the solution for inhibiting difficult boiler scale and turbine-blade deposits.

An interesting deposit was removed from a high-pressure boiler-feed-water pump supplied with lime-soda-softened water. The scale was dark brown and laminated, extremely hard and porous, containing approximately 60 per cent calcium carbonate. It was definitely identified as calcite with inclusion of organic matter and colloidal silicate in the form of "opal."

Considerable tube difficulty was experienced from a scale formation resulting in the loss of many tubes during one month of operation. The feedwater from which this scale deposited had been treated by a hot lime-soda system employing ferrous sulfate as supplementary treatment. The boilers were operated at 400 lb pressure. The deposits were identified as calcite and anhydrite with small inclusions of opal. The laminated structure of the deposit clearly indicated over- and undertreatment of the feedwater. The tube losses were due not so much to the composition and porosity of the scale as to loose bonding to the tubes, permitting steam blanketing of the metal in restricted areas with resulting poor rate of heat transfer and localized overheating.

The most interesting scales identified by petrographic analyses and other refined testing procedures[8] are those composed largely of analcite.

These silicates are similar to zeolite groups and have many peculiar properties. According to C. Doelter,[9] analcite, $Na_2O \cdot Al_2O_3 \cdot SiO_2$, will form whenever the components exist in a closed system between temperatures of 180 and 430°C (350 to 800°F). The temperature of the boiler water from which these scale deposits were taken by the author was 500°F, but the temperature of the tube surface was not known. Above the upper temperature limit of 430°C (796°F), analcite is said to pass into an anhydrous compound (nepheline) of the formula $Na_2O \cdot Al_2O_3 \cdot 2SiO_2$. Below the lower limit of 180°C, analcite passes into natrolite, another zeolite, $2H_2O \cdot Na_2O \cdot Al_2O_3 \cdot 3SiO_2$.

Analcite is not a rare mineral in regions of volcanic rocks. It is frequently found in fissures and cavities in lava, often in fine crystals. Its origin is in the hot aqueous solution that marks the final phase of volcanic activities after the magmatic solutions have reached a moderate temperature and pressure. Artificially, analcite has been made in various ways, such as heating sodium hydroxide or silicate in a sealed aluminum glass tube at 180 to 190°C (356 to 362°F) for 18 hr.[10]

A. de Schulten[11] prepared synthetic analcite by mixing solutions of sodium silicate and sodium aluminate in such proportions that the amounts of alumina and silica were the same as in analcite, adding a suitable amount of lime water, and heating in a sealed copper tube at 180°C (356°F). After 18 hr he obtained crystals of analcite of the following compositions:

Constituent	Artificial analcite, %	Natural analcite, %
Silica	55.4	54.4
Alumina	22.1	23.3
Soda (difference)	14.3	14.1
Water	8.2	8.2
Total	100.0	100.0

It is to be noted that the lime water has no other function than that of facilitating the crystallization.

Viewed through a microscope, the surface of analcite is seen to consist entirely of well-defined faces, showing typical cubes and trapezohedrons. These conditions are illustrated in Fig. 5. According to Dr. Charles Milton,[12] "When the solution is very concentrated and alkaline, the crystals form trapezohedrons, under other conditions cubes."

It is seldom that a single deposition of crystals will occur and boiler deposits, as a rule, are conglomerated mixtures with one or more substances predominating. At times, deposits are laid down as conglomerates, several deposits occurring simultaneously. This phenomenon is clearly the result of solubility equilibria and of occluded substances.

Several growths of dendritic metallic copper crystals were found in the deposit, the presence of which, in the amount of 2.5 per cent of copper oxide, had previously been determined by chemical analysis. A reproduction of these crystals is shown in Fig. 6. It was believed that they

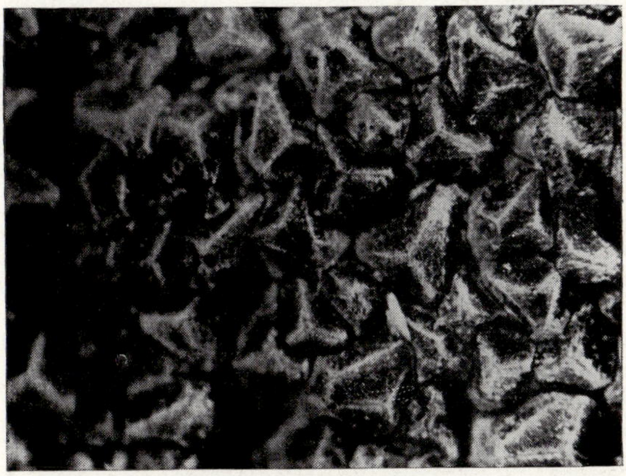

Fig. 5. Thin section of analcite scale showing dominant trapezohedral faces.

Fig. 6. Photograph of a boiler deposit showing dendritic copper crystals.

were due to copper salts in the boiler-water salines, which were reduced to a metallic state under the influence of boiler conditions. The only apparent explanation for the presence of copper salts in the water was the corrosion of condenser tubes and brass fittings. It is conceivable that, in contact with the iron surfaces of the boiler, electrochemical deposition of metallic copper could occur with an equivalent solution of iron.

In Table 3 are given the results of analyses of a dense scale laid down on the drums and tubes of a 400-lb pressure boiler, the feedwater for which was softened with intermittent cold lime-soda treatment. This scale accumulated slowly and caused no serious inconvenience. The deposit was essentially calcite with small amounts of analcite and anhydrite.

X-ray identification of numerous compounds had been used as an analytical technique for many years. Early procedures were relatively costly and required highly trained personnel. Recently, the equipment for such work has been greatly simplified and X-ray diffraction apparatus and procedures are now available for rapid routine, as well as special, determinations. The constituents in the majority of boiler scales, turbine deposits, and other accumulations in power plant equipment can now be readily and accurately identified by analysts after reasonable training.

TABLE 3. CHEMICAL ANALYSIS OF MUD DRUM DEPOSIT IDENTIFIED BY PETROGRAPHIC ANALYSIS AS A CALCITE, ANHYDRITE, AND A FEW CRYSTALS OF ANALCITE

Constituent	*Per Cent*
Iron and aluminum oxide ($Al_2O_3 \cdot Fe_2O_3$)	11.48
Silica (SiO_2)	2.80
Calcium sulfate ($CaSO_4$)	1.45
Calcium carbonate ($CaCO_3$)	76.95
Magnesium carbonate ($MgCO_3$)	3.72
Total	96.40

The diffraction apparatus (Fig. 7) consists of a highly stabilized source of low-energy X rays, a principal instrument for supporting the sample under analysis, and an X-ray detector to measure and record the diffracted radiation. The sample is normally in a powdered state. The powder is compacted into a shallow tray and placed in the instrument for analysis.

When the sample under analysis is properly inserted in the X-ray beam, it diffracts the rays at specific angles characteristic of the material. Identification is based upon the fact that the diffraction pattern is specific and reproducible for each compound. The X-ray pattern also reveals the percentages in which the various constituents are present.

There are three techniques for comparing the pattern of an unknown with that of a known material: (1) photography, (2) mechanical scanning, and (3) X-ray fluorescence.

By allowing the diffracted X rays to strike a photographic film, a picture of the characteristic diffraction is obtained. Identification is thus a method of comparing the photographic pattern of an unknown material with that of a known material. Through the joint and cooperative efforts of X-rayologists and the ASTM, an extensive library of diffraction patterns for identification of unknown materials is now available. These data in the form of indexed cards can be purchased from the ASTM.*

* American Society for Testing Materials, Philadelphia, Pa.

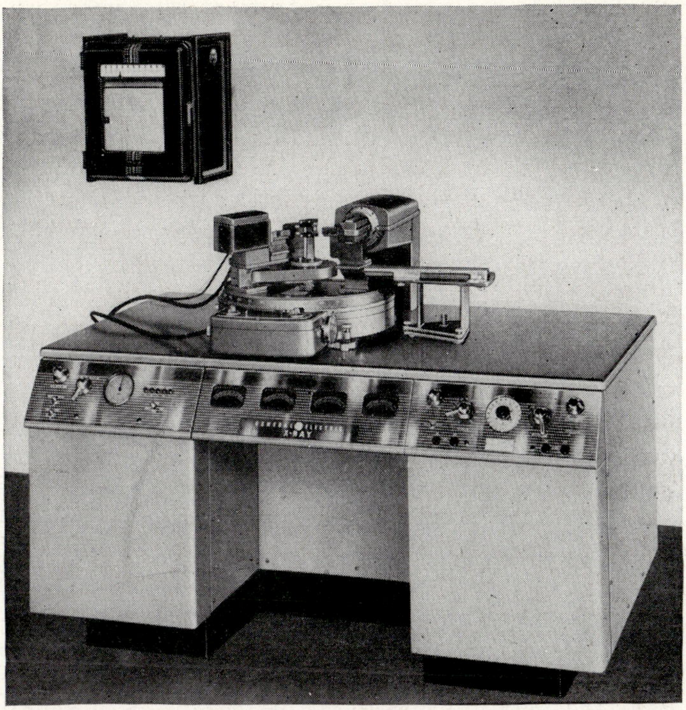

Fig. 7. X-ray diffraction equipment for identification of deposits from boilers, turbines, and miscellaneous equipment. *(Courtesy of General Electric Company.)*

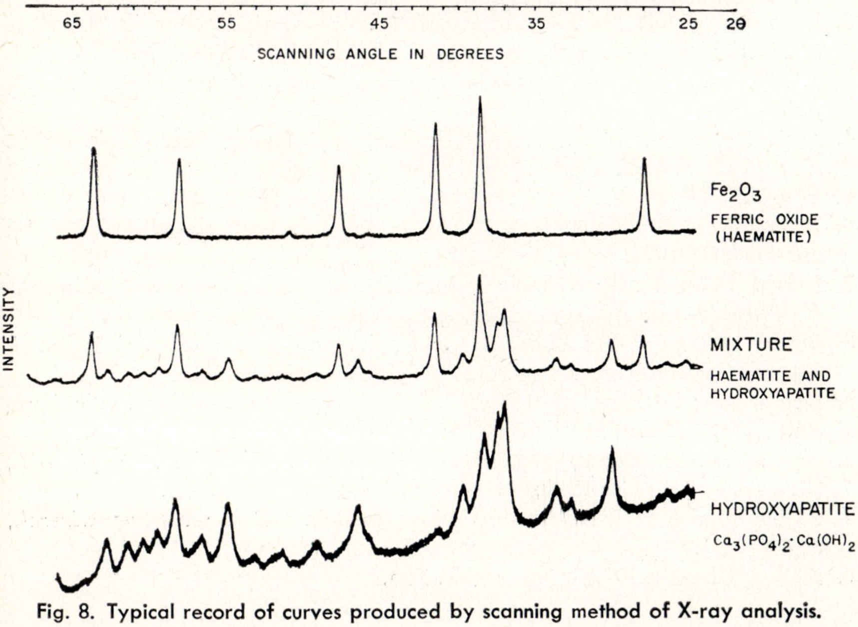

Fig. 8. Typical record of curves produced by scanning method of X-ray analysis.

The second method consists of an instrument attachment which permits scanning through a desired angular portion of the diffraction pattern. The scanning assembly contains a detector and a recorder to draw a curve of the intensity of the diffracted rays at different angles in the field scanned. Recorded lines for a typical X-ray diffraction chart for iron oxide, hydroxyapatite, and a mixture of these two compounds are shown in Fig. 8. By comparing the characteristic curves of materials being analyzed with established standards, identifications are made in a manner similar to that of the photographic method. However, the scanning method is more rapid and is readily adaptable to routine work. The entire chart has not been reproduced in Fig. 8, but a section has been shown merely to indicate the type of the recorded pattern.

A related analysis technique is X-ray fluorescence; by this method, elements having atomic numbers from 22 to 92 may be analyzed. Both X-ray diffraction and X-ray fluorescence are applicable to metallurgical studies; therefore, these techniques are of interest to those engaged in power plant operation. Studies of alloys by phase and by element can also be made.

Bibliography

1. Nordell, E.: "Water Treatment for Industrial and Other Uses," p. 49, Reinhold Publishing Corporation, New York, 1951.
2. Hall, R. E., and others: "A Physico-chemical Study of Scale Formation and Boiler Water Conditioning," Bull. 24, Carnegie Institute of Technology, Pittsburgh, Pa., 1927.
3. Dewrance, John: The Generation of Steam, presented before the Institute of Mechanical Engineers, London and reported in *Engineer*, Feb. 18, 1921, p. 182, under the title Corrosion of Boilers and in *Power*, Mar. 29, 1921, p. 525.
4. Partridge, E. P.: Formation and Properties of Boiler Scale, *Univ. Mich., Eng. Research Bull. No.* 15, June, 1930.
5. Editorial, The Formation of Scale, *Engineering*, Jan. 30, 1931.
6. Rogers, A. F.: Collophane—A Much Neglected Mineral, *Am. J. Sci.*, No. 16, pp. 269–276, 1922.
7. Larsen, Esper S.: Microscopic Determination of the Non-opaque Minerals, *U.S. Geol. Survey Bull.* 679, 1921.
8. Powell, Sheppard T.: Water as an Engineering and Industrial Material, Edgar Marburg Lecture, *Proc. ASTM*, Vol. 34, Part II, 1934.
9. Doelter, C.: IV Minerogenese und Stabilitatsfelder der Minerale die Analcimbildung, *Mineralog. u. petrog. Mitt.*, **25**: 111 (1906).
10. de Schulten, A.: Sur la reproduction artificielle de l'analcime, *Compt. rend.*, **90**: 1493 (1880); also, *Bull. soc. franç. mineral.*, **3**: 150 (1880).
11. de Schulten, A.: Sur la reproduction de l'analcime, *Bull. soc. franç. mineral.*, **5**: 7 (1882).
12. Milton, Charles: Private communication.

21 CHEMICAL CLEANING OF POWER PLANT EQUIPMENT

INTRODUCTION

Many kinds of power plant equipment require periodic cleaning to remove water-insoluble deposits from their surfaces. The accumulated deposits may be from corrosion products, from water or steam, or from oil or other suspended matter contaminating the liquids in contact with the metal.

Because of the wide variety of materials accumulating and the complex design of the equipment, various cleaning means have been devised: (1) mechanical, (2) chemical, or (3) thermal means or a combination of them. Each process has its specific adaptability and usefulness.

Possibly the oldest cleaning method is to use manually operated or power-driven tools that remove deposits by mechanical means. Thermal cleaning, such as by steam or direct flame, is another method, but it has only limited application. Under some conditions, thermal methods, however, are practical and suitable.

Cleaning by chemicals is probably the most widely adopted procedure. It has been used to some extent for many years. Recently, chemical cleaning processes have been greatly improved and are now widely employed throughout the world.

It has long been common practice to clean equipment contaminated by oils, greases, or other saponifiable materials with various alkaline detergents. More recently, organic degreasing compounds have come into use for this type of cleaning. In general, cleaning agents suitable for removing grease and dirt are likely to be totally ineffective for removing metallic films such as mill scale, incrustations due to corrosion, or adherent scale laid down by water or steam.

Mechanical methods still find wide acceptance and are relatively satisfactory in some cases, but these are unsuitable or ineffective for cleaning complex power plant equipment, many parts of which are inaccessible.

As a result, various chemical methods have been developed. It must be remembered, though, that there are many types of deposit which cannot be removed by a single acid or alkali but require a combination of chemicals.

ACID CLEANING

The use of acids has become widespread because acids, when properly compounded with inhibitors and wetting agents, are capable of, and suitable for, dissolving or loosening many incrustations which cannot be removed by other means.

Acid pickling processes, as an example, have long been employed for the removal of mill scale from sheets, plates, tubes, and other products in mills making steel and copper alloys. Great quantities of acid are used for such services. Similarly, the economy and speed gained in removing deposits by solution or chemical action have made profitable many other applications, such as the cleaning of oil wells, water mains, heat exchangers, boilers, and other equipment. The extensive acidizing of oil wells to improve yields by removing deposits and opening formations did much to attract interest and suggest the use of acid treatment for other applications.

Scale removal from boiler surfaces was a logical development after the practical application of acidizing to other equipment was demonstrated. This form of cleaning has now become a highly specialized art. Trained personnel and complex equipment are required to assure satisfactory results and to avoid many hazards involved in handling acids and other corrosive materials.

Thousands of boilers and miscellaneous equipment have been cleaned by acid during the last few years, and a majority of these treatments have been successful. Under ordinary conditions, where an effective inhibitor has been used, the corrosion of the surfaces is relatively minor. In fact, considering the large number of boilers which have been acid-cleaned, noticeable corrosion has taken place in only a relative few. There is no accurate record, however, indicating the extent to which acid-treated equipment may have been corroded. It is a recognized fact that, even though inhibitors may be used, they do not completely prevent corrosion but merely reduce the rate and extent of the attack.

Where an effective inhibitor has been used, the temperature satisfactorily controlled, the contact time limited, and subsequent washing out and neutralization of remaining acid properly performed, the extent of attack on the metal of the equipment has been minor. Repeated acidizing may be undertaken without seriously damaging equipment when the treatment is properly carried out.

Although cleaning of boilers by acidizing is most widely practiced in steam-generating plants, the process is also applicable to many other types of equipment.[1] These may be classified broadly as follows:

1. Boilers, economizers, superheaters
2. Deaerators, vent condensers, stage heaters, and heat exchangers
3. Surface condensers
4. Feedwater-treating equipment
5. Valves and miscellaneous equipment
6. Piping

Each of these applications requires special study, and the treatment must be selected and applied to meet the conditions encountered.

EFFECT OF ACID CLEANING ON BOILER STEEL

Studies of metal which has been repeatedly acid-cleaned show that some corrosion occurs each time. The rate of attack is generally not sufficiently rapid, even after many cleaning cycles, to cause serious concern. In some instances, acidizing has been blamed for active attack on the metal, when this treatment has not actually been responsible. The metal surface has been so well cleaned that it reveals corrosion previously unnoticed because the pits and surface etching were filled and coated over with iron oxide or incrustations.

The effect of acidizing may be best evaluated by meticulously cleaning a portion of the surfaces before the treatment either by wire brushing or local application of weak acid and by inspecting the metal for pitting, using low magnification and tangential illumination. The same surface should be similarly cleaned and inspected after the entire unit has been acidized. Pitting shows up on stressed metal; hence for these inspections, it is best to select the rolled ends of boiler tubes, the heads or nuts of highly strained bolts, or the bends of angle irons, welded zones, etc. In most cases, such scrupulous before-and-after inspections have not been made because the equipment has had to be returned to service as quickly as possible after the cleaning has been completed or because the operators were unaware of the procedure to be followed.

It is a well-established fact that the rate of attack on metals by acid or any other electrolyte is greatly affected by the internal stresses in the material. Examination of boilers which showed corrosion after acid cleaning demonstrated that the most severe attack was on metal which had been greatly stressed. This condition was particularly pronounced on the flared end of tubes where stresses of large magnitude occurred as the result of cold-rolling. This is illustrated in Fig. 1, which shows the "honeycomb" type of attack that is characteristic in units where corrosion developed during acid cleaning.

A photomicrograph of a section of tube which has been attacked by acid is shown in Fig. 2. The honeycomb or lace form of deep pitting is clearly shown by the photomicrograph. Inspection of numerous boilers has further demonstrated that internal appliances within the drum which have been cold-worked are readily attacked by inhibited acid, even when the unstressed metal has shown no corrosion.

Fig. 1. Corrosion resulting from acid attack on stressed metal at the flared end of a cold-rolled tube. Note honeycomb type of attack.

This type of attack has been observed by the author in a number of installations. Some of the boilers had been acid-cleaned, and the temperature of the acid solution was relatively high. In one case, the attack was probably due to the effect of an acid contaminant in the feedwater and not to acid cleaning. The pattern of the corrosion, however, was identical with the type and pattern of attack noted in acid-cleaned boilers. This occurrence tends to demonstrate that all corrosion of tube ends by acid may not be the result of acid cleaning but may have resulted from some contaminant present in the feedwater. This observation suggests the need for an

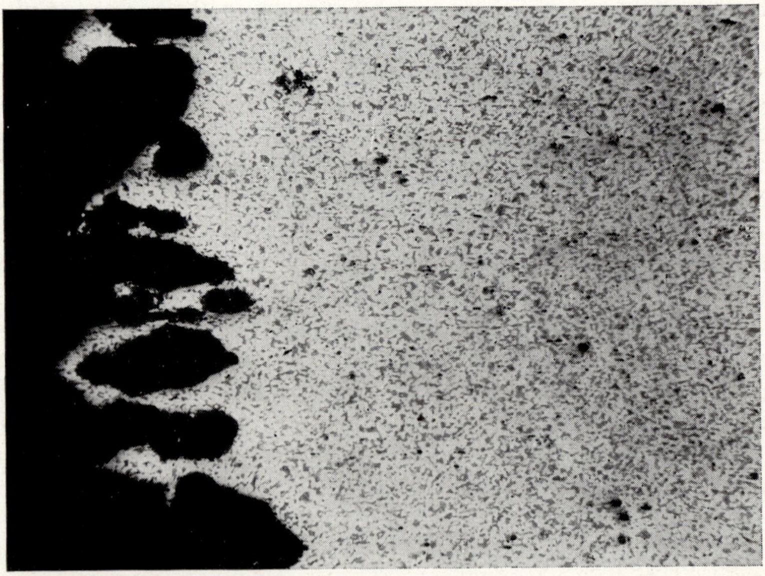

Fig. 2. Photomicrograph of a section of boiler tube taken from the rolled end, showing honeycomb attack resulting from acid cleaning.

intensive investigation to determine all conditions which may be responsible for corrosive attack and for the avoidance of unwarranted criticism of acid-cleaning, which, in the final analysis, may not be responsible.[2]

Some concern has been expressed as to possible intercrystalline attack due to the penetration of hydrogen into the steel during, or subsequent to, acid cleaning. Although relatively large amounts of hydrogen may be generated during the cleaning period, investigations have shown no attack on the crystalline structure nor any evidence of embrittlement attendant on such form of corrosion. Although such investigations have not been extensive, it seems reasonably safe to predict that embrittlement will not occur if the temperature and strength of acid used are kept within safe limits, if the acidized units are thoroughly washed, and if the acid adhering to the surfaces is completely neutralized.

ATTACK ON GASKET AND PACKING

Some damage to gasket material has been caused by acid cleaning, and special attention should be given to these matters when chemical cleaning of equipment is considered. Pressed types of asbestos gaskets are usually not affected by properly compounded acid-cleaning solvents. The principal change which sometimes appears to occur in such gaskets is that the material is likely to shear or tear readily after contact with acid. However, when this type of gasket material is compressed between steel plates, as in boiler drums, it is not noticeably affected by acid.

On the other hand, woven-asbestos gasket material is often rotted in contact with certain kinds of acid. When the gasket has been so attacked, it cannot be reused once it is removed. Handhole gaskets constructed of asbestos with metal shield, are not appreciably affected, since the acid in this case may not come directly in contact with the asbestos. If gaskets are badly corroded, they frequently leak after chemical cleaning of boilers or other power plant equipment, and in such cases, it becomes necessary to replace gaskets.

Where fluoride salts are compounded with the acid, as they often are to disintegrate silicate scales, the fluoride rapidly attacks the asbestos filler. When this occurs, it is necessary to replace the gaskets after the chemical cleaning has been completed. Ordinarily, however, boilers and other equipment can be cleaned by inhibited acid without serious attack on gaskets. One of the economic advantages of such a form of cleaning is that the removal of a large number of handhole plates and gaskets is avoided.

COST OF CLEANING

The cost of cleaning of boilers and other power plant equipment varies widely, depending on the type of equipment, the nature and extent of the deposits to be removed, the accessibility of the plant, and a number of other factors. It is impossible, therefore, to give any specific cost figures regarding such forms of cleaning. However, this type of cleaning is usually cheaper than cleaning by mechanical means, when all factors are taken into consideration. In attempting to evaluate the relative merits of chemical cleaning in contrast to those of cleaning by mechanical means, consideration and credit should be given to chemical cleaning for reduced outage of equipment, minimum replacements, and, usually, less labor. When all these factors are given their rightful importance, chemical cleaning will generally result in a marked economy. Further, there are certain types of equipment that cannot be adequately cleaned by any mechanical means yet devised, and in such cases, cleaning by chemical means is of great value.

TECHNIQUE FOR CLEANING EQUIPMENT

Prior to initiating the cleaning process, the equipment should be inspected and all conditions investigated and evaluated. The treatment required is governed by the chemical characteristics, density, porosity, and thickness of the incrustations. To evaluate these factors properly requires both chemical and microscopical examination and identification of the compound by X-ray diffraction patterns. Chemical cleaning cannot be adequately and economically carried out without such preliminary investigations and, in fact, can prove ineffective and costly, requiring a second cleaning.*

After this testing procedure has been completed and the results interpreted, the strength and quantity of acid required should be estimated and the nature and type of the supplementary chemicals and wetting agents likewise selected.† When the cleaning is to be done by a contractor specializing in this type of work, the prepared acid or other cleaning agent is usually brought to the site by specially designed tank wagons.

The solutions are heated to the proper temperature and added to the boiler in the correct proportions. Prior to delivering the acid solution to the boilers or other equipment, the temperature of the metals must be reduced so that the solution will not come into contact with too hot surfaces. The desired temperature is usually 150 to 160°F or slightly higher. The water is then drained from the equipment, and the solution is pumped into the unit. It is obvious that, before actually starting the acid cleaning, the equipment must be completely isolated from all other equipment.

* Two-stage treatment is required at times because a single cleaning with acid cannot assure removal of deposits.

† The inhibitor should be put in the acid before the solution is brought to the site of the equipment to be cleaned.

In the case of a boiler, the prepared solution is pumped in through the bottom connection until it reaches the desired point. The solution may then be circulated continuously or intermittently, or in most cases, circulation is omitted, and the solution is permitted to soak into the scale with no movement throughout the unit. There is always some circulation due to thermal currents and to differences in the specific gravity of the acid solution and the water.

Samples of the cleaning solution are removed from time to time at various points and tested for the concentration of the active constituents. The acid solvent is neutralized by contact with the incrustation which it dissolves. The degree of activity of this spent acid solution for destroying scale can be determined by the reduction in the strength of the acid. When, in the judgment of the operator, the process is completed, the solution is drained from the unit and the residual solution adhering to the metal is removed by rinsing several times with hot water.

After the acid is washed from the boiler, any remaining acid is neutralized by washing with a 0.5 to 1.5 per cent alkaline solution, usually of soda ash or caustic soda. It is imperative that all products which have been loosened by the treatment but which have not been entirely dissolved be completely removed by flushing and, where required, by supplementary mechanical means. Failure to remove such deposits adequately has resulted in severe damage to boilers.

The final step in the cleaning process is very important. This relates especially to the need for complete removal, or purging, of hydrogen gas from the equipment which has been cleaned. Such purging should be carried out with extreme care, and the equipment should not be opened until there is definite assurance that all hydrogen, including that which may have become pocketed in the superheater or elsewhere in the boiler, has been thoroughly removed from the unit.

When it is imperative to clean nondraining superheaters chemically, each tube should be individually flushed out with clean water, then neutralized with alkali, and again purged with clean water after the cleaning is completed. Where economizers are to be chemically cleaned, it is preferable to pump in the solution, to hold the solvent in the economizer for a given period of time, and then to flush and clean out in the manner previously recommended.

Difficulty from adherence of the organic inhibitor has been experienced in some cases after the cleaning is completed. Some operators prefer to destroy these inhibitor films by boiling at one-half designed operating pressure for from 12 to 24 hr. The total alkalinity during such operation should be maintained at between 2,500 and 5,000 ppm by the addition of soda ash, caustic soda, or trisodium phosphate.

DETAILED PROCEDURE FOR ACID CLEANING OF BOILERS

Generally, it is preferable to have the work performed by a trained and experienced contractor rather than by the local operators, unless personnel have been trained to employ the proper technique and to take precautions against the industrial hazards involved in such work.

The program of operation just covered in some detail for cleaning boilers by the soaking method can be summarized as follows:

1. Cool, drain, and inspect the boiler.

2. Install a vent pipe to connect the top of the steam drum or an equivalent highest accessible point to a safe location outdoors, being careful to avoid pocketing of gases.

3. Connect the acid fill pipes to the bottom wall headers or to other lowermost points in the boiler so that all lower parts will first fill with acid.

4. Make any necessary adjustment to the metal temperature either by cooling further or by warming to the desired point.

5. Fill the boiler to the desired level with dilute inhibited acid of the required strength and composition and at the desired temperature.

6. Allow to soak for the predetermined period, usually 6 to 8 hr or longer, as conditions require.

7. Make necessary tests to determine the strength of the acid in samples taken from available sampling points. If the strength of acid in the boiler has fallen rapidly, it may be necessary to drain and refill with fresh acid. The concentration of dissolved materials should be checked in the acid drained from the unit.

8. When the treatment has been completed, drain the unit as quickly as possible and test a sample of the spent acid.

9. Fill rapidly with fresh water and drain two or three times to remove remaining acid, mill scale, and iron salts.

10. Refill with an alkaline solution containing from 0.5 to 1.0 per cent (5,000 to 10,000 ppm) alkali, such as caustic soda, soda ash, or sodium phosphate. The boiler should first be filled above the point occupied by the acid and then drained down to normal working level.

11. Heat with a low fire and raise pressure sufficiently high to obtain circulation of the alkaline water. The desired pressure will vary, depending on the design and capacity of the unit.

12. Cool and drain the boiler completely.

Whenever possible, boilers should be drained and inspected after acidizing and a record kept of the condition of the surfaces for future reference and comparison with subsequent inspections.*

* A few experiences are reported which show that the acid may loosen, but not remove, all scale. When the boiler is returned to service, the heat breaks the scale loose, and it lodges in a header or tube. This stifles circulation and results in over-

There may be many deviations from the foregoing operation, but the general plan is indicative of the steps which should be followed to lessen the danger of corrosion.

In no case should any manhole covers or other plates be opened for inspection until after the unit has been completely purged of hydrogen and of acid on the surface, neutralized with alkali, and flushed with fresh water.

FACTORS INFLUENCING EFFECTIVENESS AND SAFETY OF ACID CLEANING PROCEDURES

There are numerous conditions which require control during the acidizing of any equipment in order to insure effective removal of deposits from the surfaces and a minimum of corrosive attack on the metal. It is not possible to list all the conditions which may be encountered, since there are a multitude of variables which can alter the effectiveness of the acid cleaning process. In general, however, the items which should be considered and controlled, if possible, are the following:

1. Composition and concentration of acid
2. Nature and composition of inhibitor
3. Temperature of the metal and of the solutions
4. Circulation, whether mechanically induced or by diffusion
5. Composition of the metal
6. Physical and chemical composition of the scale to be removed
7. Hazards of the process
8. Design and location of equipment
9. Procedure followed, from taking equipment out of service, through removal of acid, washing down, neutralization of acid remaining on surfaces after treatment, to final washing prior to returning to service

It is appropriate to elaborate somewhat on the foregoing items because this procedure is of interest to mechanical engineers and others responsible for acidizing of power plant equipment.

TYPE AND CONCENTRATION OF ACID SOLUTION

It is a common misconception that the cleaning of equipment requires merely the addition of a mineral acid together with a chemical inhibitor for the purpose of slowing down the rate of attack on metal surfaces. In some cases, such treatment may effect the desired results, provided tem-

---

heating and failure of the tubes. If such a condition occurs, the deposits should be completely removed before the boiler is put into service. Where evidence of this condition is obtained, it is desirable to wash out, and, if necessary, to run a turbine cleaner through, all tubes in which plugging might occur. Mud drums and lower wall headers must also be cleaned of loose deposit.

perature control and other conditions are adequately maintained. There are, however, certain types of incrustations occurring in boilers, condensers, and heat exchangers which resist solution by muriatic, sulfuric, and other acids. For instance, silicate, sulfate, and certain other scales, either alone or combined with other types of incrustation, are not appreciably soluble in muriatic acid. Where silicate compounds are encountered, ammonium bifluoride or other fluorides must be used to ensure their removal.[3] Organic deposits baked on tubes require special treatment for their removal.

Certain metallic deposits also resist solution, and where they are present, specific solvents must be employed. If oils or greases are encountered, appropriate solvents are necessary for their removal, and effective results cannot be secured without first removing such materials.[4] Experience has demonstrated that, in addition to the inhibiting chemicals, other compounds improve the effectiveness of the acid cleaning process. Most of these treatments involve trade secrets, and their use and applications are the outgrowth of research and development coupled with practical experience over a wide range of conditions encountered. Frequently it is also desirable to use a wetting agent.

The strength of acid used should be governed by the time available for cleaning the equipment and the by chemical composition of the acidizing solution. In the past, the strength of the acid employed has been as high as 15 per cent. Recently, however, there has been a marked trend to the use of acid of lower concentration, even though the cleaning takes more time. Concentrations now generally used are from 5 to 7 per cent.

EFFECT OF TEMPERATURE

Possibly the most controversial factor with regard to acidizing is the temperature at which the treatment should be carried out. There is

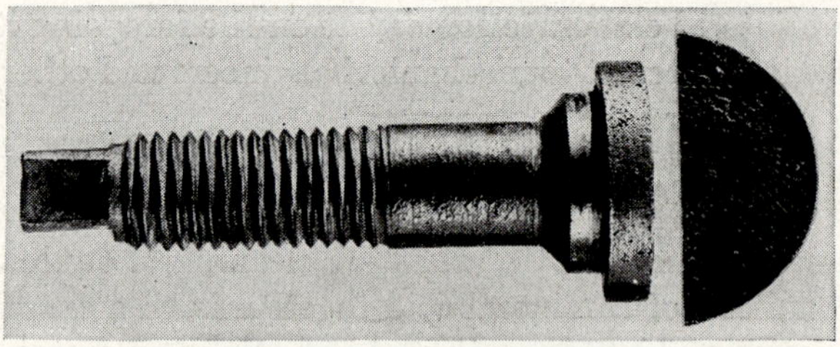

Fig. 3. Corrosion on face and side of cap showing honeycomb type of attack.

evidence that corrosion has occurred in a number of instances where temperatures have been high during cleaning. Many boilers have been cleaned at temperatures ranging from 170 to 180°F. In one case investigated by the author, the temperature was well above the boiling point of the solution. This temperature resulted in considerable damage to the boiler. It is a known fact that the inhibiting effect of most chemicals used for retarding corrosion in acid solutions is rapidly reduced as the temperature is elevated. As the result of corrosion caused by acidizing equipment at too high temperatures, there is now a trend toward maintaining the solution temperature between 150 and 160°F, or lower, during the cleaning period. If sufficient time is allowed for the cleaning, proper removal of scale and deposits can ordinarily be effected at the lower temperatures.

In Fig. 3 is shown the extent of corrosion which occurred on the face and edge of handhole caps in waterwall headers. It is probable that relatively high temperature was responsible for the condition.

EFFECT OF CIRCULATION

Various procedures have been employed which require special equipment for applying the cleaning solution to the boiler. In some instances, the operators merely inject the chemical into the top drum, whereas others prepare the solution in a tank and pump it into the boiler, recirculating from the boiler back to the tank. At present, there is insufficient evidence to prove whether controlled circulation of acid by means of pumps or circulation induced merely by diffusion and thermal currents produces the most satisfactory cleaning of equipment. Both procedures have been followed when acid treating boilers without active corrosion in either case.

Recirculation during cleaning of boilers is not widely used now because of the difficulty experienced in repumping the acid solution and from nonuniform circulation throughout the equipment due to the many parallel passes.

Only centrifugal pumps should be used for pumping the acid solution because of the damage that may result from pocketing and compressing hydrogen in reciprocating pumps. The pumps must be constructed of a special material which depends upon the severity of the service.

EFFECT OF METAL COMPOSITION

The composition and the cleanness of the metal are important factors in retarding or accelerating corrosion during acidizing of equipment. It has been observed repeatedly that low-grade steel containing slag inclusions or other impurities is readily attacked during acid cleaning while drums and tubes are not attacked. In some installations, the tops of all the acorn nuts used in the fabrication of the internal baffles have been destroyed, and similar conditions have been noted when cleaning equipment. To what extent the acid was responsible has not been fully determined, but it undoubtedly accelerated the rate of attack. It is general knowledge, however, that the metal from which such nuts are fabricated is usually not of the best quality.

Ordinarily, black iron pipe, especially if cold-worked, is readily corroded by acid, even under conditions where better grades of metal show no visible attack. The type and extent of the attack are illustrated in

Fig. 4. Nipple corroded during acidizing and illustrating the effect of acid on low-grade steel containing slag and other impurities.

Fig. 4. This shows a nipple used as a spacer in the construction of internal baffles in a boiler which has been acid-cleaned. An identical type of corrosion was observed in a number of other units employing the same grade of material.

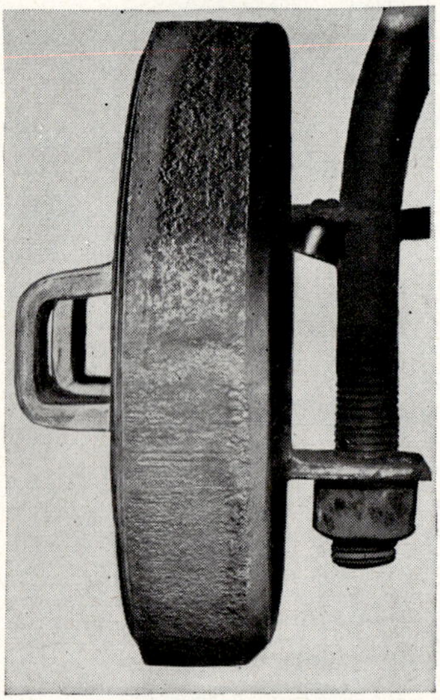

Fig. 5. Corrosion of manhole cover showing deep surface corrosion by acid attack.

The author observed active corrosion of manhole covers in two boilers operating under different conditions. The pattern of the attack is shown in Fig. 5. In one instance, it was indicated that the metal may have been fairly hot when the boiler was acidized. In the second case, the unit was relatively cold, but the feedwater has been contaminated with an organic acid compound prior to the acid treatment. It is believed that the condition of the surface of the metal may have made it more susceptible than other portions of the boiler. Such susceptible conditions are often noted in boilers supplied by different manufacturers.

COMPOSITION OF DEPOSITS

Table 1 shows various types of scale which have deposited in boilers and appurtenant power plant equipment. Frequently such deposits are composed of several compounds, each having specific solubility characteristics. Furthermore, the composition and extent of the deposit may vary

TABLE 1. TYPES OF DEPOSITS WHICH HAVE BEEN FOUND IN BOILERS, AUXILIARIES, AND MISCELLANEOUS POWER PLANT EQUIPMENT

Principal ion	Common name	Mineralogical name	Formula determined by X-ray analysis
Group A. Boiler and superheater			
Barium	Barium carbonate	Witherite	$BaCO_3$
	Barium sulfate	Barite	$BaSO_4$
Calcium	Calcium carbonate	Calcite	β-$CaCO_3$
	Calcium sulfate	Anhydrite	$CaSO_4$
	Calcium sulfate	Hemihydrate	$CaSO_4 \cdot 0.5H_2O$
	Calcium hydroxide	....	$Ca(OH)_2$
	Calcium phosphate	Hydroxyapatite	$Ca_{10}(OH)_2(PO_4)_6$
	Calcium phosphate	Whitlockite	β-$Ca_3P_2O_3$
Magnesium	Magnesium hydroxide	Brucite	$Mg(OH)_2$
Sodium	Sodium sulfate	Thenardite	$Na_2SO_4(V)$
Silicate	Silicon oxide	Quartz	SiO_2
	Sodium-iron-silicate	Acmite	$Na_2O \cdot Fe_2O_3 \cdot 4SiO_2$
	Sodium-alumina-silicate	Analcite	$Na_2O \cdot Al_2O_3 \cdot 4SiO_2 \cdot 2H_2O$
	Sodium-alumina-silicate	Natrolite	$Na_2Al_2Si_3O_{10} \cdot 2H_2O$
	Sodium-alumina-silicate	Noselite	$Na_3Al_6Si_6O_{24} \cdot SO_4$
	Sodium-calcium-silicate	Pectolite	$Na_2O \cdot 4CaO \cdot 6SiO_2 \cdot H_2O$
	Calcium silicate	Wollastonite	β-$CaSiO_3$
	Calcium silicate	Xonotlite	$5CaO_5SiO_2 \cdot H_2O$
	Calcium-alumina-silicate	Cancrinite	$4Na_2O \cdot CaO \cdot 4Al_2O_3 \cdot 2CO_29SiO_2 \cdot 3H_2O$
	Magnesium silicate	Serpentine	$3MgO \cdot 2SiO_2 \cdot 2H_2O$
Iron	Iron oxide	Ferrous oxide	FeO
	Iron oxide	Hematite	Fe_2O_3
	Iron sulfide	Triolite	FeS
Group B. Condenser, evaporator, and heat exchanger			
Calcium	Calcium sulfate	Anhydrite	$CaSO_4$
	Calcium sulfate	Gypsum	$CaSO_4 \cdot 2H_2O$
	Calcium carbonate	Aragonite	$CaCO_3$
	Calcium-sodium-silicate	Pectolite	$4CaO \cdot Na_2O \cdot 6SiO_2 \cdot H_2O$
	Calcium-magnesium-carbonate	Dolomite	$CaCO_3 \cdot MgCO_3$
	Calcium fluoride	Fluorite	CaF_2
Copper	Copper sulfide	....	CuS
	Copper carbonate	Malachite	$CuCO_3 \cdot Cu(OH)_2$
	Copper iron sulfide	....	$CuFeS$
	Copper oxide	Cuprite	Cu_2O
Magnesium	Magnesium silicate	Forsterite	Mg_2SiO_4
	Magnesium hydroxide	....	$Mg(OH)_2$
	Magnesium silicate	Serpentine	$3MgO \cdot 2SiO_2 \cdot 2H_2O$
Iron	Iron oxide	Magnetite	Fe_3O_4
	Iron oxide	Geothite	$Fe_2O_3 \cdot H_2O$
Sodium	Sodium-alumina-silicate	Noselite	$Na_3Al_6Si_6O_{24} \cdot SO_4$

TABLE 1. TYPES OF DEPOSITS WHICH HAVE BEEN FOUND IN BOILERS, AUXILIARIES, AND MISCELLANEOUS POWER PLANT EQUIPMENT.—(*Continued*)

Principal ion	Common name	Mineralogical name	Formula determined by X-ray analysis
Group C. Steam and condensate line			
Ammonia....	Ammonium bicarbonate	Teschemacherite	NH_4HCO_3
Iron.........	Iron carbonate	Siderite	$FeCO_3$
Sodium......	Sodium sulfate	....	Na_2SO_4(III)
Group D. Miscellaneous power plant equipment			
Calcium.....	Calcium carbonate	Calcite	β-$CaCO_3$
	Calcium carbonate	Aragonite	λ-$CaCO_3$
	Calcium aluminate	....	$3CaO \cdot Al_2O_3 \cdot 6H_2O$
	Calcium phosphate	Phosphate	$Ca_3P_2O_3 \cdot H_2O$
Magensium..	Magnesium hydroxide	Brucite	$Mg(OH)_2$
Group E. Turbine			
Copper......	Copper sulfite	Chalcocite	Cu_2S
Nickel.......	Nickel oxide	Bunsenite	NiO
Silicate......	Silicon oxide	Quartz	SiO_2
	Silicon oxide	Cristobalite	SiO_2
	Silicon oxide	Opal	SiO_2
Sodium......	Sodium sulfate	Thenardite	Na_2SO_4(V)
	Sodium-carbonate-sulfate	Burkeite	$Na_2CO_3 \cdot 2Na_2SO_4$
	Sodium chloride	Halite	$NaCl$

widely in different parts of the same equipment. This condition can be demonstrated by analysis of material removed from tubes in two sections of the same unit.

COMPOSITION OF INHIBITOR

The technical literature on the subject of acid cleaning and pickling of metal contains references to many organic and inorganic substances which have been used as inhibiting agents to repress the attack on the metal but permit the solution or removal of oxides or other deposits from the surface. A partial list of some of the materials which have been used is shown in Table 2. This table includes only a small percentage of the compounds which are available or have been employed. Actually, there are a great number of organic and inorganic substances which are known to have inhibiting properties. The relative effectiveness of these products varies widely, and they have specific adaptability for different services. Most inhibitors are harmless, but some are hazardous and should not be used. In this group are compounds of arsenic, cyanide, and sulfide and some of the coal-tar by-products. At least one fatality has occurred from the use of an arsenic inhibitor. Operators should be warned against the use of such materials and informed of their potential danger.

TABLE 2. PARTIAL LIST OF MATERIALS USED OR INVESTIGATED FOR INHIBITING CORROSION OF METALS BY ACIDS

Niter cake	Aniline
Sumac leaves	Pyridine
Formaldehyde	Coal-tar extracts
Glue	Slaughterhouse wastes
Flour	Ethyl iodide
Starch	Paramethylaminophenyl sulfate
Gelatin	Carbozole
Glucose	Cyanide
Thiourea	Quinoline ethiodide
Arsenious oxide	Barium salts

NOTE: The inhibiting effect of the compounds listed and of many other substances varies greatly, as does their cost and availability. Some of these materials possess slight inhibiting value, and some are dangerous to use. Many inhibitors are sold under trade names, and their composition is not readily obtainable. No inhibitor should be employed without specific information as to its effectiveness and the potential hazard involved in its use.

FUNCTION OF INHIBITORS

There are two general concepts regarding the mechanism by which inhibitors protect steel or other metals when in contact with acid solutions. One group of investigators[5] suggests the following action: iron goes into solution at the anodic regions, forming ions, while hydrogen is discharged in equivalent amount at the cathodic areas. Since most inhibitors are bases or positively charged colloids they travel to the cathodic areas and are deposited and adsorbed on the surface. The electrochemical process will stop when the reaction at either pole is stifled, a deposit of the inhibitor on the cathode area will do just that. It will stifle the reaction and retard corrosion. It has also been claimed that the presence of the inhibitor raises the hydrogen overvoltage above any potential that can exist in the acid-metal system.

A second explanation is that the inhibiting compounds serve to blanket the entire metal surface with a protective layer of the organic compound used. This occurs as simple adsorption of the inhibitor ions on the metal surface. The nitrogen atom, a component in most organic inhibitors, has a considerable influence on how this adsorption proceeds.

It has been claimed that the action of inhibitors can be explained neither by the electrochemical and overvoltage theories nor by the formation of an impervious film on the oxide-free metal surface. As proof, test results are cited. They indicate that inhibitors are adsorbed simultaneously on both metal and oxide surfaces. However, the formation is heavier on the clean metal surface. The film so formed increases the electrical resistance from liquid to metal, but this film is porous, being built up of molecule aggregates, or micelles, spaced closely together as in a pincushion filled with pins. The theory developed from these studies is that the long-chain molecules thus form long, narrow channels similar to capillaries or gel structures. This condition is reflected in reduced diffusion and migration velocities of the hydrogen and the acid-radical ions (in the case of sulfuric acid, the acid-radical ion is sulfate), which are active in the corrosion of iron in an acid solution. The net result is a reduction in the corrosion of the metal, and inasmuch as a more concentrated film is present at the clean metal surface, the oxide is attacked faster than the metal. This results in the removal of the oxide and preservation of the metal surface.

It has been contended that the corrosion of some boilers during cleaning with inhibited hydrochloric acid is due largely to the formation of ferric chloride. This, in itself, is a highly aggressive salt against which common inhibitors are not effective. According to this theory,[5] the inhibitors will retard the electrochemical reaction involving the displacement of the hydrogen from HCl by iron, but they will not prevent the oxidation-reduction reactions by which ferric chloride oxidizes metallic iron and copper to the ionic form. If metallic copper is in the sludge, ferric chloride attacks it as it does iron, producing cupric chloride, which is also an oxidizing agent and a secondary source of trouble. It is held that the plating out of copper from solution to form a chloride depends upon concentrations of the various ions present.

To limit the concentration of ferric chloride, it has been suggested that the solution be constantly circulated, thereby avoiding raising the ferric chloride concentration at localized areas to a level where active attack is possible. There is no assurance, however, that circulation will prevent corrosion in this type of treatment. If ferric chloride pockets or otherwise concentrates at surfaces and remains in a quiescent or semiquiescent state, it may react more slowly if only diffusion is available to carry away and remove the products of the reactions.

To support the foregoing theory, corrosion has been produced experimentally with dilute solutions of ferric chloride containing no hydrochloric acid except that resulting from hydrolysis.

It should be pointed out that, if inhibitors retard corrosion by covering the entire surface and preventing contact with the acid, they should be equally effective against solutions of ferric or cupric chloride. If the impervious-film theory of inhibitor mechanism is correct, any specific vulnerability of the film to certain solutions would seem doubtful. But if the film leaves certain areas exposed, they could be attacked by oxidizing agents in the absence of an opposite electrode reaction. It is a known fact that inhibitors do not give complete protection, and the actual overall attack which occurs may, no doubt, be aggravated by oxidizing agents. Whatever the role of ferric chloride and copper may be, it is true that, with the proper care, equipment can be acidized repeatedly without hazard.

Hall and Kaufman[6] have presented the following interesting theory regarding corrosion and inhibition of acid attack on metals:

Acid used for the process is supplied with an inhibitor, that is, a substance whose purpose is to minimize or prevent attack of the acid on the boiler steel, while not interferring with its dissolving effect on other substances, such as the calcium carbonate and phosphate, the iron oxide, etc., comprising the boiler sludge and scale. . . . Thus one can depend upon the inhibitor to protect against direct attack of the metal by the acid . . . if the products of reaction of the acid with the boiler sludge or scale are not of themselves potent in their attack on the metal, regardless of the inhibitor.

As a matter of fact, these products of reaction divide into two classes, one of which is innocuous as regards the boiler metal; the other of which is highly aggressive in attack on the metal, even with any usual inhibitor present in the acid solution. Materials which give rise to the first class comprise the calcium and magnesium salts of the boiler sludges and scales; those giving rise to the second class are the iron oxides, either magnetic or hematite. Whether formed in the feedline by the action of dissolved oxygen and thence entering the boiler, or whether formed in the boiler itself by reaction of boiler metal with the bonded oxygen of the boiler water, these oxides represent high oxidizing capacity. Concentrated bit by bit as the boiler is operated and held harmless by the insolubility of the oxides in the boiler water, they await only the trigger action of the acid solution which dissolves them to bring this concentrated oxidizing capacity into play against the boiler metal. Metallic copper, so frequently associated with the iron oxides, may enter into but cannot initiate the oxidation reactions, nor can it contribute thereto. In fact, by plating out on the metal it may even be protective in some measure. . . .

Fundamentally the aggressiveness of solutions containing ferric ion can be explained on the basis of the following reaction:

$$2Fe^{+++} + Fe \rightleftharpoons 3Fe^{++} \tag{1}$$

It will be noted that this process does not intrinsically involve the disappearance or formation of an acid or a base. Therefore, it might be presumed that the inhibitors customarily employed to stifle acid attack on steel would not be well adapted to minimize the reaction of ferric ion with metal. Laboratory tests have confirmed this point. . . .

In the removal of deposits from boilers by acid-cleaning, when magnetic iron oxide, for example, goes into the circulating acid, what quantitative potential has it for producing damage? If it is assumed that 1000 gallons of cleaning acid builds up to contain (homogeneously distributed in solution) 1 per cent or 83.3 pounds of material originally Fe_3O_4, twenty (20) pounds of iron can be dissolved if Reaction (1) goes to completion. If, for example, the dissolution of iron is assumed to take place evenly within three inch tubing, the general loss of metal will amount to less than 0.001 inch. If, however, the ferric ion is not homogeneously distributed in the acid solution, but rather is selectively concentrated in localized areas rich in magnetic oxide, then serious pitting or wastage in such areas may be expected.

Avoidance of such corrosion we believe is best effected by steadily maintained circulation of the acid-cleaning fluid so that localized high concentrations of ferric chloride cannot occur, and by limiting concentration of ferric chloride in the acid-washing solution through adequate blowdown and replenishment of acid solution, thus protecting against too much loss of metal in general corrosion of the surfaces by even the homogeneously distributed ferric chloride which develops mainly at the surfaces or in particular spots in the boiler. Satisfactory control of ferric ion concentration can be established by analysis.

PHOSPHORIC ACID CLEANING AND NITROGEN BLANKETING

Until quite recently practically all acid cleaning of boilers was effected by the use of inhibited hydrochloric acid. Frequently wetting agents and ammonium bifluoride were added to the hydrochloric acid to penetrate scale deposits or to remove adherent siliceous formations. Hydrochloric acid, along with the various constituents added, has been effective in removing deposits; however, this type of inhibited acid has certain limitations. As previously pointed out, the temperature at which the acid cleaning is performed must be limited to a maximum of about 160°F. Further, this acid gives off corrosive fumes and objectionable aftercorrosion takes place on the clean metal surfaces upon exposure to air.

In order to avoid these difficulties, the following measures have been taken:

1. Nitrogen blanketing of the internal surfaces of the boiler during draining subsequent to hydrochloric acid cleaning.

2. The use of inhibited phosphoric acid in place of inhibited hydrochloric acid.

3. A combination of methods 1 and 2, involving the acid cleaning with inhibited hydrochloric acid, nitrogen blanketing during the draining of the boiler, and a final treatment with very dilute phosphoric acid to prevent aftercorrosion.

The procedure for employing inhibited phosphoric acid, which was successfully used at the Duquesne Light Company, is described by Purcell and Whirl.[7] Nitrogen blanketing was utilized effectively by Call and Webb[8] during chemical cleaning of boilers to remove copper deposits. *

Dowell, Inc., has developed a technique which employs both nitrogen blanketing and phosphoric acid to prevent or minimize aftercorrosion. The boiler is acid-cleaned in the usual manner with inhibited hydrochloric

* Nitrogen blanketing is now used by many companies to inhibit rusting of equipment after acid cleaning.

acid; during draining, bottled nitrogen is delivered through a vent at the top of the boiler to displace the spent acid. The residual acid adhering to the boiler metal is partially removed by filling the boiler with clean hot water through a blowdown connection. The nitrogen is dispelled through the vents on the top of the boiler. Again the boiler is drained, and nitrogen is allowed to displace the flushing water. A solution of 0.15 per cent phosphoric acid is then used to fill the boiler and displace the nitrogen. After draining out the phosphoric acid solution, any acid remaining in the boiler is neutralized by use of an alkaline boilout using trisodium phosphate.

In commenting on their experience with phosphoric acid cleaning, Purcell and Whirl[7] state:

> The principal advantage of phosphoric over hydrochloric acid is that it can be boiled in the unit by direct firing of the boiler with negligible attack of the metal. This is possible because the inhibitor is effective at these temperatures. In addition to the beneficial washing action, the resulting natural circulation promotes distribution and provides a sufficiency of acid at the desirable locations at all times. Furthermore, the boiling operation sends solutions into certain regions at the top of the boiler which otherwise could not be reached except by recirculation with an elaborate system of internal baffling.

Purcell and Whirl further point out that hydrochloric acid has no properties of passivating or forming films over the metal surfaces, whereas phosphoric acid definitely passivates the metal and avoids the objectionable rusting of the metal after treatment.

For a time, phosphoric acid appeared to be preferable to hydrochloric acid for cleaning boilers, regardless of the higher cost. More experience with phosphoric acid cleaning, however, has indicated that there are a number of limitations to its use, and the treatment now appears to be less desirable than hydrochloric acid.

DISPOSAL OF SPENT ACID SOLUTION

With regard to discharge of the chemical solution after cleaning, operators should give special attention to the manner in which this procedure is carried out. There are many laws and ordinances which prohibit the discharge of highly concentrated acids or alkalies into city sewers. Likewise, some laws require that such materials be neutralized before they are discharged into streams. If neutralization is impractical, it is sometimes possible to secure a permit for disposal by discharging the waste liquors into holding tanks and then releasing them at a slow rate of discharge so that, when mixed with other waste waters, no damage to city sewers or surface-water streams will occur. This latter restriction is one of considerable importance and is receiving more and more attention from regulatory groups throughout the country.

HAZARDS TO PERSONNEL AND EQUIPMENT

Since mineral acids are corrosive to most metals and hazardous to personnel, especially at high concentrations, it is obvious that care should be exercised to protect both the persons performing the work and the equipment being cleaned. If the treatment is carried out by experienced operators who are cognizant of the dangers involved and who take the necessary precautionary measures, the hazards involved should be negligible. Aside from the danger of the corrosive characteristics of the acid, the treatment involves a definite hazard from fire and explosion if the hydrogen generated is not effectively removed and dissipated. Measures should be taken to prevent the discharge of hydrogen into a confined space. Extreme care should also be used to prevent hydrogen from pocketing in any equipment which is being cleaned by acid. Open flames and sparks should not be allowed near equipment that is being acidized.

How many accidents have occurred from explosion of hydrogen during, or subsequent to, acid cleaning is not known, but some accidents of this kind have been reported. In one instance, workmen were killed and extensive damage resulted from an explosion during the cleaning of a boiler. The accident was caused by failure to vent the generated gas from the boiler drum. The hydrogen was ignited by a spark from a flashlight when the boiler was opened for inspection. This accident demonstrates the extreme danger in acid cleaning by persons not thoroughly informed of the reactions which occur and of the hazards prevalent when acid is in contact with steel. Other explosion accidents from ignition of accumulated hydrogen gas have been reported. Although the explosions were not fatal to personnel, appreciable damage to equipment occurred.

An operator cleaning a boiler with inhibited acid was killed by arsenious oxide released from the inhibitor, which contained an arsenic compound. The use of toxic inhibitors should be prohibited even though they are quite effective for retarding corrosion.

MOBILE ACIDIZING EQUIPMENT

There are companies which specialize in acid cleaning of equipment. It was observed earlier that it is desirable to employ the services of such companies especially when complex equipment is to be acidized. The general type of mobile unit used for transporting and delivering acid to the equipment to be cleaned is shown in Fig. 6.

REMOVAL OF TURBINE-BLADE DEPOSITS

Deposition of solids on the blading of steam turbines has been an operating problem for many years. Prior to the development of high-pressure boilers, these solids, which would carry over with the steam, had consisted for the most part of ordinary boiler-water salt. To a large extent, this problem was completely eliminated or conveniently controlled by improvements in boiler design and by the development of effective steam scrubbers. With the advent of high-pressure boilers and turbines, new types of deposit resulting from steam contamination were encountered. Usually siliceous deposits are the most prevalent. Some of these are more water-soluble than others and can be removed when the condensate contacts the blading. The insoluble type, found most frequently

Fig. 6. Equipment employed for acid cleaning showing the storage tank, circulating pumps, heater, and other appliances. (*Courtesy of Dowell, Inc.*)

in the lower-pressure stages of high-pressure turbines, is the most difficult to remove.

Extensive research has been conducted in an effort to determine ways and means for controlling the deposits, principally those of silica, but up to the present time, no totally adequate control procedure has been developed. It is obvious, however, that so conditioning the water as to eliminate all silica completely and preventing contamination by condenser leakage would be a satisfactory solution, but such control is almost impossible of attainment. Straub and Grabowski[9] reached the following conclusion after extensive experimentation and experience:

> The plant tests have indicated that when the silica is kept below 5 ppm. and the pH of the boiler water around 11.0, no appreciable silica deposits form in turbines. When the silica in the steam becomes greater than 0.1 ppm., deposits will form. . . . Studies have been made in the laboratory relative to reducing the soluble silica in boiler water by the addition of chemicals which will precipitate the silica as an insoluble sludge. . . . The addition of the magnesium reduced the soluble silica from 82.5 ppm. to around 2 ppm.

Although the maintenance of the conditions as set forth above will greatly limit the rate of build-up of silica deposit and, in some cases, totally control the formation of silica scale, such a scale has been known to accumulate even when silica in very low concentrations has been present in the boiler waters. The accumulation of deposits on turbine blading reduces the area of the steam passages and may also reduce the load-carrying capacity of the unit. Other serious operating difficulties may develop. Warren and Howard[10] have stated these dangers concisely and are quoted herewith verbatim:

The hazards associated with the formation of deposits are sometimes not fully understood. Often the operators are governed entirely by the loss in unit capacity and when that becomes great enough to make action necessary, corrective steps are taken. This criterion may be safely used only when the deposit formations are uniform throughout all stages of the turbine and are about of equal magnitude on the rotating and stationary elements. Perhaps the most common situation is for the deposits to collect in some particular location within the turbine. Such deposits have a marked detrimental effect on turbine efficiency as well as capacity. Extensive non-uniform plugging can result in thrust failure, wheel rubbing, and other serious troubles.

A similar warning has been given by Buckland.[11]

Where the deposits become sufficiently heavy to cause marked loss of capacity, the turbine should be cleaned. This can be done by removing the casing and cleaning by (1) mechanical means, (2) washing with water, and (3) caustic washing.

It is obvious that dismantling a turbine is very costly and is to be avoided except where it is imperative to do so for mechanical failure or for scheduled inspections.

Where the accumulated deposits are soluble in water, there are several procedures which may be used. Warren and Howard[10] have given instructions for procedures to remove such deposits. They bring to the reader's attention that water-soluble materials are often washed out of the unit following a short idle period of two or three days. Reduction of temperature due to reduced load frequently advances the moisture region nearer the inlet and produces a washing effect over the area where deposits form under the higher load conditions. They describe another method whereby atomization and injection of water into the steam supply is so controlled as to insure a gradual and uniform reduction in the temperature of the turbine-inlet steam. As the steam approaches saturation, the contained moisture washes soluble deposits into the condenser.

In modern, high-pressure boiler-turbine-generator units, it is possible to lower the pressure at the throttle and thus to effect some washing. Although water-soluble materials can be removed by the methods discussed, silica deposits are generally difficult to wash from the turbine, and in such cases, it may be necessary to rely upon a caustic soda wash.

There are certain hazards involved in this type of operation, and the reader is referred to Warren and Howard's article for details regarding the proper procedure and suggested washing cycle. It is important that the precautions outlined be followed if future difficulties with the turbine are to be avoided.

Bibliography

Specific References:

1. Powell, Sheppard T.: Acid Cleaning of Boilers and Auxiliary Equipment, *Trans. ASME,* **68**: 905 (1946).
2. Farmer, H. C.: Investigation of Acid Attack on Boilers and the Effect of Repeated Acid Cleaning on the Metal, *Trans. ASME,* **69**: 405 (1947).
3. Alquist, F. D., C. H. Groom, and G. F. Williams: Chemical Removal of Scale from Heat Exchange Equipment, *Trans. ASME,* **65**: 719 (1943).
4. Brines, M. E.: Scale Removal by Chemical Means, *Power Plant Eng.,* **44** (5): 47 (1940).
5. Chappell, E. L., B. E. Roetheli, and B. Y. McCarthy: The Electrochemical Action of Inhibitors in the Acid Solution of Steel and Iron, *Ind. Eng. Chem.,* **20**: 582 (1928).
6. Hall, R. E., and C. E. Kaufman: Discussion of paper, Acid Cleaning of Boilers and Auxiliary Equipment, by Sheppard T. Powell, *Trans. ASME,* **68**: 912 (1946).
7. Purcell, T. E., and W. F. Whirl: Phosphoric Acid for Cleaning High-pressure Boilers, *Power Eng.,* **54** (8): 72 (1950).
8. Call, R. G., and W. L. Webb: Chemical Removal of Copper from Boilers, Paper 50-SA-34, presented at the American Society of Mechanical Engineers Semi-annual Meeting, 1950, and abstracted in *Mech. Eng.,* **72**: 746 (1950).
9. Straub, F. G., and H. A. Grabowski: Silica Deposition in Steam Turbines, *Trans. ASME,* **67**: 309 (1945).
10. Warren, G. G., and T. W. Howard: The Removal of Deposits from Steam-turbine Steam Passages, *Trans. ASME,* **69**: 213 (April, 1947).
11. Buckland, B. O.: Effect of Nozzle and Bucket Deposits on Turbine Shell Pressures, Capacity and Efficiency, *Power Plant Eng.,* **46** (9): 52 (1942).

General References:

Speller, F. N., E. L. Chappell, and R. P. Russell: Removal of Rust from Pipe Systems by an Acid Solvent, *Trans. Am. Inst. Chem. Engrs.,* **19**: 165 (1927).

Wasco, J. L., and F. N. Alquist: Chemical Removal of Calcium Sulfate Scale, *Ind. Eng. Chem.,* **38**: 394 (1946).

Straub, F. G.: Cause and Prevention of Turbine Blade Deposits, FSP-57-14, *Trans. ASME,* **57**: 447 (December, 1935).

Feller, E. W., and G. F. Williams: Chemical Cleaning, *Power,* **90**: 586, 659 (1946).

22 DECONCENTRATION OF SOLIDS IN BOILER WATER

Concentration of soluble and suspended solids in boiler waters must be controlled to prevent steam contamination, scale, and heavy sludge deposits. All manufacturers of modern boilers specify the upper limit of the concentration of the total solids required to meet the guarantee of steam quality. The term "steam quality" as used here refers to entrained solids rather than to percentage of moisture carried by the steam. With modern steam-scrubbing equipment built into the boiler drums, a maximum of 1 ppm is guaranteed by boiler manufacturers.*

In addition to limiting total dissolved solids, it is often necessary to control total alkalinity, sulfates, silica, and other constituents; thus the quantity of water blown down may be governed specifically by any of the constituents present in the water. It is not uncommon to blow down solely for maintaining a maximum permissible concentration of silica. This is especially necessary in high-pressure boilers where a limit of 3 to 5 ppm of silica is sometimes the safe maximum necessary to avoid siliceous turbine deposits.

The term "continuous blowdown," or "blowoff," is used to describe the process of continuously discharging a small stream of concentrated boiler water from boilers for removing suspended and dissolved solids which have concentrated during operation. Boiler water is withdrawn at a slow, constant rate; it is, therefore, possible to regulate and control boiler-saline concentration more uniformly than by intermittently blowing down with conventional, manual blowdown valves.

It is required by law that all boilers be equipped with blowdown valves to permit blowing down intermittently. Continuous blowing down is not required by code regulations but has been incorporated in recent boiler

* Limits of concentration of boiler-water solids required to meet guaranteed steam quality are established by the American Boiler Manufacturers' Association and are given in Chap. 14.

design for more effective and economic regulation of boiler-water concentrations; however, not all high-, medium-, and low-pressure units are so equipped.

Water blown from boilers may be wasted with or without heat recovery. In high-pressure installations and wherever steam or hot water can be economically utilized, equipment for such recovery is provided. All designers of modern steam-generating stations recognize the importance of this factor in heat-balance studies; therefore, heating equipment is adjusted to conserve, as far as possible, available Btu from this source.

A number of ingenious devices incorporated into continuous blowdowns are now on the market. These have varying degrees of merit. In low-pressure units, the blowdown water is merely flashed, and the steam is delivered to heaters or, in some plants, to hot-process reaction tanks. At both high- and low-pressure stations, the Btu available from this water is often used for feedwater heating in accordance with a scheme which fits into the heat balance of the station.

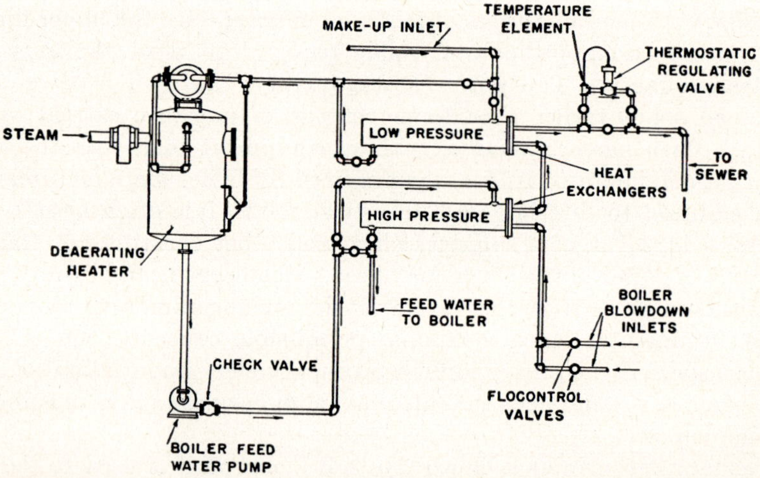

Fig. 1. Coil-type heat exchangers with automatic blowoff control. (*Courtesy of Cochrane Corporation.*)

One arrangement is essentially a continuous blowdown system in which the heat of the blowdown is utilized in an apparatus which is practically a two-stage heater. The outlet from the feedwater heater (Fig. 1) is passed through coils in the high-temperature stage of the blowdown heat exchanger; this imparts to the feedwater a portion of the heat from the blowdown water passing over the coils. The temperature of the cold feedwater is raised several degrees before discharging into the boiler. Partially cooled blowdown water passes from the high-temperature stage of the heater to the low-temperature stage. Temperature of the cold-water make-up is raised by passing over the coils in this second stage.

The available heat in the blowdown water is thus effectively recovered.

Amount of water blown down will vary with the condition of the raw-water make-up, the operation of the boiler, and other conditions. The apparatus may be installed wherever it is convenient to the piping system and boilers. It is equipped with a thermostatic control valve and a special meter. It is necessary to clean the apparatus from time to time to remove the solids which have been deposited from the blowdown water.

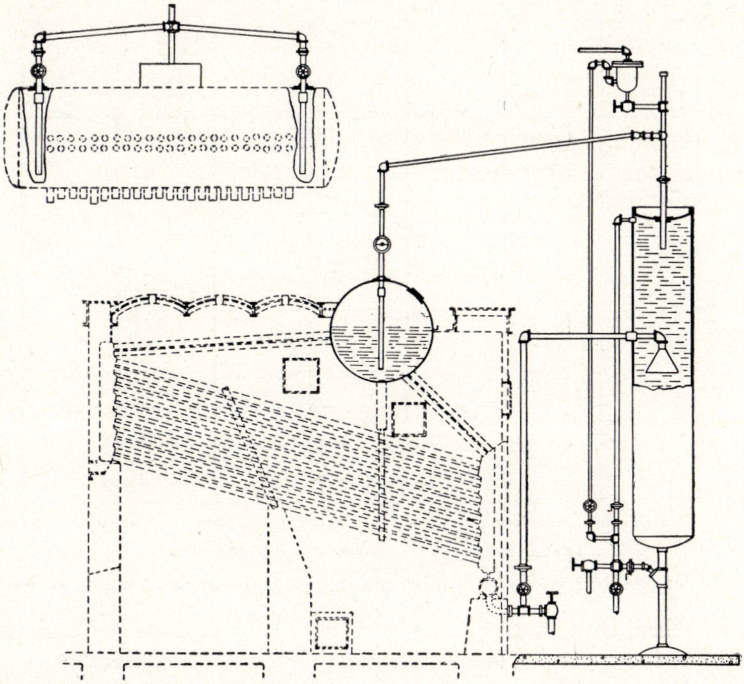

Fig. 2. Blowdown system providing for heat recovery. *(Courtesy of Elgin Softener Corporation.)*

Frequency of cleaning will depend upon the character of the raw make-up water and the rate of operation of the boilers.

DECONCENTRATORS

A popular type of deconcentrator for low-pressure boilers provides for the discharge of the blowdown water to a tank in which suspended solids can settle and from which the partially clarified water is returned to the boiler.

Another system applicable to low-pressure boilers combines internal treatment and continuous blowing down. The apparatus consists of an outside, closed subsidence tank in which the insoluble solids from the blowdown water are removed and from which the partially clarified water is returned to the boiler. By this method, scale and sludge formation can be retarded; some economy is effected by saving heat of the blowdown water. A typical installation of this kind is shown in Fig. 2.

RECOVERY OF HEAT FROM BOILER BLOWDOWN WATER

Where the necessary amount of boiler blowdown is appreciable, heat is recovered in the form of flashed steam or by the use of a heat exchanger in the feedwater cycle or by both methods. The fraction of the blowdown

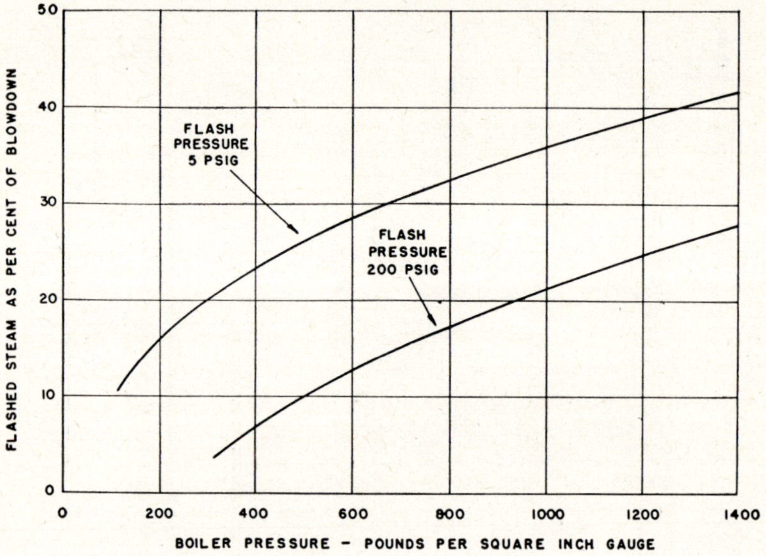

Fig. 3. Per cent of blowdown flashed into steam by reduction of boiler pressure.

which will flash into steam is shown in Fig. 3 for various boiler pressures from 100 to 1,000 psig. The continuous blowdown is delivered to a flash tank which is usually vented to the deaerating heater.

As shown in Fig. 4, a large fraction of the total heat in the blowdown is recoverable in this manner. The residual water is usually discharged to the sewer unless the total amount is relatively large.

In high-pressure boiler plants, 900 psi to 2,500 psi, two or more flash tanks are used to reduce the temperature stepwise and to recover the heat of the continuous blowdown. The flashed steam is piped to low-pressure headers. A typical arrangement for such a conventional system is shown in Fig. 5.

CALCULATION FOR DETERMINING THE AMOUNT OF BLOWDOWN REQUIRED

In any type of evaporating system, the solids in the feedwater become concentrated in the circulating water. This occurs in cooling towers and evaporative condensers as well as in boilers.

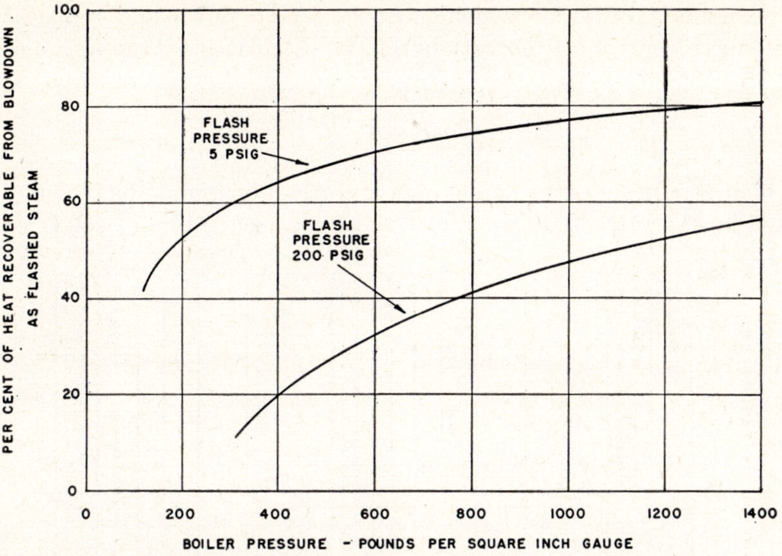

Fig. 4. Per cent of heat recoverable from blowdown as flashed steam.

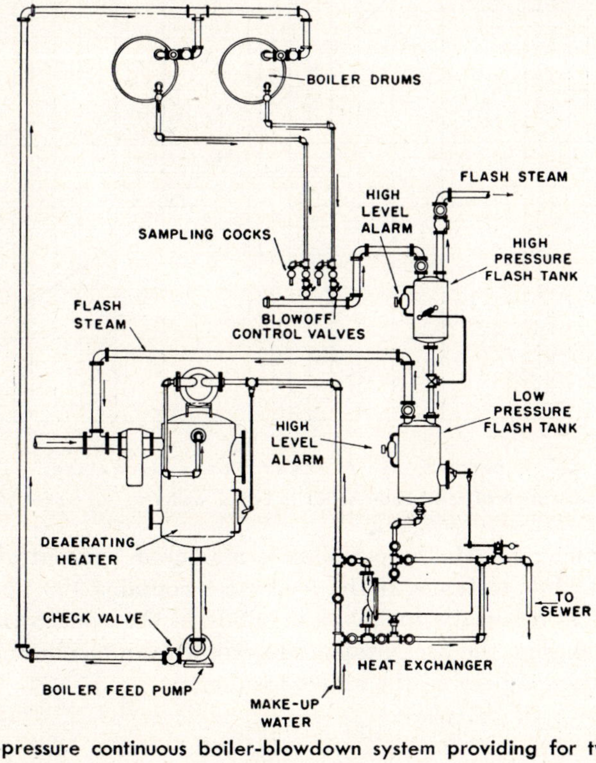

Fig. 5. High-pressure continuous boiler-blowdown system providing for two-stage heat recovery.

The amount of water which must be blown down and replaced by make-up to control concentrations can be calculated by the formula

$$B = \frac{F}{M - F} \times 100$$

in which B represents the percentage of water which must be blown down (based on the rate of evaporation); F, the concentration of solids in the

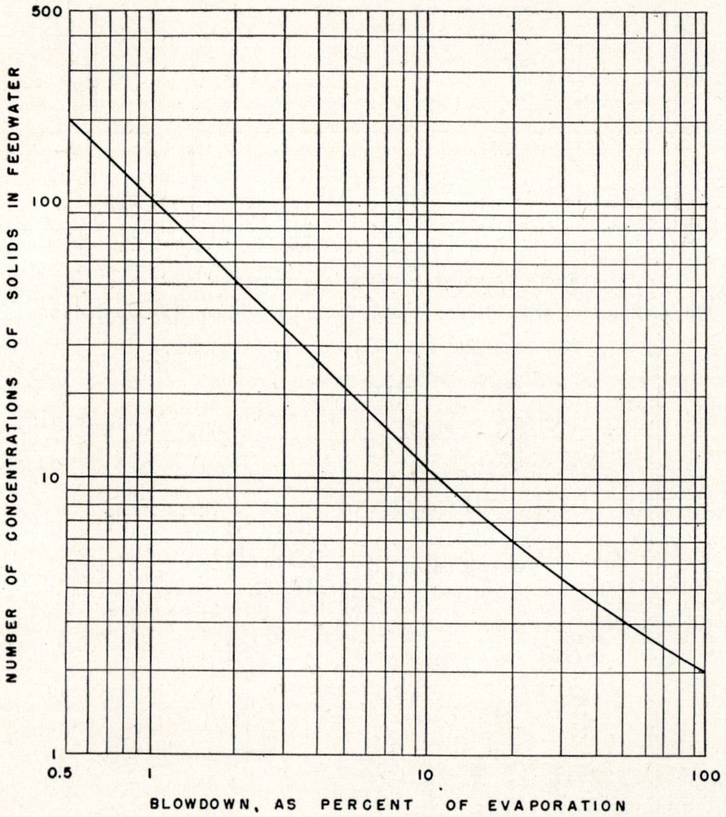

Fig. 6. Concentration of feedwater solids in boiler water versus boiler blowdown.

boiler feedwater; and M, the permissible maximum concentration in the boiler water. For instance, if the feedwater contains 100 ppm of total solids and it is necessary to hold the solids in the concentrated boiler water at 1,500 ppm, the percentage of blowdown can be obtained by substitution of these figures in the above formula

$$B = \frac{100 \times 100}{1{,}500 - 100} = 7.15$$

The number of times that the feedwater solids will be concentrated in the boiler in this case is obviously 15, but the same information can be calculated for any given percentage of blowdown without knowing the total solids. If C is the number of times the feedwater solids are concentrated, then

$$C = \frac{100 + B}{B}$$

In the above example

$$C = \frac{107.15}{7.15} = 15$$

In making these calculations, it is necessary to bear in mind that the total solids in the feedwater are not necessarily the same as the total solids in the make-up but that the latter will have to be multiplied by the percentage of make-up used. Actually, "100 per cent make-up" is more or less a figure of speech, since in all cases of direct-contact heating, steam is condensed, which dilutes the soluble salts.

Figure 6 is a graph showing the amounts of concentration of feedwater solids in the boiler water for varying rates of blowdown.

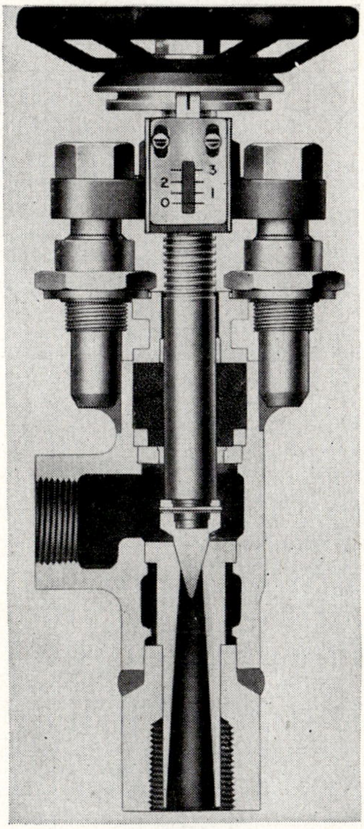

Fig. 7. Calibrated high-pressure orifice—and—plug valve for boiler-blowdown control (1,500 psig.) (*Courtesy of Hancock Valve Division, Manning, Maxwell & Moore, Inc.*)

REGULATION OF BLOWDOWN BY FLOW CONTROL VALVES

In practically all blowdown systems, whether they are of the specialized or conventional type, flow control valves are essential in establishing the volume of water which passes from the boiler through the heat-reclaiming apparatus. These valves are designed to minimize plugging, resist high temperatures, and withstand the erosive and corrosive action of the blowdown. Design of the blowdown flow control valves has become quite specialized. Companies offering such valves supply charts or nomographs from which it is possible to select the size of valve required to deliver the volume of blowdown necessary to deconcentrate the boilers at different pressures.

If an appreciable back pressure is imposed on the downstream side of the valves, a correction factor must be made in calculating the pressure drop across the valve.

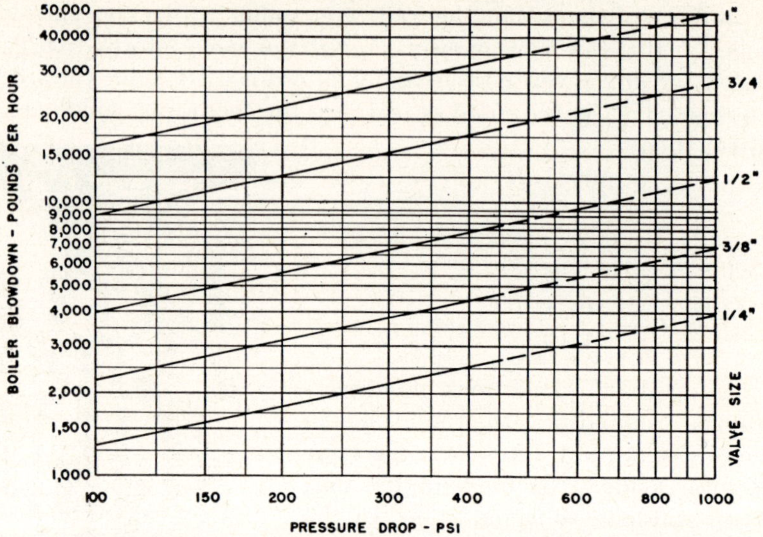

Fig. 8. Curves for obtaining capacity of continuous-blowdown valves in wide-open position, V-port type.

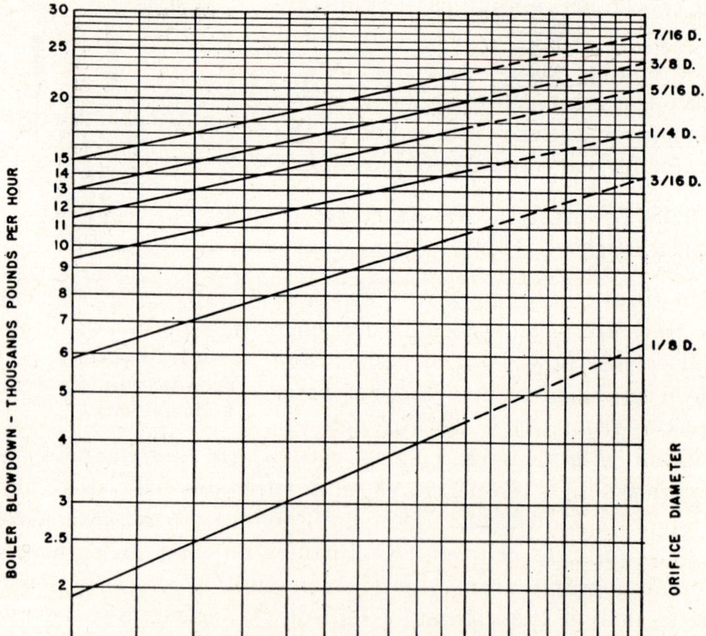

Fig. 9. Curves for indicating capacity of continuous-blowdown valves in wide-open position, orifice and tapered-plug type.

Two types of valve have been used for continuous blowdown control: one employs two or four V ports; the other, a fixed orifice with a tapered plug, as shown in Fig. 7. The discharge from the V-port valve varies directly with the number of turns beyond the "just opened" position. The orifice-type valve does not have a straight-line discharge curve but is more suitable for higher pressures where erosion and wire drawing are troublesome. Figures 8 and 9 show the maximum capacities of both types of valve at various pressure drops.

USE OF HEAT EXCHANGERS FOLLOWED BY ORIFICE BOXES TO CONTROL FLOW OF COOLED BLOWDOWN

This system for controlling continuous blowdown is shown in Fig. 10. It is applicable to both high- and low-pressure boilers and is capable of

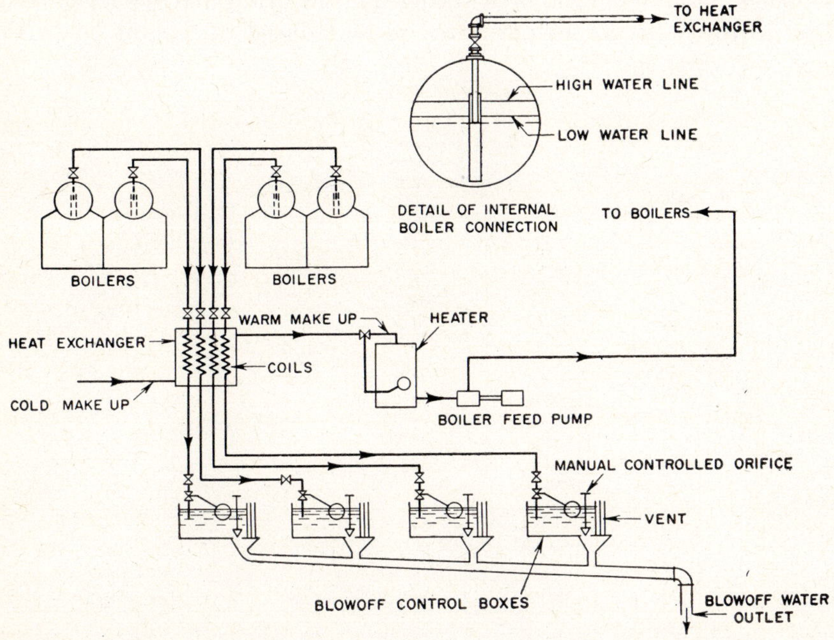

Fig. 10. Arrangement for the continuous blowing down of boilers utilizing heat from blowdown water for feedwater heating.

many modifications to conform to permissible heat-recovery use, as may fit into the heat balance of a station. With this arrangement, the water blown from the boiler is utilized to heat the feedwater.

A continuous stream of the concentrated boiler water is discharged at a fairly constant rate through the heat exchanger. From here it passes through the blowoff control box and thence into the sewer. The heat exchanger, which is practically a closed heater, consists of a cast-iron shell with bolted flanged heads through which helical copper coils pass. Leakage is prevented by stuffing glands into the heads of the heat exchanger. There are individual coils for each boiler and a control or weir box connected with each coil. By this means, the water blown from each boiler may be controlled independently of the other units. The blowoff-control weir box consists of an open steel tank with an adjustable opening in the bottom of the tank. A constant head of water in the box over the weir opening is maintained by a float valve on the inlet side, which discharges boiler water from the coil of the heat exchanger.

Piping is so arranged that the water may be drawn either from the surface, as a surface blowdown, or from the lower portion of the boiler. In the latter case, the continuous blowdown lines are connected into the main blowoff. Water is drawn from the upper portion of each drum of the boiler, and the blowoff is so arranged in the drum that the level of the water in the boiler is not permitted to be reduced to a point below the usual low-water line.

23 MISCELLANEOUS TREATMENTS

There are now available many forms of feedwater treatment which, if intelligently selected to conform to specific boiler plant conditions, will insure satisfactory results. Regardless of the many reliable processing methods from which to select, ineffective and, in some cases, dangerous treatments are extensively used. A thriving business is done through vague and unethical advertising methods and sales promotion based on pseudoscientific claims by salesmen and advertising literature. As a general criterion, no treatment that the seller surrounds with secrecy should be used. No material should be used whose composition the seller is unwilling to divulge completely. No process or equipment offered should be used whose operating details cannot be revealed. The author has investigated many processes which have been proved to be worthless or pure frauds.

These comments will have the support of all informed persons and should be taken at face value by all who are responsible for the selection of feedwater-conditioning treatments.

CONTROL OF ALGAE AND BACTERIA

There is public awareness of the need for chemical treatment to control algae growth and bacteria in water used for general sanitary purposes. Such treatments are now recognized as equally important for conditioning water for power station services and miscellaneous processing requirements.

From earliest antiquity,[1] man has known of the powerful algaecidal value of copper for destroying algae and similar types of organism. However, the widespread use of copper salts for the treatment of municipal water supplies did not occur until the publication of the results of the comprehensive studies of Moore and Kellerman.[2] These two investigators demonstrated, in work carried on at the U.S. Department of Agriculture, the practical application of copper sulfate treatment. The assembled data showed that different copper salts when used in very

TABLE 1. COPPER TREATMENT FOR ALGÆ*
Quantity of Copper Sulfate Required for Different Organisms

Organisms	Ppm	Lb per million gal of water
Diatomaceæ:		
Asterionella	0.10	0.8
Fragilaria	0.25	2.1
Melosira	0.30	2.5
Synedra	1.00	8.3
Navicula	0.07	0.6
Chlorophyceæ:		
Cladophora	1.00	8.3
Conferva	1.00	8.3
Hydrodictyon	0.10	0.8
Scenedesmus	0.30	2.5
Spirogyra	0.20	1.7
Ulothrix	0.20	1.7
Volvox	0.25	2.1
Zygnema	0.70	5.8
Microspora	0.40	3.3
Draparnaldia	0.30	2.5
Raphidium	0.30	2.5
Coelastrum	0.30	2.5
Cyanophyceæ:		
Anabæna	0.10	0.8
Clathrocystis	0.10	0.8
Coelosphærium	0.30	2.5
Oscillaria	0.20	1.7
Microcystis	0.20	1.7
Aphanizomenon	0.15	1.2
Protozoa:		
Euglena	0.50	4.2
Uroglena	0.05	0.4
Peridinium	2.00	16.6
Glenodinium	0.50	4.2
Chlamydomonas	0.50	4.2
Cryptomonas	0.50	4.2
Mallomonas	0.50	4.2
Dinobryon	0.30	2.5
Synura	0.10	0.8
Schizomycetes:		
Beggiatoa	5.00	41.5
Cladothrix	0.20	1.7
Crenothrix	0.30	2.5
Leptomitus	0.40	3.3

* From G. C. Whipple: "Microscopy of Drinking Water," p. 255, John Wiley & Sons, Inc., New York, 1914.

small amounts killed algae and many other types of vegetation and microscopic life present in water.

The dosage required to destroy such life varies with the type of organism present and other chemical and physical conditions of the water. A list of common algae and allied aquatic life together with the lethal doses of copper sulfate required for their destruction is shown in Table 1.

Copper sulfate treatment is used extensively for control of such growths in circulating water, in cooling-tower installations, and in reservoir water supplies and is used wherever algae growths are troublesome for power station water uses.

Technique for Applying Copper Sulfate. The algaecide may be applied intermittently or continuously. For the treatment of large reservoirs, it is common practice to place the chemical in a cloth bag, such as a gunny sack, and drag it through the water behind a boat. The frequency of treatment varies widely: in some cases, application is required only once or twice during warm weather; in others, the treatment must be carried on daily.

When treating circulating cooling-tower water supplies, it is preferable to treat the water returned to the tower in order to destroy growths which develop on the wooden surfaces. This is desirable for killing growths at the top of towers and also for permitting adequate time for reaction with the water in the cooling-tower basins, where the insoluble copper carbonate formed in the water can settle out and not be carried into the circulating water supply. Other methods of treatment are used, depending on the type of system and the ingenuity of the operators.

Copper sulfate does not kill many types of bacteria and, in fact, is much less effective as a bactericidal agent than are chlorine and other chemicals. Actually, the destruction of algae by copper sulfate results in a great increase in some bacteria. This condition has been observed many times. In studies made by the author many years ago,[3] an enormous increase in the number of bacteria in the water of shallow reservoirs was noted after copper sulfate treatment.

Copper sulfate treatment, although effective for algae control, has limitations and should be used advisedly in the conditioning of industrial water supplies.

CHLORINE TREATMENT

As is well known, chlorine treatment is extensively used for control of algae and for prevention of bacterial slimes on condenser tubes. The chemical solution is applied either continuously or intermittently, as the size of the station and other local conditions dictate to be effectively or economically necessary.

Industrial plants or power-generating stations may have either separate or combined sanitary, process, and cooling-water systems. A polluted water source may require sterilization to assure a safe, potable water supply, or a boiler feedwater, or a cooling water free from slime-forming organisms. Chlorination is the accepted practice to meet one or all of the foregoing requirements. Efficient performance of water-purification equipment, such as flocculators, precipitators, filters, softeners, and aerators, universally requires a freedom from microorganisms capable of

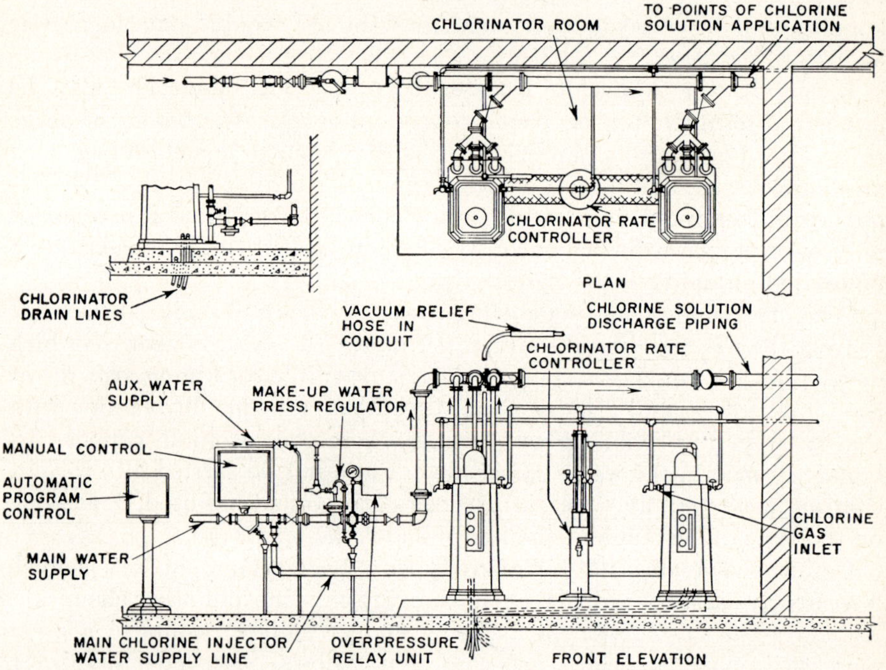

Fig. 1. Arrangement of master vacuum chlorinator installation. (*Courtesy of Wallace & Tiernan Co., Inc.*)

building up organic accumulations or capable of gas formation or of anaerobic bacterial corrosion. A process known as "break-point chlorination"* may be employed for destruction of organic matter and for removal of ammonia from boiler feedwater.

Chlorination. All waters commonly used for cooling purposes are infected with slime-forming microorganisms whose incubation and multiplication are accelerated at the warm water-contact surfaces of steam or vapor condensers and in low-temperature heat exchangers. Mechani-

* The addition of chlorine in sufficient quantities to produce and maintain free available chlorine residuals to the exclusion or near exclusion of combined available chlorine residuals. This technique is different from that used in marginal chlorination, where little attention is given to the type of residual produced or to its persistence.

cal entrapment of foreign materials in the slime complicates and swells the accumulation on the heat-exchange surfaces. The rate of accumulation of organic slimes depends upon the type of bacteria, the temperature, the food materials, and the oxygen concentration in water. Periodic cleaning requires that the unit be out of service and nonproductive, whereas preventive continuous or intermittent chlorination treatment may be used to maintain continuous operation at optimum efficiency.

The chlorinator shown in Fig. 1 is a solution-feed vacuum-type machine. The chlorine gas flow is regulated and metered under a partial vacuum created by an aspirator-type injector wherein the chlorine is

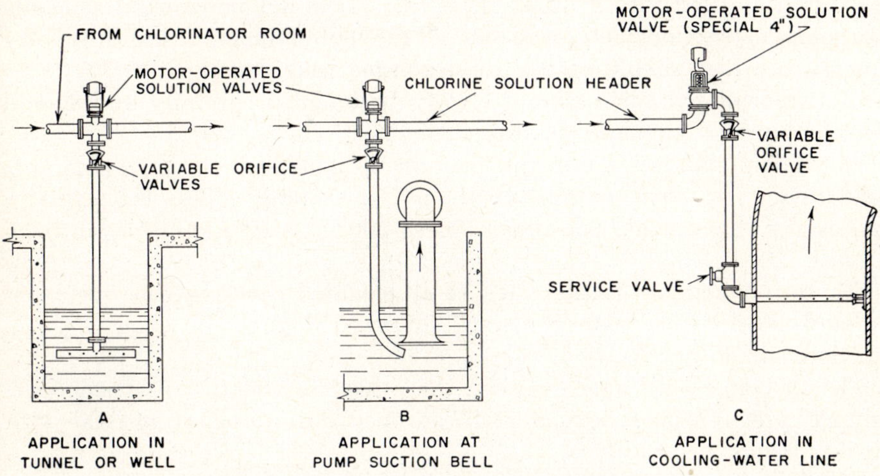

Fig. 2. Points of chlorine-solution application.

mixed with water before being discharged to the point of application. Operation may be manually controlled, semiautomatic, or intermittently programmed to start and stop, or it may consist of fully automatic proportioning, as in deammoniation of boiler feedwater.

Three methods of applying chlorine solution in open conduits or pipelines are shown in Fig. 2.

Instruments have been developed for indicating and monitoring absolute pressures in surface condensers; there are also instruments for recording the performance of the chlorination processes.

Chlorine Compounds. There are now available on the market many compounds which are effective algaecides.* Their particular merit is that they can be applied by apparatus less costly than that required for applying chlorine. On the other hand, these compounds are relatively expensive and may be uneconomical when used, if there are high losses due to blowing down towers or to other forms of wastage.

* Refer to Chaps. 11 and 13 for further discussion of these compounds.

The algaecides such as the pentachlorophenates[4] and similar compounds are fairly extensively sold and are marketed under various trade names. Bromine has been advocated recently for algae and bacterial-slime control in place of chlorine. Although a few large installations have been made, especially on the West Coast, experience with this type of treatment is still limited.

Handling Chlorine to Avoid Hazard. Considering the enormous tonnage of chlorine used annually, there are surprisingly few accidents. However, chlorine is a poisonous gas and must be handled carefully. There are a number of precautions which can be taken whenever chlorine is stored and fed into a water supply.[5,6] It is not pertinent to include detailed instructions here, but some of the more important precautionary measures which should be taken are summarized below:

1. Chlorine cylinders should always be handled carefully and should not be bumped violently or dropped.

2. When cylinders are installed in an upright position, rugged guards should be provided to prevent them from falling. This is especially necessary when small cylinders (100 to 200 lb) are located on platform scales.

3. If cylinders are to be raised to an elevated position, this operation should be done by safety lifting clamps. Magnets or sling hoisting devices should not be used.

4. All connecting lines between the cylinders and the discharge pipe should preferentially be made of approved flexible tube material with male fittings.

5. Dry chlorine gas can be conveyed in iron, steel, copper, or copper alloys. Chlorine solution should be carried in rubber or rubber-lined pipes or other approved materials which will withstand the aggressive chlorine water.

6. Ordinary lubricants should not be used. The preferred lubricant for such service is graphite.

7. No solder except silver or pure lead should be used.

8. An approved gas mask should always be provided. It should be located outside the room housing the chlorine cylinder and the feeding equipment. The mask should be tested at intervals to be certain that it is in good condition.

9. The chlorine room should be provided with an exhaust fan of sufficient capacity to change the air in the room at least 60 times per hour. The inlet of the fan should be located at floor level, and the exhaust should discharge outside the building in a manner which will avoid damage to personnel or property in the area.

10. The temperature of chlorine cylinders should never be raised above 100°F, and if it is imperative to do so, it should be done by circulating

warm air. Heat from a fire or from any other source of flame is extremely dangerous applied to a cylinder containing liquid chlorine and should never be permitted. More specific instructions should be obtained from equipment vendors and chemical companies manufacturing chlorine and chlorine compounds.

ROUGHING FILTERS AND STRAINERS

Many appliances are sold for the removal of oil or suspended matter from feedwater.* Some of the devices offered for this service, being

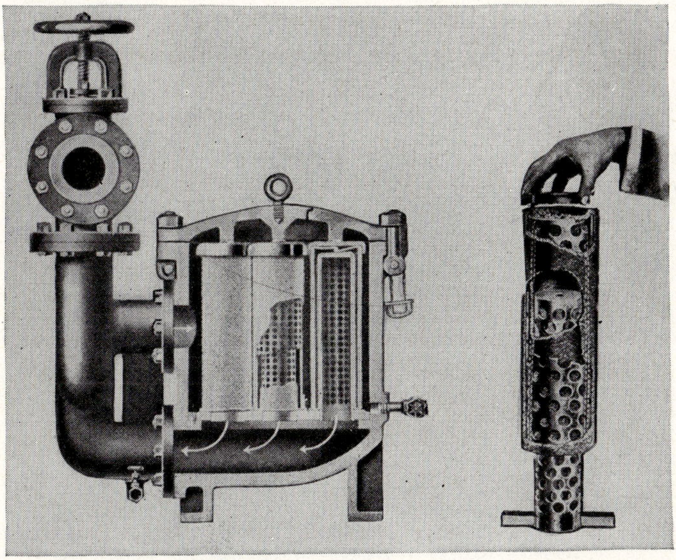

Fig. 3. Feedwater filter and grease extractor and a sectional view of the filter cartridge. (*Courtesy of Blackburn-Smith Mfg. Co., Inc.*)

nothing more than coarse strainers, are of little value and accomplish little or no purification. There are on the market, however, several efficiently designed filters and grease extractors which remove most of the coarse suspended solids. The filter medium used extensively for this purpose is terry cloth or other closely woven material. The cloth is frequently wound around a cylinder or cartridge, the water going through a number of layers in passing through the apparatus. These extractors are built as single- or twin-unit types. Where twin filters are employed, one unit is held in reserve, and it is put into service while the other is being cleaned.

These appliances may, in the absence of other forms of feedwater purification, markedly improve the quality of the raw water. Their effective-

* Refer to Chap. 5 for a discussion of filter equipment designed for this purpose.

ness in the removal of grease is generally recognized. They should not be looked upon as complete purification systems, since they do not remove the soluble scale-forming solids. When used for clarification of turbid surface waters or waters containing suspended solids from trade-waste pollution, their use is justified either alone or in connection with more complete methods of treatment.

In Fig. 3 is shown a feedwater filter and grease extractor. The filter medium of this unit is composed of terry cloth built up in sections over perforated cartridges. The filter compartment is divided by partitions

Fig. 4. Sectional view of Lagonda filter and grease extractor. (*Courtesy of Elliott Company.*)

into two chambers. The water enters the upper, or filtering, chamber, passes to the lower or effluent chamber through the series of filtering cartridges that are set in the partition, and then goes through the effluent to the feedwater line as indicated by the arrows. The number of cartridges in each filter varies with the size and capacity of the unit. A lower end of the inner cylinder is fastened into the partition that separates the inlet from the outlet chambers. The lower end of each outer cylinder is supported on a finished surface, and all the outer cylinders are secured in place at the top end by projections from the underside of the top cover. These projections press down upon the caps when the top cover is bolted into place.

When the cartridges become fouled by suspended solids, they may be removed, and new ones may be inserted in their places.

The apparatus in Fig. 4 consists of a shell containing a filter spool. The filtering medium is special terry cloth wound around the spool with alternate layers of a link spacer mat (Fig. 5). On the end of the filtering cloth, hooks are provided which are fastened on the filter-spool core. When the winding is completed, the whole is held in place by a small rod inserted through the filtering medium and the flanges on the spool. The filter spool is inserted in a perforated cylinder which is held in place in the chamber by cap screws. The filtering element fits into a recess in the bottom casting of the filter unit. These units have two individual filtering chambers. Only one chamber is in operation at a time, the other being held in reserve for cutting into service while the alternate strainer is being cleaned. To clean the filter, the element is withdrawn from the filtering chamber. The terry cloth may be washed with soap and hot water to which soda ash has been added in order to remove any oil that may have been entrained by the cloth. Under some conditions, it is desirable to clean the cloth by gasoline or other solvents.

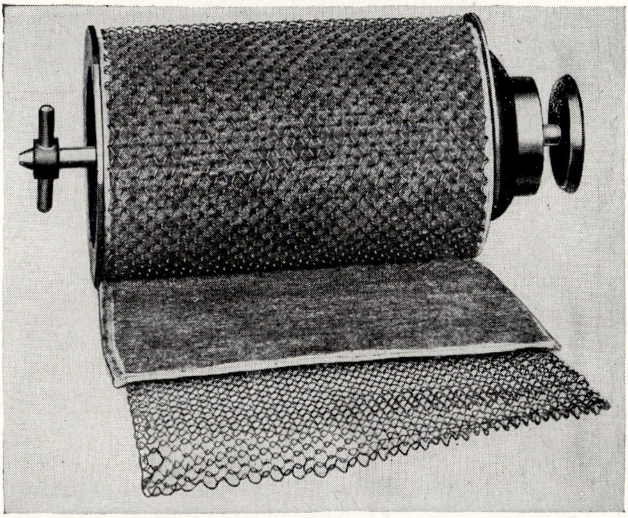

Fig. 5. Filter spool from a Lagonda filter and grease extractor showing position of terry cloth and spacer mat.

The frequency of cleaning the filter will depend on the character of the raw water passing through the unit.

Bibliography

1. Baker, M. N.: "The Quest for Pure Water," American Water Works Association, New York, 1948.
2. Moore, G. T., and K. F. Kellerman: Copper as an Algicide and Disinfectant in Water Supplies, *U.S. Bur. Plant Ind. Bull.* 76, 1905.

3. Powell, S. T.: The Effect of Algae on Bicarbonates in Shallow Reservoirs, *J. AWWA*, **2:** 703 (1915).
4. Turner, H. J., Jr., D. M. Reynolds, and A. C. Redfield: Chlorine and Sodium Pentachlorophenate as Fouling Preventives in Sea Water Conduits, *Ind. Eng. Chem.*, **40:** 450 (1948).
5. Logan, J. O.: Handling Chlorine Safely, *Water & Sewage Works*, **95:** 333 (1948).
6. Hedgepeth, L. L.: Preparing and Packaging Liquid Chlorine and Allied Products for Water Plants, *J. New Engl. Water Works Assoc.*, Vol. 52, No. 1, 1938.

24 ANALYTICAL PROCEDURES

ANALYTICAL TEST METHODS FOR WATER-TREATMENT CONTROL

The satisfactory control of water quality in steam power or other plants is dependent upon numerous factors, among which frequent control tests and their interpretation are of primary importance. The extent of the test work will vary with the design of the plant, the water-treatment facilities provided, and the chemical controls required.

The analytical procedures given in this chapter are representative of those used in the control of chemical treatments applied to treated process water, make-up water, cooling water, feedwater, and boiler water. Space requirements do not permit the inclusion of all standard and alternate methods used in current practice. For additional methods, reference should be made to procedures published by the American Society for Testing Materials and the American Public Health Association and to other texts.

The frequency schedule for testing the water and the selection of methods to be used require consideration in each plant.

Facilities should be provided for sampling water throughout the feedwater-steam-condensate cycle. These test points should be located to permit sampling of raw and treated water, boiler water, saturated and superheated steam, condensates from steam turbines and stage heaters, and deaerated feedwater.

COLLECTION OF SAMPLES

An important part of the analytical work is the taking of representative samples, which are preserved in their original state until the analyses can be made. An accurate representative sample is one which truly represents the water or steam in that part of the water cycle from which it is taken. Perfection in this respect may not always be possible, but in so far as may be practical, the samples should be taken with this object in mind.

Water and steam conditions will change from time to time and sometimes momentarily. Therefore, it is necessary not only to take the samples carefully but to consider the influence of operating conditions at the time so that the analytical results may be properly evaluated; or, in some cases, the individual samples should be taken during average load or "on load" conditions.

For the most part, water samples are to be taken in resistant glass containers of pyrex quality. If a number of tests are to be made on one sample, it is desirable to work from a single 500- to 1,000-ml bottle full of water, keeping the stopper in place when not removing a part of the sample. For the most accurate silica determinations, rubber, plastic,* or selected metallic containers are recommended. Regardless of the nature of the sample container, it should be the general rule to make analyses as soon as possible after sampling.

In special cases, as in determining the soluble-solids content of steam-condensate samples, containers of corrosion-resistant metal or other insoluble containers are required.

Before taking a sample from an infrequently used sampling line, the sample line should be not only flushed to remove all stagnant water but given additional flushing to remove any sediment or other impurity that may be present.

Samples should be taken from locations as close to the boiler as possible, and the sample lines and coolers should be free of any traps that will collect suspended matter before reaching the sample bottle.

In power stations, samples are classified into five groups: steam condensate, concentrated boiler water, make-up water, circulating water, and bearing cooling water.

Steam Condensate. These samples include those of saturated and superheated steam, evaporator vapor condensate, and condensate from condensers and boiler feedwater in low make-up plants where the feedwater is practically all condensate.

Concentrated Boiler and Evaporator Water. Such samples are those obtained from the continuous- or intermittent-boiler-blowdown lines and from the continuous-blowdown lines on the evaporators. Long exposure of these samples to air will change the hydroxide alkalinity to carbonate, and the pH value will be lowered; so the time interval between sampling and starting the analyses should not be unduly long.

Boiler-feed Make-up Water. Normally, the raw water which is used as make-up for boilers is processed before it enters the feedwater. Samples should be taken at frequent intervals for tests and control of the

* Plastic bottles are not satisfactory as sample containers to be used for exact iron determinations because iron compounds are absorbed by the container.

softening and evaporating processes. Such samples should be taken during normal flow conditions.

Circulating Water. The circulating water is used to cool the main turbine condensers and the auxiliary coolers for condensate. The sampling points are located at the inlet and outlet of the main turbine condensers. Samples of such water are taken to show the effect of periodic chlorination or other type of chemical treatment.

The determination of free chlorine and dissolved oxygen and pH should be made on the spot immediately after sampling.

Bearing Cooling Water. Such samples are required for cooling auxiliary equipment bearings. At times the main cooling-water circuit furnishes such water, or a completely separate source of treated or untreated water supply is provided. Sampling and testing of such water are similar to those for the main circulating service water.

TEMPERATURE REGULATION

The regulation of the water temperature for flowing samples to conductivity and pH recording or indicating equipment is of special concern in obtaining comparable results. The ideal temperature is 77°F, or 25°C, but if this is not possible, efforts should be made to hold the sample water at the nearest point to 77°F as may be feasible.

ALKALINITY DETERMINATION

The alkalinity results obtained by this volumetric control method are used for the regulation of chemical treatment, for the maintenance of desired pH values in the water, and for examination of water samples generally. Frequently, the pH value of the water is also measured in order to aid in interpretation of the alkalinity results and to determine whether the water is acid or alkaline to some specified pH value.

The determinations of alkalinity to the phenolphthalein and methyl orange (P and M) end points are carried out with the same sample of water and in consecutive steps. If the sample is alkaline to phenolphthalein, it will turn pink when the indicator is added. This indicates the presence of alkaline hydroxides, carbonates, phosphates, and silicates. If the sample is alkaline to methyl orange only, that is, no pink color develops when phenolphthalein is added, but the sample turns yellow after addition of methyl orange, the sample contains no alkaline hydroxides or carbonates, and only the less alkaline bicarbonates, phosphates, etc., are present. The methyl orange test represents total alkalinity; the phenolphthalein test, partial alkalinity.

Samples containing suspended alkaline salts present as insoluble carbonates, phosphates, etc., should be filtered before starting the test. In this procedure, 100-ml samples are used, but for routine purposes, smaller samples, such as those of 50 or 25 ml, may be substituted, and the results may be calculated using a factor to compensate for the smaller samples used in the test. As a matter of convenience, the alkalinity results are expressed as equivalent calcium carbonate. Other forms of alkalinity may be calculated as required.

Reagents:

Phenolphthalein indicator solution

Dissolve 5 grams of phenolphthalein in 1 liter of a 50 per cent solution of ethyl alcohol in distilled water. Neutralize with dilute NaOH solution to a faint pink color.

Methyl orange indicator solution

Dissolve 1 gram of methyl orange in 1 liter of distilled water.

Standard sulfuric acid, 0.02*N*

Procedure for P Alkalinity (Carbonate and Hydroxide): To a measured 100 ml of clear sample placed in a white casserole or an Erlenmeyer flask over a white background, add four drops of neutral phenolphthalein indicator. If a pink color is developed, titrate the sample with the 0.02*N* sulfuric acid solution from a burette having graduation marks for each 0.1 ml. The acid should be added slowly and the sample stirred or agitated until the pink color just disappears. Near the end point add only one or two drops of acid at a time.

Calculation:

$$\text{ppm } P \text{ alkalinity, as } CaCO_3 = \frac{\text{ml acid} \times 1{,}000}{\text{ml sample}}$$

Example: If a 50-ml sample of water is used and 4 ml of acid is required, the *P* alkalinity is

$$P \text{ alk.} = \frac{4 \times 1{,}000}{50} = 80$$

Procedure for M Alkalinity (Total): To the same sample used for the *P* alkalinity determination, add 2 to 4 drops of methyl orange indicator or until a distinct yellow color is produced.

Without refilling the burette, continue to titrate slowly with the 0.02*N* sulfuric acid until a definite pink tinge is seen in the yellow color of the sample. A strong pink color indicates overtitration of the end point. In case of difficulty in observing the end point add a second sample containing the same amount of indicator to a second casserole or flask and place this next to the sample being titrated. By comparison, the first definite change in color can be seen.

Calculation: Read from the burette the total milliliters of acid which have been used for both the *P* and *M* titrations.

$$\text{ppm } M \text{ alk., as } CaCO_3 = \frac{\text{ml acid} \times 1{,}000}{\text{ml sample}}$$

Example: If a 50-ml sample of water is used and 7 ml of acid is required, the *M* alkalinity is

$$M \text{ alk.} = \frac{7 \times 1{,}000}{50} = 140$$

If it is desired to show the results as equivalent sodium carbonate, as required for calculating ASME embrittlement ratios, the results as calcium carbonate are multiplied by the factor 1.06. For example, 100 ppm, as $CaCO_3$, are equal to 106 ppm, as Na_2CO_3.

HYDROXIDE ALKALINITY: BARIUM CHLORIDE METHOD

Application and Description. An exact volumetric method for the determination of hydroxide ion in water is not available. The interpretation of the results of any titration procedure is complicated by the presence of such materials as carbonate, phosphate, silicate, aluminate, and organic matter. If the water is practically free of silicates, aluminates, and organic matter and if these materials are present in relatively small amounts as compared with the hydroxide ion, the method described below may be used. This method covers the titration of water to which barium chloride is added to remove carbonate ions which interfere with the determination of the hydroxide ion. Sulfates are also precipitated on addition of barium chloride. Normally, this method is accurate to within 10 per cent of the total hydroxide present. If an accuracy greater than 1 to 2 per cent is required, a more refined method should be used.*

Reagents:

Hydrochloric acid, 0.02*N*
Barium chloride solution

Dissolve 100 grams of barium chloride, $BaCl_2.2H_2O$, in 1 liter of distilled water.

Phenolphthalein indicator solution

Procedure: To a 100-ml or suitable-sized sample in a titration flask or white porcelain casserole, add 5 ml of barium chloride solution to precipitate the carbonate and sulfate ions.†

* The strontium chloride method, APHA or ASTM D-514, is recommended.

† If there are only very small amounts of carbonates and/or sulfates present, it will be desirable to add a small percentage of sodium carbonate to the barium chloride

Add 2 to 3 drops of the phenolphthalein indicator solution and titrate with the standard 0.02N hydrochloric acid solution until the pink color of the indicator just disappears. Sulfuric acid is less desirable because of the production of barium sulfate and the possibility of occlusion of hydroxide in the precipitate.

Calculation:

$$\text{ppm of hydroxide, OH} = \frac{\text{ml acid} \times 340}{\text{ml sample}}$$

Example: Suppose 5.0 ml of 0.02N hydrochloric acid were required to titrate 100 ml of sample, then by substituting in the above formula

$$\frac{5 \times 340}{100} = 17 \text{ ppm OH}$$

TOTAL HARDNESS BY TITRATION WITH SOAP SOLUTION

Application and Description. The determination of total hardness by titration with standard soap solution is used in the analysis of raw-water supplies, in the control of the chemical treatment of raw water for boiler- or evaporator-feed purposes, and, in some instances, for the detection of condenser leakage or other raw-water leakage into the boiler feedwater.

The method is based on the ability of the soap solution used for titration to form calcium and magnesium soaps which do not form a stable lather on shaking in the test bottle. When an excess of soap solution has been added, a stable lather will be formed.

Reagents:

Phenolphthalein indicator solution
Sodium hydroxide solution, 0.1N (*carbonate free*)

Store in a 1-oz dropping bottle.

Sulfuric acid solution, 0.1N

Store in a 1-oz dropping bottle.

Standard hard water

A calcium chloride standard solution may be used for the standard hard water. However, if magnesium is present, the accuracy of the test may be improved by mixing the calcium chloride standard solution with the magnesium sulfate standard solution in the ratio of calcium to magnesium in the sample. For average water, the mixture is made up

solution. The resulting precipitate will provide nuclei for precipitation of carbonate in the sample.

of 80 ml of calcium chloride standard solution and 20 ml of magnesium sulfate standard solution.

1. *Calcium chloride standard solution*

Dissolve 0.5 gram of pure calcium carbonate in a beaker with a little dilute hydrochloric acid, being careful to avoid splattering. Wash down the sides of the beaker with freshly boiled and cooled distilled water. Neutralize the excess acid with ammonium hydroxide, using litmus paper of phenolphthalein as the indicator. Make up to 500 ml with freshly boiled and cooled distilled water, and store in a tightly stoppered glass bottle. One milliliter equals 1 mg of hardness, as calcium carbonate.

2. *Magnesium sulfate standard solution*

Dissolve 0.6 gram of anhydrous magnesium sulfate in 500 ml of freshly boiled and cooled distilled water. Store in a tightly stoppered pyrex glass bottle. One milliliter equals 1 mg of hardness, as calcium carbonate.

Standard soap solution

Dissolve with vigorous shaking approximately 100 grams of pure powdered castile soap in 1 liter of 80 per cent ethyl alcohol. Let this solution stand at least overnight, then decant. The stock soap solution is approximately 9 to 10 times stronger than the dilute standard soap solution. Dilute a portion of the stock solution with 80 per cent ethyl alcohol until, when titrated against calcium chloride solution, 1 ml of the resulting dilution is equivalent to 1 ml of the standard calcium chloride solution. The adjustment of the strength of the dilution should include an allowance for the lather factor of the adjusted soap solution from at least five determinations. One milliliter soap solution − lather factor = 1 mg $CaCO_3$.

Lather Factor: The lather factor of the soap solution should be determined by adding small increments of soap solution to 50 ml of freshly boiled and cooled distilled water until a stable lather is formed after shaking. Since the soap solution may change, the lather factor should be checked each week.

Add 50 ml of the prepared cooled water to the 8-oz test bottle and adjust the pH value to the phenolphthalein end point, as in the sample. Titrate with the soap solution in the manner described until a permanent lather is formed. Record the milliliters of soap used as the lather factor.

Procedure: In order to obtain reproducible results the samples should be brought to about the same pH value of approximately 8.3. The temperature of the samples should be between 60 and 100°F. Samples which contain appreciable amounts of suspended matter should be settled or filtered clear.

Add 50 ml of the sample to be tested to a clean, clear-glass, round or square 8-oz bottle, which is fitted with a glass or rubber stopper. Add three drops of phenolphthalein indicator and adjust the pH value to the end point of the indicator by adding drops of 0.1*N* sodium hydroxide or of 0.1*N* sulfuric acid solution as required. For example, if the sample is pink, add from a dropping bottle enough acid to remove all but a faint pink in the sample. If an excess of acid is added and no pink shows, add drops of the 0.1*N* sodium hydroxide until the first pink color appears.

To the neutralized sample, add standard soap solution from a burette in small increments of, say, 0.2 ml at a time, and shake the bottle vigorously after each addition. When a permanent lather begins to form, reduce the increments of soap solution to 0.1 ml until, with further shaking, a lather will persist over the surface of the water for a period of 5 min. At this time, the bottle should be lying on its side. Record the milliliters of soap solution used.

When this point is reached, test for a false end point by adding 0.2 ml more of the soap solution, and shake. If the lather disappears, the titration should be continued to a final end point. The presence of magnesium in the sample may be the cause of the false end point, and this point should be recorded.

If the sample is known to contain only small amounts of hardness, the additions of soap solution should be in increments of 0.05 ml at a time. If the sample requires more than 8 ml of soap solution, it is preferable to take 25 ml of sample and dilute to 50 ml with distilled water. In this case, the factor for calculating hardness is doubled.

Calculation:

Total hardness, as ppm $CaCO_3$

$$= (\text{ml soap solution} - \text{ml lather factor}) \times \text{hardness factor}$$

Example: If 50 ml of sample requires 1.5 ml of soap solution to produce a stable lather, the lather factor is 0.2 ml, and the hardness factor is 20. Then $20 \times (1.5 - 0.2) = 20 \times 1.3 = 26$ ppm total hardness, as $CaCO_3$.

Hardness Factor: The factor used for calculating the hardness should be determined each week in order to note any change in the soap solution.

To make this test, add 5 ml of standard hard water to 45 ml of distilled water such as is used for the lather factor. Adjust the mixture to the phenolphthalein end point, and titrate with the soap solution until a lather which is stable for 5 min is obtained.

Calculation:

$$\text{Hardness factor} = \frac{5 \times 20}{(\text{ml of soap} - \text{ml lather factor})}$$

Example: If 5 ml of the standard hard water requires 4.1 ml of soap solution and the lather factor is 0.2 ml, then the soap solution consumed is

$$4.1 - 0.2 = 3.9 \text{ ml}$$

This is equivalent to the 5 mg of calcium carbonate in the standard hard water. Therefore, the hardness factor may be calculated as

$$20 \times \frac{5}{3.9 \text{ (ml of soap used)}} = 25.6$$

This hardness factor of 25.6 will be used for future hardness calculations, instead of the factor of 20 mentioned in the foregoing example.

A new acceptable method*,† which is presently being widely used for determining hardness depends on the use of an organic titrating agent which has the ability to sequester calcium and magnesium and an indicator which changes color when hardness is removed from the sample.

Low-hardness Control Test (Zeolite Effluent): This test is made to determine when it is necessary to regenerate a softener and when the water is again soft after brining and rinsing. The soap solution in this determination should be dilute in order to use a reasonable volume.

For collection of sample, use a 100-ml oil-sample bottle with 40-ml graduating marks. Adjust the alkalinity with either standard acid or alkali so that the sample is pink to phenolphthalein. Add three drops of B & B soap solution.‡ Stopper the bottle, shake vigorously, and place on its side. If a lather is formed which persists for 5 min, the water is soft. If additional soap solution is required, add a few drops at a time, shaking vigorously after each addition until a permanent lather is formed.

CHLORIDES, AS SODIUM CHLORIDE, NaCl

Application and Description. The results of the chloride tests are frequently used for chemical control, detection of leakage of raw water into the feedwater system, as at the main turbine condenser, and estima-

* G. Schwarzenbach, W. Biedermann, and F. Bangerter, Complexons VI: New Simple Titrating Methods for Determining the Hardness of Water, *Helv. Chim. Acta*, **29:** 811 (1946); U.S. Patents 2,583,890 and 2,583,891. Hall Laboratories, Inc., holds nonexclusive licenses under these patents with the right to grant sublicenses to others who purchase the several reagents.

† J. D. Betz and C. A. Noll, *J. AWWA*, **42:** 49 (January, 1950).

‡ Dissolve 100 grams of pure soap in 1,600 ml of 90 per cent ethyl alcohol. Dilute with distilled water (800–1,000 ml will be required) until 2.4 ml of the soap solution give a lather with 40 ml of solution of calcium chloride, $CaCl_2$, containing 0.25 gram per liter. Three drops should give a permanent lather with distilled water.

tion of total solids in water samples when the chloride content increases and decreases with other materials in the water.*,†

The determination of chlorides is accomplished by titration of the sample with standard silver nitrate solution in the presence of potassium chromate indicator. A red color is developed when all the chloride has been precipitated by the silver nitrate.

For normal routine testing of water samples, the concentration of chloride may be determined within satisfactory limits by the procedures given in this method. If a high degree of accuracy and reproducibility is required, this method may be modified to include the following:

1. Use a yellow light or yellow goggles in observing the end point.
2. Make adjustments for indicator blank.
3. Remove color and any interfering reducing material, such as sodium sulfite, by boiling the sample with chloride-free hydrogen peroxide, filtering, and cooling.
4. Adjust the pH value of the sample to 8.3 ± 0.5.
5. Use a standard 50-ml volume of water for titration. If the chloride concentration is high, it may be necessary to use a smaller sample and dilute to 50 ml with chloride-free distilled water so that the concentration of chloride ion is preferably between 5 and 200 ppm.

Reagents:

Standard silver nitrate solution (*1 ml* = *1 mg* of *NaCl*)

Dissolve 2.9068 grams of dried, reagent-grade silver nitrate, $AgNO_3$, in 1 liter of chloride-free distilled water.

Sodium chloride standard solution

Dissolve 5.0000 grams of fused NaCl in 500 ml of chloride-free distilled water. Dilute 10 ml of this solution with chloride-free distilled water to a final volume of 100 ml. One milliliter of this solution is equivalent to 1 mg of NaCl.

Potassium chromate indicator solution

Dissolve 50 grams of neutral potassium chromate, K_2CrO_4, in 100 ml of distilled water. Add silver nitrate solution until a slight red precipitate is formed. After letting the solution stand overnight, filter to

* Condensate samples may contain only traces or very small amounts of chloride, and determination of chlorides by the foregoing method may not be possible.

† I. M. Kolthoff and H. J. Yutzy, Nephelometric Determination of Chloride, *J. ACS*, **55**: 1915 (1933). See also N. H. Furman, and G. W. Low, Comparison of Several Electrometric and Nephelometric Methods for the Estimation of Small Amounts of Chloride, *J. ACS*, **57**: 1588 (1935).

remove the precipitate and dilute the filtrate to 1 liter with distilled Water.

Hydrogen peroxide, 30 *per cent*
Phenolphthalein indicator solution

Dissolve 5 grams of phenolphthalein in 1 liter of a 50 per cent solution of ethyl alcohol.

Sulfuric acid, 2 *per cent*

Procedure: To 50 ml of filtered sample in a white porcelain casserole, add four drops of phenolphthalein indicator. If a pink color develops, add 2 per cent sulfuric acid with stirring just sufficient to remove the pink color. If no pink color is developed, neutralize to phenolphthalein pink with sodium carbonate solution, and then just destroy the pink color by adding a little sulfuric acid. If the P and M alkalinities of boiler water have already been determined, the sample may be used for the chloride determination without further neutralization.

Add 1 ml of potassium chromate indicator to the neutralized sample, and titrate slowly with standard silver nitrate while stirring. Each small addition of silver nitrate will produce a brick-red color, which will disappear less and less rapidly as the end point is approached. At the same time, the solution will become more turbid owing to the precipitation of silver chloride. Stop the titration when the last drop of silver nitrate produces a slight reddish-orange or pink color which remains after stirring.

Blank: Determine the indicator blank by titrating with silver nitrate the same volume of chloride-free distilled water, using 1 ml of indicator solution. The effect of the indicator blank can be canceled from the calculations by titrating a second sample containing half the normal volume and subtracting the two titration results before calculating the chloride concentration.

Calculation:

$$\text{ppm chloride, as NaCl} = \frac{(\text{ml silver nitrate} - \text{blank}) \times 1{,}000}{\text{ml sample}}$$

Example (using distilled water blank): Suppose that 50 ml of water required 2.7 ml of silver nitrate and that the blank determination required 0.2 ml, then

$$\frac{(2.7 - 0.2) \times 1{,}000}{50} = 50 \text{ ppm chloride, as NaCl}$$

Example (using different-size samples): Suppose 50 ml of water required 2.7 ml of silver nitrate and that 25 ml of water required 1.5 ml, then

$$\frac{(2.7 - 1.50) \times 1{,}000}{(50 - 25)} = 48 \text{ ppm chloride, as NaCl}$$

SULFATES IN WATER BY THE BENZIDINE METHOD

Application and Description. The most accurate method for the determination of sulfates is that of gravimetric analysis. This is time-consuming and impractical for most routine control procedures. The results of sulfate determinations are useful in the regulation of sulfate concentrations in cooling-tower water and boiler water. The sulfate data are also used in the estimation of the total solids in the water. In acid solution, benzidine hydrochloride precipitates sulfates as benzidine sulfate. The solubility of benzidine sulfate in water is significant. The benzidine method described minimizes this effect and is sufficiently accurate for general use.

Reagents:

Benzidine solution

Dissolve 11.2 grams of benzidine hydrochloride in 400 ml of distilled water, and mix with 100 ml of 5 per cent hydrochloric acid solution. Saturate with benzidine sulfate and filter. If benzidine base is used, weigh out 8 grams and rub with 8 ml concentrated hydrochloric acid. Then add 85 ml of water and 17 ml of concentrated hydrochloric acid, dilute to 500 ml, and filter.

Phenolphthalein indicator solution

Standard sodium hydroxide solution, 0.1*N*

$$\text{One milliliter NaOH} = 4.803 \text{ mg } SO_4$$

Procedure: Measure 50 ml of filtered water sample into a graduated cylinder, and pour into a 250-ml Erlenmeyer flask. Add 10 ml of the benzidine solution from a 10-ml graduated cylinder or pipette. Thoroughly mix the contents of the flask by giving it a whirling motion, and allow the mixture to stand for 10 min.

If sulfates are present in the sample, they will be precipitated by the benzidine solution. Mix the contents of the flask and filter off the benzidine sulfate precipitate, using a small funnel and quantitative filter paper pulp. In case the filtered solution is clear, add a drop of benzidine; if not all the sulfates have been precipitated, the clear solution will become cloudy. In this case, add 2 ml of the benzidine solution, mix, let stand a few minutes, and filter again. Repeat this process until a clear filtrate is obtained that will not turn cloudy on addition of a drop of benzidine.

It is important to use special care in washing the precipitate. Rinse the empty flask with 25 ml of distilled water saturated with benzidine sulphate, and pour this rinse water into a graduated cylinder. Pour 5 ml over the filter paper, revolving the funnel so that the water runs over the entire surface of the paper. Allow the water to run through completely so that none is retained in the funnel, and repeat the process with an additional 5-ml portion. In this manner, wash the paper five times, allowing the wash water to drain completely after each application. Discard the wash water. Then carefully remove the filter paper pulp containing the washed precipitate from the funnel, and place it in the original flask. Rinse the funnel with approximately 25 ml of distilled water, and add to the original flask.

Before beginning the titration, place a rubber stopper in the flask, and shake violently in order to disintegrate the filter paper completely. Do not begin to titrate until the paper has been reduced to a pulp. Then add two to four drops of phenolphthalein indicator, and titrate with 0.1*N* sodium hydroxide, a few drops at a time, shaking violently after each addition. The titration is complete when the resulting pink color remains for a period of ½ min. This will give the hydroxide solution an opportunity to become thoroughly mixed with the contents of the flask and is a further assurance that all the filter paper has been completely disintegrated.

The number of milliliters of standard sodium hydroxide used in the titration multiplied by 142 gives the sulfate content of the water, expressed as parts per million of sodium sulfate.

In case the sodium sulfate content of the water is less than 200 ppm, however, 100 ml of the sample should be measured for the test instead of 50 ml. In this case, the number of milliliters of standard sodium hydroxide multiplied by 71 gives the sodium sulfate content of the water.

In case the sodium sulfate is over 1,700 ppm, use a 25-ml sample, and multiply the titration by 284 to obtain the sodium sulfate. The above procedure will apply to all waters containing up to 3,400 ppm of sodium sulfate.

Blank: Errors due to failure to wash out the acid benzidine reagent completely cannot be eliminated entirely, and each operator should run a blank sufficiently often to determine the extent of the error in his procedure. If consistent results are obtained, the blank need be run only once a week. To establish the necessary correction, 10 ml of benzidine should be added to a 50-ml portion of distilled water, and the sample should then be titrated with 0.1*N* sodium hydroxide. The quantity of sodium hydroxide required should be recorded on the report forms as the sulfate blank, and this number of milliliters should be deducted from all quantities of 0.1*N* sodium hydroxide used in the sulfate determination before calculating the results.

Calculation:

$$\text{ppm } SO_4\text{, as } Na_2SO_4 = (\text{ml NaOH for sample} - \text{ml NaOH for blank}) \times 142$$

Example: Suppose that 50 ml of boiler water require 2.5 ml of sodium hydroxide and that the blank consumed 0.3 ml, then

$$2.5 - 0.3 = 2.2$$
$$2.2 \times 142 = 312 \text{ ppm } Na_2SO_4$$

A second acceptable method for determining sulfates by titration is based on the precipitation of sulfate ion by barium chloride in an alcoholic solution in the presence of THQ (tetrahydroxyquinone) indicator. The description and procedure may be found in "Standard Methods for the Examination of Water and Sewage," published by the American Public Health Association.

PHOSPHATE AS PO_4: COLORIMETRIC METHOD

Application and Description. The following colorimetric control method is for the determination of phosphate, as PO_4, in boiler or evaporator waters which are not too highly colored. If considerable color is present, steps should be taken to decolorize the sample or to balance out the color effect when making the color comparisons. The test results are used to regulate the chemical treatment of the water and to interpret the effect of phosphates on the boiler-water alkalinity.

The method is based on the formation of a phosphomolybdate when the molybdic acid solution is added and the production of a blue color when the phosphomolybdate is reduced with stannous chloride solution. The intensity of the blue color is proportional to the amount of phosphate in the sample.

Apparatus: The procedure which follows is based on the use of an a-c electrophotometer, but color comparators, such as Nessler tubes, the Duboscq-type colorimeter, or comparators* with color standards, are also employed. If comparators with prepared color standards are used, the reagents recommended by the manufacturer should be substituted for those given in this method.

Reagents:

Standard phosphate reagent

Stock solution: dissolve 7.1638 grams of dried potassium phosphate, KH_2PO_4, in 1 liter of distilled water. This solution has a concentration of 5,000 ppm of PO_4, and 1 ml = 5 mg PO_4.

Dilute solution: dilute 20 ml of the stock solution to 1 liter with distilled water. This dilute solution has a concentration of 100 ppm of PO_4, and 1 ml = 0.1 mg PO_4.

* W. A. Taylor Co., Baltimore, Md.; LaMotte Chemical Products Co., Baltimore, Md.

Molybdic acid solution

Dissolve 10 grams of reagent-grade, 85 per cent, molybdic acid in 500 ml of distilled water. Add slowly with stirring 47 ml of concentrated reagent-grade sulfuric acid. Dilute to 1 liter with distilled water.

Stannous chloride solution

Stock solution: dissolve 12 grams of stannous chloride dihydrate, $SnCl_2 \cdot 2H_2O$, in 100 ml of concentrated hydrochloric acid. Keep in tightly stoppered glass bottle.

Dilute solution: dilute 2.5 ml of stock solution to 100 ml with distilled water. Keep in tightly stoppered glass bottle. Always make up new dilute solution each day.

Procedure: Turn on the light switch for the photometer and insert the light filter which is to be used. Filters of 420 to 650 μ wave length are satisfactory. Filters of higher wave length are more sensitive to color change, but it may be necessary to dilute the sample in order to remain within the range of the photometer.

To a 25-ml test tube, add 5 ml of clear, filtered sample and 10 ml of the molybdic acid solution. Mix the contents of the tube by inserting a rubber stopper and inverting several times. Add 2.5 ml of the dilute stannous chloride solution and mix again. Allow the mixture to stand for 2 min for full development of the blue color.

Prepare a blank by adding to a second test tube 5 ml of sample, 10 ml of molybdic acid, and 2.5 ml of distilled water, and mix. Transfer the blank to the color comparison tube of the photometer. Then with the color filter in place and the scale reading at zero, balance the galvanometer to the zero mark by turning the light-adjustment knob.

Remove the blank, transfer the sample in which the blue color has developed to the color comparison tube, and insert it in the photometer. With some instruments, matched comparison tubes are supplied and it is merely necessary to shift the tube with the blank out of the light path and substitute the tube with the sample in its place. Now zero the galvanometer by turning the knob on the color density scale, and at this point, record the reading on the scale.

Read the parts per million of phosphate in the sample from the standard calibration curve, on which the scale readings have been plotted as a function of the parts per million of phosphate in standard phosphate solutions for the light filter which is being used.

Preparation of Standard Curve: The standard curve should be prepared by repeating the foregoing test, using prepared standard solutions of 10, 20, 30, 40, 50, and 60 ppm of phosphate and plotting on graph paper the photometer readings against the parts per million of phosphate in the standards. The curve should be nearly a straight line.

Precautions:

1. Keep all glassware clean. The comparison tubes must be kept very clean and free from finger marks, otherwise the readings will be in error.
2. Do not use old dilute stannous chloride solution. Keep this solution lightly stoppered, and renew once a day.
3. Check the standard calibration curve once a week with a standard phosphate solution of, say, 20 ppm concentration.
4. If the light bulb in the photometer is changed, the standard calibration curve should be rechecked.

SODIUM SULFITE IN BOILER WATER

Application and Description. The following volumetric method is used to determine the concentration of sodium sulfite in boiler water and to guide the rate of feeding sodium sulfite solution.

The method is based on the oxidation of sodium sulfite in the acidified sample when it is titrated with standard potassium iodate solution in the presence of starch indicator and an excess of potassium iodide. The potassium iodate releases free iodine in the acidified sample, which becomes fixed as soluble iodide during the oxidation of sodium sulfite. The addition of any excess of potassium iodate liberates an excess of iodine, which causes the starch indicator to turn blue in color, indicating the end point of the titration.

Depending on the temperature of the sample and the sensitivity of the starch solution used, there will be a blank correction of varying magnitude, which must be determined. The blank is especially important in the case of samples containing only small amounts of sulfite.

Reagents:

Potassium iodate standard solution

Dissolve 0.5661 grams of reagent-grade potassium iodate, KIO_3, in about 100 ml of distilled water, add 1.0 gram of sodium carbonate, and dilute to 1 liter with distilled water.

Potassium iodide solution

Dissolve 10 grams of iodate-free potassium iodide, KI, and 0.2 gram sodium carbonate in 100 ml of distilled water. Store in a stoppered bottle fitted with a dropper pipette to transfer 1 ml to the sample. If a color develops, discard and make up fresh.

Hydrochloric acid, 1:1

Concentrated reagent-grade hydrochloric acid is diluted with an equal volume of distilled water. It is helpful to provide a pipette with rubber bulb for the acid bottle for transfer of 3 ml of acid to the sample.

Starch indicator solution

The starches referred to here should be of tested sensitivity to small amounts of iodine in water. Make a water paste with 6 grams of root starch, such as potato or arrowroot, and after dispersing any lumps, dilute to 1 liter with distilled water, and stir. Add 20 grams of potassium hydroxide, KOH, and dissolve with stirring. Let stand for 2 hr to dissolve the starch. Add 27.5 ml of concentrated hydrochloric acid, HCl, and 6 ml glacial acetic acid, and stir. Adjust to pH 4, using potassium hydroxide or hydrochloric acid as required. Use reagent-grade chemicals for the above. Keep in stoppered bottles. This solution should keep for appreciable periods.

Sulfite-free boiler water for blank

Prepare synthetic boiler water of average composition, or oxidize the sulfite in an average sample of boiler water with a small amount of bromine water, and boil off all excess bromine.

Preparation of Samples: Since the sodium sulfite is oxidized by the air and the rate of oxidation increases rapidly with increase in temperature, the sample should be cooled to 80°F, or less. Cooling the sample to 50 or 60°F will increase the sensitivity of the starch indicator. Air is excluded while the sample is collected and stored prior to titration. The recommended procedure is to insert the end of the sample cooler outlet nipple or an extension in the form of a rubber hose into a 250-ml Erlenmeyer widemouthed flask or widemouthed bottle, which should be filled without splashing from the bottom. The flask should be allowed to overflow so as to remove the air. Finally, the full flask is stoppered, excluding air, until the titration is to be made. It is preferable to use a flask with a 200-ml calibration mark instead of a bottle.

Procedure: In the following steps, proceed rapidly to prevent loss of sulfite due to oxidation in the air.

Pour out the excess water from the flask to the 200-ml calibration mark, and place over a white background.* Add 3 ml of 1:1 hydrochloric acid, 1 ml of starch solution, and 1 ml of potassium iodide solution. While stirring, titrate the sample with the standard potassium iodate solution. As the titration proceeds, the starch will develop a blue color which will

* If there is difficulty in seeing the end point, a white casserole can be substituted for the flask and white background.

fade out more and more slowly until the addition of one more drop will produce a permanent light-blue color. Record the milliliters of potassium iodate solution used to reach the endpoint. Make a blank determination on 200 ml of sulfite-free boiler water following the same procedure given for the sample. The milliliters of potassium iodate required to produce a blue color is the blank.

Calculation: Each milliliter of titrating solution used for the sample in excess of the blank for boiler water containing no sulfite is equivalent to 5 ppm of sodium sulfite.

Example: Suppose a 200-ml sample required the addition of 1.2 ml of the iodate solution for titration and that the blank is 0.2 ml, then

$$\text{ppm sodium sulfite} = (1.2 - 0.2) \times 5 = 5$$

pH VALUE (HYDROGEN-ION CONCENTRATION)

Definition. The pH value of the water is defined as the logarithm of the reciprocal of the gram equivalents per liter of the hydrogen ion. It may be expressed by the equation

$$\text{pH} = \log \frac{1}{\text{H}^+}$$

where H^+ designates hydrogen-ion concentration in moles per liter.

Pure water at 25°C has a concentration of H^+ ions of 10^{-7} and a concentration of OH^- ions of 10^{-7}, in moles per liter. This is equal to a pH value of 7 and pOH value of 7. It will be noted that, when the hydrogen-ion concentration is expressed as a negative power of 10, the pH value is equal to the exponent without the minus sign.

The product of the concentrations of the hydrogen ion, H^+, and hydroxyl ion, OH^-, is equal to 10^{-14}, and the sum of the pH and pOH values is equal to 14. Since the pOH value equals 14 − pH, the pH value only is reported.

More simply explained, the pH value is a number between 0 and 14 which represents the degree of alkalinity or acidity of the water. When the pH is 7, the water is said to be neutral, since it contains the same concentration of H^+ ions as OH^- ions. The H^+ ions are acid ions, and the OH^- ions are alkaline ions. When the pH is less than 7, the water is acid, since in that case, the H^+ ions exceed the OH^- ions; and when the pH is more than 7, the water is alkaline.

Application. The pH value of water is generally used in the control of chemical treatments given to the raw water, boiler feedwater, and boiler water. The data are also useful in the control of vents on heaters and in the interpretation of conductivity and alkalinity tests.

Since the pH value is a measure of the activity of the alkalinity or acidity in the water, it is of special interest in the investigation and inhibiting of corrosion or embrittlement of the metal in the system. It gives an indication of whether or not certain chemical reactions have taken, or will take, place in the water.

Methods of Determination. For accurate results, the pH value is determined by the electrometric method, using glass and calomel electrodes immersed in the sample and a potentiometer, known also as a "pH meter," which is calibrated to read in terms of pH. Depending on the pH meter used, the current may be supplied from batteries or from a power pack. The latter is preferred for general plant work.

Less accurate results are obtained by the color-comparator method. With this method, a suitable indicator is added to the sample, and the resulting color is compared with fixed color standards for that indicator. Comparators, indicators, and color standards covering certain ranges of pH are supplied. Since the indicator solution will, in many cases, affect the pH of the sample, the results are likely to be in error. There are other sources of error, such as the accuracy of the color standards and poor color matching. However, the data may be used in an empirical manner to show marked changes in the pH of the water.*,†

DISSOLVED OXYGEN IN FEEDWATER

Application and Description. The basic method for the determination of dissolved oxygen in water was developed by Winkler and has been recognized as the standard procedure. In this method, the dissolved oxygen in the water converts manganous ion to manganic ion, which, in turn, releases iodine from potassium iodide in amounts equivalent to the dissolved oxygen. The iodine is measured by titration with standard sodium thiosulfate solution. Starch indicator is used to determine the end point.‡

In applying the method to the determination of small concentrations of oxygen in feedwater (less than 0.01 ml per liter), it has been necessary to modify the basic Winkler procedure to minimize errors resulting from sampling, interfering ions, variations in chemical reagents, and operator technique. The following method is the Schwartz-Gurney modification of the Winkler procedure.

* Electrometric apparatus and equipment for making colorimetric tests can be purchased from many laboratory-supply companies. There are a number of types of testing methods which may be used.

† Color standards and instructions for determining the pH value of liquids can be purchased from a number of laboratory-supply companies.

‡ Under certain conditions, such as lack of low-temperature cooling water, the titration end point may be determined more accurately by the use of electrometric indicator equipment.

Apparatus:

One 500-ml glass sampling vessel
Two 250-ml glass sampling vessels
Three 100-ml oil bottles for fixing solution fitted with 2-ml pipettes on rubber stoppers
One 750-ml white porcelain casserole and stirring rod
Two microburettes, 10-ml capacity, with 0.05-ml graduations

Reagents:

Manganous sulfate solution

Dissolve 480 grams of $MnSO_4 \cdot 4H_2O$ in distilled water, and make up to 1 liter.

Alkaline potassium iodide solution

Dissolve 360 grams of sodium hydroxide (or 500 grams of potassium hydroxide) and 100 grams of potassium iodide in 1 liter of water.

Hydrochloric acid, concentrated
Sodium thiosulfate solution

Dissolve 24.82 grams of $Na_2S_2O_3 \cdot 5H_2O$ in 1 liter of distilled water to make a $0.1N$ stock solution. Dilute the $0.1N$ stock solution, as required, to make a $0.005N$ solution for titration. Weaker solutions may be used, but when starch is used, the end point will not be so easy to see.

Starch indicator

Grind 0.5 grams of root starch (arrowroot or equivalent) in a mortar with a little distilled water, and pour into 100 ml of boiling distilled water. Allow to cool and settle, and use the clear supernatant solution. This solution may be preserved for a short time by adding 0.12 grams of salicylic acid. It should be made up fresh every few days.

Potassium iodate solution

Dissolve 3.25 grams of $KIO_3 \cdot HIO_3$ or 3.5667 grams of KIO_3 in 1 liter of distilled water to make a $0.1N$ stock iodate solution. Dilute the stock iodate solution as required to make $0.005N$ solution for titration.

Preparation of Samples: Care must be exercised to prevent contamination of the sample from the atmosphere during collection and from addition of reagents. The sample should be obtained at a temperature as close to 21°C as possible (Fig. 1). Thorough flushing of all sample lines is essential, and the rate of water flow should be adjusted to from approximately 750 ml to 1 liter per minute. The water should flow through the sample vessel for not less than 15 min. Sampling vessels may be Erlenmeyer flasks with three-hole rubber stoppers, modified gas sampling tubes, or glass reagent bottles with ground-glass stoppers. In

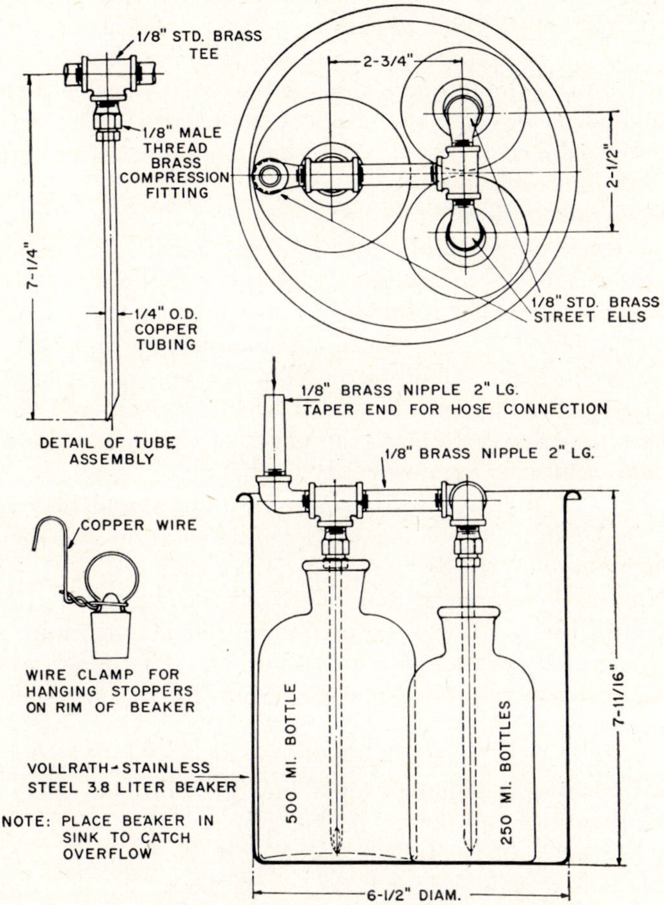

Fig. 1. Apparatus for dissolved-oxygen test.

all cases, thorough flushing and freedom from gas bubbles is absolutely essential.

If Erlenmeyer flasks are used, the sample flows first through a 250-ml bottle and then a 500-ml bottle. The stopper should not be inserted until the flask is well flushed. If gas sampling tubes are used, the sample is divided, part flowing to a 250-ml bottle and part to a 500-ml bottle. If reagent bottles are used, a three-arm manifold of copper or stainless-steel tubing is required to fill a 500-ml bottle and two 250-ml bottles simultaneously. The outlet ends of the three tubular arms should have openings of such size as to distribute the flow of sample water proportionately between the three sample bottles. Three reagent bottles are placed in copper or corrosion-resistant containers of appropriate size, approximately 6½ in. in diameter by 8 in. in height. The assembly for the copper manifold and the pail, as shown in Fig. 1, is such that the overflow of water from the container will leave the tops of the bottles submerged. The reagent bottles should have glass stoppers with conical tips, and the stoppers should be wired to the necks of the bottles. The bottles must be adjusted in capacity to hold 500 ml and 250 ml of water by filling the bottoms with nonreactive solid material.

Allow the water to flow through the bottles until it overflows from the container. Raise the manifold until all arms disengage from the bottles and then lower it into the container. Allow the container to overflow for not less than 15 min or for a sufficient time to purge the container of all oxygen-contaminated water. Raise the manifold and reinsert the arms into the three bottles. Allow to overflow for not less than 20 min to flush out thoroughly any of the water which first entered the bottles. Lift the manifold slowly to clear the necks of the bottles, and lower it into the pail. Samples are now ready for fixing.

Procedure: Stopper one of the 250-ml. samples, and remove it from the pail. Add 2 ml of the manganous sulfate solution and then 2 ml of the alkaline potassium iodide fixing solution to each of the two remaining bottles. Insert the stoppers, and mix by inverting the bottles several times. Return the bottles to the pail, and after about 5 min, add 2 ml of the concentrated hydrochloric acid solution. Restopper, and mix as before until all the precipitated material from the first two solutions has been dissolved.

Remove both bottles from the pail, and pour the liquid in the two 250-ml bottles, known as the blank sample, into the white porcelain casserole used for the titrations. A 600-ml beaker with a white background underneath may be used in place of the casserole. If the fixed solution contains considerable yellow color, which is due to the liberation of iodine by the reaction with dissolved oxygen, it should be titrated with the 0.005*N* thiosulfate solution until only a light-yellow tint remains. Then add approximately ½ ml of starch indicator, which should give a blue color.

If the sample shows no yellow color and the addition of the starch indicator produces no blue color, it is likely that the amount of dissolved oxygen is small and/or the temperature of the solution is too high for the starch indicator to be sensitive. In this case, add exactly 1 ml of 0.005*N* potassium iodate solution to the mixture. This should produce the blue starch color.

Continue the titration of the blank sample with the 0.005*N* sodium thiosulfate until the blue color just disappears. Record the milliliters of standard solutions used under the symbol *B*.

Transfer the solution in the 500-ml sample bottle to the casserole and proceed with the titration in the same manner as for the blank sample. If, for example, it was necessary to add 1 ml of 0.005*N* potassium iodate to produce the blue starch color in the blank sample, the same amount should be added in this titration. Record the milliliters of standard solutions used under the symbol *S*.

It is desirable that the blank sample in the two 250-ml bottles be titrated first, since, if any of the potassium iodate solution is required, it will be evident in this sample.

Calculations:

$$\text{ppm of dissolved oxygen} = S - B \times 0.16$$

Example: If the 500-ml sample required 0.5 ml of thiosulfate and the blank in the two 250-ml bottles required 0.2 ml of thiosulfate, then

$$\text{ppm dissolved oxygen} = (0.5 - 0.2) \times 0.16 = 0.048$$

If equal amounts of 0.005*N* potassium iodate are purposely added to produce the blue color in the blank and the sample, the effect of adding the potassium iodate is canceled out.

If it is desired to express the results in terms of milliliters of dissolved oxygen per liter of sample, multiply the parts per million of dissolved oxygen by 0.698 to obtain dissolved oxygen at 0°C and 760 mm Hg pressure.

Alternative Methods:

Standard Winkler Procedure: If it is not necessary to test for extremely low values, the standard Winkler method will be satisfactory. The procedure for this test is the same as the Schwartz-Gurney modified method, but a 250-ml sample only is fixed for titration. In this case, no blank is made or subtracted.

ASTM Referee Method: The ASTM referee method, Designation D-888-49T, can be found in the ASTM books of standards. In this method, two sampling tubes of 500-ml capacity each are used to collect the sample. The sample in one tube is used as a partial blank. The end point of the titration is found by a potentiometric procedure given in the method.

Heat Exchange Institute Method: The Heat Exchange Institute Method is in most respects similar to the ASTM Referee Method, Designation D-888-49T.

Bibliography:

Schwartz, M. C., and W. B. Gurney: The Determination of Traces of Dissolved Oxygen by the Winkler Method, *ASTM Proc.*, **34** (II): 796 (1934).

Daugherty, T. H.: Technique in the Determination of Dissolved Oxygen, *ASTM Proc.*, **37** (II): 615 (1937).

"Standard Methods for the Examination of Water and Sewage," 9th ed., pp. 106–110, American Public Health Association and American Water Works Association, New York, 1946.

DISSOLVED AND SUSPENDED SOLIDS

These methods consist of evaporation of a measured sample to dryness and weighing the residue. If only dissolved solids are to be determined, the sample is filtered. The suspended solids are determined by subtracting the dissolved solids from the total solids found by evaporating the unfiltered sample. For routine control purposes, this is too time-consuming, and in the case of steam condensates, it requires special apparatus.

Total dissolved solids can be determined by measuring the specific conductance, or conductivity, of the water, expressed in micromhos, and by applying a factor. This method is preferred for rapid control work, and it is entirely satisfactory for routine control. It is the only practical control method for estimation of solids in steam.

Gravimetric Evaporation Method. If only dissolved solids are to be determined, the sample should be settled clear or filtered to remove any suspended matter. If the suspended matter is to be determined, both the filtered and unfiltered samples should be evaporated. The difference will be the suspended matter.

Heat a 150-ml pyrex beaker or an aluminum-foil or platinum evaporating dish to 130°C, dry in a desiccator, and weigh accurately. Small beakers or aluminum-foil dishes will serve for boiler water or samples containing appreciable amounts of dissolved solids, but platinum is used for steam condensate or very dilute solutions. Measure a suitable sample of the water, using 100 ml in the case of boiler water, 500 to 1,000 ml for most other water samples, and 5 to 20 liters for steam condensate.

For the determination of solids in steam condensate, the sample must be protected from contamination, such as by dust from the air, and the sample containers should be of corrosion-resistant metal, such as stainless steel, block tin, or aluminum. The tubular connections to the evaporating dish should be of the same type of material.*,†

For the determination of solids in waters of higher concentration than

* ASTM Designation D-1069-49T, "Suspended and Dissolved Solids in Industrial Water."

† See procedure for determining total solids in condensate samples, Chap. 15.

steam condensate, evaporate to dryness over a water or steam bath or on a sand bath or carefully controlled hot plate. It is imperative to avoid splattering. Evaporation under a radiant-heat lamp is useful for this purpose. Dry in an oven at 103°C for 1 hr, cool in a desiccator, and weigh. Total solids in parts per million are calculated by dividing the increase in weight of the evaporating dish (expressed in milligrams) by the volume of the sample used (expressed in liters).

Conductivity Method. In many cases, this method is the only practical means of observing continuously or at frequent intervals the changes in the concentrations of dissolved electrolytes in water, or of detecting and estimating the impurities in condensed steam.

The method is based on the fact that the electrical conductivity of water increases or decreases with the concentration of the ionized dissolved material. Different ions increase the conductivity by different amounts, but in most cases there is sufficient similarity in the composition of the water so that a direct relationship can be used. This is especially true if the high conductivity acid or hydroxide ions are neutralized. If these ions predominate, the conductivity is changed by a much greater amount than it is by equivalent concentrations of the ions of neutral salts.

The result of this test is reported in terms of specific conductivity, expressed as micromhos, at a standard temperature of 77°F, or 25°C. A conductivity of 1×10^{-6} reciprocal ohm (mho) is equivalent to 1 micromho. In this case, specific conductance, or conductivity, is defined as the reciprocal of the resistance in ohms, measured between opposing faces of a cube 1 cm on edge, of an aqueous solution at a specified temperature.

When it is required that the results in micromhos be converted to equivalent parts per million, the conversion factor may be found by evaporating and weighing the soluble solids in a number of representative clear samples for which the conductivity is known. The results in parts per million are then divided by the micromhos conductivity. The average of these results is used as the multiplying factor for converting micromhos into parts per million of solids by evaporation. For boiler water, the conversion factor should be in the range of 0.45 to 0.65, but if the caustic alkalinity is neutralized, as by addition of 5 per cent sulfuric acid or powdered gallic acid, the factor should be fairly steady at about 0.8.

For steam condensate, the conversion factor may vary over a range of 0.4 to 0.8, depending on the dissolved ionized gases and the non-ionized solids, such as silica, in the condensate.

It is possible to apply corrections to the conductivity readings by using data such as shown in Fig. 6, Chap. 15. However, it is preferable to remove dissolved gases by an efficient degasser. This is especially true when the conductivity of the steam condensate is in excess of 1 micromho.

In most instances, the conductivity readings are taken continuously on flowing samples, and the results in micromhos are shown on a conductivity-recorder chart.

In the case of laboratory tests, indicating conductivity instruments are used, and, except for boiler water or waters of similar character which do not change rapidly on exposure to air, the tests should be made on flowing samples.

Apparatus for determining the conductivity of water can be purchased from laboratory-supply companies, which furnish complete instructions for manipulation of the equipment and interpretation of the results.

COPPER, BY THE COLORIMETRIC METHOD

Application and Description. This method* is intended for the determination of small amounts of copper, either suspended or dissolved, in boiler feedwater or samples of condensate for which the determination of copper is of interest. Other samples containing small amounts of copper may also be tested by this method.

The procedure is based on the reaction of the soluble copper in the sample with diethanolamine in the presence of sulfide when the pH of the mixture is 5 to 6. This results in the development of a yellow color, the intensity of which is proportional to the concentration of copper.

Apparatus: The color may be measured by comparison with color standards or Nessler tubes, but the use of an electrophotometer such as the Fisher or Klett-Summerson is preferred. The photometer readings are converted to milligrams of copper from a calibration curve sheet on which the milligrams of copper are plotted as a function of the photometer reading. A light filter of 425 mμ or approximate equivalent is used.

Reagents:

Standard copper solution

Dissolve 0.1 gram of pure copper in 5 ml of 8*N* nitric acid, add 1 ml of concentrated sulfuric acid, and fume to dryness. Dissolve the residue in copper-free distilled water, and dilute to 1 liter. One milliliter of this solution is equivalent to 0.1 mg of copper. Dilute 50 ml of the above stock solution to 1 liter with copper-free distilled water. One milliliter of this solution is equivalent to 0.005 mg of copper.

Diethanolamine solution

Dissolve 4 grams of diethanolamine† in 200 ml of methanol. Dissolve 1 ml of carbon disulfide in 200 ml of methanol. Mix equal volumes of

* Woelfel, Warren C., Colorimetric Determination of Copper with Carbon Disulfide and Diethanolamine, *Ind. Eng. Chem.*, **20**: 722 (1948).

† Eastman 1598.

the diethanolamine and carbon disulfide solutions sufficient for one day's use only.

Hydrochloric acid, 1:1
Ammonium hydroxide, 1:1
Sodium pyrophosphate solution

Dissolve 30 grams of sodium pyrophosphate, $Na_4P_2O_7 \cdot 10\ H_2O$, and dilute to 1 liter with copper-free distilled water.

Sodium acetate solution

Dissolve 200 grams of sodium acetate trihydrate in distilled water, and dilute to 1 liter with copper-free distilled water.

Copper-free distilled water

Preparation of Samples: A sample of sufficient volume should be collected so that when it is evaporated the residue will contain between 0.0025 and 0.15 mg of copper in the 100-ml final volume of sample plus reagents. If samples of boiler feedwater are taken, it is likely that 500 to 1,000 ml will be sufficient. Frequently, both iron and copper are determined on the same sample. In this case, four-fifths of the evaporated residue may be used for the determination of copper and the balance for iron.

Collect the samples in acid-cleaned glass bottles with glass or rubber stoppers, and add 1 ml of 1:1 hydrochloric acid and 1 ml 8*N* nitric acid to each liter of sample to dissolve or keep in solution the copper and iron.

Evaporate the acidified samples on a hot plate in 600- or 1,000-ml pyrex glass beakers fitted with watch-glass covers to permit easy escape of the vapor and at the same time protect the sample from contamination. If it is found that suspended copper is dissolved with difficulty, continue the evaporation to dryness, and then take up with 2 ml of 1:1 hydrochloric acid and about 25 ml of copper-free water. If the copper is readily soluble, evaporate to about 25 ml only.

Procedure: With a rubber policeman, scrub down the sides of the 600-ml beaker with the acid residue remaining after evaporation of the sample. Transfer the residue to a 100-ml beaker, and after washing the sides of the beaker with copper-free distilled water, add the washings to the beaker. Make up the total volume to about 50 ml.*

Add 1 ml of the sodium pyrophosphate and 10 ml of acetate buffer solution.

Using a pH meter with external electrodes, adjust the pH of the above mixture to between 5.0 and 6.0 with ammonium hydroxide solution. Add

* If iron is to be determined, make up to exactly 50 ml and remove exactly 10 ml with a pipette for the iron test.

dropwise, with stirring, the 1:1 ammonium hydroxide until the desired pH of, say, 5.2 has been reached.

Set the null point of the electrophotometer, using a blank in the comparison cell. Prepare the 100-ml blank from copper-free distilled water plus the same reagents used for the sample.

Transfer the sample in the beaker to a 100-ml measuring tube and add 1 ml of diethanolamine reagent mixture. Dilute to 100 ml with copper-free distilled water, and mix. Allow to stand for 30 min in order to dissipate any color due to iron. Transfer a portion of the sample to a photometer cell,* and record the dial reading. From the photometer† reading, find the milligrams of copper in the 100 ml of mixture, as shown on a standard calibration curve.

If the color due to copper is so strong as to give a reading which is out of range of the instrument and standard curve, repeat the test with a less concentrated portion of the sample rather than dilute after developing the color.

Calculation: The parts per million of copper in the sample may be calculated by the use of the following formula:

$$\text{ppm of Cu} = \frac{\text{mg copper} \times 1{,}000}{\text{volume of original sample} \times \text{fraction of concentrated sample used}}$$

Example: Suppose 0.016 mg of copper was found by the photometer reading, the sample evaporated was 1,090 ml, and the fraction of the sample used for copper was $\frac{4}{5}$, then, substituting in the above formula

$$\text{ppm Cu} = \frac{0.016 \times 1{,}000}{1{,}090 \times \frac{4}{5}} = 0.018$$

IRON BY THE COLORIMETRIC METHOD

Application and Description. The following method‡ for the determination of small amounts of iron in condensate or feedwater samples is dependent on the development of a pink color, which is proportional to the amount of iron present. The pink color is developed because of the formation of a phenanthroline-iron complex. This occurs in the presence of hydroxylamine hydrochloride at a pH of 2.0 to 9.0.

Apparatus: The color may be measured by comparison with color standards or Nessler tubes, but the use of an electrophotometer such as

* Frequently, the large photometer cells which give 40 to 50 mm solution depth for the light path are preferred, but small cells may be used for higher concentrations of copper.

† Suitable photometers, with the necessary cells, can be purchased from all laboratory-supply companies.

‡ ASTM Designation, D-1068.

the Fisher or Klett-Summerson is preferred. The photometer readings are converted to milligrams of iron from a calibration curve sheet on which the milligrams of iron are plotted as a function of the photometer reading. A light filter of 425 mμ or approximate equivalent is used.

Reagents:

Standard iron solution

Prepare a standard iron solution by dissolving 0.7023 grams of ferrous ammonium sulfate, $FeSO_4 \cdot (NH)_2SO_4 \cdot 6H_2O$, in 100 ml of distilled water and 5 ml of 95 per cent sulfuric acid. Dilute to 1 liter with iron-free distilled water. Dilute 100 ml of this solution to 1,000 ml to obtain a standard solution containing 0.01 mg of iron per milliliter. One milliliter of this solution diluted to 100 ml gives a concentration of 0.1 ppm.

Hydrochloric acid, 1:1
Hydroxylamine hydrochloride solution

Dissolve 10 grams of reagent-grade hydroxylamine hydrochloride, $NH_2OH \cdot HCl$, in 90 ml of distilled water.

Acetate buffer solution

Dissolve 100 grams of ammonium acetate, $NH_4C_2H_3O_2$, in 100 ml of distilled water, add 200 ml of concentrated acetic acid, dilute to 1,000 ml with redistilled water, and mix.

Ammonium hydroxide, 1:1
Congo Red test paper (for use when pH meter is not available)
Phenanthroline solution

Dissolve 0.125 grams of reagent-grade orthophenanthroline in 10 ml of iron-free ethanol, and dilute to 100 ml with redistilled water.

Iron-free redistilled water

Redistill steam condensate and store in clean glass bottles.

Preparation of Samples: A sample of sufficient volume should be collected so that, when it is concentrated, the residue will contain between 0.005 and 0.1 mg. of iron in 50 ml of liquid, or a concentration of not less than 0.1 ppm.

If samples of steam condensate or boiler feedwater are being tested, it is likely that not less than 250 ml of sample will need to be concentrated to approximately 25 ml. If copper is to be determined at the same time, a 1,000-ml sample will need to be evaporated. In that case, one-fifth of the concentrate is used for the iron determination, and the balance is used for the copper determination. In any event, the final 50 ml of concentrated sample and reagents should contain between 0.005 and 0.1 mg of iron. If the comparison curve for the photometer is extended, the maximum iron may be increased to 0.5 mg.

The samples are taken from the samples lines and cooling coils in glass-stoppered glass bottles to which 1 ml of 1:1 hydrochloric acid is added. The addition of the acid will keep the iron in a soluble state and prevent it from plating out on the sides of the bottle.

For the most accurate results, the samples should be taken through stainless-steel lines and cooling coils. Copper sampling equipment may be used if the copper determination is to be omitted.

The acidified sample is then evaporated on a hot plate in a clean 600-ml pyrex beaker fitted with a watch-glass cover, which will permit the vapor to escape easily. The samples are evaporated until only about 25 ml or less of residue remains. The beaker is then stored under a suitable cover to cool prior to the testing of the sample. If the sample should be evaporated to dryness, take up again with 1 ml of 1:1 hydrochloric acid, wash down the sides of the beaker with the acid, and finally wash with a little water.

Procedure: With a rubber policeman, scrub down the sides of the beaker with the acid residue remaining after evaporation of the sample. Transfer the residue to a 50-ml cylinder or Nessler tube, and after washing the sides of the beaker with iron-free water, add the washings to the tube. The total volume should now be about 40 ml. This will permit adding the reagents and still not exceed a total final volume of 50 ml.* Other final volumes may be used if the comparison cell is smaller. Cells of 40- to 50-mm light depth are preferred.

To the sample which is now in the tube add the following reagents in the order given, and mix between each addition:

1. One-half milliliter of hydroxylamine hydrochloride solution.
2. Two milliliters of phenanthroline solution.
3. Acetate buffer solution, 2.5 ml.
4. Approximately 1 to 1½ ml of 1:1 ammonium hydroxide solution, which should be just sufficient to show red color with the congo red indicator paper. If the paper is dark in color, add more ammonium hydroxide. Alternately, adjust pH to 4.2 with pH meter.
5. Iron-free water to dilute the above mixture to the 50-ml mark on the tube. Mix.

Set the null point of the electrophotometer, using a blank in the comparison cell. Prepare the blank from iron-free distilled water plus the

* If copper is to be determined, make up to exactly 50 ml and remove exactly 40 ml with a pipette for the copper test. Dilute the sample remaining in the tube with iron-free distilled water to approximately 40 ml.

reagents used for the sample, and dilute to 50 ml. Place the sample in a comparator cell, and record the dial reading indicating the intensity of the color developed. From the photometer reading, find the milligrams of iron in the sample as shown on a standard calibration curve.

If the color due to iron is so strong as to give a reading which is out of range of the instrument and standard curve, repeat the test with a less concentrated portion of the sample rather than dilute after developing the color.

Calculation: The parts per millions of iron in the sample may be calculated by the use of the formula

$$\text{ppm Fe} = \frac{\text{mg iron} \times 1{,}000}{\text{volume of original sample} \times \text{fraction of concentrated sample used}}$$

Example: If 0.015 mg of iron was found, the sample evaporated was 1,000 ml, and the fraction of the sample used for iron was ⅕, then, substituting in the above formula

$$\text{ppm Fe} = \frac{0.015 \times 1{,}000}{1{,}000 \times \frac{1}{5}} = 0.075$$

AMMONIA BY DIRECT NESSLERIZATION

The direct nesslerization method[*,†,‡] is useful for the determination of ammonia in steam condensate and boiler feedwater. Knowledge of the ammonia concentration in condensates and boiler feedwater is useful in interpreting the results of conductivity and pH measurements and in regulating the ammonia concentration. In the case of colored and raw waters, it is recommended that the ammonia be separated first by distillation methods. Organic nitrogen not present as free ammonia will not be determined by the direct Nesslerization method.

The method is based on the development of a yellowish-brown color when Nessler reagent is added to a water sample containing ammonia.

FREE CARBON DIOXIDE, BY ACIDITY TITRATION

Application and Description. Carbon dioxide may be present in water as carbonates, bicarbonates, and carbonic acid, which is also known as free carbon dioxide. The bicarbonates and carbonates are determined by the alkalinity tests. However, when the water contains free carbon

* "Standard Methods for the Examination of Water and Sewage," 9th ed , p. 64, American Public Health Association and American Water Works Association, New York, 1946.

† Symposium on Steam Purity, *ASTM Proc.*, **41**: 1261 (1941).

‡ Schwartz, M. C., W. B. Gurney, and T. E. Crossan, Determination of Ammonia in Condensed Steam, *Trans. ASME*, **62**: 723 (1940).

dioxide, it is acid to a phenolphthalein-indicator neutral point of pH 8.3. In this case, the free carbon dioxide is determined by making an acidity titration with standard alkali solution and phenolphthalein indicator.

If it is known that some other acid is present, the result should be expressed in terms of that acid rather than of free carbon dioxide. Therefore, it is necessary to know the history and treatment of the water before reaching a conclusion regarding the kind of acid in the water.

The determination of free carbon dioxide in water provides data which are of value in the chemical treatment of raw-water supplies, the regulation of vents on heaters and deaerators, and the interpretation of conductivity tests made on samples of steam condensate.

Reagents:

Neutral phenolphthalein-indicator solution

Neutralize the indicator solution with drops of $N/44$ sodium hydroxide solution until a faint pink color is seen.

Sodium hydroxide solution, $N/44$

Prepare with reagent-grade sodium hydroxide, NaOH, dissolved in freshly boiled and cooled distilled water. Standardize with $N/50$ or $N/44$ sulfuric acid, and store in a pyrex glass or plastic bottle. Protect from contact with the air by drawing off with a siphon, using an Ascarite-filled breathing tube or a wash bottle containing strong caustic solution for admitting air to the reagent bottle. Make up a fresh solution at one- or two-month intervals.

Preparation of Samples: Water samples containing more or less carbon dioxide than water in equilibrium with the atmosphere will change in carbon dioxide content when standing in contact with air. Thus, if the water is saturated with free carbon dioxide, CO_2, it will lose carbon dioxide to the atmosphere, or if the water contains very little free carbon dioxide in it will absorb carbon dioxide from the air. These facts must be considered in taking and handling the samples.

When testing raw or treated waters which have been exposed to the air prior to sampling, the samples may be taken in the usual manner, but they should be titrated as soon afterward as possible.

When testing steam condensates which contain large or small amounts of free carbon dioxide and are subject to rapid change, the sampling procedure should be more exact. In this case, the samples should be taken in 500-ml widemouthed pyrex bottles or flasks fitted with rubber or glass stoppers. Actually, these containers should have a maximum capacity of about 550 or 600 ml. The containers should have a calibration mark at 500 ml or, if that is not feasible, at 450 or 400 ml.

If the sample is hot, it should be passed through a cooling coil which is throttled at the outlet end, and it should be cooled to room temperature. The sample container should be filled through a sample tube or hose reaching to the bottom, and if any air is being drawn in through the sample connections, the connections should be tightened before proceeding. The sample flow should be continued at a rate of about 500 ml per minute until the water initially collected is replaced several times. The stopper is then washed with the overflowing sample, and after withdrawing the sample tube, the stopper is inserted in such a way as to exclude air. If the test can be made on the spot without any loss of time, it will not be necessary to insert the stopper.

Procedure: Remove the stopper from the pyrex sample flask, and pour off the water from the top until the level reaches the calibration mark. Place the flask over a white background and add five drops of neutralized phenolphthalein indicator. From a burette, add drop by drop, the $N/44$ sodium hydroxide, NaOH, solution used for titration. After each addition, stir the contents of the flask by gentle swirling. When a definite permanent pink color is developed, read and record the milliliters of sodium hydroxide used.

Calculation: Each millimeter of $N/44$ sodium hydroxide is equivalent to 1 mg of free carbon dioxide. Therefore, the formula for calculation becomes

$$\text{ppm } CO_2 = \frac{\text{ml N/44 NaOH} \times 1{,}000}{\text{ml sample}}$$

Example: If the sample is 500 ml and 3.0 ml of $N/44$ NaOH is used for the titration

$$\text{ppm } CO_2 = \frac{3 \times 1{,}000}{500} = 6$$

SOLUBLE SILICA: COLORIMETRIC METHOD

Application and Description. The colorimetric methods described are applicable to the determination of soluble (noncolloidal) silica in clear boiler water or other clear waters except those containing excessive color interference. Methods for insoluble silica are available.*

In these methods, the soluble silica in the sample first produces a yellow color when ammonium molybdate and acid are added. A blue color is then formed when a reducing agent, such as sodium sulfite, is added. The intensity of the blue color is proportional to the concentration of silica.†

* Methods by Armour Research Foundation.

† A small amount of color interference in the samples can be canceled out by the blank recommended, but if a large amount of color, as from tannin treatment, is present, gravimetric procedures are recommended.

The result is read from a standard calibration curve or by comparison with color standards.

Two colorimetric methods for the determination of soluble silica in water are described. These are:

1. Rapid control method: soluble silica concentrations in the range of 0 to 20 ppm.

2. Precise method: soluble silica concentrations in the range of 0 to 3 ppm in the presence of 0 to 150 ppm of phosphate.

The concentration of phosphate ion in the sample may be sufficient to develop some color interference. If the tests show this to be true, add oxalic acid in the prescribed amount to remove the color.

Apparatus: The color intensity may be measured by means of the following equipment: Nessler tubes, Duboscq-type color comparator, Fisher- or Klett-Summerson-type electrophotometer and spectrophotometer.*

Usually, the instruments increase in accuracy in the order listed. For example, the spectrophotometer is the most accurate for measuring small amounts of silica, since it can be operated with light of optimum wave length of about 815 $m\mu$.

The examples given in the methods which follow refer to the electrophotometer. In this case, filters in the range of 415 to 650 $m\mu$ wave length can be used successfully. The filters of higher wave length are more suitable for accurate measurements of silica in the range of 0.05 to 3 ppm. For measurement of very small amounts of silica in the range of 0.00 to 1 ppm, use the spectrophotometer in place of the filter photometer.

Reagents for Rapid Control Method (0 to 20 Ppm SiO_2):

Sodium silicate standard solution

Fuse 3.0 grams of reagent-grade sodium carbonate, Na_2CO_3, with 0.2 grams of pure dry silica in a platinum crucible. Dissolve in 200 ml of distilled water, and store in a plastic or rubber bottle. This solution should contain 1,000 ppm of silica, SiO_2; 1 ml = 1 mg of SiO_2. Alternatively, the sodium silicate solution may be made by dissolving a calculated amount of sodium metasilicate in water. The actual silica in the standard stock solution should be checked by gravimetric test. The standard solution is diluted with distilled water to make the intermediate standards required for the calibration curve.

Hydrochloric acid solution

Add 10 ml of concentrated reagent-grade hydrochloric acid to a volumetric flask, and dilute with distilled silica-free water to the 500-ml mark.

* Apparatus, with instructions for operation, can be purchased from laboratory-supply companies.

Ammonium molybdate solution

Dissolve 50 grams of reagent-grade ammonium molybdate in about 200 ml of distilled water. Dilute to 500 ml with distilled water.

Sodium sulfite solution

Dissolve 85 grams of reagent-grade anhydrous sodium sulfite or 170 grams of reagent-grade $Na_2SO_3 \cdot 7H_2O$, and dissolve in about 450 ml of distilled water. Dilute with distilled water to 500 ml.

Oxalic acid solution

Dissolve 10 grams of oxalic acid dihydrate in 100 ml of distilled water.

Procedure for Rapid Control Method: To 10 ml* of clear sample in a pyrex glass flask, add 5 ml of hydrochloric acid and 5 ml of ammonium molybdate solution. Mix and allow to stand 1 or 2 min to develop the yellow color. Then add 10 ml of sodium sulfite, and allow to stand for 2 min. If the sample contains 60 ppm of phosphate or more, add 1 ml of oxalic acid to destroy the color due to interference from this source.

Prepare a blank and set the null point of the photometer in the following manner:

If color is present in the samples in sufficient intensity to cause measurable interference, add to 10 ml of the sample 5 ml of distilled water, 5 ml of hydrochloric acid, and 10 ml of sodium sulfite solution.

If there is no appreciable color interference, mix 15 ml of silica-free distilled water, 5 ml of hydrochloric acid, 5 ml of ammonium molybdate, and 10 ml of sodium sulfite solution.

Transfer the blank to the color photometer tube, insert into the instrument with the proper light filter, and set the reading dial at zero when the galvanometer is balanced at the center mark.

Place the comparison tube containing the sample and reagents in the photometer, and record the color intensity on the photometer scale when the galvanometer is balanced.

Read the parts per million of silica, as SiO_2, in the sample from the standard calibration curve on which the photometer scale readings have been plotted as a function of parts per million of silica.

Preparation of Standard Calibration Curve: Prepare the standard calibration curve by repeating the foregoing test procedure, using mixtures of silica-free water and silica solutions containing 5, 10, 15, 20, and 30 ppm SiO_2. Then plot on graph paper the photometer scale readings against the parts per million of soluble silica in the standard solutions. The curve should be nearly a straight line.

* If the glass comparison cells of the photometer hold more than 25 ml, use 20 ml or 50 ml of sample as required and standardize accordingly. The same amounts of reagent as for the 10 ml sample are added. If the sample contains more than 20 ppm of silica, dilute as required with silica-free distilled water.

Reagents for Precise Method (0 to 3 Ppm SiO_2):

Sodium silicate standard solution
Hydrochloric acid 1:1

Dilute concentrated hydrochloric acid (sp gr 1.19) with an equal volume of distilled water.

Ammonium molybdate solution

Dissolve 30 grams of ammonium molybdate tetrahydrate, $(NH_4)_6$-$Mo_7O_{24}\cdot 4H_2O$, in 270 ml of distilled water.

Ammonium oxalate solution

Dissolve 30 grams of oxalic acid dihydrate, $H_2C_2O_4\cdot 2H_2O$, in 270 ml of distilled water.

Amino reducing solution

Dissolve 1 gram of sodium sulfite, 30 grams of sodium bisulfite, and 0.5 gram of 1-amino-2-napthol-4-sulfonic acid* in 200 ml of distilled water.

For accurate results, the reagents and distilled water should be the purest available. Store all solutions in plastic bottles.

Procedure for Precise Method: Turn on the light switch, and warm up the instrument. To 100 ml of clear sample in a pyrex flask, add 2 ml hydrochloric acid and 5 ml ammonium molybdate, wait 5 min, and then add 5 ml oxalic acid and 2 ml amino reducing solution. The final pH should be 1.1. If silica is present, the blue color should be fully developed in 10 min.

Prepare the following blank for setting the null point of the photometer: in the amounts stated, add to 100 ml of clear sample all of the foregoing reagents except the ammonium molybdate. Place the blank in a comparison cell, insert into the comparator, and set the dial scale at zero with the galvanometer balanced.

Insert the sample cell in which the blue color has been developed into the photometer and read the dial setting in terms of color intensity when the galvanometer is balanced at the center mark.

Read the parts per million of silica, SiO_2, in the sample by reference to a standard calibration curve (Fig. 2) on which the photometer readings of color intensity are plotted as a function of parts per million of silica. For accurate results in the low-concentration range, the light filter should pass light having a wave length of not less than 650 $m\mu$.

* Eastman 360.

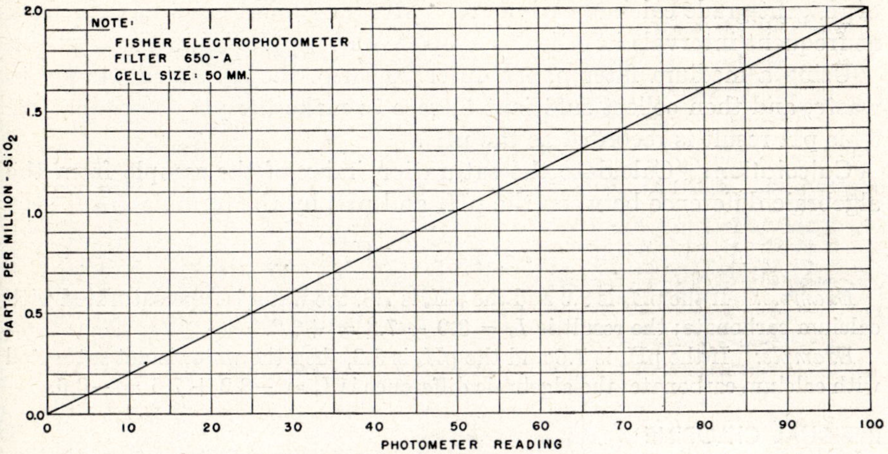

Fig. 2. Silica calibration curve.

SATURATION INDEX: MODIFIED MARBLE TEST

Application and Description. The saturation index (Is) of the water is defined as the algebraic difference between the actual pH, known as the pH_a, and the pH of the water when saturated with calcium carbonate, known as pH_s. The saturation index may be positive or negative depending on the pH_a for the water being tested. The test is made on the water at a standard temperature of 77°F, but if the saturation index is required at some higher temperature, it may be estimated.*

This test is of value when estimating whether calcium carbonate will be precipitated in the system or whether the water is likely to be corrosive to metals with which it comes into contact. For example, if the I_s is positive, the water is likely to deposit calcium carbonate scale which may be protective. If the I_s is negative, no calcium scale will be formed, and the water will be more corrosive.

Apparatus: Electrometric or colorimetric pH-determination equipment.†

Reagents:

Reagent-grade calcium carbonate

Procedure: Bring the sample close to the standard temperature of 77°F (25°C). Determine the pH of the sample using a pH meter and glass and calomel electrodes, as described under the method for pH value. This result is recorded as the pH_a.

To 200 ml of the sample in a 500-ml Erlenmeyer flask, add 2 or 3 grams

* See Fig. 2, Chap. 13.

† See Methods of Determination of pH.

of reagent-grade calcium carbonate. Mix for 5 min by swirling the flask at frequent intervals in order to saturate the sample.

Using a medium filter paper, filter 20 ml of the saturated sample to waste, and then collect sufficient filtrate to make the pH determination. This pH result is recorded as the pH_s.

Calculation: Calculate the saturation index of the sample from the algebraic difference between the pH_a and pH_s by the formula

$$I_s = pH_a - pH_s$$

Example: If the pH_a is 8.0 and the pH_s is 7.8, the water is supersaturated with calcium carbonate; the result is $I_s = 8.0 - 7.8 = +0.2$.

Example: If the pH_a is 7.6 and the pH_s is 8.2, then the water is not saturated with calcium carbonate; the algebraic difference is $I_s = -8.2 + 7.6 = -0.6$.

RESIDUAL CHLORINE

Application and Description of the Orthotolidine Method. When water is treated with chlorine or hypochlorite solutions to aid in coagulation or to prevent growth of slimes on filters, softeners, surfaces of pipes, heat-exchanger tubes, and the like, it is necessary to control the dosage by testing for residual chlorine. Frequently, an accurate titration method is preferred, but if there are no appreciable amounts of color-interfering substances present, a simple color test, using a colormetric comparator, including comparison cells, color standards, and orthotolidine solution, may be used.

Apparatus: Color standards and comparators* are available for 0.05 to 10 ppm of chlorine, but for normal treatment, color standards having a range of 0.05 to 1.0 ppm in steps of 0.05 ppm of residual chlorine are used. As in the case of other color comparisons, a standard light source is necessary in order to obtain accurate results.

Reagents:

Standard orthotolidine solution

Dissolve 1.35 grams of orthotolidine dihydrochloride in 500 ml of distilled water. Add this solution with constant stirring to 500 ml of dilute hydrochloric acid made by mixing 350 ml of distilled water and 150 ml of concentrated hydrochloric acid (sp gr 1.18 to 1.19).

Procedure: Fill one 10-ml-capacity comparison cell to the mark with the sample, and place it behind the standard of the comparator. To a second 10-ml cell, add 0.5 ml† of orthotolidine solution, fill to the mark with the sample, and mix thoroughly. Allow the mixture to stand for

* Available from laboratory-supply houses.

† Use 0.75 ml in 15-ml cells and the same ratio for other volumes.

sufficient time to develop maximum color. Normally, maximum color is developed in 1 or 2 min, but if the sample is cold or contains chloramines, a longer period may be required.

In the case of cooling water to condensers, the maximum color may appear in 1 min or less, since further standing may permit chlorine to be consumed. If the chlorine dose is applied for other purposes, as in closed systems, and residual chlorine after a longer period of reaction is required, the point of sampling and the period of standing to develop maximum color should be given further consideration.

Place the second cell in the reading position of the comparator, and using a standard light source behind the comparator, compare the color developed in the treated sample with the color standards until the color of the sample is matched. If the color match lies between two color standards, the result is estimated. In making the color comparison, the comparator should be at such a distance from the eye that the two fields are brought into clear juxtaposition.

If it is uncertain when maximum color will be developed, take readings at frequent intervals, and note any increase or decrease in depth of color in the sample. The maximum reading should be recorded.

The comparison cells and the comparator must be maintained in a clean condition at all times.

OIL IN WATER

Application and Description. When it is likely that water in the power plant or elsewhere will contain oil which will cause objectionable conditions, it is desirable to determine the amount or presence of the oil and to examine a sufficient number of samples to find the source of the oil. The oil may be determined by either qualitative or quantitative means.

Methods of Determination: If there is sufficient oil in the water, it will be visible as an oil slick on the surface of the water or as an emulsion which appears as turbidity. It may be distinguished from other materials by further tests.

Another qualitative test is to place some of the water in a dish and add small shavings of camphor, which will float on the surface. If no appreciable amount of oil is present, the camphor particles will spin around on the surface. If there is a small amount of oil in the water, the camphor will show no activity. Care should be taken to clean the sample containers thoroughly.

Quantitative tests consist of turbidity or Tyndall-effect measurements, which require standardization with known concentrations of oil in the water, or extraction methods using the following wet extraction method:

Shake the sample with a sufficient number of portions of diethyl ether, petroleum ether, or chloroform. The selection of solvent will depend upon the type of oil contamination present in the water. Extract with each solvent layer, and transfer into a previously dried and weighed light evaporating dish or beaker before again treating the sample with a further portion of solvent.

Evaporate the accumulated extractions in a steam bath or its equivalent, and then weigh the residue. More refined methods have been developed*,† and should be used when justified.

Calculation:

$$\text{ppm of oil in sample tested} = \frac{\text{mg of residue} \times 1{,}000}{\text{ml of original sample}}$$

* American Petroleum Institute, Division of Refining, New York, N.Y.

† American Society for Testing Materials, "Tentative Method of Test for Chloroform-extractable Matter in Industrial Water," Designation, D 1178-51T. ASTM Standards, Part 7, 1952, p. 1152.

APPENDIX

TABLE 1. DECIMAL PARTS OF A MILLIMETER EXPRESSED IN DECIMAL PARTS OF AN INCH

Mm	In.	Mm	In.	Mm	In.	Mm	In.	Mm	In.
1.00	0.03937	0.80	0.0314960	0.60	0.0236220	0.40	0.0157480	0.20	0.0078740
0.99	0.0389765	0.79	0.0311023	0.59	0.0232283	0.39	0.0153543	0.19	0.0074803
0.98	0.0385826	0.78	0.0307086	0.58	0.0228346	0.38	0.0149606	0.18	0.0070866
0.97	0.0381889	0.77	0.0303149	0.57	0.0224409	0.37	0.0145669	0.17	0.0066929
0.96	0.0377952	0.76	0.0299212	0.56	0.0220472	0.36	0.0141732	0.16	0.0062992
0.95	0.0374015	0.75	0.0295275	0.55	0.0216535	0.35	0.0137795	0.15	0.0059055
0.94	0.0370078	0.74	0.0291338	0.54	0.0212598	0.34	0.0133858	0.14	0.0055118
0.93	0.0366141	0.73	0.0287401	0.53	0.0208661	0.33	0.0129921	0.13	0.0051181
0.92	0.0362204	0.72	0.0283464	0.52	0.0204724	0.32	0.0125984	0.12	0.0047244
0.91	0.0358267	0.71	0.0279527	0.51	0.0200787	0.31	0.0122047	0.11	0.0043307
0.90	0.0354330	0.70	0.0275590	0.50	0.019685	0.30	0.0118110	0.10	0.0039370
0.89	0.0350393	0.69	0.0271653	0.49	0.0192913	0.29	0.0114173	0.09	0.0035433
0.88	0.0346456	0.68	0.0267716	0.48	0.0188976	0.28	0.0110236	0.08	0.0031496
0.87	0.0342519	0.67	0.0263779	0.47	0.0185039	0.27	0.0106299	0.07	0.0027559
0.86	0.0338582	0.66	0.0259842	0.46	0.0181102	0.26	0.0102362	0.06	0.0023622
0.85	0.0334645	0.65	0.0255905	0.45	0.0177165	0.25	0.0098425	0.05	0.0019685
0.84	0.0330708	0.64	0.0251968	0.44	0.0173228	0.24	0.0094488	0.04	0.0015748
0.83	0.0326771	0.63	0.0248031	0.43	0.0169291	0.23	0.0090551	0.03	0.0011811
0.82	0.0322834	0.62	0.0244094	0.42	0.0165354	0.22	0.0086614	0.02	0.0007874
0.81	0.0318897	0.61	0.0240157	0.41	0.0161417	0.21	0.0082677	0.01	0.0003937

TABLE 2. DECIMAL OF AN INCH AND OF A FOOT

Fraction of inch or foot		Inch equivalent to foot fraction	Fraction of inch or foot		Inch equivalent to foot fraction	Fraction of inch or foot		Inch equivalent to foot fraction	Fraction of inch or foot		Inch equivalent to foot fraction
	0.0052	1/16		0.2552	3 1/16		0.5052	6 1/16		0.7552	9 1/16
	0.0104	1/8		0.2604	3 1/8		0.5104	6 1/8		0.7604	9 1/8
1/64	0.015625	3/16	17/64	0.265625	3 3/16	33/64	0.515625	6 3/16	49/64	0.765625	9 3/16
	0.0208	1/4		0.2708	3 1/4		0.5208	6 1/4		0.7708	9 1/4
	0.0260	5/16		0.2760	3 5/16		0.5260	6 5/16		0.7760	9 5/16
1/32	0.03125	3/8	9/32	0.28125	3 3/8	17/32	0.53125	6 3/8	25/32	0.78125	9 3/8
	0.0365	7/16		0.2865	3 7/16		0.5365	6 7/16		0.7865	9 7/16
	0.0417	1/2		0.2917	3 1/2		0.5417	6 1/2		0.7917	9 1/2
3/64	0.046875	9/16	19/64	0.296875	3 9/16	35/64	0.546875	6 9/16	51/64	0.796875	9 9/16
	0.0521	5/8		0.3021	3 5/8		0.5521	6 5/8		0.8021	9 5/8
	0.0573	11/16		0.3073	3 11/16		0.5573	6 11/16		0.8073	9 11/16
1/16	0.0625	3/4	5/16	0.3125	3 3/4	9/16	0.5625	6 3/4	13/16	0.8125	9 3/4
	0.0677	13/16		0.3177	3 13/16		0.5677	6 13/16		0.8177	9 13/16
	0.0729	7/8		0.3229	3 7/8		0.5729	6 7/8		0.8229	9 7/8
5/64	0.078125	15/16	21/64	0.328125	3 15/16	37/64	0.578125	6 15/16	53/64	0.828125	9 15/16
	0.0833	1		0.3333	4		0.5833	7		0.8333	10
	0.0885	1 1/16		0.3385	4 1/16		0.5885	7 1/16		0.8385	10 1/16
3/32	0.09375	1 1/8	11/32	0.34375	4 1/8	19/32	0.59375	7 1/8	27/32	0.84375	10 1/8
	0.0990	1 3/16		0.3490	4 3/16		0.5990	7 3/16		0.8490	10 3/16
	0.1042	1 1/4		0.3542	4 1/4		0.6042	7 1/4		0.8542	10 1/4
7/64	0.109375	1 5/16	23/64	0.359375	4 5/16	39/64	0.609375	7 5/16	55/64	0.859375	10 5/16
	0.1146	1 3/8		0.3646	4 3/8		0.6146	7 3/8		0.8646	10 3/8
	0.1198	1 7/16		0.3698	4 7/16		0.6198	7 7/16		0.8698	10 7/16
1/8	0.1250	1 1/2	3/8	0.3750	4 1/2	5/8	0.6250	7 1/2	7/8	0.8750	10 1/2
	0.1302	1 9/16		0.3802	4 9/16		0.6302	7 9/16		0.8802	10 9/16
	0.1354	1 5/8		0.3854	4 5/8		0.6354	7 5/8		0.8854	10 5/8
9/64	0.140625	1 11/16	25/64	0.390625	4 11/16	41/64	0.640625	7 11/16	57/64	0.890625	10 11/16
	0.1458	1 3/4		0.3958	4 3/4		0.6458	7 3/4		0.8958	10 3/4
	0.1510	1 13/16		0.4010	4 13/16		0.6510	7 13/16		0.9010	10 13/16
5/32	0.15625	1 7/8	13/32	0.40625	4 7/8	27/32	0.65625	7 7/8	29/32	0.90625	10 7/8
	0.1615	1 15/16		0.4115	4 15/16		0.6615	7 15/16		0.9115	10 15/16
	0.1667	2		0.4167	5		0.6667	8		0.9167	11
11/64	0.171875	2 1/16	27/64	0.421875	5 1/16	43/64	0.671875	8 1/16	59/64	0.921875	11 1/16
	0.1771	2 1/8		0.4271	5 1/8		0.6771	8 1/8		0.9271	11 1/8
	0.1823	2 3/16		0.4323	5 3/16		0.6823	8 3/16		0.9323	11 3/16
3/16	0.1875	2 1/4	7/16	0.4375	5 1/4	11/16	0.6875	8 1/4	15/16	0.9375	11 1/4
	0.1927	2 5/16		0.4427	5 5/16		0.6927	8 5/16		0.9427	11 5/16
	0.1979	2 3/8		0.4479	5 3/8		0.6979	8 3/8		0.9479	11 3/8
13/64	0.203125	2 7/16	29/64	0.453125	5 7/16	45/64	0.703125	8 7/16	61/64	0.953125	11 7/16
	0.2083	2 1/2		0.4583	5 1/2		0.7083	8 1/2		0.9583	11 1/2
	0.2135	2 9/16		0.4635	5 9/16		0.7135	8 9/16		0.9635	11 9/16
7/32	0.21875	2 5/8	15/32	0.46875	5 5/8	23/32	0.71875	8 5/8	31/32	0.96875	11 5/8
	0.2240	2 11/16		0.4740	5 11/16		0.7240	8 11/16		0.9740	11 11/16
	0.2292	2 3/4		0.4792	5 3/4		0.7292	8 3/4		0.9792	11 3/4
15/64	0.234375	2 13/16	31/64	0.484375	5 13/16	47/64	0.734375	8 13/16	63/64	0.984375	11 13/16
	0.2396	2 7/8		0.4896	5 7/8		0.7396	8 7/8		0.9896	11 7/8
	0.2448	2 15/16		0.4948	5 15/16		0.7448	8 15/16		0.9948	11 15/16
1/4	0.2500	3	1/2	0.5000	6	3/4	0.7500	9	1	1.0000	12

TABLE 3. SUMMER TEMPERATURES OF MUNICIPAL WATER SUPPLIES FOR VARIOUS CITIES USING SURFACE WATER SOURCES*

Location	Temperature at main outlet, °F			
	June	July	August	September
Atlanta	78.1	83.5	79.5	77.8
Baltimore	61.0	66.0	70.0	64.0
Birmingham	78.0	82.0	81.0	79.0
Boston	68.3	74.3	73.4	69.4
Buffalo	62.0	71.0	73.0	66.0
Chicago	55.4	68.0	69.4	62.5
Cincinnati	76.0	82.0	81.0	77.0
Cleveland	58.0	68.0	73.5	71.0
Detroit	64.0	75.0	74.0	68.0
Kansas City	84.0	93.0	91.0	85.0
Louisville	77.0	82.0	82.0	77.0
Nashville	84.0	88.0	88.0	84.0
New Orleans	86.0	89.0	90.0	90.0
Oakland	59.0	62.0	64.0	64.0
Philadelphia	71.0	79.0	77.0	72.0
Pittsburgh	75.2	80.6	80.6	75.2
Sacramento	70.7	70.7	80.6	77.0
St. Louis	77.0	85.0	83.0	75.0
Washington	43.0	67.0	73.0	75.0

* Data taken from "Surface Water and City Water Temperatures During the Summer," published in *Heating and Ventilating,* January, 1939; and U.S. Dept. of Commerce publication, "Effect of City Water and Sewerage Facilities on Industrial Markets," April, 1938.

TABLE 4. SUMMER TEMPERATURES OF MUNICIPAL WATER SUPPLIES FOR VARIOUS CITIES USING WELLS AS A SOURCE*

Location	Temperature at main outlet, °F			
	June	July	August	September
Albuquerque, N. M	72.0	72.0	72.0	72.0
Aurora, Ill	60.0	60.0	60.0	60.0
Camden, N. J	58.0	58.0	58.0	58.0
El Paso, Tex	84.0	85.0	85.0	84.0
Fresno, Calif	72.0	72.0	72.0	72.0
Houston, Tex	84.0	84.0	84.0	84.0
Jacksonville, Fla	84.8	86.3	86.7	82.4
Kalamazoo, Mich	52.0	52.0	52.0	52.0
Lafayette, Ind	53.0	53.0	53.0	53.0
Lansing, Mich	57.5	58.0	59.0	59.0
Lincoln, Neb	58.0	59.0	59.0	59.0
Lowell, Mass	50.0	50.0	50.0	50.0
Madison, Wis	53.0	52.0	52.0	53.0
Marion, Ind	54.0	54.0	55.0	55.0
Montgomery, Ala	70.0	70.0	71.0	71.0
Pensacola, Fla	70.0	70.0	70.0	70.0
Peoria, Ill	56.0	56.0	56.0	54.0
Pontiac, Mich	55.0	55.0	55.0	55.0
San Antonio, Tex	76.0	76.0	76.0	76.0
Sioux Falls, S. D	55.0	55.0	55.0	55.0

* Data taken from "Surface Water and City Water Temperatures During the Summer," published in *Heating and Ventilating*, January, 1939; and U.S. Dept. of Commerce publication, "Effect of City Water and Sewerage Facilities on Industrial Markets," April, 1938.

CENTIGRADE (ABSOLUTE)	CENTIGRADE	FAHRENHEIT	FAHRENHEIT (ABSOLUTE)	
373	100.0	212	672	BOILING
370.2	97.2	207	667	
367.4	94.4	202	662	
364.7	91.7	197	657	
361.9	88.9	192	652	
359.1	86.1	187	647	
356.3	83.3	182	642	
353.6	80.6	177	637	
350.8	77.8	172	632	
348.0	75.0	167	627	
345.2	72.2	162	622	
342.4	69.4	157	617	
339.7	66.7	152	612	
336.9	63.9	147	607	
334.1	61.1	142	602	
331.3	58.3	137	597	
328.6	55.6	132	592	
325.8	52.8	127	587	
323.0	50.0	122	582	
320.2	47.2	117	577	
317.4	44.4	112	572	
314.7	44.7	107	567	
311.9	38.9	102	562	
309.1	36.1	97	557	
306.3	33.3	92	552	
303.6	30.6	87	547	
300.8	27.8	82	542	
298.0	25.0	77	537	
295.2	22.2	72	532	
292.4	19.4	67	527	
289.7	16.7	62	522	
286.9	13.9	57	517	
284.1	11.1	52	512	
281.3	8.3	47	507	
278.6	5.6	42	502	
275.8	+ 2.8	37	497	
273.0	0	32	492	FREEZING
270.2	− 2.8	27	487	
267.4	5.6	22	482	
264.7	8.3	17	477	
261.9	11.1	12	472	
259.1	13.9	7	467	
255.2	17.8	0	460	

$F = 1.8 \times C + 32$

$C = (F - 32) \div 1.8$

ABSOLUTE TEMP.

$F = F + 460$

$C = C + 273$

Fig. 1. Centigrade-Fahrenheit thermometer scales.

Table 5. Capacity and Conversion Tables

Quantity of Water

The gallon, New York Statute measure (1829) = 8 lb of distilled water or 221.184 cu in., with barometer at 30 in., and temperature, 39.83°F. The legal gallon of United States is the old British wine gallon, containing 231 cu in., being the contents of a cylinder 7 in. in diameter and 6 in. high; it contains 8.3389 lb of water at maximum density, weighed in air at 30 in. Hg pressure, and 62°F. 1 U.S. gal = 0.1337 cu ft = 0.8327 imperial gal.

The imperial gallon contains 10 imperial lb of distilled water at 62°F, weighed in air of the same temperature, and with barometer at 30 in. 1 imperial gal. = 1.200 U.S. gal = 277.27 cu in. = 0.160 cu ft.

1 ton of water (2,000 lb) = 240 gal ±

1 drop of water, 4°C (39°F) = 1 minim (U.S.) = 0.0616 cu cm = 0.00376 cu in. = 0.0000162 gal. Experiments by E. G. Bradbury indicated that 16,000 drops of water from metallic surfaces are approximately equal to 1 gal, a somewhat larger volume per drop than given by Kuichling, who found 1 drop per sec equal to 3 gal per day.

1 in. of rainfall per month (30 days) = 0.8962 cfs per sq mile = 0.5793 mgd per sq mile.

1 in. on 1 acre = 27,154.3 gal = 3,630 cu ft = 0.0833 acre-ft.

TABLE 6. AREAS AND CAPACITIES PER FOOT OF DEPTH FOR VERTICAL CYLINDRICAL TANKS 1 TO 100 FT IN DIAMETER

Diameter	Area, sq ft, or cu ft per ft of depth	U.S. gal per ft of depth	Diameter	Area, sq ft, or cu ft per ft of depth	U.S. gal per ft of depth	Diameter	Area, sq ft, or cu ft per ft of depth	U.S. gal per ft of depth
1′	0.785	5.87	6′	28.27	211.5	28′	615.8	4,606
1′ 1″	0.922	6.89	6′ 3″	30.68	229.5	28′ 6″	637.9	4,772
1′ 2″	1.069	8.00	6′ 6″	33.18	248.2	29′	660.5	4,941
1′ 3″	1.227	9.18	6′ 9″	35.78	267.7	29′ 6″	683.5	5,113
1′ 4″	1.396	10.44	7′	38.48	287.9	30′	706.9	5,288
1′ 5″	1.576	11.79	7′ 3″	41.28	308.8	31′	754.8	5,646
1′ 6″	1.767	13.22	7′ 6″	44.18	330.5	32′	804.3	6,016
1′ 7″	1.969	14.73	7′ 9″	47.17	352.9	33′	855.3	6,398
1′ 8″	2.182	16.32	8′	50.27	376.0	34′	907.9	6,792
1′ 9″	2.405	17.99	8′ 3″	53.46	399.9	35′	962.1	7,197
1′ 10″	2.640	19.75	8′ 6″	56.75	424.5	36′	1,018	7,616
1′ 11″	2.885	21.58	8′ 9″	60.13	449.8	37′	1,075	8,043
2′	3.142	23.50	9′	63.62	475.9	38′	1,134	8,483
2′ 1″	3.409	25.50	9′ 3″	67.20	502.7	39′	1,195	8,940
2′ 2″	3.637	27.58	9′ 6″	70.88	530.2	40′	1,257	9,404
2′ 3″	3.976	29.74	9′ 9″	74.66	558.5	41′	1,320	9,876
2′ 4″	4.276	31.99	10′	78.54	587.5	42′	1,385	10,360
2′ 5″	4.587	34.31	10′ 6″	86.59	647.7	43′	1,452	10,860
2′ 6″	4.909	36.72	11′	95.03	710.9	44′	1,521	11,370
2′ 7″	5.241	39.21	11′ 6″	103.9	777.0	45′	1,590	11,900
2′ 8″	5.585	41.78	12′	113.1	846.0	46′	1,662	12,430
2′ 9″	5.940	44.43	12′ 6″	122.7	918.0	47′	1,735	12,980
2′ 10″	6.305	47.16	13′	132.7	992.9	48′	1,810	13,540
2′ 11″	6.681	49.98	13′ 6″	143.1	1,071	49′	1,886	14,110
3′	7.069	52.88	14′	153.9	1,152	50′	1,964	14,690
3′ 1″	7.467	55.86	14′ 6″	165.1	1,235	52′	2,124	15,890
3′ 2″	7.876	58.92	15′	176.7	1,322	54′	2,290	17,130
3′ 3″	8.296	62.06	15′ 6″	188.7	1,412	56′	2,463	18,420
3′ 4″	8.727	65.28	16′	201.1	1,504	58′	2,642	19,760
3′ 5″	9.168	68.58	16′ 6″	213.8	1,600	60′	2,827	21,150
3′ 6″	9.621	71.97	17′	227.0	1,698	62′	3,019	22,580
3′ 7″	10.08	75.44	17′ 6″	240.5	1,799	64′	3,217	24,060
3′ 8″	10.56	78.99	18′	254.5	1,904	66′	3,421	25,590
3′ 9″	11.04	82.62	18′ 6″	268.8	2,011	68′	3,632	27,170
3′ 10″	11.54	86.33	19′	283.5	2,121	70′	3,848	28,790
3′ 11″	12.05	90.13	19′ 6″	298.6	2,234	72′	4,072	30,450
4′	12.57	94.00	20′	314.2	2,350	74′	4,301	32,170
4′ 1″	13.10	97.96	20′ 6″	330.1	2,469	76′	4,536	33,930
4′ 2″	13.64	102.0	21′	346.4	2,591	78′	4,778	35,740
4′ 3″	14.19	106.1	21′ 6″	363.1	2,716	80′	5,027	37,600
4′ 4″	14.75	110.3	22′	380.1	2,844	82′	5,281	39,500
4′ 5″	15.32	114.6	22′ 6″	397.6	2,974	84′	5,542	41,450
4′ 6″	15.90	119.0	23′	415.5	3,108	86′	5,809	43,450
4′ 7″	16.50	123.4	23′ 6″	433.7	3,245	88′	6,082	45,490
4′ 8″	17.10	128.0	24′	452.4	3,384	90′	6,362	47,590
4′ 9″	17.72	132.6	24′ 6″	471.4	3,527	92′	6,648	49,720
4′ 10″	18.35	137.3	25′	490.9	3,672	94′	6,940	51,920
4′ 11″	18.99	142.0	25′ 6″	510.7	3,820	96′	7,238	54,140
5′	19.63	146.9	26′	530.9	3,972	98′	7,543	56,420
5′ 3″	21.65	161.9	26′ 6″	551.5	4,126	100′	7,854	58,750
5′ 6″	23.76	177.7	27′	572.6	4,283			
5′ 9″	25.97	194.3	27′ 6″	594.0	4,443			

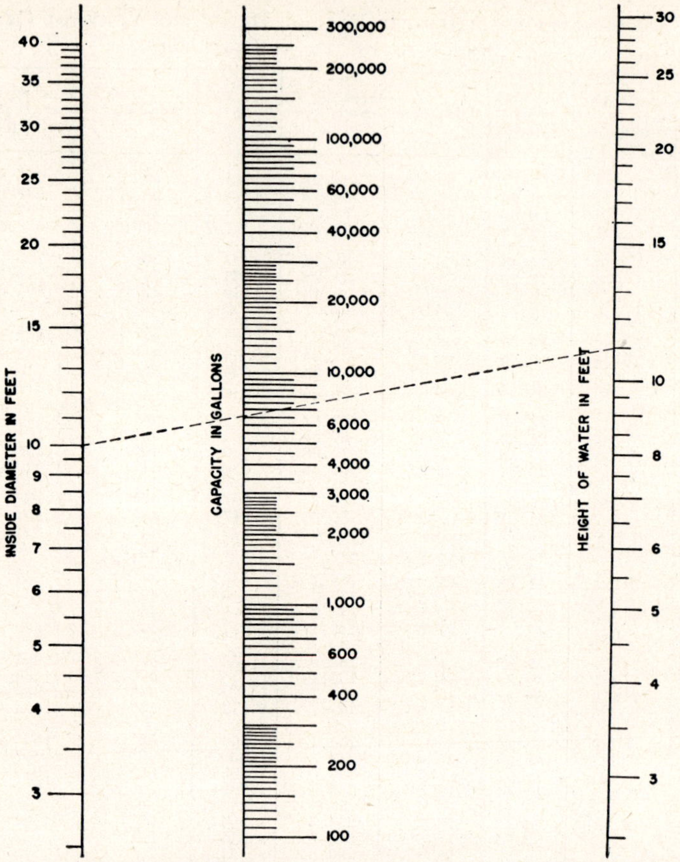

Fig. 2. Nomograph for calculating capacity of tanks.

TABLE 7. WEIGHT OF WATER PER CUBIC FOOT AT DIFFERENT TEMPERATURES

Temp, °F	Wt per cu ft, lb	Temp, °F	Wt per cu ft, lb	Temp, °F	Wt per cu ft, lb	Temp, °F	Wt per cu ft, lb	Temp, °F	Wt per cu ft, lb
32	62.42	150	61.18	260	58.55	380	54.36	500	48.7
40	62.42	160	60.98	270	58.26	390	53.94	510	48.1
50	62.41	170	60.77	280	57.96	400	53.5	520	47.6
60	62.37	180	60.55	290	57.65	410	53.0	530	47.0
70	62.31	190	60.32	300	57.33	420	52.6	540	46.3
80	62.23	200	60.12	310	57.00	430	52.2	550	45.6
90	62.13	210	59.88	320	56.66	440	51.7	560	44.9
100	62.02	212	59.83	330	56.30	450	51.2	570	44.1
110	61.89	220	59.63	340	55.94	460	50.7	580	43.3
120	61.74	230	59.37	350	55.57	470	50.2	590	42.6
130	61.56	240	59.11	360	55.18	480	49.7	600	41.8
140	61.37	250	58.83	370	54.78	490	49.2		

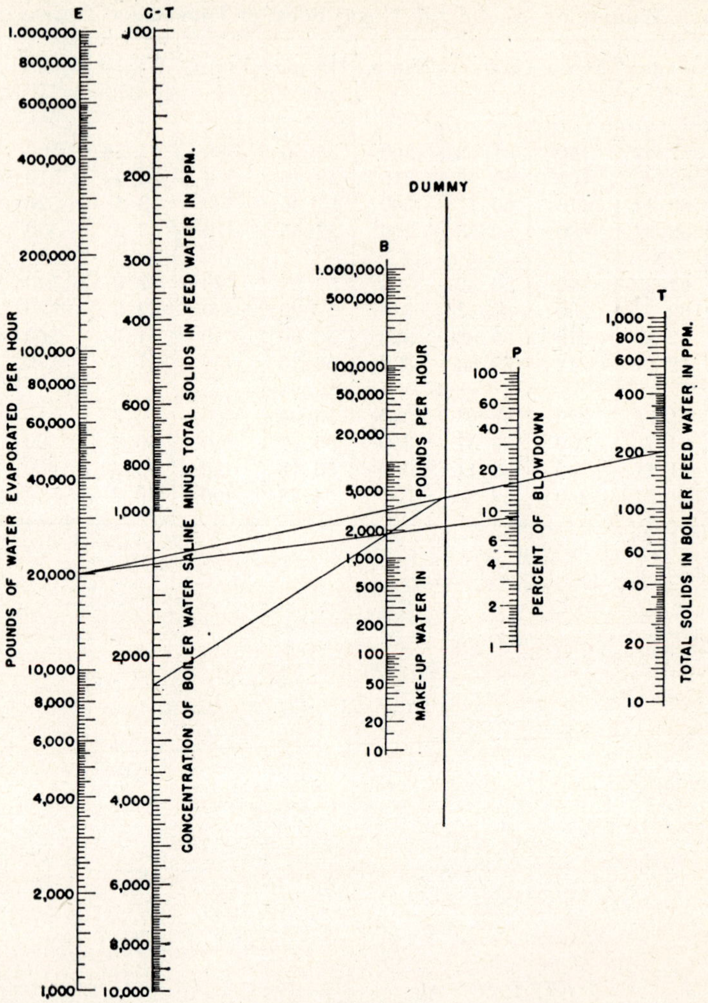

Fig. 3. Nomograph for determining per cent blowdown.

TABLE 8. EQUIVALENT QUANTITIES OF WATER

Gal per 24 hr	Gal per hr	Gal per min	Cu ft per sec	Gal per 24 hr	Gal per hr	Gal per min	Cu ft per sec
2,500,000	104,166.6	1,736.0	3.8681	144,000	6,000.0	100.0	0.2228
2,000,000	83,333.3	1,388.0	3.0945	129,600	5,400.0	90.0	0.2005
1,500,000	62,500.0	1,041.7	2.3205	115,200	4,800.0	80.0	0.1782
1,000,000	41,666.6	694.4	1.5472	100,800	4,200.0	70.0	0.1559
950,000	39,583.3	659.7	1.4697	100,000	4,166.6	69.4	0.1547
900,000	37,500.0	625.0	1.3925	86,400	3,600.0	60.0	0.1337
850,000	35,416.6	590.2	1.3152	75,000	3,125.0	52.9	0.1160
800,000	33,333.3	555.5	1.2378	72,000	3,000.0	50.0	0.1114
750,000	31,250.0	520.8	1.1604	60,000	2,500.0	41.6	0.0928
700,000	29,166.6	486.1	1.0830	57,600	2,400.0	40.0	0.0811
650,000	27,083.3	451.3	1.0057	50,000	2,083.0	34.7	0.0774
600,000	25,000.0	416.7	0.9283	43,200	1,800.0	30.0	0.0668
550,000	22,916.6	381.9	0.8510	28,800	1,200.0	20.0	0.0446
500,000	20,833.3	347.2	0.7736	25,000	1,041.6	17.3	0.0387
450,000	18,750.0	312.5	0.6961	20,000	833.3	13.8	0.0309
400,000	16,666.6	277.7	0.6189	15,000	625.0	10.4	0.0232
350,000	14,583.3	243.0	0.5415	14,400	600.0	10.0	0.0223
300,000	12,500.0	208.3	0.4642	10,000	416.6	6.9	0.0155
250,000	10,416.7	173.6	0.3868	7,200	300.0	5.0	0.0111
200,000	8,333.3	138.8	0.3094	5,000	208.3	3.4	0.0077
150,000	6,250.0	104.1	0.2321	1,440	60.0	1.0	0.0022

TABLE 9. EQUIVALENTS OF VARIOUS MEASURES OF RATE OF FILTRATION

Rate	U.S. gal, millions per acre per 24 hr	Imperial* gal, millions per acre per 24 hr	U.S. gal per sq ft per hr	Imperial* gal per sq ft per hr	Cu ft per sq yd per hr	Vertical velocity, in. per hr	Vertical velocity, mm per hr	Vertical velocity, m per 24 hr = cu m per sq m per 24 hr
1 million U.S. gal per acre per 24 hr	1.0	0.833	0.96	0.80	1.15	1.53	39	0.935
1 million imperial gal per acre per 24 hr	1.200	1.0	1.15	0.96	1.38	1.84	46.8	1.122
1 U.S. gal per sq ft per hr	1.045	0.870	1.0	0.83	1.20	1.60	40.7	0.978
1 imperial gal per sq ft per hr	1.255	1.045	1.20	1.0	1.44	1.92	48.9	1.174
1 cu ft per sq yd per hr	0.869	0.724	0.83	0.69	1.00	1.33	33.9	0.813
1 lin in., vertical velocity, per hr	0.652	0.543	0.62	0.52	0.75	1.0	25.4	0.610
100 lin mm, vertical velocity, per hr	2.566	2.139	2.46	2.05	2.95	3.94	100.0	2.400
1 lin m, vertical velocity, per 24 hr = 1 cu m per sq m per 24 hr	1.069	0.891	1.02	0.85	1.23	1.64	41.7	1.0

* Sometimes called British.

TABLE 10. SPECIFICATIONS FOR FILTER SAND AND ANTHRACITE COAL

Specifications. The following specification is typical of good practice in specifying sand for rapid sand filters:

"Filter Sand. All filter sand shall be composed of hard, durable grains, either sharp or rounded, free from clay, loam, dirt and organic matter, with not more than one per cent flat or micaceous particles. After immersion for 24 hours in warm concentrated hydrochloric acid, the sand, crushed and powdered, shall not have lost more than five (5) per cent of its original weight. The sand shall not contain more than one and one half (1½) per cent calcium and magnesium taken together and calculated as calcium carbonate ($CaCO_3$); and not more than two (2) per cent by weight shall be lost on ignition of the crushed and powdered sand.

"Filter Sand Size. The sand shall have an effective size of not less than 0.55 of a millimeter and a uniformity coefficient of not more than 1.65. Not more than two-tenths (0.2) per cent shall be finer than two-tenths of a millimeter, not more than one (1) per cent shall be finer than twenty-five hundredths of a millimeter, and at least 90 per cent shall be finer than eight-tenths of a millimeter. The diameters of sand grains shall be computed as the diameters of spheres of equal volume and all percentage calculated by weight.

"The Contractor may place in the filter tanks sand which does not strictly conform to the specifications for effective size and uniformity coefficient provided that he shall, after placing, thoroughly wash the sand by means of wash water introduced by the strainer system at a satisfactory rate, and he shall remove the fine sand remaining upon the surface of the filter after washing by scraping, if necessary. He shall repeat this process until the sand remaining is of the required depth, size and uniformity.

"Placing Filter Sand. Especial care shall be taken in transporting and placing filter sand to prevent contamination of any sort, and sand which may have become dirty, either before or after placing in the filters, shall be washed or removed and replaced by clean sand in a satisfactory manner. All sand contaminated by organic matter shall be rejected. After sand is placed in the filters, it shall be washed at least ten (10) times with the washing devices connected with the filters, and fine sand appearing at the surface shall be removed by scraping. Sufficient sand shall be placed in each filter to secure a layer at least thirty (30) inches in depth after the sand has been washed and scraped, and conforming to the sizes above shown.

"Testing and Sampling. The Contractor shall furnish for tests, samples of sand consisting of not less than one (1) quart, taken under the direction of the Engineer. Each sample shall be labeled, stating whether it has been washed or screened, or both, and the place from which it is taken. At least one sample shall be furnished for each fifty (50) cubic yards of filter sand. All samples shall be furnished in tight receptacles. The sand will be carefully tested by the Engineer to determine its conformity with the specifications."

Method of Examination of Filter Sand. As the result of some experimental work for the Massachusetts State Board of Health, the late Allen Hazen published a paper in 1892, which established two factors, "effective size" and "uniformity coefficient," which are frequently used in an examination of a sand to be used for filtration purposes. The term "effective size" was adopted by Hazen to apply to that size of sand in millimeters below which 10 per cent of the sample was finer and above which 90 per cent of the sand was larger. "Uniformity coefficient" was the term applied to the ratio between the size in millimeters such that 60 per cent of the sand is finer than it and the effective size.

With sieves about 5½ inches in diameter, 100 grams is the greatest amount to be used for a screen analysis. Some laboratories prefer to take exactly 100 grams so that the grams retained on each sieve is also the per cent retained.

Quoting the "Manual of Water Works Practice":*

"The dried sand is put in the top of the nest of sieves, each fitting into the one below with a cover at the top and a pan at the bottom. The sieves may be shaken by hand for occasional field tests, but generally a shaking machine is employed.

"The machine frequently used is geared so that 150 turns of the handle give 700 single shakes to the sieves. This is enough ordinarily."

A hand test on one sample requires from 20 to 30 min of most careful painstaking work. Usually 15 min sieving is specified for the standard commercial forms of shaking machines. The amount of sand retained on each sieve is weighed and recorded. The accompanying table gives the mechanical analysis of a sample of filter sand and Fig. 4 shows the method of plotting and determining the effective size and uniformity coefficient.

There is considerable difference of opinion whether round grains are superior to sharp grains. Sand used for filtration purposes may be obtained from the seashore, from river beds, or from sand banks. It should consist mainly of quartz grains but also may contain hard silicates. As it occurs in nature, it is frequently mixed with clayey or other fine particles which must be removed from it by washing before it is used.

Anthracite Coal. Crushed anthracite coal has been used as a subsitute for sand in a number of places, including Denver, Cumberland, and Harrisburg. The filters at Cumberland were made up of anthracite coal before 1914.

Anthracite coal has an average specific gravity of 1.5 as against 2.65 for silica sand. The weights per cubic foot are approximately 55 lb and 100 lb, respectively. A ton of anthracite coal will fill almost twice as much filter volume as a ton of sand. Of course, the cost per ton is a great deal more than that for sand. The principal advantages claimed for anthracite coal over sand are:

1. Only about one-half the backwashing velocity as for sand is required for anthracite of the same size in order to produce equal expansion (not necessarily equal washing). This is due to the lower specific gravity of the anthracite.
2. Anthracite with the same effective size as sand has a greater porosity.

Sieve Analysis of a Sample of Filter Sand*

Mesh	Sieve opening, mm	Weight retained, grams	Per cent finer than
10	1.651	0	100
14	1.168	0.4	99.6
20	0.833	18.9	80.7
24	0.701	22.6	58.1
28	0.589	26.7	31.4
32	0.495	16.1	15.3
35	0.417	11.4	3.9
42	0.351	2.8	1.1
50	0.279	1.1	0
100	0.140	0	0

* Size of sample, 100 grams.

Reproduced by permission of the American Water Works Association; from "Water Quality and Treatment," American Water Works Association, New York, 1940, p.178.

* American Water Works Association: "Water Quality and Treatment," AWWA Manual, 2d ed., 1950, p. 257.

SOURCE OF MATERIAL ________________

SIEVE	U.S. STD. NO.	100	80	70	60	50	40	30	20	16	10	8
	OPENING-MM.	0.149	0.177	0.210	0.250	0.297	0.42	0.59	0.84	1.19	2.00	2.38
GRAMS RETAINED ON SIEVE												
✓												
TOTAL PER CENT PASSING SIEVE												
✓												

AMOUNT EXAMINED ________ GRAMS. DATE ________

A. 60% FINER THAN ______ MM. B. 10% FINER THAN ______ MM. = EFFECTIVE SIZE
✓ ______ " ✓ ______ . " AVERAGE ______ MM.

UNIFORMITY COEFFICIENT $\frac{A}{B}$ ______ ✓ ______ AVERAGE ______

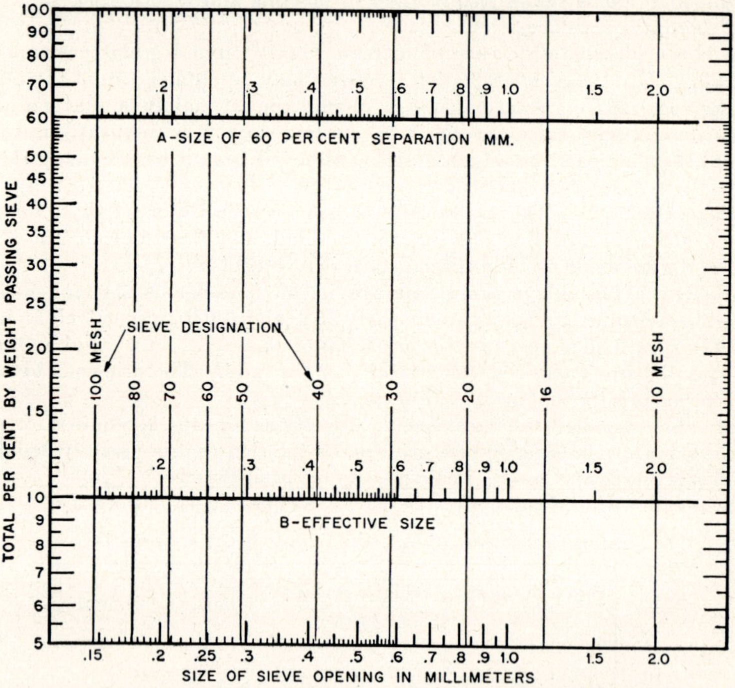

Fig. 4. Chart for mechanical analysis of filter medium.

Table 11. Depth of Layers of Filter Medium and Sizing of Material*

Sand and gravel:
- 27 in. of fine sand, 0.45 to 0.5 mm effective size
- 3 in. of coarse sand, 0.8 to 1.2 mm effective size
- 3 in. of $\frac{1}{4}$- to $\frac{1}{12}$-in. gravel
- 4 in. of $\frac{3}{4}$- to $\frac{1}{4}$-in. gravel
- 5 in. of $1\frac{1}{2}$- to $\frac{3}{4}$-in. gravel
- 6 in. of $2\frac{1}{2}$- to $1\frac{1}{2}$-in. gravel

Anthrafilt and gravel:
- 30 in. of No. 1 anthrafilt, 0.68 to 0.72 mm effective size
- 3 in. of $\frac{1}{4}$- to $\frac{1}{12}$-in. gravel
- 4 in. of $\frac{3}{4}$- to $\frac{1}{4}$-in. gravel
- 5 in. of $1\frac{1}{2}$- to $\frac{3}{4}$-in. gravel
- 6 in. of $2\frac{1}{2}$- to $1\frac{1}{2}$-in. gravel

Anthrafilt:
- 30 in. of No. 1 anthrafilt, 0.68 to 0.72 mm effective size
- 2 in. of No. 2 anthrafilt, $\frac{3}{16}$ to $\frac{3}{32}$ in.
- 3 in. of No. 3 anthrafilt, $\frac{5}{16}$ to $\frac{3}{16}$ in.
- 3 in. of No. 4 anthrafilt, $\frac{9}{16}$ to $\frac{5}{16}$ in.
- 4 in. of No. 6 anthrafilt, $1\frac{5}{8}$ to $\frac{13}{16}$ in.
- 6 in. of No. 7 anthrafilt, $2\frac{3}{8}$ to 2 in.

* Nordell, Eskel: "Water Treatment for Industrial and Other Uses," p. 309, Reinhold Publishing Corporation, New York, 1951.

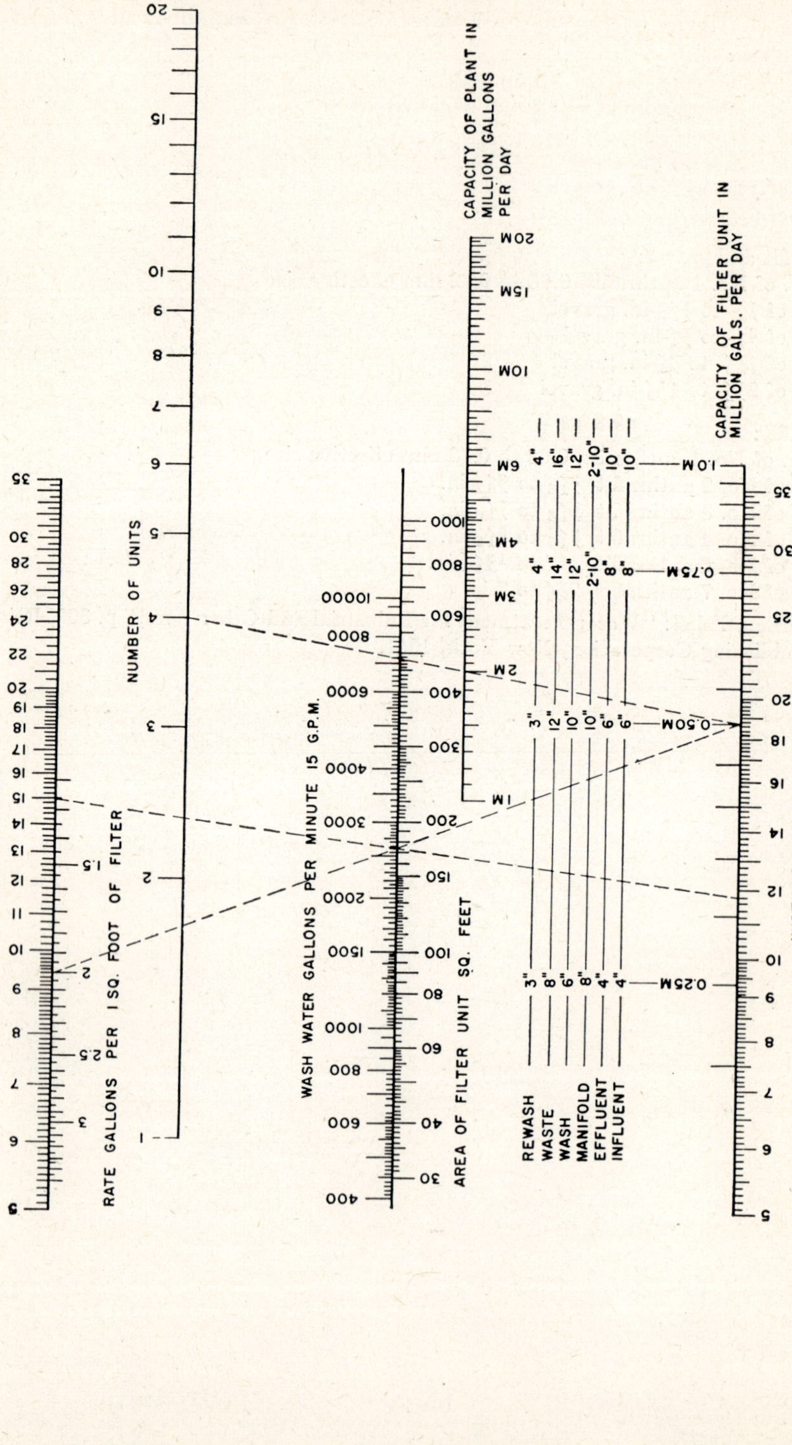

Fig. 5. Nomograph relating dimensions of concrete gravity filters to capacity, backwash requirements, and valve sizes.

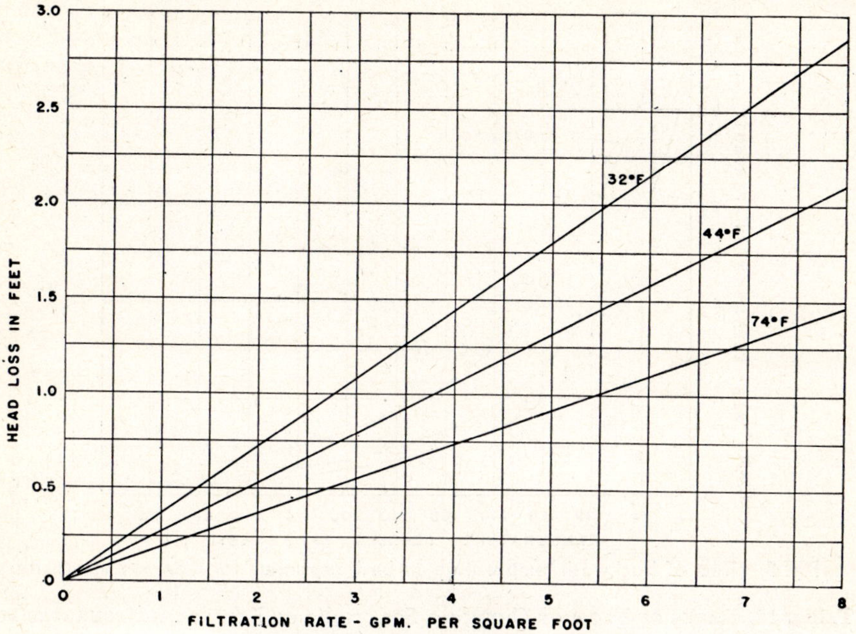

Fig. 6. Initial head loss in gravity sand filters.

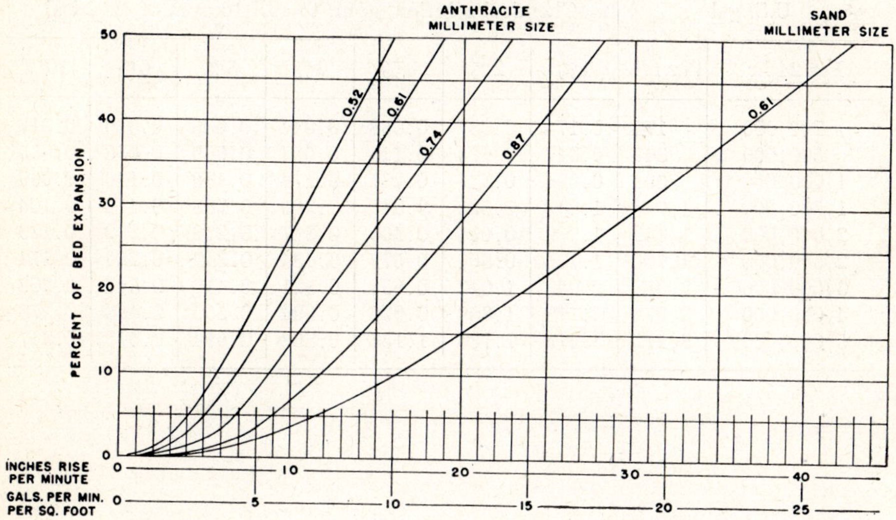

Fig. 7. Relation between per cent of bed expansion and backwash rates at 65°F.

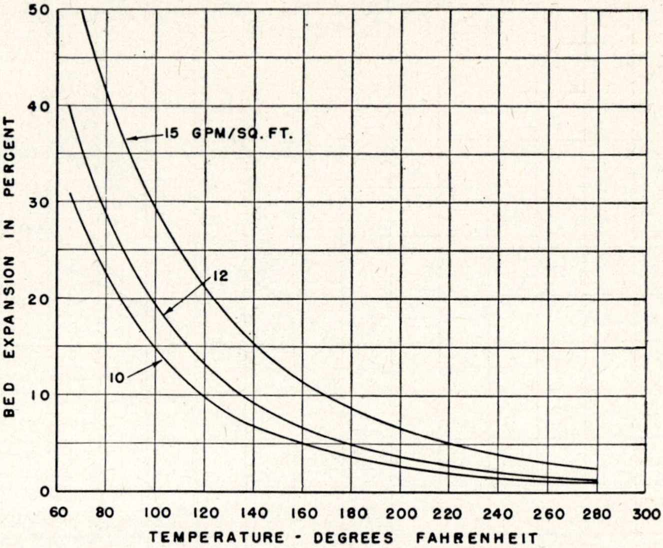

Fig. 8. Effect of backwash temperature on bed expansion for 0.74-mm anthracite.

TABLE 12. RATES OF FLOW OF CHEMICAL SOLUTIONS OF DIFFERENT STRENGTHS FOR DIFFERENT RATES OF FILTRATION

U.S. gal		Chemical solution, gpm, to add 10 ppm of chemical						
Per 24 hr	Per min	1%	2%	3%	4%	5%	6%	10%
250,000	174	0.174	0.087	0.058	0.044	0.035	0.029	0.017
500,000	347	0.347	0.174	0.116	0.087	0.069	0.058	0.035
1,000,000	694	0.694	0.347	0.231	0.174	0.139	0.116	0.069
1,500,000	1,042	1.042	0.521	0.347	0.260	0.208	0.174	0.104
2,000,000	1,389	1.389	0.694	0.463	0.347	0.278	0.232	0.139
2,500,000	1,736	1.736	0.868	0.579	0.434	0.347	0.289	0.174
3,000,000	2,083	2.083	1.042	0.694	0.521	0.416	0.347	0.208
4,000,000	2,778	2.778	1.389	0.926	0.694	0.556	0.463	0.278
5,000,000	3,472	3.472	1.736	1.157	0.868	0.694	0.579	0.347

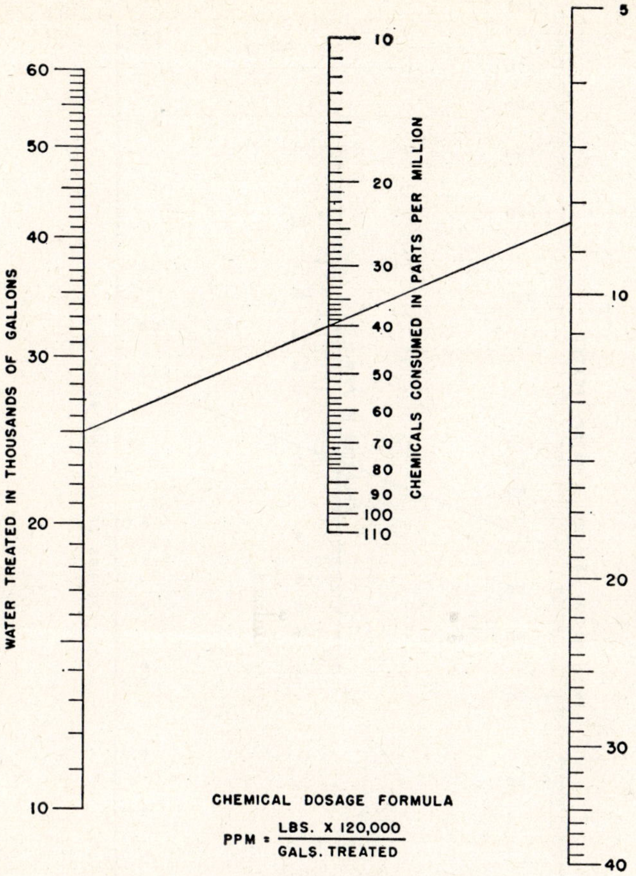

Fig. 9. Nomograph for calculation of chemical dosage.

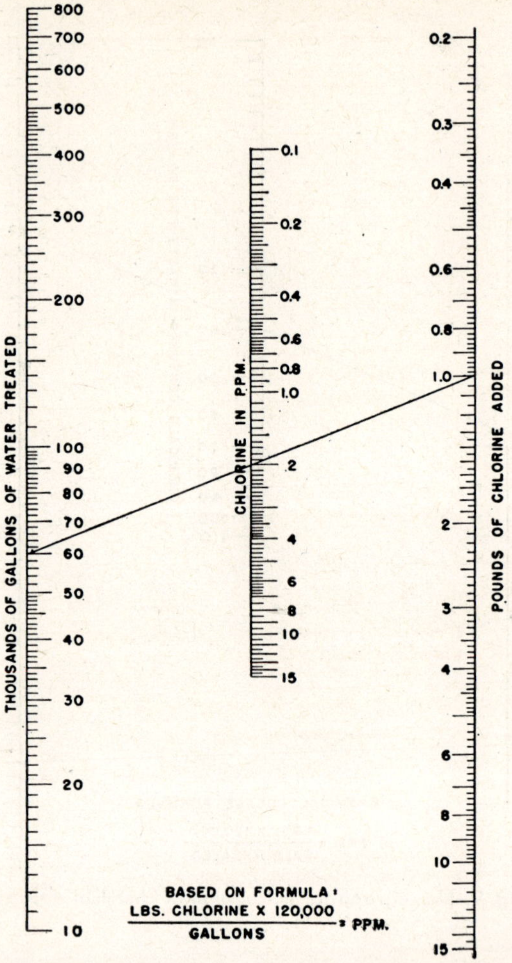

Fig. 10. Nomograph for determining chlorine dosage.

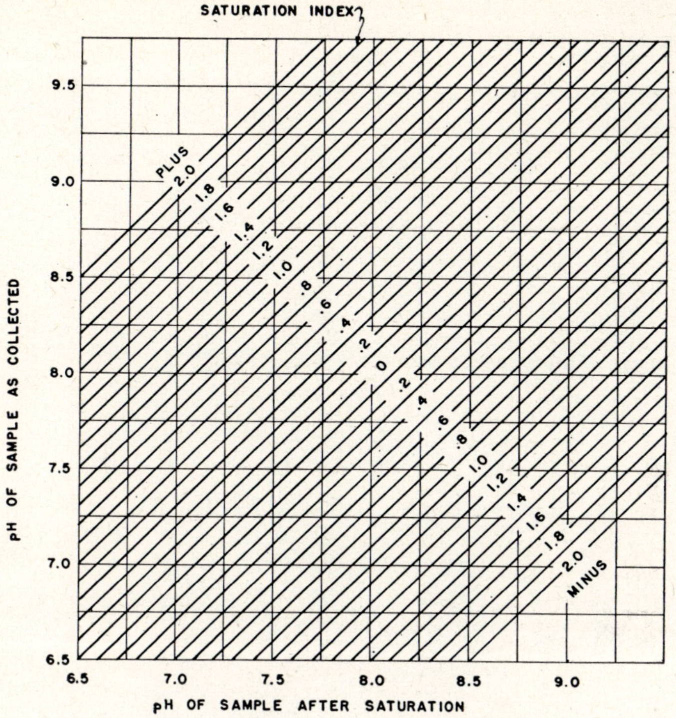

Fig. 11. Saturation index by marble test.

INDEX